DRAFTING
IN A COMPUTER AGE

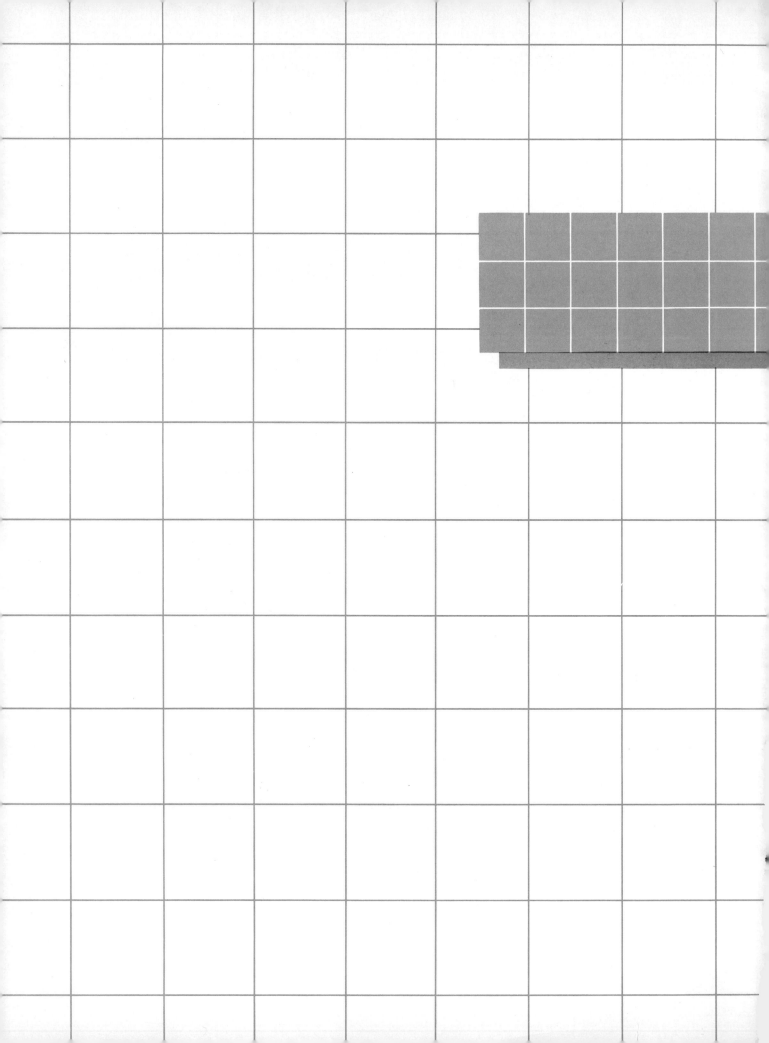

DRAFTING
IN A COMPUTER AGE

Paul Ross Wallach and
Dean Chowenhill

DELMAR PUBLISHERS INC.®

NOTICE TO THE READER

Publisher does not warrant or guarantee any of the products described herein or perform any independent analysis in connection with any of the product information contained herein. Publisher does not assume, and expressly disclaims, any obligation to obtain and include information other than that provided to it by the manufacturer.

The reader is expressly warned to consider and adopt all safety precautions that might be indicated by the activities described herein and to avoid all potential hazards. By following the instructions contained herein, the reader willingly assumes all risks in connection with such instructions.

The publisher makes no representations or warranties of any kind, including but not limited to, the warranties of fitness for particular purpose or merchantability, nor are any such representations implied with respect to the material set forth herein, and the publisher takes no responsibility with respect to such material. The publisher shall not be liable for any special, consequential or exemplary damages resulting, in whole or in part, from the readers' use of, or reliance upon, this material.

Cover credits (clockwise from upper left): General Dynamics; McDonnell Douglas; Mike Bruton/Hinkel Photographics, Inc.; Pontiac

Delmar Staff
Associate Editor: Cameron O. Anderson
Editing Manager: Gerry East
Project Editor: Christine E. Worden
Design Coordinator: Susan C. Mathews
Production Coordinator: Linda Helfrich

For information, address Delmar Publishers Inc.
2 Computer Drive West, Box 15-015
Albany, New York 12212

COPYRIGHT © 1989
BY DELMAR PUBLISHERS INC.

All rights reserved. No part of this work covered by the copyright hereon may be reproduced or used in any form or by any means—graphic, electronic, or mechanical, including photocopying, recording, taping, or information storage and retrieval systems—without written permission of the publisher.

Printed in the United States of America
Published simultaneously in Canada
by Nelson Canada,
A division of The Thomson Corporation

10 9 8 7 6 5 4 3

Library of Congress Cataloging in Publication Data

Ross Wallach, Paul.
    Drafting in a computer age/Paul Wallach and Dean Chowenhill.
       p. cm.
    Includes index.
    ISBN 0-8273-2925-3. ISBN 0-8273-2926-1 (instructor's guide)
    1. Mechanical drawing. 2. Computer graphics. I. Chowenhill,
Dean. II. Title.
T353.R7962 1989
604.2'4—dc19
                                    88-20379
                                        CIP

CONTENTS

PREFACE

1 INTRODUCTION TO CONTEMPORARY DRAFTING • 1

History/1 Drafting Today/2 Drafting Fields/5

2 INTRODUCTION TO COMPUTER-AIDED DRAFTING SYSTEMS • 7

Evolution of Drafting Tools/7 CAD Classifications/8 CAD System Components/10

3 DRAFTING EQUIPMENT, SUPPLIES, INKING • 22

Introduction/22 Conventional Drafting Supplies/22 Conventional Drafting Equipment/27 Inking/39 CAD Equipment/43

4 SKETCHING AND LETTERING • 48

Introduction/48 Supplies/48 Working Drawing Sketches/48 Sketching Guidelines and Procedures/51 Computer-aided Sketching/56 Computer-aided Lettering/57

▲ *The World of CAD*

5 CAD PROCEDURES • 63

Personal CAD System/63 Turning On (Booting) the System/63 Starting a Drawing/64 Creating a Simple Drawing/68 Saving and Ending a Drawing/68 To Edit an Existing Drawing/69

▲ *The World of CAD*

6 DRAFTING CONVENTIONS AND FORMATS • 73

Standardization/73   Line Conventions/74   Drafting Conventions/75   Drawing Formats/77   CAD Line Conventions/81   CAD Formats/82

7 GEOMETRIC CONSTRUCTION • 86

CAD Versus Drawing Instruments/86   Geometric Forms/86   Conventional Geometric Construction/88   Computer-aided Geometric Construction/93

▲ *The World of CAD*

8 MULTIVIEW DRAWINGS • 99

Multiview Orthographic Projections/99   Selection of Views/99   Planes of Projection/104   Angles of Projection/106   Visualization/107   Inclined and Oblique Surface Projections/108   Dimensioning/111   Drawing Procedures/112   Computer-aided Multiview Drawings/117

▲ *The World of CAD*

9 DIMENSIONING CONVENTIONS AND SURFACE FINISHES • 135

Introduction/135   Systems of Dimensioning/137   Dimensioning Elements/137   Dimensioning Guidelines/140   Special Dimensioning Procedures/143   Surface Control/147   Dimensioning with CAD/152

10 TOLERANCING AND GEOMETRIC TOLERANCING • 158

Introduction/158   Tolerance Dimensions/158   Tolerance Practices/159   Mating Parts/161   Geometric Tolerancing/164   Tolerancing with CAD/169

11 SECTIONAL VIEWS • 178

Introduction/178   Cutting Plane/178   Section Lining/179   Sectional View Types/182   Computer-aided Sectional Views/186

12 AUXILIARY VIEWS AND REVOLUTIONS • 192

Introduction/192   Foreshortening/192   Auxiliary Planes/194   Projection/195   Primary Auxiliary Views/196   Secondary Auxiliary Views/196   Completeness of Auxiliary Views/197   Procedures/199   Revolutions/201   Computer-aided Auxiliary Views/203

▲ *The World of CAD*

Contents vii

13 DESCRIPTIVE GEOMETRY • 213

Introduction/213   Points in Space/213   Geometric Planes/214   Geometric Lines/215   Computer-aided Descriptive Geometry/218

 The World of CAD

14 DEVELOPMENT DRAWINGS • 223

Introduction/223   Surface Forms/224   Pattern Drawing Terminology/225   Parallel Line Development/227   Radial Line Development/228   Computer-aided Development Drawings/235   CAD Parallel Line Development/235

 The World of CAD

15 PICTORIAL DRAWINGS • 247

Introduction/247   Axonometric Drawings/249   Oblique Drawings/261   Perspective Drawings/264   Computer-aided Pictorial Drawings/273

16 FASTENERS • 284

Introduction/284   Threaded Fasteners/284   Drafting Procedures/288   Types of Threaded Fastening Devices/290   Fastener Templates/293   Threaded Permanent Fasteners/297   Computer-aided Fastener Drawings/298

 The World of CAD

17 DRAFTING PROCEDURES AND STORAGE SYSTEMS • 307

Introduction/307   Functional Drafting/307 Revisions/310   Storage Systems/311   CAD Storage Systems/312   Data Bases/312

 The World of CAD

18 WORKING DRAWINGS • 323

Introduction/323   Working Drawings/323   Working Drawing Dimensions/330   Computer-aided Working Drawings/337

19 WELDING DRAWINGS • 363

Introduction/363 Welding Processes/364 Welding Symbols/365 Data Selection/369 Welding Type Symbol Applications/370 Welded Joints/376 Weld Symbols with CAD/377

▲ *The World of CAD*

20 GEARS AND CAMS • 385

Introduction/385 Gear and Drive/385 Cam Mechanisms/398 Drawing Gears with CAD/404 Drawing Cams with CAD/408

21 PIPING DRAWINGS • 416

Introduction/416 Pipe Connections/416 Pipe Standards/418 Piping Drawings/418 Computer-aided Piping Drawings/422

▲ *The World of CAD*

22 ELECTRONICS DRAFTING • 431

Introduction/431 Electronics Industry Standards/431 Electronics Drawing Symbols/432 Types of Electronic Drawings/433 Standard Symbols Charts/438 Drawing Schematic Diagrams/438 Drawing Connection Diagrams/440 Drawing Logic Diagrams/440 Computer-aided Electronics Drafting/441

23 TOOL DESIGN DRAFTING • 448

Introduction/448 Jigs and Fixtures/448 Standard Parts/449 Jig and Fixture Design/455 Computer-aided Jig and Fixture Design/458

 The World of CAD

24 ARCHITECTURAL DRAFTING • 465

Introduction/465 Gathering Information/465 Residential Styles/468 Design Considerations/473 Design Sequence/479 Working Drawings/481 Support Documents/494 Other Types of Buildings/498 Ergonomics/502 Computer-aided Architectural Drafting/504

Appendices • 510

Tables/512 Abbreviations/542 Glossary/546

Index • 555

PREFACE

The authors would like to welcome you to this new concept for an engineering drawing textbook. It was designed and written for your classes. With the increasing development of our industry and the ever increasing new breakthroughs of new technology there will always be a large demand for qualified engineers and drafters. With this demand there must be qualified schools and teaching materials. The authors strove to fulfill these needs with the textbook DRAFTING IN A COMPUTER AGE.

An excursion through each chapter will teach basic knowledge and drafting skills. This information then is used to instruct how to perform the same tasks with a computer-aided drafting (CAD) system. It is important that skills and knowledge be taught first with drawing instruments before using the CAD system—otherwise the CAD system could become a very expensive teaching aid.

STANDARDS

The language of drafting is a uniform and standardized system that is used throughout the world. The standards for the U.S. customary system is developed by the American National Standards Institute (ANSI). The standards for the metric system is developed by Systems International (ISO). Careful attention was given to the dimensioning and tolerancing chapters (ANSI Y14.5M).

DRAFTING IN A COMPUTER AGE fulfills the need for an instructional drafting text that will teach the fundamentals of engineering drawing with new concepts. In summary, the following are features of the text:

1. Each chapter concentrates on clear, concise text.
2. There are over 1300 illustrations and photographs to help clarify the text, and to aid the students in reading two-dimensional drawings and visualizing in three dimensions.
3. There are many step-by-step procedural illustrations taking the students through exercises in design, layout, and finished drawings.
4. At the end of every chapter the drafting concept is redefined, and it is explained how to accomplish it on a CAD system.
5. The basic CAD operations are completely explained throughout the book, starting with turning on the system.
6. The explanations for the computer-generated art are kept simple, so students can get a clear overview of how most CAD systems operate. Students will be well-versed in CAD upon completing this book, and should have no fears about learning any CAD system.
7. Each chapter will develop and strengthen specific technical concepts, and make the student proficient in solving drafting problems.
8. Key terms are boldfaced within the text, and listed at the end of every chapter for easy identification.

ORGANIZATION

DRAFTING IN A COMPUTER AGE has the chapters organized in a logical sequence; however each chapter is a stand-alone unit of instruction. This permits easy reorganization of the chapters to fit any instructor's drafting program. No previous drafting knowledge is needed to use this textbook.

STUDENT EXERCISES

Student exercises are presented at the end of each chapter. The exercises start very simply and become progressively more complex. The drawing exercises are consistent within each chapter. The general order of the exercises is as follows:

1. sketching exercises
2. instrument drawings
3. engineering change orders (ECOs)
4. design problems
5. CAD drawings
6. drawing layouts and listing data charts with the cartesian coordinate system

Most exercises and instructional materials are dimensioned as in industry today, with the inch-decimal and millimeter (metric) systems. A few exercises will use the inch-fraction system.

CONTENT

Chapter 1 presents an introduction to modern industry. Specific careers related to drafting also give the students insight into their occupational options for the future.

Chapter 2 is an introduction to computer-aided drafting. General information about CAD hardware and software is covered in enough detail to insure that the student understands CAD and its applications.

Chapters 3 through 7 give the students the background needed to learn and draw the basic drafting concepts (instruments, supplies, sketching, lettering, CAD, conventions, and formats). The latest ANSI standards are used where appropriate in these chapters and throughout the text.

Chapters 8 through 15 teach the students the concepts of mechanical drafting required to design and draw finished technical working drawings (multiview drawings, dimensioning, tolerancing, geometric tolerancing, sectional drawings, auxiliary drawings, revolutions, descriptive geometry, developments, and pictorial drawings). Each of these technical drawing areas is presented using traditional drafting instruments and supplies as well as using CAD.

Chapters 16 through 24 teach the students how to prepare finished technical working drawings required for production (fasteners, drafting systems, working drawings, welding drawings, gears, cams, piping drawings, electronics, drawings, jigs, fixtures, and architectural drawings). CAD procedures for each of these subjects is examined in detail.

The World of CAD, special interest articles at the end of selected chapters, highlight a variety of uses for CAD now and in the future. Profiles of individuals who use CAD in their careers are also included.

The appendices have the necessary tables needed for most working drawings; A comprehensive glossary and abbreviations are included.

DRAFTING IN A COMPUTER AGE is an outstanding up-to-date and complete introductory drafting text. This book gives beginning students the skill and knowledge required for drafters in today's complex industry. We welcome you to the new challenges in drafting as we approach the twenty-first century.

ACKNOWLEDGMENTS

The authors would like to thank the following drafting teachers from across the nation for their extensive time and effort in making this the best textbook possible:

Wade McCarty
Lanier H. S., TX

Nate Moore
William Christian H. S., MO

Jess Kruegier
Lincoln Way H. S., IL

James Fox
Three Rivers H. S., MI

Ronald Rosenlof
Apollo H. S., AZ

Guy Forsythe
Westerville South H. S., OH

Dante Galiozzi
Tracy H. S., CA

Eugene J. Magi
Trumbull H. S., CT

Herb Luhr
Guilderland H. S., NY

Harvey Felt
Rio Vista H. S., CA

George Martens
Winter Park H. S., FL

Larry Boucher
Dover-Sherborn Regional H. S., MA

Dr. Jule Scarborough
Northern Illinois University, IL

Special thanks to the following companies for their technical support:

AutoCAD®
Versacad
IBM
Apple Computer
Hewlett-Packard
Cascade Graphics Systems

Chapter opening photos are credited as follows: Chapters 1, 5, 6, and 17—Hewett-Packard; Chapter 2—IBM Corporation; Chapter 3—Kneuffel & Esser Co.; Chapter 15—General Motors Corporation, Engineering Standards.

The authors would like to dedicate this book to Ann Ross Wallach and Cathy Chowenhill whose encouragement, help, and support were invaluable.

AutoCAD is registered in the U.S. Patent and Trademark Office by Autodesk, Inc.

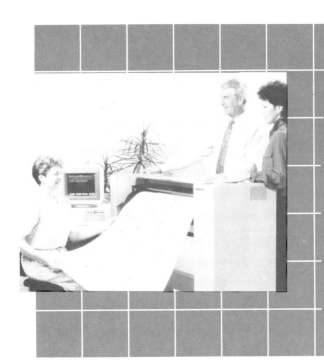

chapter 1
INTRODUCTION TO CONTEMPORARY DRAFTING

OBJECTIVES

The student will be able to:

- relate the historical development of drafting
- state the importance and need for drafting as a technical communication skill
- identify the role and responsibilities of various drafting specialists
- state how drafting is used in different fields as the major source of communication

HISTORY

Drawings have been used throughout history as an art form and a method of communication. Early humans drew crude pictures of animals on cave walls, **(Fig. 1–1)** and by 4000 B.C. stone tablet drawings **(Fig. 1–2)** were used to define the outline of buildings. By 2000 B.C. drawings were prepared on parchment. Later when the Egyptian pyramids and Greek Parthenon were designed and constructed, from 500 to 1000 B.C., detailed construction drawings were widely used **(Fig. 1–3)**.

By the first century A.D., Romans were using detailed and dimensional drawings as guides in

FIGURE 1–1 Early humans drew crude pictures on cave walls.

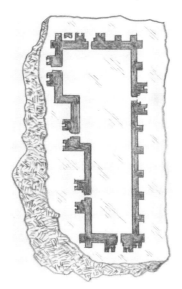

FIGURE 1–2 By 4000 B.C. stone tablets were being used to define the outline of fortifications.

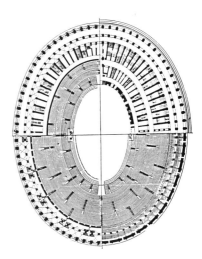

FIGURE 1–3 Detailed construction drawings were used to build the Egyptian pyramids, the Greek Parthenon, and the Roman Coliseum (shown).

2 Drafting in a Computer Age

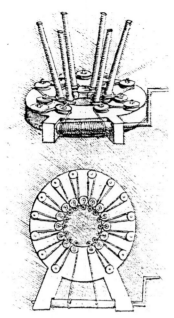

FIGURE 1–4 Leonardo Da Vinci used pictorial and two-dimensional working drawings to develop his inventions.

FIGURE 1–5 Humans communicate with written words, body movements, and drawings. (Courtesy of The Turner Corporation, Photograph by Michael Sporzarsky)

FIGURE 1–6 Describe this spacecraft in detail using only words. (Courtesy of NASA)

building their structures and roadways. But not until the fifteenth century were flat orthographic projection planes used for product drawings, as shown in Leonardo da Vinci's sketch in **Fig. 1–4**.

The Language of Drawing

Humans now routinely communicate with each other through the use of verbal sounds, written words, body movements, and many kinds of drawings **(Fig. 1–5)**. Written, verbal, or body language is very effective in communicating personal and social ideas or emotions. Drawings are effective in describing the precise size and shape of objects. To illustrate this point, try to describe the object shown in **Fig. 1–6** using words alone. This exercise clearly illustrates the old Chinese proverb, "one picture is worth a thousand words."

DRAFTING TODAY

Drawing is now the universal language of industry and construction. Through the use of standardized symbols and projection methods, drawings help overcome international language barriers by describing manufactured products, buildings, and processes without words. Today, drawing is the primary method of communication between designers and clients, architects and builders, engineers and production personnel, and between advertisers and customers. A drawing, when used to show the material, size and shape of a product, is known as a technical drawing, engineering drawing, architectural drawing, or mechanical drawing. All manufactured products or buildings, regardless of size, can begin either on the drawing board or computer. In fact, every part of every product you will use or see today started as a technical drawing.

The design and manufacture of small, simple products, such as the paper clip shown in **Fig. 1–7** still requires technical drawings for its production. Larger and more complex products may require hundreds or many thousands of technical drawings to be properly manufactured. This is because technical drawings are required for each manufactured part at each step of the design, production, and assembly process. The following are just a small sampling of the many products our industries manufacture today. Each must have many technical drawings in order to be manufactured:

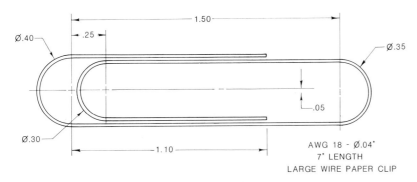

FIGURE 1-7 The design and manufacture of a simple paper clip requires technical working drawings.

trucks (**Fig. 1-8**)
spacecraft and aircraft (**Figs. 1-9A through H**)
large structures (**Figs. 1-10A and 1-10B**)

Related Careers

Drafting skill and knowledge is required in the preparation of original design drawings, detailed working drawings from which products are manufactured. They are also required for graphic instructions in finishing, assembling, transporting, and operating products. Drafting skills are also needed in the creation of drawings that help sell products. Because drafting is used in so many fields for so many purposes, a thorough knowledge of drafting is essential in all technical fields.

Engineers, architects, and designers use drafting skills to communicate technical ideas. Drafters or drafting technicians use the principles of mathematics and science to translate ideas and preliminary design sketches into finished technical drawings. Drafters specialize in many

FIGURE 1-8 Truck (Courtesy of Freightliner)

FIGURE 1-9A Space shuttle being piggybacked (Courtesy of NASA)

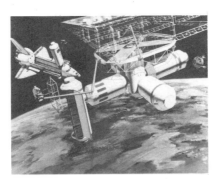

FIGURE 1-9B Space shuttle servicing a space station (Courtesy of NASA)

FIGURE 1-9C Space shuttle with cargo hatches open (Courtesy of NASA)

FIGURE 1-9D Space shuttle take off (Courtesy of NASA)

FIGURE 1-9E Spacecraft fly-by of Saturn (Courtesy of NASA)

FIGURE 1-9F Turboprop engine for aircraft (Courtesy of Allied Signal Aerospace Company)

4 Drafting in a Computer Age

FIGURE 1–9G Pictorial cutaway of the F–16B aircraft (Courtesy of General Dynamics)

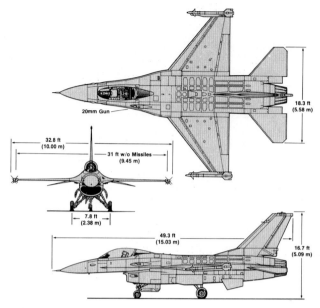

FIGURE 1–9H Multiview drawing for the F–16A aircraft (Courtesy of General Dynamics)

FIGURE 1–10A Large ground structures require many working drawings. (Courtesy of NASA)

FIGURE 1–10B Ship builders constructing an ocean drilling rig (Courtesy of D. E. Gorter, Marine Engineers, Holland)

different fields such as aeronautical, architectural, electrical, electronic, civil, topographic, or mechanical drafting.

In addition to these areas of specialization, drafting and drafting-related professions and occupations are classified by levels of responsibility and duties performed.

Engineers design products, including the selection of materials, standard components, and manufacturing methods. Engineers also supervise the preparation of all drawings required for production.

Architects design structures for specific needs and supervise the preparation of all architectural drawings needed for presentation and construction.

Designers create product design drawings under the direction of an engineer or architect.

Chief drafters supervise all drafting personnel and set drafting standards, practices, and procedures.

Senior drafters create basic layout drawings or sketches and supervise the work of junior drafters or detailers.

Junior drafters complete final technical drawings under the direction of a senior drafter.

Detailers prepare specialized final detailed drawings from which products are manufactured or built.

Checkers review each drawing to insure dimensional accuracy, verify correct use of symbols and standards, and check agreement with related drawings.

Tracers do routine drafting work usually involving changes and corrections on existing drawings.

Technical illustrators are part artists, part drafters who prepare pictorial renderings of engineering or architectural drawings for client presentation or product marketing. An example of the drawings, done by an illustrator can be seen in **Fig. 1-11**

In large companies drafting duties are divided into very small areas of specialization and levels of responsibility. In smaller companies one person may be responsible for many different drafting tasks and functions. Regardless of company size or organization, a thorough knowledge of drafting principles and practices is essential for success.

DRAFTING FIELDS

Although the aforementioned drafting positions are similar in many industries, there are unique job differences among drafters depending on the materials and processes employed in each industry. All drafters

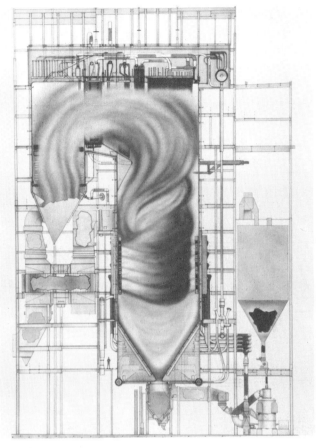

FIGURE 1-11 The technical illustrator is part artist and part drafter when illustrating this steam generator. (Courtesy of Combustion Engineering, Inc.)

must possess a broad knowledge of the drafting basics covered in this text. In addition, drafters must gain specific knowledge of the manufacturing standards and procedures practiced in the industry in which they work. Industries which employ the largest number of drafters include the automotive, aerospace, electronic, plastics, wood products, machine tool, and construction industries.

In the automotive and aircraft industries, drafters involved in the design and manufacture of engines need in-depth knowledge of foundry, diemaking, and machining processes. Drafters working with transmissions, power trains, fuel, lubricating and brake systems need additional training in pneumatics, hydraulics, and cams and gears. Drafters who prepare drawings for steering and suspension systems need additional grounding in statics, dynamics, and strength of materials. Drafters who prepare autobody or aircraft surface drawings must become specialists in pattern or developmental drafting.

In the electronics industry drafters must understand electronic and electrical circuits, and the methods used to design, develop, and manufacture printed circuit boards. Since electronic consumer products are housed in a chassis made of wood, metal, or plastic, a knowledge of manufacturing with these materials may also be required.

Drafters in the machine tool industry must possess in-depth knowledge of physical metallurgy, foundry operations, machine tool operations, tool design, metalworking fasteners and fastening methods, cams and gears, computerized numerical control, and finishing methods.

Drafters in the wood products or furniture industry must understand wood machine tool processes, mass production methods, properties of woods and related synthetic products, wood joint applications, wood fasteners, and finishing methods.

Drafters in the construction industry are broadly classified as architectural drafters. However, most architectural drafters work for companies that specialize in a building type such as residential or commercial light construction, medium-rise commercial construction, high-rise structural steel buildings, highways, bridges, or parks. In the design of small structures one drafter may prepare many or all of the drawings required for construction. In contrast, larger projects require the use of more drafting specialists. For example, drafters who draw high-rise structures or bridges are known as structural drafters or detailers. They must be thoroughly grounded in statics, dynamics, and strength of construction materials. Structural drafters must also be familiar with steel shapes and the many methods used to join structural materials.

Landscape architectural drafters who prepare park, highway, and site development drawings must possess a detailed understanding of surveying practices, geotechnical properties of soils, hydraulics, solar energy, and environmental codes and standards.

In very large construction projects drafting specialists parallel the role of many building subcontractors. In these cases drafters specialize and may prepare only the drawings for site development, framing, foundations, electrical, plumbing, heating, ventilating, or air-conditioning systems. In addition some drafting personnel specialize in the preparation of schedules, specifications, contracts, budgets, and related legal documents.

Regardless of the level of responsibility or drafting specialization, all drafters must understand the basic principles of drafting and design. They must be able to communicate ideas graphically through the use of conventional drafting tools and/or computer-aided drafting systems.

Today over one-half million people work in drafting, design, and related positions. As more products are developed and manufactured, and as more buildings are designed and constructed, the need for personnel with high levels of drafting skill and knowledge will continue to increase.

EXERCISES

1. Research one of the professional positions that you find interesting and write a short paper on the education, training, and the type of work it involves.
2. Select one of the drafting positions described in this chapter. Research a short paper on the education, training, and the type of work it involves.
3. Do a short interview with an engineer, technician, or manager from any industry in your area. Take notes and write a short report on his/her background, education, training, and work performed.
4. Take a field trip to an industry with a CAD system. Watch a demonstration and interview the CAD operator. Take notes and write a short report on his/her background, education, training, and work performed.
5. Looking at the picture of the space shuttle, try to imagine the different engineers, technicians, and support staff required to go from the conception of the design to the completion of the first test flight. Make a list of those involved.
6. Interview a university counselor and list the high school prerequisites recommended for acceptance into an engineering school.
7. Interview a vocational or community college counselor and list the high school prerequisites recommended for acceptance into a technical program.

chapter 2
INTRODUCTION TO COMPUTER-AIDED DRAFTING SYSTEMS

OBJECTIVES

The student will be able to:

- differentiate between different computer-aided drafting (CAD) systems
- describe the role of CAD in industry today
- list the prerequisite skills for operating a CAD system
- identify the major components of a microcomputer-based CAD system
- describe the function of major CAD components

EVOLUTION OF DRAFTING TOOLS

Early drafters measured distances with crude scales and made freehand sketches. Later drafters used simple wooden straightedges as an aid to drawing. For centuries drafters used T squares, triangles, and compasses **(Fig. 2–1A)** to draw more accurate lines, arcs, and circles. Later, the development of the drafting machine and parallel slide systems **(Fig. 2–1B)** enabled drafters to draw much faster and with greater accuracy. Today, computer systems give drafters the capability of drawing faster and much more accurately **(Fig. 2–1C)**.

Development of CAD

Computer systems designed to perform drafting functions are known as **computer-aided drafting** systems, or **CAD**. Computer systems with the capacity to graphically solve more complex design problems are called **computer-aided drafting and design** systems or **CADD**. Drafters interact with CAD or CADD systems to rapidly produce drawings that are extremely accurate, consistent, and easily changeable.

FIGURE 2–1A Handheld drawing instruments (Courtesy of Teledyne Post)

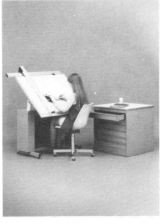

FIGURE 2–1B Track drafting machine (Courtesy of Teledyne Post)

FIGURE 2–1C Computer-aided drafting (CAD) system (Courtesy of Cascade Graphics Systems)

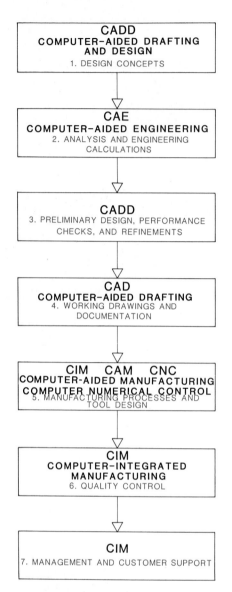

FIGURE 2-2 Flow chart for various CAD software processes

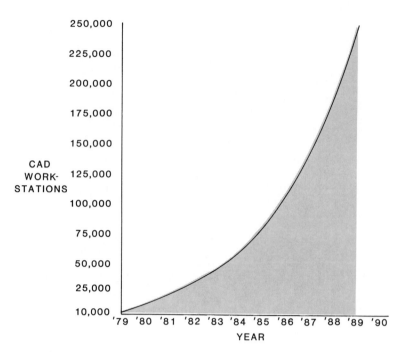

FIGURE 2-3 CAD growth forecast

When computer systems are designed to integrate technical drawings with manufacturing processes, they are called **computer-integrated manufacturing (CIM)** or **computer-aided manufacturing (CAM)** systems. **Fig. 2-2** shows the common acronyms used to describe these different computer systems.

Early CAD systems of the 60s, like all early computer systems, were extremely large, expensive, complicated, and difficult to use. Some systems filled entire rooms and cost millions of dollars. Today's **personal computer** systems (**PCs**) are smaller, faster, less expensive, easier to use, and can perform more complex tasks. These improvements and changes were made possible by the development and continual refinement of the **integrated circuit (IC)** chip. This led to an enormous expansion in the use of CAD systems in all phases of drafting, design, engineering, manufacturing, and construction as shown in **Fig. 2-3**. The developmental process continues with breakthroughs in system size, speed, accuracy, and ease of use.

CAD, the Drafting Tool

The most advanced and sophisticated CAD systems can't draw or design without human input. Although faster, more accurate, and more consistent than the T square, triangle, drafting machine, and parallel slide, a CAD system is simply an electronic drafting tool. Always remember the middle word in CAD is *aided*. CAD systems *aid* the drafter in designing, drawing, and calculating faster, easier and with more accuracy.

Just as a knowledge of mathematics is necessary to effectively use a calculator, a knowledge of orthographic projection, dimensioning, sectioning, symbology, and descriptive geometry is necessary to effectively use a CAD system. Otherwise, a CAD system is not a tool, but only a toy.

CAD CLASSIFICATIONS

There are three general classifications of CAD systems: mainframe, minicomputer, and microcomputer systems.

Mainframe computer systems process information extremely quickly, contain a vast amount of storage capacity (data base), and can rapidly solve very complex engineering problems. Mainframe systems often consist of a large main computer with many terminals (workstations) that access the main computer memory (**Fig. 2-4**). Separate workstations can be located some distance from the main computer and often include two display screens (monitors)—one for graphics and

Chapter 2 Introduction to Computer-Aided Drafting 9

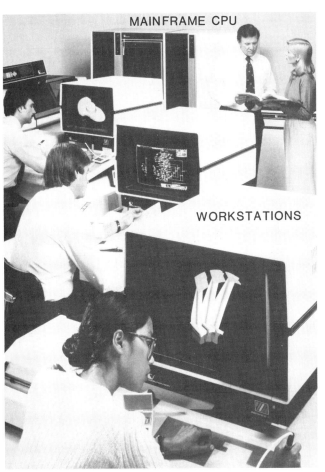

FIGURE 2–5 A dual monitor workstation for a large CAD system (Courtesy of Cascade Graphics Systems)

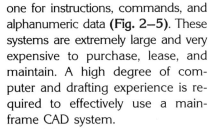

FIGURE 2–4 Multiple workstations (terminals) of a mainframe CAD system (Courtesy of Computervision Corporation)

FIGURE 2–6 A minicomputer CAD system may run approximately 3 to 20 workstations. (Courtesy of IBM Corporation)

one for instructions, commands, and alphanumeric data **(Fig. 2–5)**. These systems are extremely large and very expensive to purchase, lease, and maintain. A high degree of computer and drafting experience is required to effectively use a mainframe CAD system.

Minicomputer systems are similar to mainframe systems except that the speed is slower and the memory capacity is smaller. Three to twenty separate workstations **(Fig. 2–6)** can be connected. Minicomputers are vanishing due to the increased power, sophistication, and economy of microcomputers.

The newest and fastest growing CAD systems are **microcomputer** systems, often referred to as personal computers (PCs). Microcomputer systems are relatively inexpensive **stand-alone** systems **(Fig. 2–7)**. That is, they process information without a link to a mainframe computer. The microcomputer systems of today are much faster and have more memory capacity than mainframe systems of just a few years ago.

Networking

Stand-alone microcomputer workstations can be connected to a mainframe or minicomputer. This is known as **networking** and allows smaller PCs to use the main computer's larger data base and share information among the networked terminals **(Fig. 2–8)**. Networking also saves power. If the main computer

FIGURE 2–7 The microcomputer is a stand-alone system. (Courtesy of Radio Shack, a division of Tandy Corporation)

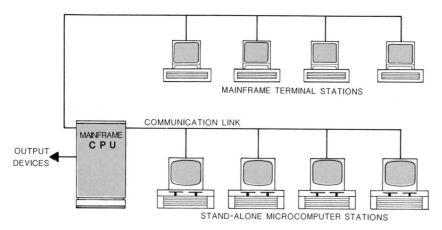

FIGURE 2-8 Workstations for the mainframe computer system and individual stand-alone microcomputers may be connected together (networking).

is down, all terminals will be inoperable, but the PCs can continue to function by storing information on floppy disks. When the mainframe is running again, the PCs can bring files stored through the mainframe network up to date.

CAD SYSTEM COMPONENTS

The components of a CAD system are divided into hardware and software items. **Hardware** includes the computer, monitor, and the **peripherals** which include the input, output, and storage devices. **Fig. 2-9** shows the relationship of peripherals to a computer and computer software.

CAD **software** is a set of computer instructions that cause the computer to act as a drafting system. The instructions are stored on a computer disk or magnetic tape, and are loaded into the computer when needed.

Central Processing Unit

The **central processing unit (CPU)** consists of a computer and a display screen (monitor). The CPU is the brain of the system. Here, all operator input is translated into electrical pulses, the computer's language, and stored in the memory storage system for immediate or later use.

Input Devices

Input is all the information fed into a CAD system by an operator. Input includes graphic, alphabetic, and numerical data. Devices used to input data into the CPU differ depending on the data used and output required.

Keyboards (Fig. 2-10) are similar to standard typewriter keyboards. In addition to normal typing functions, CAD keyboards are used to select drafting tasks. Tasks such as line, arc, circle, erase, etc. are assigned numerical codes which are selected by typing (entering) the numerals with the keyboard. Keyboards can also be used to draw lines by entering the grid coordinates of lines (see Chapter 5, CAD Procedures). However, this method of drawing is very slow and therefore rarely used in CAD operations.

A **mouse (Fig. 2-11)** is a handheld input device. It contains a ball that rolls on a flat surface. As the ball rotates it moves a **cursor**, a bright marker on the monitor, to locate drawing points (coordinates) and/or

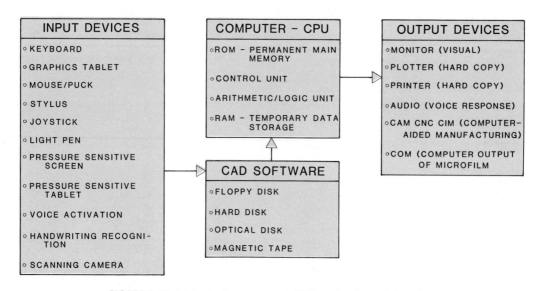

FIGURE 2-9 A block diagram of a CAD system's peripherals

FIGURE 2–10 An alphanumeric keyboard (Courtesy of Hewlett-Packard, AM Bruning Easy Draft 2)

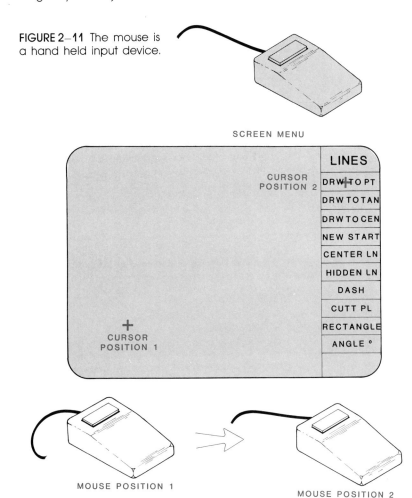

FIGURE 2–11 The mouse is a hand held input device.

FIGURE 2–12 The mouse controls the positioning of the cursor on the monitor.

point to (pick) drawing commands on the screen **(Fig. 2–12)**.

A **puck (Fig. 2–13)** is similar to a mouse but has a flat bottom and is moved over a graphics tablet to control the cursor on the monitor.

Stylus pens (Fig. 2–14) are shaped like a pen or pencil and operate like a mouse or puck.

A **graphics tablet (Fig. 2–15)** is a flat surface upon which the operator draws by touching points with a mouse, puck, or stylus.

A **joystick (Fig. 2–15)** is a spring-loaded stick that tilts and moves a cursor on a monitor. When the stick is tilted up or down, the cursor on the screen moves up or down. Likewise when the stick is moved right or left, the cursor moves right or left.

Light pens (Fig. 2–16) are electronic pointing devices that sense light from a monitor. With a light pen a CAD operator can point directly to the screen to pick menu items or to draw.

On pressure sensitive screens or tables, the cursor is controlled by slight pressure from a finger or a small pointed object.

Voice activation controls the computer through commands spoken into a microphone interface.

Handwriting activation controls the computer through written commands on a small electronic pad that recognizes handwriting.

Optical scanning cameras are the quickest way to enter existing drawings and text into the computer's memory. Scanning cameras convert conventional drawings into digitized CAD drawings.

A **menu** is not an input device, but an aid used to select desired input commands. A graphic menu is similar to a restaurant menu. It is a list of choices. Menu items may be drawing functions such as line, circle, arc, text, etc. Menus also include other drafting commands that manipulate drawings such as erase, move, copy, zoom-in, zoom-out, change scale, store, etc. There are

12 Drafting in a Computer Age

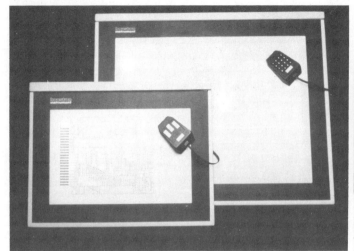

FIGURE 2-13 The puck is used to move the cursor on the screen and to locate coordinate points (digitize). (Courtesy of Houston Instruments, a division of AMETEK)

FIGURE 2-16 The light pen operates directly on the monitor. (Courtesy of Texas Instruments, Inc.)

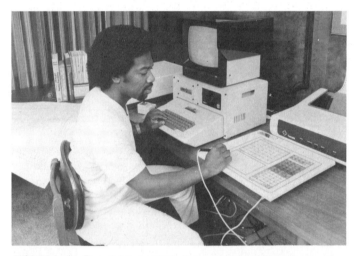

FIGURE 2-14 The stylus controls the cursor and selects drawing commands from the menu. (Courtesy of Cascade Graphics Systems)

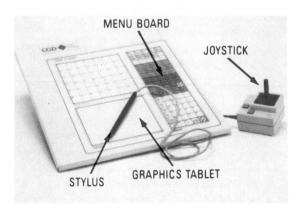

FIGURE 2-15 The joystick controls the cursor on the monitor. (Courtesy of Cascade Graphics Systems)

two common types of menus. One type is displayed on the monitor **(Fig. 2-16)**. The second type is affixed to a graphics tablet **(Fig. 2-17)**. Input devices such as the mouse, puck, or stylus are used to activate tasks shown on a menu.

Output Devices

In conventional drafting the output is an ink or pencil drawing prepared on a paper or vellum surface. The output of a CAD system may be the same. However, the drawing is first viewed on a monitor before a plotter or printer electromechanically transfers the drawing onto a paper or vellum surface. Output devices in CAD systems includes monitors, plotters, and printers.

A **monitor (Fig. 2-18)** or screen is a **cathode ray tube (CRT)**. It is the most common CAD output device. CAD systems include one or two monitors. Monitors may be **monochrome,** usually green or amber, or may have full color capabilities.

The sharpness of the monitor is called the **resolution.** The picture on a monitor screen is comprised of a series of dots. Each dot is called a **pixel.** A CAD monitor should have a higher resolution than a monitor used

Chapter 2 Introduction to Computer-Aided Drafting 13

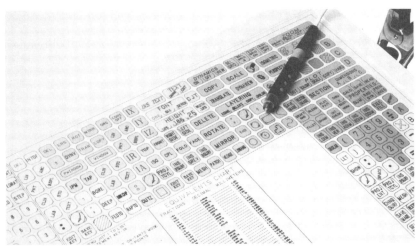

FIGURE 2–17 A menu selection sheet affixed to a graphics tablet (Courtesy of Computervision Corporation)

FIGURE 2–18 All CAD systems have a monitor for visual output. (Courtesy of Motorola, Inc.)

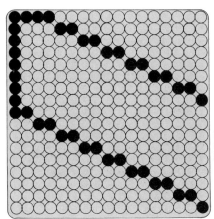
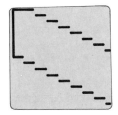

PIXELS AS THEY APPEAR ON THE MONITOR'S SCREEN SHOWS "JAGGIES" FOR ANGULAR AND CURVED LINES

FIGURE 2–19 The line display on a monitor with a low number of pixels will have poor resolution.

for word processing because of the precision and clarity required for reading the drawing lines. An acceptable resolution for a 13" color monitor used for CAD is 680 × 520. This means that there are 680 pixels horizontally across the screen and 520 pixels vertically on the screen. This display will have 353,600 pixels or spots of light on the monitor's screen.

Monitors with fewer pixels produce very jagged lines (low resolution) **(Fig. 2–19)**. Monitors with more pixels produce smoother lines (high resolution) **(Fig. 2–20)**. However, the final appearance of a drawing depends on the quality of the plotter, not on the resolution of the monitor.

Plotters convert images shown on a CAD monitor to ink drawings on paper or vellum. These drawings are known as **hard copy.** Plotters with multiple pens can produce multicolor drawings depending on how the input was entered into the computer. Different line weights can also be drawn with different pens. There are two kinds of plotters—flatbed and drum.

Flatbed plotters (Fig. 2–21) contain a flat surface and pen carriage. Drawing paper is placed on the flat surface and the pen carriage moves over the surface, drawing lines similar to hand-held pens, but much faster, more accurately, and more consistent in weight.

Drum plotters include a cylinder and pen carriage. Drawing paper is placed on the cylinder which moves back and forth to produce lines on the vertical axis. The pen carriage moves along the horizontal axis of the cylinder to produce lines on the opposite axis **(Fig. 2–22)**.

Plotter sizes range from A size **(Fig. 2–23)** to E size. They can plot drawings as small as 8 $\frac{1}{2}$" x 11" or as large as 26" x 48". The scale of the finished drawing can also be adjusted on the plotter.

Printers are output devices that transform monitor images into hard copy in a few seconds. Although

14 Drafting in a Computer Age

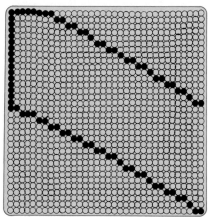

WITH THE ADDITION OF MORE PIXELS ANGULAR AND CURVED LINES WILL HAVE SMALLER "JAGGIES"

FIGURE 2-20 The line display on a monitor with a high number of pixels will have high resolution.

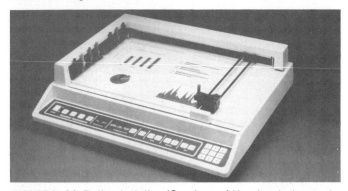

FIGURE 2-21 Flatbed plotter (Courtesy of Houston Instruments, a division of AMETEK)

FIGURE 2-22 Drum plotter (Courtesy of Hewlett-Packard)

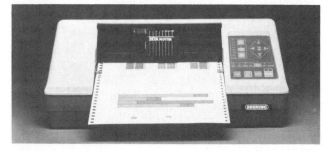

FIGURE 2-23 A small desktop drum plotter for A size drawings (Courtesy of Bruning Computer Graphics)

printer speeds are faster, the line quality is inferior to drawings produced on a **plotter**. Printers reproduce precisely the degree of resolution shown on the monitor. Printed hard copy is known as a screen dump. CAD-system printers are usually used to produce hard copy prints for preliminary checking purposes.

Printers are either dot matrix or laser operated. **Dot matrix printers (Fig. 2-24)** are inexpensive and fast. However, because of the space between the dots, line resolution is poor. The newest dot matrix printers are adding more **dots per inch (DPI)** and improving resolution. **Laser printers** are more expensive but produce lines of higher resolution and are very fast. A comparison of a wood Jorgensen clamp drawing produced on a plotter, dot matrix, and laser printer is shown in **Fig. 2-25**. Note the difference in line qualities and resolution.

Other computer output devices not always associated with CAD systems include electronic typewriters, audio output, and computer-generated microfilm. These devices may become more popular as CAD systems become more sophisticated.

The ultimate output of a CAD system is a finished product, not just a drawing of a product. Through the development of computer-integrated manufacturing (CIM) systems this type of output is now a reality, although currently in limited use.

FIGURE 2-24 Dot matrix printer (Courtesy of Cascade Graphics Systems)

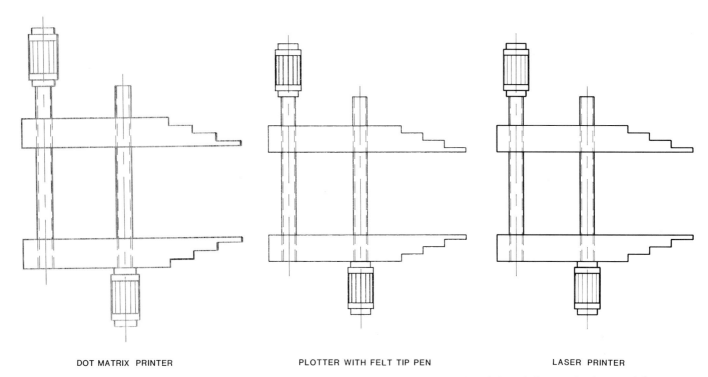

DOT MATRIX PRINTER   PLOTTER WITH FELT TIP PEN   LASER PRINTER

FIGURE 2-25 A comparison of various CAD-generated drawings for a dot matrix printer, plotter, and a laser printer.

Data Storage

One of the greatest benefits of CAD over manual drafting is not in the preparation of original single drawings, but in the use of stored data. CAD storage capacity enables drafters to draw a detail or component only once, store it in a computer library, then retrieve it to be used or altered any number of times. This elimination of repetitive tasks saves thousands of drafting hours on many large projects. Storage capacity continues to increase as IC chips get smaller, and the capacity of CAD storage devices is correspondingly increased.

There are two types of computer memory systems—permanent and temporary. Permanent memory is known as **read only memory (ROM)**. ROM is the data that allows the computer to read the software programs. Temporary memory is known as **random access memory (RAM)**, and includes the information input from the operator and the software program being used.

CAD data storage devices include magnetic tapes, floppy disks, microdisks, hard disk drives, and optical disks.

Magnetic tapes used for data storage may be large reel-to-reel systems or small cassettes. Magnetic tape storage is relatively slow and has been replaced by other types of disk storage systems.

Floppy disks (Fig. 2–26) are the most commonly used storage device for CAD data. Floppy disks are either 8" or 5.25" square. Data is stored on and read from the disk through the head window on the disk. The disk spins at approximately 300 RPMs and is read by a magnetic head in the disk drive.

Microdisks are smaller and bound in hard plastic so they are less fragile than floppy disks. They are 3.5" square and can store more data than a standard 5.25" floppy disk.

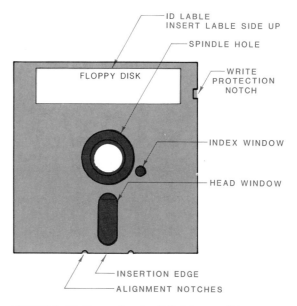

FIGURE 2–26 Example of a 5.25" floppy disk

Hard disk drives (Fig. 2–27) may be a peripheral unit or built directly into the CPU. A hard disk holds much more data than a microdisk, and stores and retrieves data more quickly. A hard disk spins at speeds in excess of 3600 RPMs. The amount of data stored on hard disks is measured in megabytes (10^6) soon to be gigabytes (10^9).

Optical disks can store much more data than any other storage device. One side of an optical disk can store a full set of encyclopedias. Optical disks are used only for permanent data because with laser storage data cannot be changed. Once a breakthrough is made to allow the manipulation of data, the optical disk may become a major CAD data storage system.

Software

A Stradivarius violin would be useless with inferior strings or in the hands of an amateur musician. Likewise, the most sophisticated computer hardware is ineffective without quality software and a knowledgeable operator.

Computer software for CAD is the set of instructions given to a computer. Instructions are written in a computer language such as C, ASSEMBLER, or BASIC by computer programmers. The instructions are known as a computer program. Drafters don't write or create computer programs, but use existing programs for operating CAD systems. Computer programs are kept on a data storage device such as a floppy or hard disk. Quality software includes a wide variety of drafting tasks and options and is relatively easy to use.

CAD software is either operational or applications software. Operational software controls the operation of the computer and allows the communication between hardware components. Applications software is the CAD program. It may be a general purpose drafting program or an application specially written for engineering drawing, civil drafting, architectural drawing, and structural drafting.

Improvements in software programs to expand the usability of CAD include increased speed, increased number of functions, expansion of calculations performed, and the execution of complex tasks with minimal input. Other improvements include refinement of automatic functions such as dimensioning and sectioning, and increased color capability **(Figs. 2–28A** through **2–28M).** Expansion of three dimensional options **(Fig. 2–29)** and ergonomic design capabilities have also extended the use of CAD into areas not formerly used. **Ergonomics** is planning with human dimensions to insure that people will work more efficiently with the designed product. CAD systems now enable handicapped persons to successfully function as drafters **(Fig. 2–30A** and **B).** Other experimental programs now allow drafters to input graphic commands by voice activation.

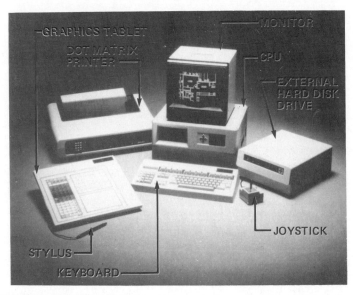

FIGURE 2–27 External hard disk drive and other peripherals (Courtesy of Cascade Graphics Systems)

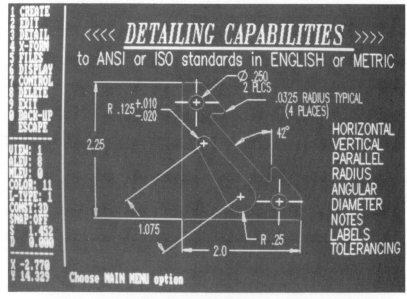

FIGURE 2–28A Detailed working drawing (Courtesy of CADKEY)

Chapter 2 Introduction to Computer-Aided Drafting 17

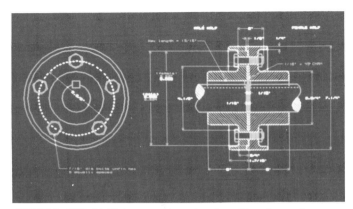

FIGURE 2–28B Multiview working drawing (Courtesy of VersaCAD)

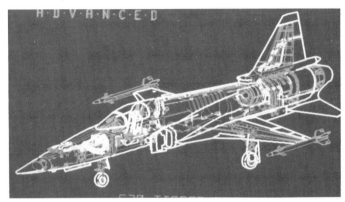

FIGURE 2–28C Pictorial section (Courtesy of VersaCAD)

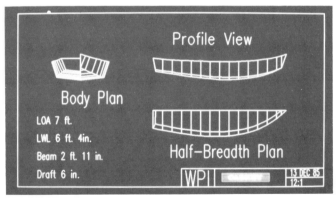

FIGURE 2–28D Multiview drawing (Courtesy of CADKEY)

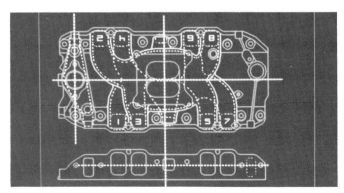

FIGURE 2–28E Multiview working drawing (Courtesy of VersaCAD)

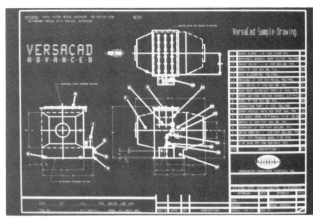

FIGURE 2–28F Multiview working drawing (Courtesy of VersaCAD)

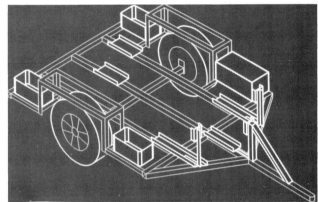

FIGURE 2–28G Assembly drawing (Courtesy of CADKEY)

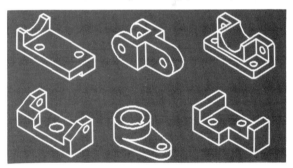

FIGURE 2–28H Pictorial drawings (Courtesy of CADKEY)

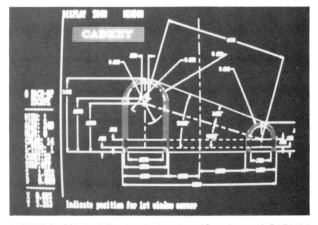

FIGURE 2–28I Architectural drawing (Courtesy of CADKEY)

18   Drafting in a Computer Age

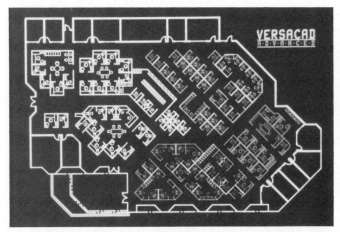

FIGURE 2–28J Architectural drawing (Courtesy of VersaCAD)

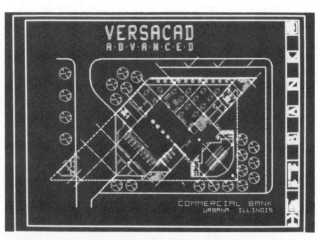

FIGURE 2–28L Site plan (Courtesy of VersaCAD)

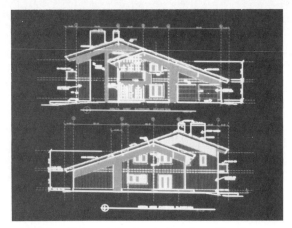

FIGURE 2–28K Architectural elevation (Courtesy of VersaCAD)

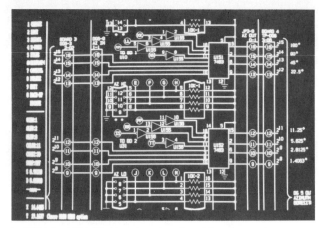

FIGURE 2–28M Electronics drawing (Courtesy of CADKEY)

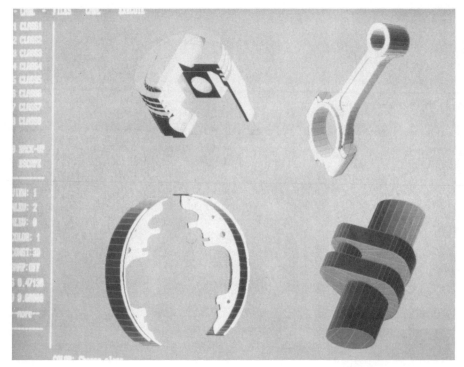

FIGURE 2–29 Shaded surface three-dimensional images help CAD designers develop their concepts. (Courtesy of CADKEY)

As CAD software is improved, the ease of use (user friendliness) is increased. Poor software is difficult to use because options are limited. Good software is easier to use because more options are provided and more help for the operator is programmed into the software. **Fig. 2–31** shows an example of a student CAD output.

A

B

FIGURE 2-30A and B CAD systems now enable the handicapped to function successfully as drafters. (Courtesy of Cascade Graphics Systems)

20 Drafting in a Computer Age

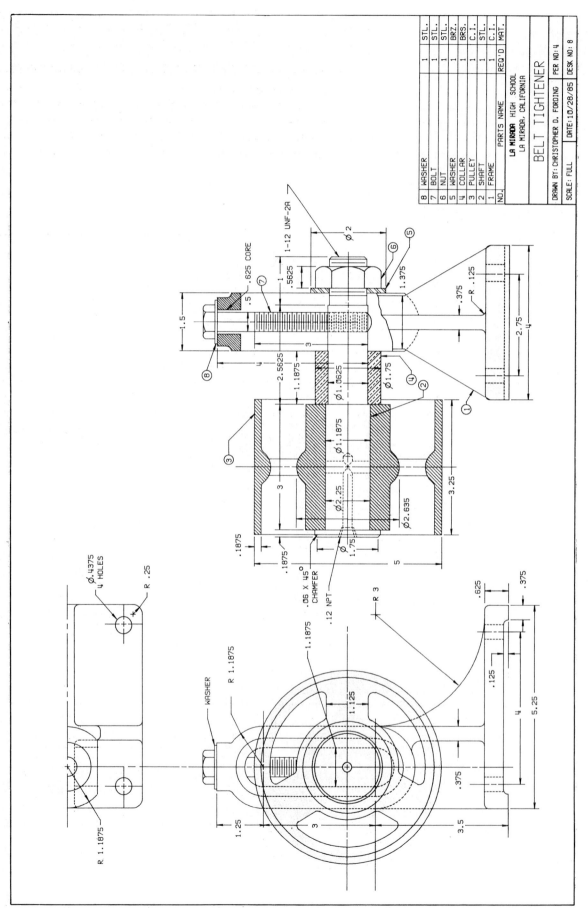

FIGURE 2–31 CAD working drawing (Courtesy of VersaCAD)

EXERCISES

1. Take a field trip to a vendor selling CAD systems. Make a list of various systems and the manufacturers' addresses. Write for additional information.
2. Take a field trip to an industry or office with a CAD system. Observe its operation. List the features of the software observed.
3. Compare prices of CAD hardware and software.
4. Make a sketch of a CAD system and label all the hardware and peripherals.

KEY TERMS

Cathode ray tube (CRT)
Central processing unit (CPU)
Computer-aided drafting (CAD)
Computer-aided drafting and design (CADD)
Computer-aided engineering (CAE)
Computer-aided manufacturing (CAM)
Computer-integrated manufacturing (CIM)
Cursor
Dot matrix printer
Dots per inch (DPI)
Drum plotter
Ergonomics
Flatbed plotter

Floppy disk
Graphics tablet
Hard copy
Hard disk drive
Hardware
Integrated circuit (IC)
Joystick
Keyboard
Laser printer
Light pen
Magnetic tapes
Mainframe
Menu
Microcomputer
Microdisk
Minicomputer

Monitor
Monochrome
Mouse
Networking
Optical disks
Peripherals
Personal computer (PC)
Pixel
Puck
Random-access memory (RAM)
Read-only memory (ROM)
Resolution
Software
Stand-alone
Stylus pens

chapter 3
DRAFTING EQUIPMENT, SUPPLIES, INKING

OBJECTIVES

The student will be able to:

- understand the function of all manual drafting instruments and supplies
- complete a pencil drawing using a full set of drafting instruments
- measure with a civil engineer's scale
- measure with a mechanical engineer's scale
- measure with a metric scale
- measure with an architect's scale
- complete an ink drawing using drafting instruments

INTRODUCTION

Because of the many advantages of CAD systems **(Fig. 3–1)** over manual drafting systems, the number of computer-generated technical drawings in manufacturing and construction is growing at a rapid rate. However, a successful drafter, designer, engineer, or architect must also be able to create drawings using conventional drafting instruments and supplies due to the following reasons:

1. A CAD system may not always be available.
2. Not all companies can or will supply CAD systems to every drafter.
3. Field drawings and instructional sketches must be completed in environments hostile to computers.
4. Creative design in many areas is still done in pencil.

Whether a drawing is prepared as a pencil sketch, inked instrument drawing, or computer-generated hard copy **(Fig. 3–2)**, the final output must be orthographically correct, dimensionally accurate, symbologically true, and easily readable.

CONVENTIONAL DRAFTING SUPPLIES

Conventional drafting supplies consist of surface media and marking devices. Surface media upon which pencil, ink, or CAD-plotted drawings are drawn include a variety of materials and surfaces such as **tracing paper, opaque drawing paper, vellum, film (Fig. 3–3)**, and preprinted **grid papers (Fig. 3–4)**. All drafting surfaces are held in place with drafting tape.

Marking devices used to make lines on surface media include pencils, pens, transfer film, and tapes.

FIGURE 3–1 Computer-aided drafting systems are gradually replacing manual drafting. (Courtesy of the College of San Mateo)

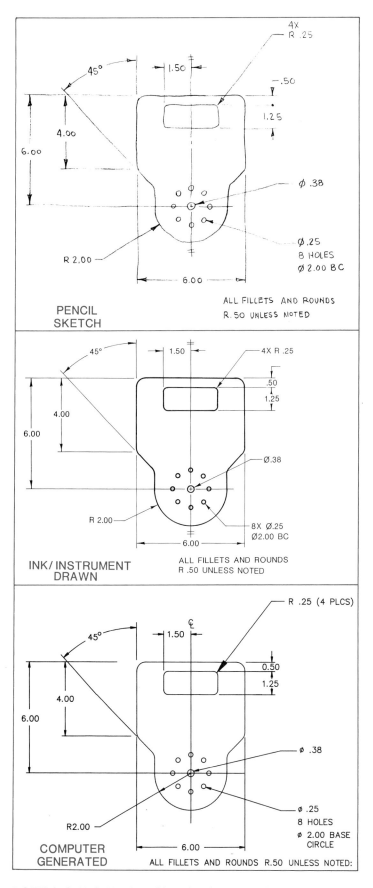

FIGURE 3–2 All finished working drawings must be easily read.

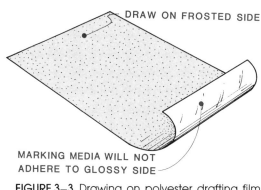

FIGURE 3–3 Drawing on polyester drafting film

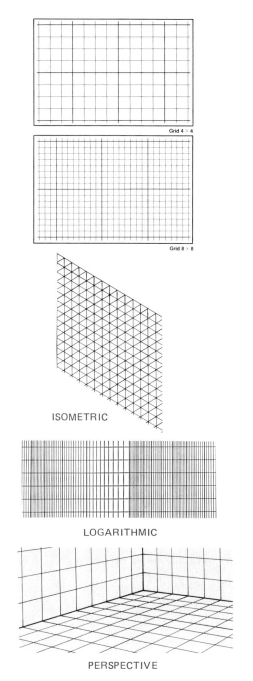

FIGURE 3–4 Examples of grid paper

24 Drafting in a Computer Age

FIGURE 3–5 Typical wood-encased drafting pencil (Courtesy of J. S. Staedtler, Inc.)

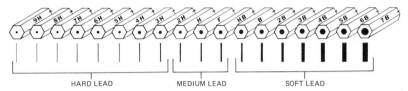

FIGURE 3–6 The various degrees of drafting pencils and their graphite lead widths

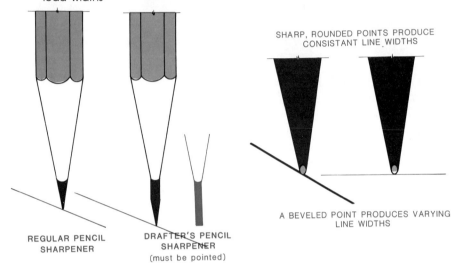

FIGURE 3–7 Pointing the drafting pencil

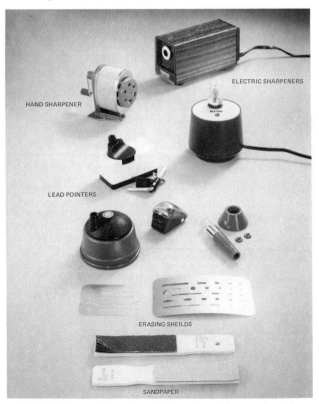

FIGURE 3–8 Examples of pencil and lead sharpeners (Courtesy of Teledyne Post)

Wood pencils are the oldest type of drafting marking device (Fig. 3–5).

Pencils and Lead Holders

Drafting pencils are classified by degree of lead hardness (Fig. 3–6). Hard leads are used for fine layout work, medium leads for most object lines, and soft leads for lettering and rendering. To insure line weight consistency, pencils must be sharpened evenly (Fig. 3–7). Sharp points produce fine lines while round or dull points produce broad lines. Sharpening devices for drafting pencil points include hand or electric rotating sharpeners, cylindrical lead pointers, and sandpaper for hand pointing leads (Fig. 3–8).

Many conventional drafters prefer mechanical **lead holders** (Fig. 3–9). These leads are not sharpened with a sharpener but pointed with a mechanical pointer or on a sandpaper block. Fine line leads and matching holders (Fig. 3–10) don't require either sharpening or pointing. Leads for mechanical holders are available in all grades; fine line leads are available in thicknesses of 0.3, 0.5, 0.7, and 0.9 mm. Drafting leads are made of graphite, not lead. However, some leads are made of plastic, specially designed for use on drafting film.

After drawing with any graphite lead, remember to brush the surface periodically with a dusting brush to remove the accumulation of foreign matter (Fig. 3–11).

Technical Pens

A **technical pen** (Fig. 3–12) is an instrument used to ink lines and lettering on a technical drawing. Inking procedures are similar to pencil drawing except care must be taken to not smear ink lines before they dry. See inking procedures on page 40. Technical pens are designed to be used with a straight edge. Care must be taken to avoid touching the

Chapter 3 Drafting Equipment, Supplies, Inking 25

FIGURE 3-9 The lead holder grips the drafting lead. The lead can be sharpened on sandpaper, a file, or a mechanical sharpener. Leads are available in all drawing grades. (Courtesy of Koh-I-Noor)

FIGURE 3-10 Fine line mechanical pencils do not require sharpening. Leads are available in all drawing grades. (Courtesy of Koh-I-Noor)

FIGURE 3-11 A dusting brush should be used to remove erasures and other foreign matter from the drawing's surface (Courtesy of Teledyne Post)

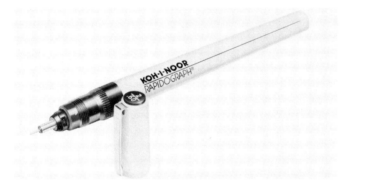

FIGURE 3-12 Technical inking pen (Courtesy of Koh-I-Noor)

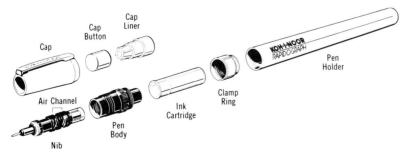

FIGURE 3-13 The parts of a technical pen (Courtesy of Koh-I-Noor)

straight-edge base with the pen to keep ink from running under the surface. Once dry, ink lines can take much abuse compared to graphite lines which can smudge at any time.

Parts of a technical pen are shown in **Fig. 3-13**. The heart of the technical pen is the point. The following are types of inking points:

1. A stainless steel point is a general purpose point for drawing on inking vellum. It is reasonably priced.
2. A jewel point is a hard point used on abrasive surfaces such as drafting film. It is moderately priced.
3. A tungsten-carbide point is an extremely hard point made to last while inking on abrasive surfaces. It is very expensive.

Technical pens are available in sizes 0.1 to 0.9 mm. The line widths are coded in several ways, depending on the manufacturer. **Fig. 3-14** shows typical line widths and two coding systems.

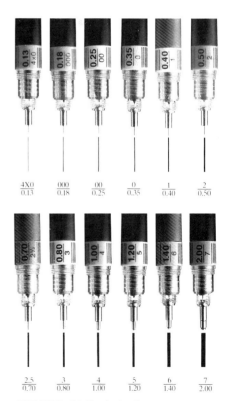

FIGURE 3-14 Technical pen identification designating line widths (Courtesy of Faber-Castell)

Transfer Letters and Tape

A wide variety of film **transfer letters,** numerals, symbols, and lines are available for use on technical drawings. Film transfers are applied to a drawing by rubbing the surface with a hard, smooth stylus until the letter, numeral, symbol, or line is transferred to the drawing surface. **Fig. 3–15** shows the use of this transfer film. **Fig. 3–16** shows some of the thousands of symbol transfer films and **Fig. 3–17** shows a sample of line transfer film. Transfer film lines are available in different widths, and with a variety of spacing and intersecting patterns.

FIGURE 3–15 Position the letter and rub with a hard/smooth tip and letters will transfer onto the drawing. Many styles and sizes are available. (Photo by Ann Ross Wallach)

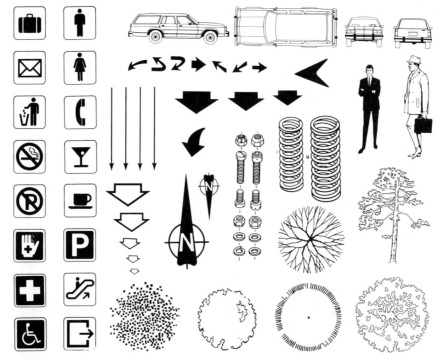

FIGURE 3–16 Many hundreds of different symbol appliques are available, or may be designed and printed onto transfer sheets.

FIGURE 3–17 Preprinted lines, textures and shading may be applied to drawings. Hundreds of various patterns are available.

Lines can also be drawn with tapes which are manufactured in a variety of widths and patterns **(Fig. 3–18)**. Taped lines are effective for very thick lines or borders.

FIGURE 3–18 Line tapes are available in various widths. Rolls are 50' long. (Courtesy of Zipatone, Inc.)

Erasers

Layout and construction lines often need to be removed, drafting mistakes need to be corrected, and graphite smudges must be erased from drawing surfaces. An electric eraser **(Fig. 3–19)** is the most efficient line remover since a gentle touch removes graphite or ink without damaging the surface. A medium-hard eraser with no abrasives is best for most drafting work. To erase ink from drafting film, a drop of water or commercial ink remover can be used. Very small ink areas can be removed by gently scraping with a sharp knife.

Softer graphite lines can be removed with a rubber or vinyl eraser, and smudges can be reduced by sprinkling the drawing surface with a powder cleaner **(Fig. 3–20)**. When using any eraser, always use an erasing shield **(Fig. 3–21)** to insure that only the desired line area is removed.

FIGURE 3–19 The electric eraser is the most efficient cleaner. (Courtesy of Koh-I-Noor)

CONVENTIONAL DRAFTING EQUIPMENT

Quick preparation of accurate manual drawings will be difficult without the aid of conventional drafting equipment.

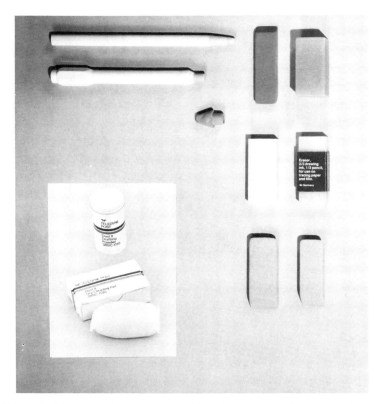

FIGURE 3–20 There is a large variety of hand erasers and powder cleaners. (Courtesy of Teledyne Post)

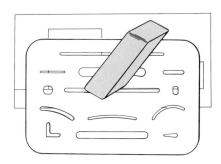

FIGURE 3–21 The erasing shield protects lines not to be erased.

Drafting Machines

Track drafters (Fig. 3–22), are very efficient manual drafting machines. A track drafting machine includes a vertical track that slides left and right on a horizontal track **(Fig. 3–23)**. A protractor head is attached to the vertical track and allows the vertical scale (arm) and horizontal scale (arm) to be rotated to any angle. Fig. 3–24 shows the parts of a protractor head. Although the protractor head can be moved to any angle, an indexing button allows the arms to be automatically stopped every 15°. These are the angles commonly used in drawings **(Fig. 3–25)**.

Elbow drafting machines include a protractor head similar to the track drafter. The elbow machine **(Fig. 3–26)** is connected at the top of a drawing board. It cannot cover an area as large as that covered by a track drafting machine, however, the elbow machine is just as effective.

FIGURE 3–22 The track drafter is a very efficient manual drafting machine. (Courtesy of Vemco Corporation)

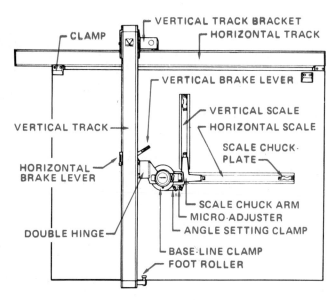

FIGURE 3–23 The track drafting machine and its parts (Courtesy of Mutoh America, Inc.)

1. CENTRAL INDEXING BUTTON
2. INDEXING RELEASE LEVER
3. DOUBLE WING BRAKE LEVER
4. EXCLUSIVE DOUBLE-LENGTH VERNIER
5. OPTICALLY CLEAR CURSOR WINDOW
6. LARGE, FULLY VISIBLE PROTRACTOR
7. SINGLE-LEVER ADJUSTMENT
8. MICROMETER ADJUSTMENT SCREW

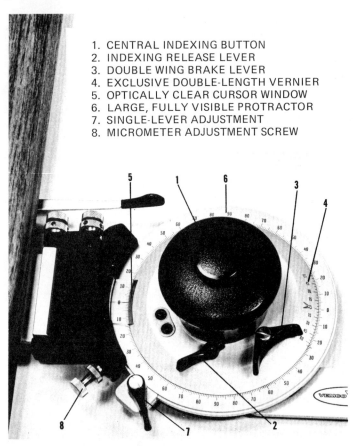

FIGURE 3–24 Parts of a protractor head for a drafting machine (Courtesy of Vemco Corporation)

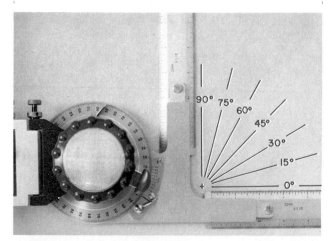

FIGURE 3–25 The indexing button will lock at 15° intervals.

Chapter 3 Drafting Equipment, Supplies, Inking   29

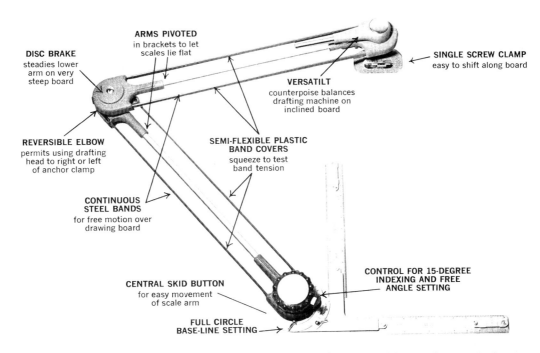

FIGURE 3–26 An elbow drafting machine and its parts (Courtesy of Vemco Corporation)

Large-scale drawings with continuous lines longer than the length of track or elbow machine arms are cumbersome to draw with an elbow or track drafter. For this reason, many large architectural drawings are made with a **parallel slide** drafting system **(Fig. 3–27)**. In a parallel slide system, the horizontal slide is used to draw all horizontal lines. Triangles are placed on the horizontal slide and used for all vertical and angular lines. Some parallel slide and track systems are portable and integrated with a drawing board **(Fig. 3–28)**. These systems are excellent for field-generated drawings.

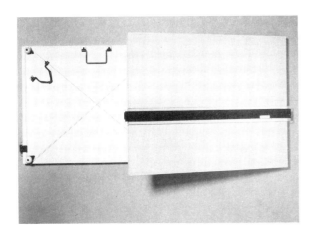

FIGURE 3–27 The parallel slide is permanently attached to the board. The blade slides up and down, and functions like the blade of a T square. (Courtesy of Teledyne Post)

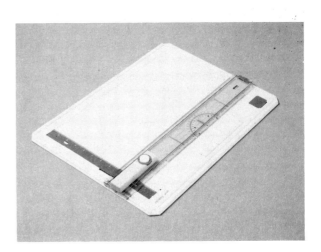

FIGURE 3–28 Portable drawing unit with parallel slide (Courtesy of Koh-I-Noor)

T Square and Triangles

The use of the **T square (Fig. 3–29)** and **triangles** is one of the oldest drafting systems but tends to produce inaccurate lines if the T square is not rigidly held perpendicular to an edge at all times. **Fig. 3–30** and **3–31** show the recommended method of mounting drawing paper onto a drawing board. When used with a T square or parallel slide, triangles **(Fig. 3–32)** are used to draw 15°, 30°, 45°, 60°, 75°, and 90° (vertical) lines. A 45° triangle is used to draw 90° (vertical) and 45° lines **(Fig. 3–33)**. A 30°–60° triangle is used to draw 90°, 30°, and 60° lines **(Fig. 3–34)**. By combining triangle arrangements 15° and 75° lines can be drawn **(Fig. 3–35)**. For drawing angles other than at 15° increments, the adjustable triangle can be used **(Fig. 3–36)**.

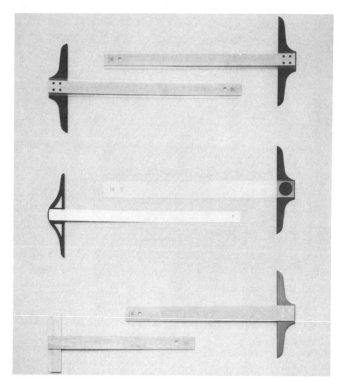

FIGURE 3–29 Various T squares (Courtesy of Teledyne Post)

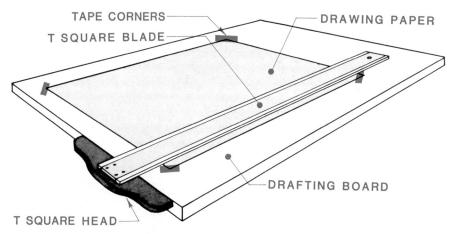

FIGURE 3–30 Line up drawing paper with T square on the drawing board as shown. Tape down corners of drawing media.

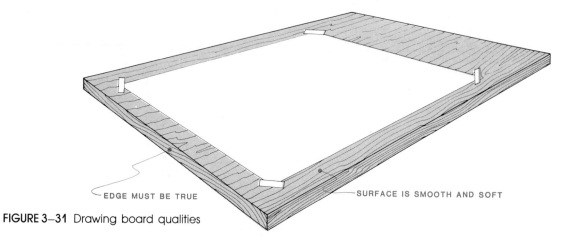

FIGURE 3–31 Drawing board qualities

Chapter 3 Drafting Equipment, Supplies, Inking   31

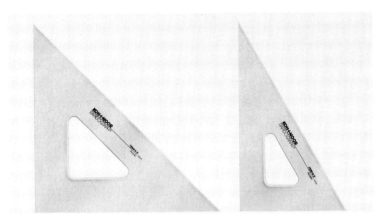

FIGURE 3–32 The 45° and 30–60° triangles (Courtesy of Koh-I-Noor)

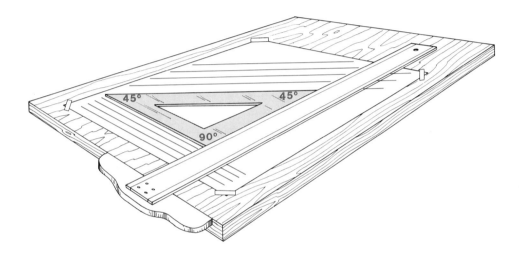

FIGURE 3–33 Drawing 45° and vertical lines

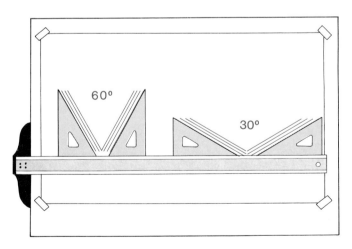

FIGURE 3–34 Positioning triangles to draw 30° and 60° lines

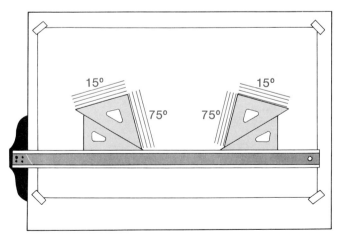

FIGURE 3–35 Positioning triangles to draw 15° and 75° lines

32 Drafting in a Computer Age

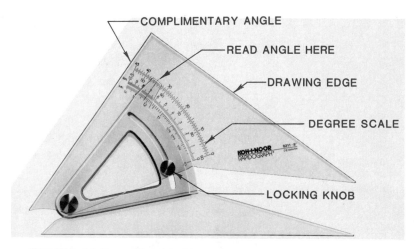

FIGURE 3-36 The adjustable triangle may be set to any angle. (Courtesy of Koh-I-Noor)

Arc and Circle Instruments

All drawings are comprised of three kinds of lines; straight lines, arcs and circles, and irregular curved lines. Drafting machines can only be used to draw straight lines. Regular curved lines which radiate from a center point are drawn with a compass or circle template. There are three types of compasses used in drafting—bow, beam, and friction. **Bow compasses (Fig. 3-37)** are used for small circles up to 6" in diameter, **friction compasses** are used for medium circles from 4" to 10" in diameter, and **beam compasses** are used for large circles over 10" in diameter.

The bow compass is the most commonly used compass. It should be tilted and revolved between the thumb and index finger for best results **(Fig. 3-38)**. **Fig. 3-39** shows the correct method of setting and sharpening a compass point and lead to produce the clearest and most consistent line quality.

Dividers (Fig. 3-40) are similar to compasses except both tips are points. Dividers are used to step off or transfer distances on a drawing.

Irregular curves on technical drawings are drawn with the aid of an irregular curve template, sometimes called a French curve **(Fig. 3-41)**. When using a French curve, follow pre-plotted points in gradual overlapping progression **(Fig. 3-42)**.

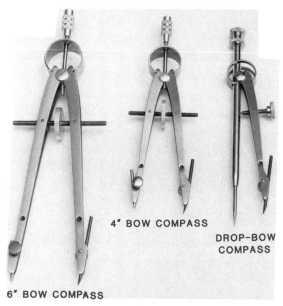

FIGURE 3-37 Bow compasses are used for drawing circles and arcs. (Courtesy of Vemco Corporation)

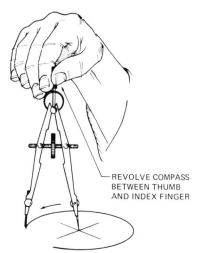
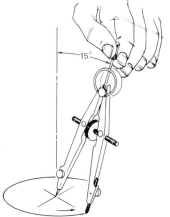

FIGURE 3-38 Drawing procedure for handling a bow compass (Reprinted from TECHNICAL DRAWING AND DESIGN by Goetsch and Nelson, © 1986 Delmar Publishers Inc.)

Chapter 3 Drafting Equipment, Supplies, Inking 33

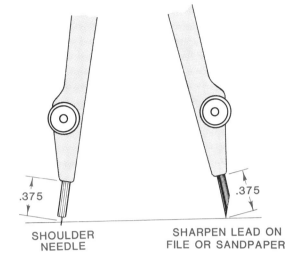

FIGURE 3-39 The compass point and the compass lead

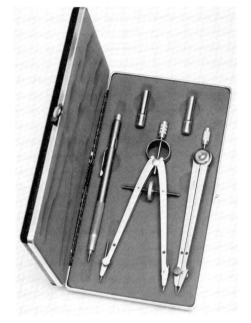

FIGURE 3-40 The divider is used to transfer measurements, and to increase or divide line measurements. Included in the drafting set is a bow compass, a lead holder, and two small containers of spare parts and leads. (Courtesy of Vemco Corporation)

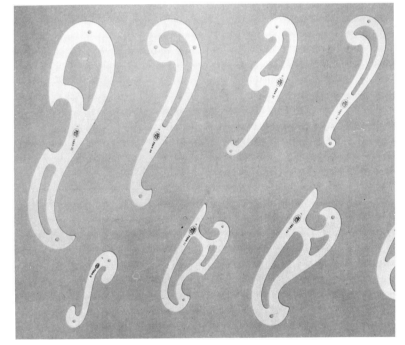

FIGURE 3-41 A French curve is used to draw irregular curves (Courtesy of Teledyne Post)

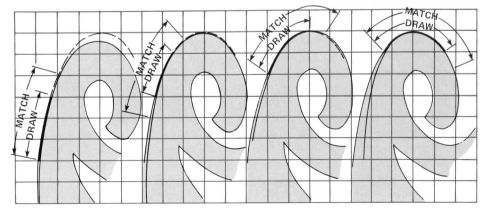

FIGURE 3-42 The procedural steps to draw an irregular curve with a French curve (Reprinted from MECHANICAL DRAFTING by Madsen, Shumaker, and Stewart, © 1986 Delmar Publishers Inc.)

Another device used to draw irregular curves is a **flexible curve (Fig. 3–43)**. A flexible curve can be bent into smooth contours, then held in position while tracing its outline with a pencil or technical pen.

Protractor

When a protractor head on a drafting machine, adjustable triangle, or adjustable T square is not available, a **protractor** can be used to measure and layout angles **(Fig. 3–44)**.

Mechanical Lettering Aids and Templates

For standard circles, ellipses, and arcs, a **template** as shown in **Fig. 3–45A** may be preferable to using a compass.

Templates also have the additional advantage of providing guides for drawing complete symbols, letters, and numerals **(Fig. 3–45B)**. A template used for lettering guide lines is shown in **Fig. 3–46**. Guide line templates are used by sliding the template along the top of a T square, slide, or blade with a sharp pencil inserted into a hole. The guides can be positioned to draw guide lines for different heights of lettering.

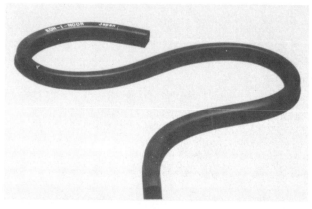

FIGURE 3–43 A flexible curve may be shaped and the desired curve traced along its edge. (Courtesy of Koh-I-Noor)

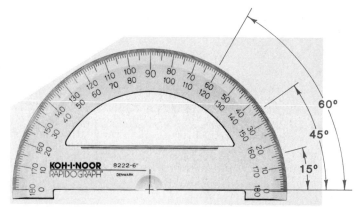

FIGURE 3–44 A semicircular protractor is used to measure angles in degrees. (Courtesy of Koh-I-Noor)

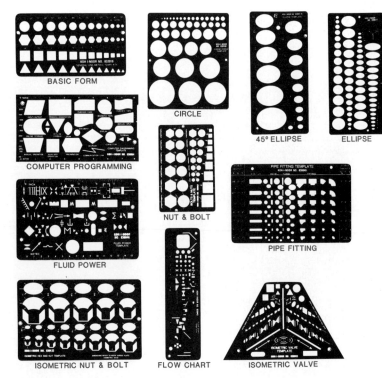

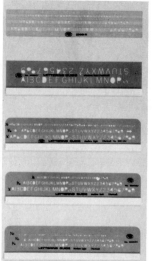

FIGURE 3–45A Sampling of some typical drawing templates (Courtesy of Teledyne Post)

FIGURE 3–45B Sampling of some typical lettering templates (Courtesy of Teledyne Post)

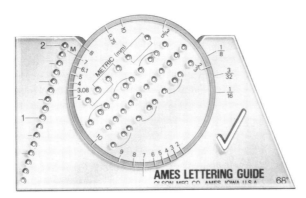

FIGURE 3-46 A guide line template is used by sliding the template along the top of the T square blade with a sharp pencil inserted into a hole. Various spaced guide lines may be drawn. (Courtesy of Olson Manufacturing & Distribution, Inc.)

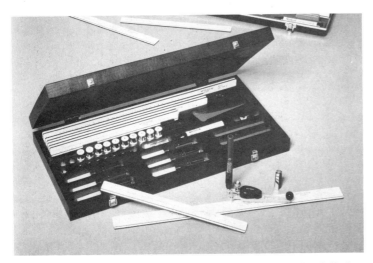

FIGURE 3-47 The Leroy lettering set uses a scribe with the lettering templates to letter with ink or pencil. (Courtesy of Teledyne Post)

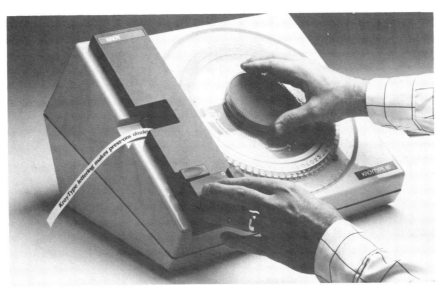

FIGURE 3-48 The Kroy lettering machine prints a single strip of text that will adhere to the drawing. (Courtesy of Teledyne Post)

Other mechanical aids to lettering include the Leroy lettering set **(Fig. 3-47)** which is a template guide with a special technical pen point. Lettering machines such as the Kroy **(Fig. 3-48)** print labels on transparent tape which can be affixed to drawings without blocking out lines. Computer-guided lettering machines draw letters, symbols, or numerals directly on the drawing after they are keyed in.

Drafting Scales

Whether drawing with manual equipment or with a CAD system, an understanding of different scales is essential. Scales are available in several different shapes **(Fig. 3-49)**. Scales allow drafters to create drawings that are proportioned to the actual size of the object being drawn. Scales are based on either the **U.S. customary system** or the **metric system.** Scales used in drafting include the mechanical engineer's scale, the civil engineer's scale, the architect's scale, and the metric scale.

The **mechanical engineer's scale** uses a U.S. customary inch-fraction unit of measure **(Fig. 3-50)**. Scale subdivisions include $1/32''$, $1/16''$, $1/8''$, $1/4''$, $3/8''$, $3/4''$, $1/2''$, and $1''$ units.

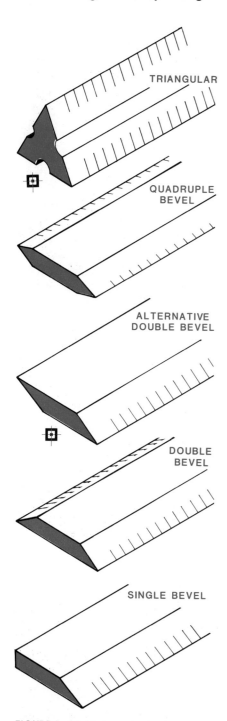

FIGURE 3–49 Various scale shapes

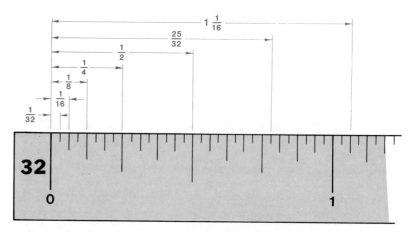

FIGURE 3–50 The mechanical engineer's scale uses inch-fraction units of measure.

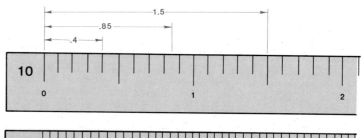

FIGURE 3–51 The civil engineer's scale uses inch-decimal units of measure.

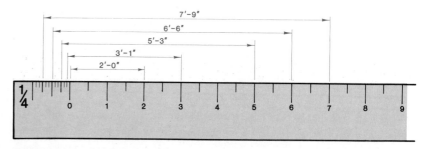

FIGURE 3–52 The architect's scale uses foot-inch units of measure. The most often-used architect's scale is $1/4'' = 1'-0''$

The **civil engineer's scale** uses a U.S. customary inch-decimal unit of measure **(Fig. 3–51)**. The inch-decimal unit is used by most industries using the U.S. customary system. Civil engineer's scale subdivisions include 10, 20, 30, 40, 50, and 60 parts per inch.

The **architect's scale** is used to prepare plans for structures. It is divided into twelve parts, so inches on a drawing can equal the actual inch or feet dimensions of a building **(Fig. 3–52)**. For example, most architectural drawings are prepared to a scale $1/4'' = 1'-0''$. This means that a $1/4''$ line on a drawing represents one foot on a building. Thus, at $1/4'' = 1'-0''$ scale, an eight-foot wall would appear 2″ long on the drawing. The architect's scale subdivisions are $3/32''$, $3/16''$, $1/8''$, $1/4''$, $3/8''$, $3/4''$, $1/2''$, 1″, $1\text{-}1/2''$, and 3″.

The basic unit of measure for metric scales is the **millimeter (mm)** **(Fig. 3–53)**. In metric drawings the abbreviation, mm, is not used since all dimensions are in millimeters. All of the world except the United States uses the metric system for technical drawing and manufacturing.

Drafting scales are either **open divided** or **full divided (Fig. 3–54)**. Only one major unit of open-divided scales is graduated with a full-divided unit. It is adjacent to the zero. Full-divided scales contain full subdivision lines throughout the entire length of the scale. In selecting the proper scale for each drawing, the drafter must consider the amount of space available, readability of the finished drawing, and ease of drawing. **Fig. 3–55** provides some basic guidelines for proper scale selection.

In selecting a scale remember that a decimal or fractional part of an inch can be made equal to any unit of measure such as an inch, foot, yard, or mile.

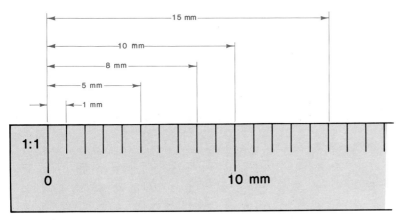

FIGURE 3–53 The basic unit of measure for metric scales is the millimeter (mm).

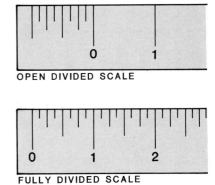

FIGURE 3–54 Drafting scales are either open divided or full divided.

SCALE SELECTIONS FOR ENGINEERING DRAWING			
Types of Drawings	**Mechanical Engineer's Scale**	**Civil Engineer's Scale**	**Metric Scale**
The object to be drawn must be smaller than the drawing format. All the drawings, dimensions, and notations will fit at actual size.	Full Size (1:1)	1:1	1:1
The object to be drawn is larger than the drawing format and must be reduced in half to fit the format.	½" = 1"	1:2	1:2
The object to be drawn is much larger than the drawing format and must be reduced eight to ten times in size to fit the drawing format.	⅛" = 1"	1" = 10" 1:10	1:10
The object to be drawn is very large, such as a building, and must be reduced at least fifty times in size to fit the drawing format.	¼" = 1' – 0" (1:48)	1:50	1:50
The object to be drawn is small and cannot easily be drawn full size. Doubling the drawing size makes the drawing easier to drawn and interpret.	1" = ½" (2:1)	2:1	2:1
The object to be drawn is very small. For ease of drawing and interpreting, the original size must be increased eight or ten times.	1" = ⅛" (8:1)	10:1	10:1
When an industrial product is extremely small, such as circuitry chips, the drawings must be drawn approximately 100 to 500 times larger.	Not used	100:1 500:1	100:1 500:1

FIGURE 3–55 Typical scale selection for engineering drawings

Drafting Furniture and Lighting

Since drafters spend many hours drawing each day, drafting furniture **(Fig. 3–56)** must be adjustable to individual preferences and sizes (ergonomic design) and be maneuverable to enable the drafter to reach any part of a large drawing comfortably. Local lighting **(Fig. 3–57)** must also be easily adjustable, provide sufficient illumination, and produce a minimum of shadows.

Economic Considerations

When purhasing scales and other drafting instruments **(Fig. 3–58)** always select the highest quality affordable. Purchasing sets of instruments **(Fig. 3–59A** and **B)** is often more economical than purchasing individual instruments. **Fig. 3–60** shows a set of basic drafting supplies and equipment recommended for a beginning student.

FIGURE 3–56 Drafting tables and chairs are ergonomically designed in a very large number of sizes and styles. This drafting table is mounted with a track-drafting machine. (Courtesy of Teledyne Post)

FIGURE 3–57 A drafting lamp will insure the proper lighting for drafting. (Courtesy of Vemco Corporation)

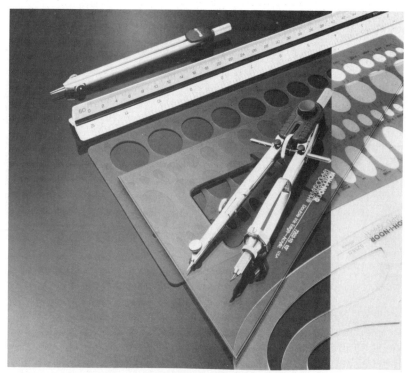

FIGURE 3–58 Many considerations must be made when purchasing drafting instruments. (Courtesy of Koh-I-Noor)

Chapter 3 Drafting Equipment, Supplies, Inking   39

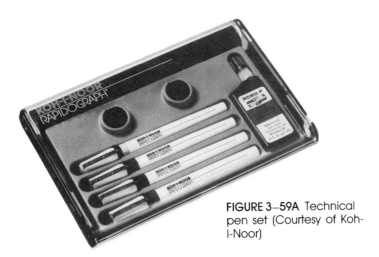

FIGURE 3–59A Technical pen set (Courtesy of Kohl-Noor)

INKING

When a drawing requires a sharper and more precise line width than can be produced with a graphite pencil, inking is the solution. Ink is very dense and black, insuring superior prints and photographic qualities needed for reproduction and microfilming. **Fig. 3–61** compares the qualities of ink and pencil drawings.

Ink for technical pens is stored in an ink reservoir inside the barrel of the pen. To start the ink flow, shake the pen. A weighted wire inside the inking tip will feed the ink through the opening. Draw the first line on a scrap surface to eliminate any hardened ink particles before proceeding. When the pen is not in use, seal the cap snugly so the ink will not dry and harden inside the point **(Fig. 3–62)**.

Cleaning

Cleaning a technical pen may be done by running hot water over the tip. Do not remove the wire from the tip. If this fine wire bends, the point will not function well **(Fig. 3–63)**. If hot water does not clean the pen point, use an **ultrasonic pen cleaner** **(Fig. 3–64)**, or soak the point in liquid pen cleaner for an hour.

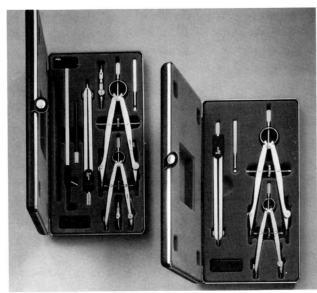

FIGURE 3–59B Complete sets of drafting instruments and supplies may be purchased. (Courtesy of Teledyne Post)

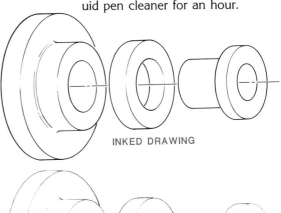

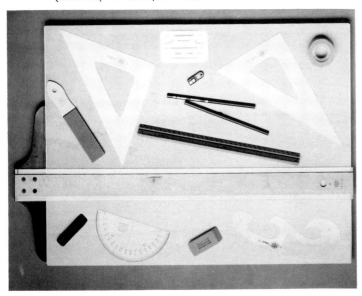

FIGURE 3–60 Basic drafting instruments required for the beginning student (Courtesy of Teledyne Post)

FIGURE 3–61 Compare the line quality of the ink and pencil drawing.

Dry Seal 1 Cap threaded firmly to pen seals cap liner to pen body.

Dry Seal 2 Open end of writing tube seals against pliable base of liner.

No humidification required.

No spring-tension point seal to jam with dried ink.

FIGURE 3–62 A tightened cap prevents the ink from drying in the tip. (Courtesy of Koh-I-Noor)

FIGURE 3–63 A weighted wire controls the ink flow through the inking tip. (Courtesy of Koh-I-Noor)

FIGURE 3–64 An ultrasonic technical pen cleaner (Courtesy of Keuffel & Esser Company, Rockaway, NJ)

FIGURE 3–65 Good quality inked lines will adhere tightly to the drawing media and will not feather out.

Ink

Ink used in technical pens may be a black waterproof or nonwaterproof drawing ink (also called **india ink**). Ink will penetrate vellum and adhere. Drafting film will not allow ink to penetrate; therefore, a special ink with good adhering qualities must be used for plastic film. Quality drawing ink must

1. flow smoothly,
2. dry quickly,
3. adhere to drawing media,
4. not chip or peel, and
5. dry very opaquely.

Inking Surfaces

Ink drawings are prepared on inking vellum or plastic drafting film. Inking surfaces must not be porous or the ink will feather out. **Fig. 3–65** shows various inked lines. The use of good quality inking media permits re-inking of good inked lines over erased areas without feathering.

Be certain to draw on the correct side of the drafting film. Ink on the matte side, not on the glossy side. When inking on drawing medias with printed grids, ink on the side opposite the printed grid lines since ink will not adhere well to the printed grid lines.

Before inking, the surface of the drawing media should be cleaned with pounce powder. Rubbing a small amount of pounce powder over the drawing surface will clean smudges and oil from fingerprints, allowing the ink to adhere to the drawing surface **(Fig. 3–65)**.

Inking Procedures

To produce a quality ink drawing follow the steps outlined in **Fig. 3–66**:

Step 1. Draw a fast and accurate layout with pencil and paper.
Step 2. Attach inking vellum or film over the pencil layout.

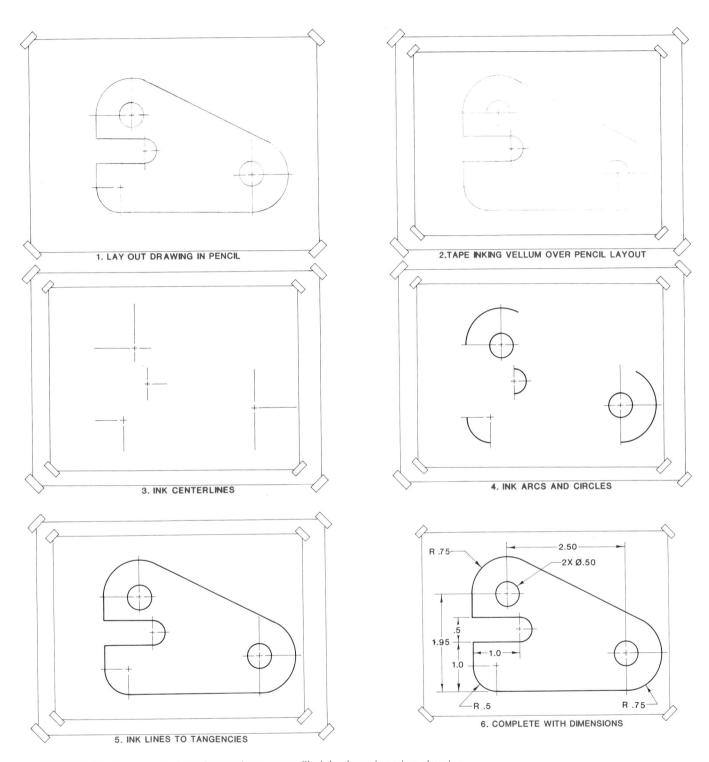

FIGURE 3-66 Procedural steps to produce a quality inked engineering drawing.

Step 3. Ink the centerlines.
Step 4. Ink circles and arcs.
Step 5. Ink tangencies.
Step 6. Complete line work.
Step 7. Dimension and add notes.

When inking, follow these guidelines:

1. When inking horizontal lines, start at the top and work down.
2. When inking vertical lines, start at the left, if righthanded, or at the right if lefthanded and work across the paper.
3. Keep the technical pen vertical for consistent line width **(Fig. 3-67)**.
4. Wipe any ink from the outside of the pen point to insure consistent line width **(Fig. 3-68)**.
5. Keep instruments from touching the inking surface to eliminate **capillary smearing (Fig. 3-69)**.

42 Drafting in a Computer Age

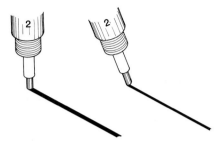

FIGURE 3–67 Keeping the technical pen vertical will consistently produce the standard line.

FIGURE 3–68 Before inking, wipe the tip clean to insure consistent line widths. (Courtesy of Koh-I-Noor)

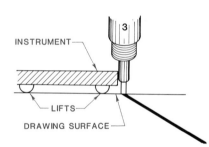

FIGURE 3–69 Lifting drawing instruments slightly off the drawing's surface will prevent capillary smearing.

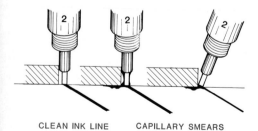

FIGURE 3–70 If ink touches an instrument, capillary action will cause an ink smear.

6. Keep the inking tip from touching the drawing instrument's edge to prevent capillary ink smears **(Fig. 3–70)**.
7. Shiny ink lines are wet. Do not touch with instruments or fingers until dry. The ink will appear flat when dry.
8. Move the technical pen at a slow, steady pace to allow a smooth flow of ink onto the line.
9. Practice making smooth tangencies by first drawing arcs, then lines to tangency points **(Fig. 3–71)**.

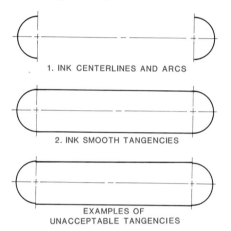

FIGURE 3–71 Drawing procedures to ink tangencies

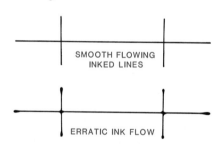

FIGURE 3–72 Quality inked drawings must have smooth, crisp, and consistent line widths.

10. If ink lines do not flow smoothly, change to fresh ink, clean the pen, and/or add a new replacement point **(Fig. 3–72)**.
11. Ink circles with a circle template or inking compass **(Fig. 3–73)**.
12. Ink letters or numerals with a lettering template or lettering device **(Fig. 3–74)**.

Fig. 3–75 shows several examples of inked drawings used for technical illustration.

FIGURE 3–73 Inking compass (Courtesy of Koh-I-Noor)

FIGURE 3–74 Inking notes and dimensions (Courtesy of Koh-I-Noor)

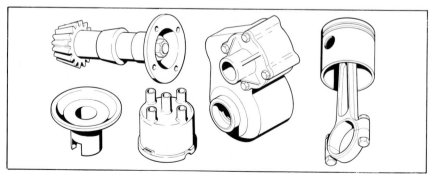

FIGURE 3–75 Examples of inked pictorial drawings

COMPUTER-AIDED DRAFTING & DESIGN

CAD EQUIPMENT

CAD systems consist of a central processing unit and supporting input, output, and data storage devices **(Fig. 3–76)**. Supplies needed to make this equipment functional include floppy disks for the computer, pens, and plotting media for the plotter.

Floppy Disks

The most common floppy disk is the 5.25" square disk, referred to as a five-and-a-quarter floppy **(Fig. 3–77)**. Data stored on a floppy disk is measured in bytes. One byte holds one character of digital data. The range of data storage depends on the type of floppy disk used.

There are several types of 5.25" disks—**single-sided** or **double-sided**. Single-sided means only one side of the disk may be used for data storage. These disks hold 180KB of data (K = 1024 bytes of digital data). Double-sided double-density means both sides can be used for data. These are the most common disks used on CAD systems. They can

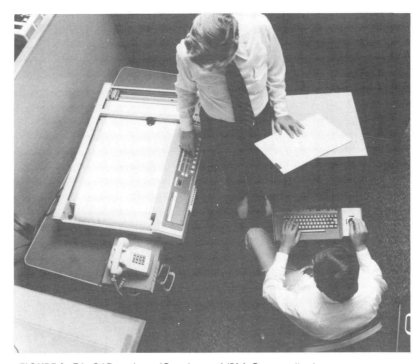

FIGURE 3–76 CAD system (Courtesy of IBM Corporation)

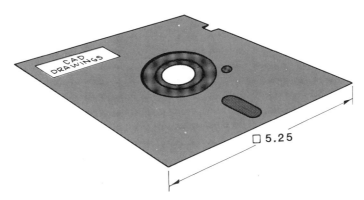

FIGURE 3–77 Typical $5\frac{1}{4}$" floppy disk

44 Drafting in a Computer Age

store 360KB of data. Double-sided high-density disks are capable of storing 1.2MB of data (M = 1,048,576 bytes).

It is important that the proper disk is used in the computer; the computer reference manual will specify which type to use. Proper disk care is very important. If a disk is misused the data may become scrambled or the disk may become completely unuseable. The chart in **Fig. 3-78** lists some tips on the proper care of disks. It also is important not to place a disk on the monitor, computer, or on the graphic tablet, as data may be scrambled by their magnetic fields.

Plotter Pens

The most advanced CAD systems **(Fig. 3-79)** cannot produce professional-level drawings without the use of quality plotter pens **(Fig. 3-80)**. Plotter pens are available in ink, felt-tip, or ballpoint in a variety of sizes and colors.

The selection of CAD plotting pens is more difficult than that of manual drafting pens. Plotter pens are not standardized; therefore, it is important that the pens specified in the plotter manual be used. Fig. 3-80 shows some of the shapes of the various plotter pens. Note that the main difference is in their physical shape. Even though pen shape varies, the tips are the same. The categories of plotter pen points **(Fig. 3-81)** include the following:

1. Refillable liquid ink—The tips on these pens are similar to those of

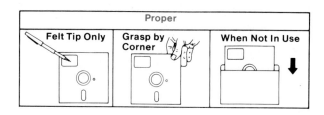
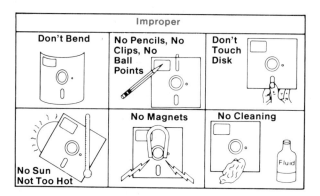

FIGURE 3-78 Floppy disk care (Courtesy of IBM Corporation)

FIGURE 3-79 An advanced CAD system (Courtesy of Cascade Graphics Systems)

FIGURE 3-80 Plotter tips are available with ink, felt-tip, and ballpoint, in an assortment of colors. (Courtesy of Koh-I-Noor)

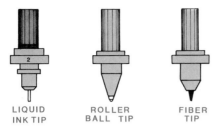

FIGURE 3–81 Plotter pen points

Characteristics	Pen Recommendations			
Type of Plot:	Refillable Liquid-Ink	Disposable Liquid-Ink	Disposable Fiber-Tip	Disposable Ballpoint
Business Graphics	No	Good	**Best**	Good
Check Plot	Good	**Best**	Good	**Best**
Design/Engineering	**Best**	**Best**	No	No
Fill-In	No	Good	**Best**	No
Final Plot	**Best**	**Best**	No	No
Multiple Precision Line Widths	**Best**	No	No	No
High Accuracy	**Best**	No	No	No
High Speed	**Best**	**Best**	No	No
Archival/Reproducible	**Best**	**Best**	No	No
Waterproof/Fade-Resistant	**Best**	**Best**	No	No
High Quality/Maintenance Free	No	**Best**	No	No

FIGURE 3–82 Usage chart for plotter pens (Reprinted courtesy of Koh-I-Noor)

Plotter Pen Media Guide	
Vellum	Very high quality paper. Can be used with all types of pens.
Tracing Paper	Very similar to bond or vellum.
Bond Paper	Good grade paper. Use with all types of pens.
Chart Paper	Low cost paper. Best used with fiber tips. Can be used with DPP™.
Clay-Coat Paper	Coated paper. Excellent for colored graphs. Use with fiber pens only.
High Gloss Paper	Same characteristics as clay-coat.
Matte Paper	Excellent for colored graphs. Best with fiber tips. Some brands can be used with liquid-ink pens and DPP™
Matte Film	Highest quality surface. Best dimensional stability. For archival plots, and final plots. Use liquid-ink pen designed for use on film (or both paper and film).
Overhead Film	For overhead transparencies. Use fiber-tip pens with a film ink.

FIGURE 3–83 Drawing medias for plotters (Reprinted courtesy of Koh-I-Noor)

inking pens used in manual drafting.

2. Disposable liquid ink—These produce the same high quality line as refillable liquid ink pens, but are maintenance free. However, in the long run, they are more expensive.
3. Disposable fiber-tip—These are relatively inexpensive, do not produce high-quality crisp lines, but are excellent for filling large areas in charts.
4. Disposable ball or roller points—These pens do not produce a high-quality line, but are good for check prints. They are durable and inexpensive, but will skip if operated at high speed.

The variety of plotter points enables a CAD drawing to be plotted with line-weight contrasts or in multicolors. **Fig. 3–82** is a usage chart for plotter pens.

Plotting Media

The selection of plotting media (drawing surfaces) is just as important as the selection of plotter pens, and depends on the intended use of the drawing. The chart in **Fig. 3–83** lists nine types of media used for plotting and their primary uses. The two most commonly used plotting media are vellum and **bond**. Vellum is used when blueline or blueprints are to be made from the plotted drawings, and bond is used when the drawings are to be duplicated using a photocopy machine.

Exercises

1. Divide a "B" (11" x 17" or 12" x 18") drawing format into 11 parts. Practice sketching the line work with drafting pencils, **Fig. 3–84**.
2. Divide a "B" drawing format into 11 parts. Practice drawing with instruments and drafting pencils, **Fig. 3–84**.

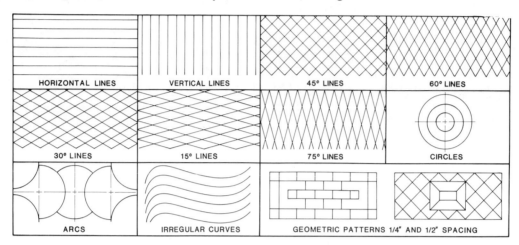

FIGURE 3–84 Divide four B-size drawing formats into 11 parts as shown. Practice the line work with pencil sketching, drawing instruments and pencil, drawing instruments and ink, and a CAD drawing.

1 ───
2 ───
3 ──
4 ──────────────────────────────────────
5 ─────────────────
6 ──────────────────────────────────
7 ────────────────────────
8 ──────────
9 ──────────────
10 ────────────────────
11 ──────
12 ──
13 ──────────────
14 ────────
15 ──────────────────
16 ──────────────────────────────
17 ──
18 ──────────────────────────────────────
19 ──
20 ──
21 ──────────
22 ──────────────
23 ──────────────────────────────
24 ──
25 ──────────────────

FIGURE 3–85 Measure each line with as many different scales as are available.

3. Divide a "B" drawing format into 11 parts. Practice inking **Fig. 3-84**.
4. With a CAD system, practice laying out the line work in **Fig. 3-84**.
5. Measure each line in **Fig. 3-85** with these scales:
   - full size—inch-decimal
   - full size—inch-fraction
   - $1/2'' = 1''$
   - $1'' = 10'$
   - full size—metric (millimeters)
   - $1/4'' = 1'-0''$
6. With a CAD system draw **Fig. 3-86**.
7. Make a hard copy of the CAD work for Exercise #6 with a plotter and/or a printer.

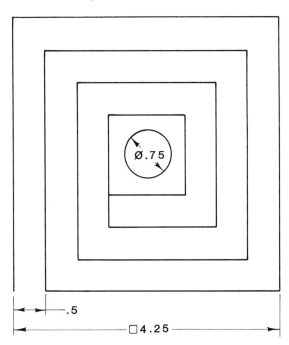

FIGURE 3-86 Draw this figure with a CAD system.

KEY TERMS

Architect's scale
Beam compass
Bond
Bow compass
Capillary smear
Civil engineer's scale
Dividers
Double-sided floppy disk
Elbow drafter
Film
Flexible curve

Friction compass
Full-divided scale
Grid paper
India ink
Irregular curve
Lead holder
Mechanical engineer's scale
Metric system
Millimeter (mm)
Opaque drawing paper
Open-divided scale
Parallel slide

Protractor
Single-sided floppy disk
Technical pen
Template
Tracing paper
Track drafter
Transfer letters
Triangles
T square
Ultrasonic cleaner
U.S. customary system
Vellum

chapter 4
SKETCHING AND LETTERING

OBJECTIVES

The student will be able to:

- use proper line weights to sketch and letter
- complete a two-dimensional sketch of simple objects
- complete a three-dimensional sketch of simple objects
- file a record of sketched ideas
- select and sharpen sketching and lettering pencils
- keep sketches in proportion
- dimension a sketch
- shade a sketch
- letter clearly on a sketch

INTRODUCTION

Highly developed manual drafting skills are not required for the preparation of CAD drawings (Fig. 4–1); however, a competent CAD operator must possess a thorough understanding of the principles of drafting and be skilled in technical sketching. The ability to quickly and legibly sketch ideas and products is needed by CAD operators. Observe how the sketch shown in Fig. 4–2 is used to describe the actual object shown in Fig. 4–3. This is another example of the proverb, "One picture is worth a thousand words." CAD operators use sketches to

1. record ideas immediately when a CAD system is not available,
2. help organize new designs,
3. help to formulate new ideas,
4. make certain that others understand ideas and requests,
5. explain complex drawings prior to developing a CAD drawing,
6. make changes and update designs prior to developing a CAD drawing,
7. aid in creative design,
8. plan the sequences to be used in preparing CAD drawings, and
9. record design alternatives for future reference.

SUPPLIES

Sketching supplies consist of a wide variety of pencils, erasers, and drawing media (see Chapter 3). Soft pencils in the range shown in Fig. 4–4 are used for sketching. Pencil points (Fig. 4–5) are rounded for sketching compared to the sharp points used for instrument drawing. However, sharp points on soft pencils are used for layout and lettering guide line sketching. Chisel points are used for shading. Sketching pencils are sharpened with a mechanical sharpener, or with a knife and file, or sandpaper to shape the point.

Although many types of drawing media are used for technical sketching, grid drawing paper is the most practical and easy to use (Fig. 4–6). Grids provide guide lines for rough scaling and proper alignment of perpendicular and parallel lines. Use light-colored surfaces if sketches are to be photocopied and use translucent surfaces if diazo prints are to be made. Translucent surfaces are also helpful if progressive design sketches are to be traced.

WORKING DRAWING SKETCHES

Working drawing sketches are used as the major design reference in manufacturing and construction. Sketches are also used as working drawings when time and conditions preclude the preparation of instrument or CAD working drawings. These sketches are usually multiview, orthographic drawings, but are sometimes prepared as pictorial drawings. An **orthographic** drawing is a drawing of several views of an object on a drawing surface that is perpendicular to both the view and the lines of projection. A **pictorial**

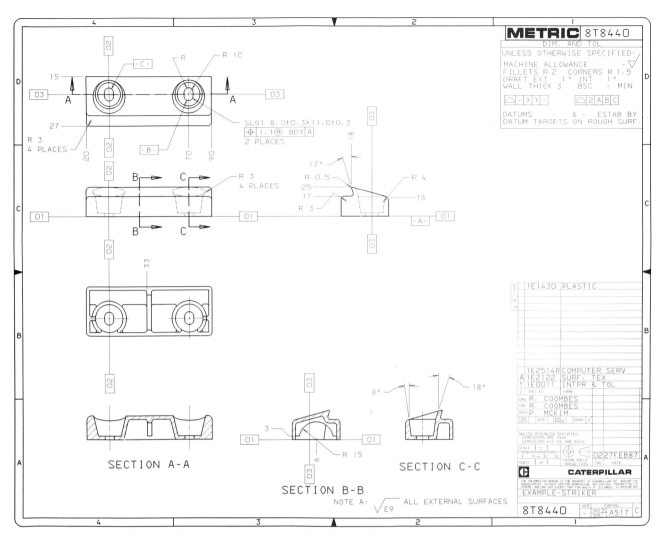

FIGURE 4-1 No manual drafting skills were required for this CAD-generated working drawing. (Courtesy of Caterpillar Inc.)

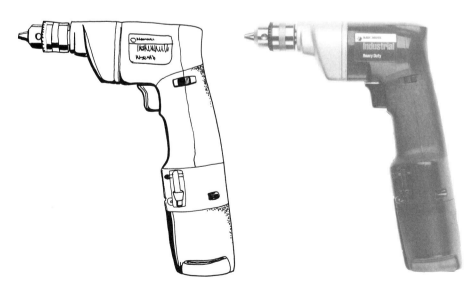

FIGURE 4-2 A freehand sketch is worth a thousand words.

FIGURE 4-3 An industrial product (Courtesy of Black and Decker U.S. Power Tools Group)

drawing shows an object's depth; three sides of an object can be seen in one view.

Orthographic Multiview Sketches

Multiview sketches provide the greatest amount of detail for manufacturing and construction. **Figure 4-7** shows the steps recommended to complete a **multiview** sketch. As with instrument drawing (see Chapter 18), blocking in the overall outline before completing the internal details is the key to maintaining correct scales, angles, and proportions.

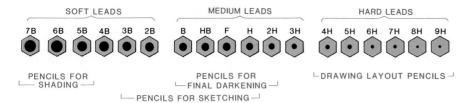

FIGURE 4–4 Recommended pencil grades for sketching

Pictorial Sketches

Blocking in the basic outline is also the recommended procedure for pictorial sketching. There are three types of pictorial sketches: isometric, oblique, and perspective.

Isometric sketches are prepared by establishing a vertical corner line and projecting receding lines 30° **(Fig. 4–8)**. As with multiview sketches, always block in the entire outline of the object before cutting corners, or adding surface details such as holes and projections.

Oblique sketches are pictorials that recede on only one side of an orthographic view. In preparing an **oblique** sketch, first sketch a front view of the object **(Fig. 4–9)**. Then extend lines from each corner upward at 45° or 30°. Connect the ends of these lines with lines that are parallel to the lines of the front view. Oblique sketches are easy to dimension since the width and length of the front view is sketched to actual scale. Only the depth dimension is foreshortened by the receding lines.

Perspective sketches are pictorials that contain receding sides designed to approximate the actual appearance of an object. There are three types of perspectives: one-point, two-point, and three-point.

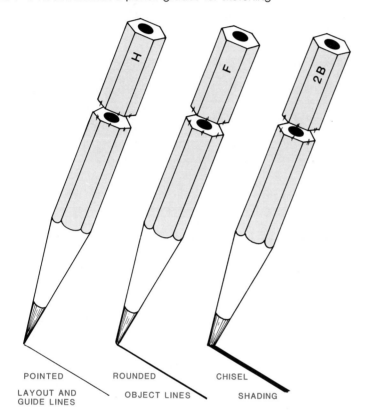

FIGURE 4–5 Recommended pencil points for sketching

1. **One-point perspective** sketches are similar to oblique drawings, except the receding lines do not follow a consistent angle. Receding lines are connected to a vanishing point. A vanishing point represents the point at which all receding lines appear to come together. It is similar to road or train tracks disappearing on the horizon. In sketching a one-point **perspective**, first outline the front view, as in **oblique** sketching. Establish a vanishing point and sketch lines from each corner to this point **(Fig. 4–10)**. Sketch lines representing the back of the

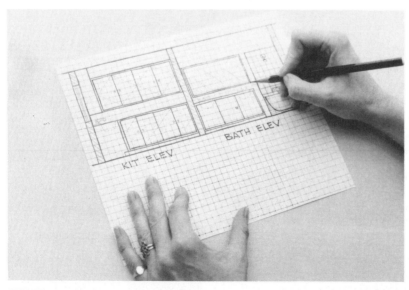

FIGURE 4–6 Sketching on grid paper is faster and neater. (Photo by Ann Ross Wallach)

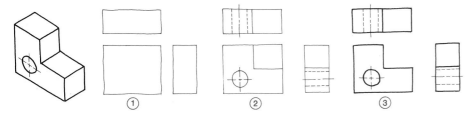

FIGURE 4-7 Steps to sketch a multiview drawing

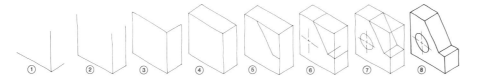

FIGURE 4-8 Steps to sketch an isometric drawing

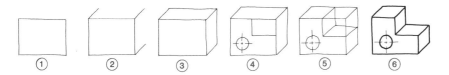

FIGURE 4-9 Steps to sketch an oblique drawing

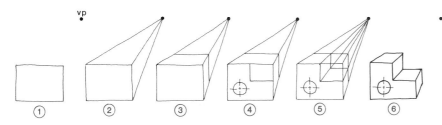

FIGURE 4-10 Steps to sketch a one-point perspective drawing

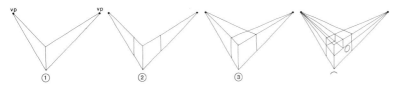

FIGURE 4-11 Steps to sketch a two-point perspective drawing

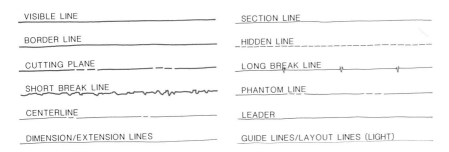

FIGURE 4-12 Line conventions for sketching working drawings

object parallel to the lines on the front view. This blocks in the sketch. Now any angular cuts and details can be added.

2. **Two-point perspective** sketches are similar to **isometric** sketches, except side lines recede to two vanishing points. First sketch the front vertical corner (**Fig. 4–11**). Establish two vanishing points above or below the front corner line, and connect the top and bottom of the front corner line to each of these points. Vanishing points on two-point perspective sketches must always be located on a horizontal line representing the horizon. Next, establish the depth of the sides and connect the back corner lines also to the vanishing points. In two-point perspectives, vertical lines are all parallel.

3. On three-point perspectives, vertical lines are projected to a third vanishing point which is aligned with the vertical front corner line. Three-point perspectives are usually used for architectural rendering, and rarely used for technical sketching.

SKETCHING GUIDELINES AND PROCEDURES

Line conventions and lettering for technical sketches are similar to those used for instrument drawings (**Fig. 4–12**). Sketching standards differ only in the degree of line raggedness. Lines are dark and wide for object lines, dark and thin for dimension and center lines, and thin and light for layout lines and guide lines. **Figure 4–13** shows the application of line conventions to a technical sketch.

All sketches are comprised of straight lines, circles and arcs, irregular curves, and letters and numerals (**Fig. 4–14**). Sketches cannot be prepared with the precision and accuracy of an instrument or CAD drawing; however, care must be taken

52 Drafting in a Computer Age

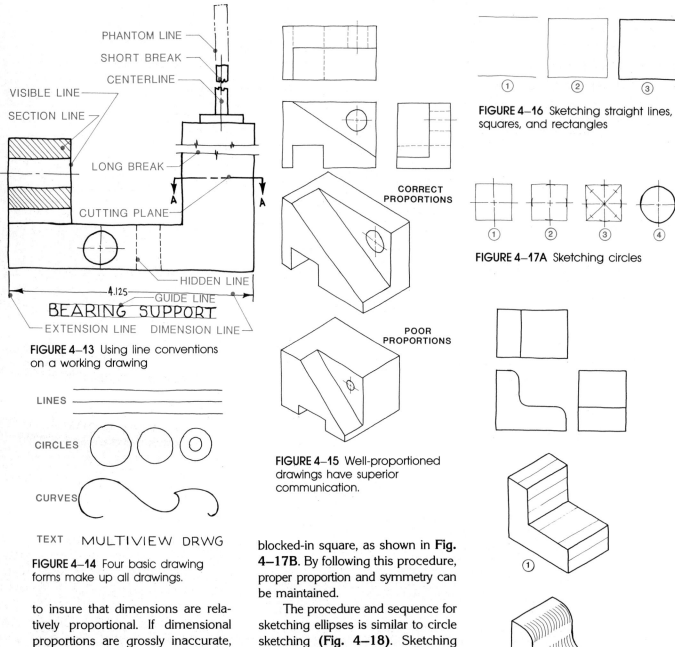

FIGURE 4–13 Using line conventions on a working drawing

FIGURE 4–14 Four basic drawing forms make up all drawings.

FIGURE 4–15 Well-proportioned drawings have superior communication.

FIGURE 4–16 Sketching straight lines, squares, and rectangles

FIGURE 4–17A Sketching circles

FIGURE 4–17B Sketching fillets and rounds

to insure that dimensions are relatively proportional. If dimensional proportions are grossly inaccurate, the sketch will misrepresent the actual appearance of the object **(Fig. 4–15)**.

When sketching straight lines, squares, and rectangles, use short strokes. Do not attempt to draw continuous lines. Right angle lines, unless sketched on grid paper, should be laid out and sketched **(Fig. 4–16)**.

Circles and arcs can be accurately and symmetrically sketched by following the sequence shown in **Fig. 4–17A**. Just as the circle was derived from the square in Fig. 4–17A, all fillets and rounds should first be blocked-in square, as shown in **Fig. 4–17B**. By following this procedure, proper proportion and symmetry can be maintained.

The procedure and sequence for sketching ellipses is similar to circle sketching **(Fig. 4–18)**. Sketching accurate angles, other than right angles (90°), can be difficult without using a triangle or protractor; however, by estimating and dividing a right angle into even angles, an acceptable level of accuracy can be achieved **(Fig. 4–19)**.

When sketching, hold the pencil comfortably **(Fig. 4–20)** and pull the pencil **(Fig. 4–21)**, never push it. To maintain a consistently rounded point and to avoid wearing a flat chisel point, rotate the pencil frequently **(Fig. 4–22)**. When erasing soft pencil sketches, use a good quality medium soft eraser.

Shading

Surfaces exposed to a major light source will appear bright. Conversely, surfaces not directly exposed to a light source will be shadowed; therefore, adding shadow to

Chapter 4 Sketching and Lettering   53

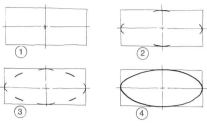

FIGURE 4–18 Sketching ellipses

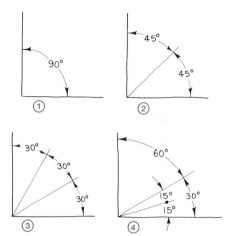

FIGURE 4–19 Sketching and estimating angles

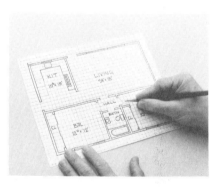

FIGURE 4–20 Use a comfortable pencil grip for sketching. (Photo by Ann Ross Wallach)

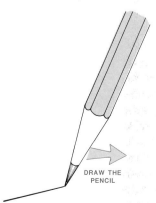

FIGURE 4–21 Always draw the pencil—never push it.

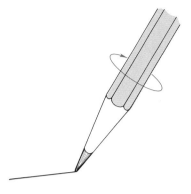

FIGURE 4–22 Rotate the pencil frequently for rounded points.

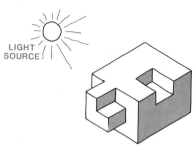

FIGURE 4–23 Shade (darken) the surfaces opposite the light source.

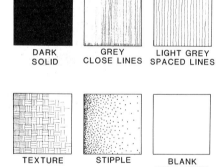

FIGURE 4–24 Shading techniques

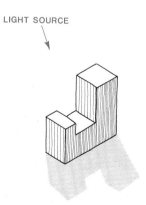

FIGURE 4–25 The light source dictates the position of the shadows.

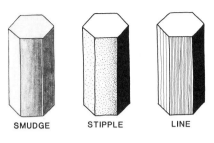

FIGURE 4–26 Freehand shading of sharp corners

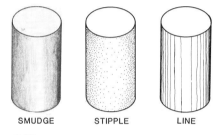

FIGURE 4–27 Freehand shading of rounded surfaces

sketches creates realism **(Fig. 4–23)**. In Fig. 4–23, the light source is located above and to the left of the object. The opposite areas are shadowed because direct light is blocked from these surfaces. **Figure 4–24** shows techniques for shadowing (**shading**) these areas.

Surfaces are rarely totally hidden from a light source. Some surfaces are very light, very dark, or appear in a variety of shadow grades depending on the position, intensity, and number of light sources. **Figure 4–25** shows an object with three light levels: light, medium, and dark. In addition, Fig. 4–25 shows a separate shadow cast by the object.

Light travels in a straight line and cannot bend around corners unless reflected; therefore, light intensity on surfaces always changes at the corner of objects. **Figure 4–26** shows several methods of sketching these differences to add realism to a sketch. On objects without corners, such as spheres and cylinders, light and dark areas change gradually. **Figure 4–27** shows several methods of shadowing a cylinder sketch to add realism.

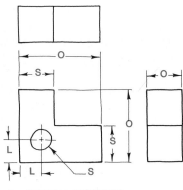

O = OVERALL DIMENSION
L = LOCATION DIMENSION
S = SIZE DIMENSION

FIGURE 4–28 Types of dimensions

Dimensioning

When sketches are used for instruction, manufacturing, or construction, dimensions are usually necessary to adequately describe the object. When sketches are dimensioned, the overall width, depth, and height dimensions are placed on the outside of the **location dimensions (Fig. 4–28)**. These are known as **overall dimensions**. Dimension lines are sketched parallel to object lines and connected to the object with extension lines and arrows. **Location dimensions** show the location of parts of an object. **Size dimensions** show the size of any hole or projection on the object and are placed between the object and the overall dimension lines. Dimensions are usually shown on multiview sketches. Pictorial sketches **(Fig. 4–29)** are normally used only to show the general appearance of products, and do not require dimensions.

Lettering

Some information is best communicated with words or numerals. Almost every sketch or drawing contains notes, labels, and dimensions which must be legible and consistent. In **Fig. 4–30**, the title, material,

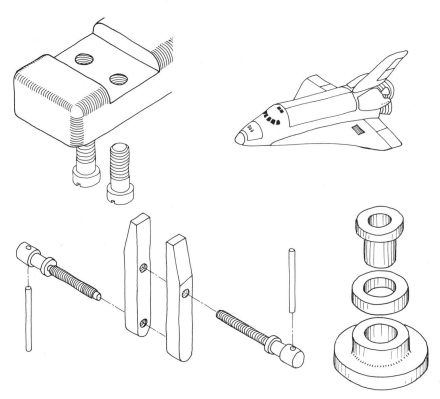

FIGURE 4–29 Examples of industrial sketches

and hole size are described with a lettered notation. Each numeral used for dimensioning must be distinct from all other numerals. The letter D which looks like an O, or the numeral 3 which looks like an 8, can be easily misread and create costly manufacturing or construction mistakes. For this reason, the **American National Standards Institute (ANSI)** style of letters and numerals **(Fig. 4–31)** is used by most American industries. **Figure 4–32** shows the most efficient and quickest method of making these letters and numerals. A slanted form of this style **(Fig. 4–33)** is acceptable, but is rarely used or recommended for engineering drawings.

To letter most efficiently and legibly, always use guide lines **(Fig. 4–34)**. Guide lines keep letters consistent and contained in the area selected. As a general rule, most lettering is $1/8''$ high with $3/16''$-high title headings. Drawings to be microfilmed should be $3/16''$ or $1/4''$ high, using the microfont style shown in **Fig. 4–35**.

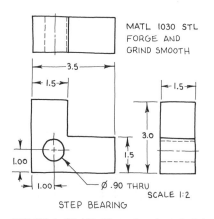

FIGURE 4–30 Working drawing sketch with dimensions and notations

ABCDEFGHIJK
LMNOPQRS
TUVWXYZ
1234567890

FIGURE 4–31 Single-stroke vertical gothic lettering (ANSI standards)

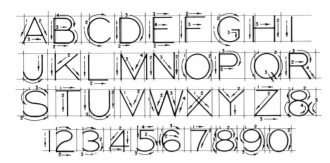

FIGURE 4-32 Recommended lettering strokes

ABCDEFGHIJKLMNOPQRSTUVWXYZ
1234567890

FIGURE 4-33 Single-stroke slant (68°) gothic lettering

ABCDEFGHIJ KLMNOPQRSTUVWXYZ

FIGURE 4-34 Guide lines will aid freehand lettering.

ABCDEFGHIJKLMNO
PQRSTUVWXYZ.,:;
1234567890=÷+−±

FIGURE 4-35 Microfont lettering is recommended if drawing is to be microfilmed.

DRAFTING IN A COMPUTER
AGE, DELMAR PUBLISHER

FIGURE 4-36 The area between letters should be made as equal as possible.

DRAFTING IN A COMPUTER
AGE, DELMAR PUBLISHER

FIGURE 4-37 Inconsistent areas between letters creates reading difficulties.

Correct spacing between letters is also needed to produce lettering that is most readable. **Figure 4–36** shows lettering which is consistently spaced. By contrast, **Fig. 4–37** shows wide and inconsistent spacing of letters. Notice how difficult it is to quickly read Fig. 4–37 compared to Fig. 4–36. Spacing between words and sentences is also important for effective reading. The space between words should be approximately equal to the height of the letters. The space between sentences should be twice the height of the letters (**Fig. 4–38**).

Fractions are one of the most frequent causes of dimensional errors. If a numerator or denominator is misread as a whole number, the result can be disastrous. For this reason, the ANSI spacing for fractions (**Fig. 4–39**) should be used to avoid confusion.

When all of the procedures and guidelines for sketching are combined, the results can be a readable and functional technical sketch as shown at the left in **Fig. 4–40**. If these guidelines are ignored, the confusing series of lines and letters shown at the right in Fig. 4–40 can result.

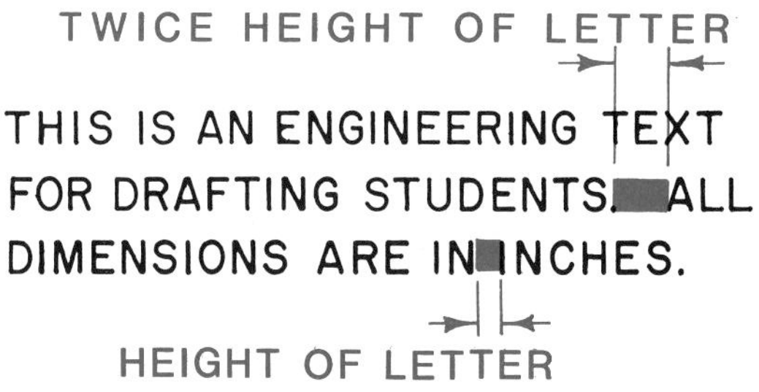

TWICE HEIGHT OF LETTER

THIS IS AN ENGINEERING TEXT
FOR DRAFTING STUDENTS. ALL
DIMENSIONS ARE IN INCHES.

HEIGHT OF LETTER

FIGURE 4-38 Spacing between words and sentences

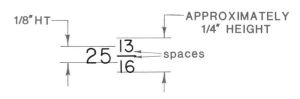

FIGURE 4-39 Fractions on the working drawing

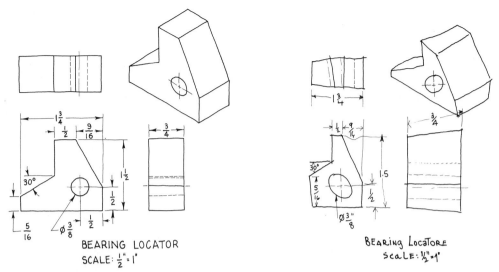

FIGURE 4-40 Which sketch gives the best communication?

COMPUTER-AIDED DRAFTING & DESIGN

COMPUTER-AIDED SKETCHING

Although most sketching is done by hand, it is possible to sketch on a CAD system. Sketching on a CAD system is usually not practical since sketching involves creating a series of short straight lines **(Fig. 4–41)**. Since each line is stored as a separate piece of data, CAD sketching uses excessive memory space with limited productivity. To sketch on the CAD system, follow these steps:

Step 1. Type or select the SKETCH task from the menu.

Step 2. Record the sketch interval. This is the desired length of each short line comprising the sketch. **Figure 4–42** shows the difference between sketches with long and short intervals.

Step 3. Turn off SNAP and ORTHO. SNAP command is when the cursor will snap or locate itself to the closest grid mark on the monitor. ORTHO command will keep all lines vertical (90°) or horizontal (180°).

Step 4. Sketch a line with a puck. It takes practice to sketch with skill.

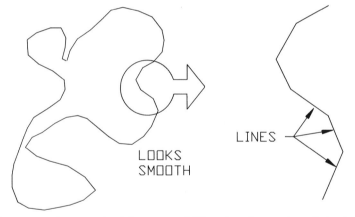

FIGURE 4-41 A sketched line on a CAD system is a series of short, straight lines.

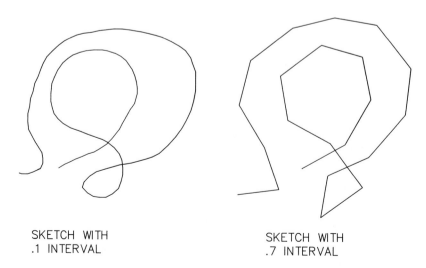

FIGURE 4-42 The short line interval for sketching may be adjusted in length.

Chapter 4 Sketching and Lettering 57

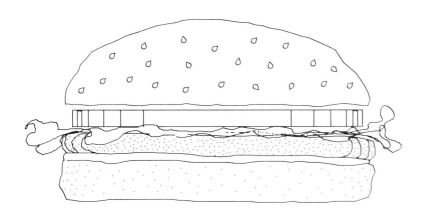

FIGURE 4-43 A drawing of a hamburger using the CAD sketch mode.

THIS IS THE NORMAL
TEXT — IT LOOKS
COMPUTER LIKE

THIS STYLE LOOKS
MORE LIKE HAND
LETTERING

THIS IS FANCY
LETTERING CALLED
COMPLEX

*THIS IS ITALIC
it looks best in
lowercase letters*

FIGURE 4-44 A few examples of the many lettering fonts available with a CAD system

CAD sketching is usually restricted to objects without much precise definition. For example, **Fig. 4-43** shows a CAD sketch of a hamburger which requires little accuracy.

COMPUTER-AIDED LETTERING

Unlike CAD sketching, CAD lettering represents one of the great advantages of CAD drafting over manual drafting. When using a CAD system, all notes, dimensions, labels, and material and parts lists are keyboarded rather than hand lettered. Lettering on a CAD system not only increases the speed of the lettering process but also produces dense, consistent letters and numerals in a variety of styles and sizes. **Figure 4-44** shows several styles of CAD type and **Fig. 4-45** shows some of the configurations possible.

As with other CAD lines and symbols, CAD letters and numerals can also be erased, moved, copied, and rotated. In addition to standard lettering, CAD systems can easily produce standard title blocks from memory **(Fig. 4-46)**. The speed of the CAD lettering process is determined by the keyboarding proficiency of the operator. Keyboarding is therefore a necessary new skill for the contemporary drafter. To produce letters or numerals on a CAD system follow these steps:

Step 1. Type or select the TEXT command from the menu.

Step 2. Select the starting point for the text using the puck.

Step 3. Type (enter) the height of lettering desired.

Step 4. Type (enter) the desired rotation angle; (O) is the rotation angle for horizontal lettering.

Step 5. Type the text that is desired. Press RETURN and the text appears on the screen.

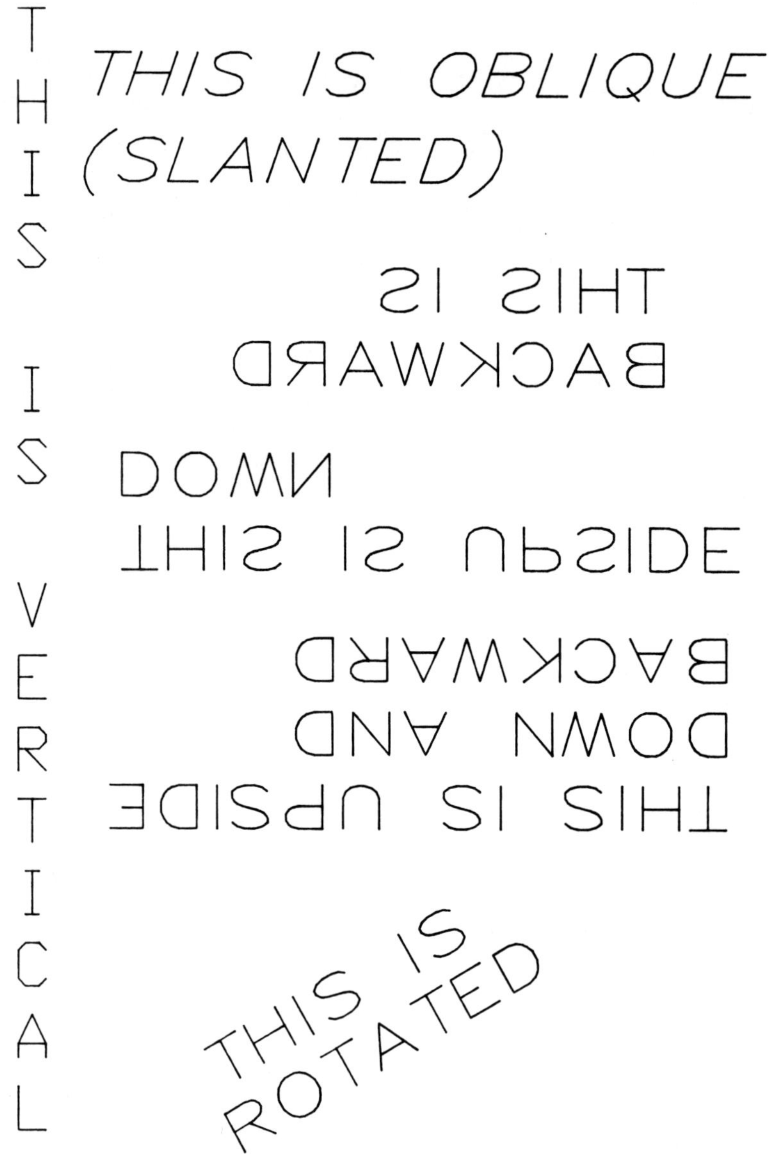

FIGURE 4–45 Examples of different lettering configurations that are possible with a CAD system

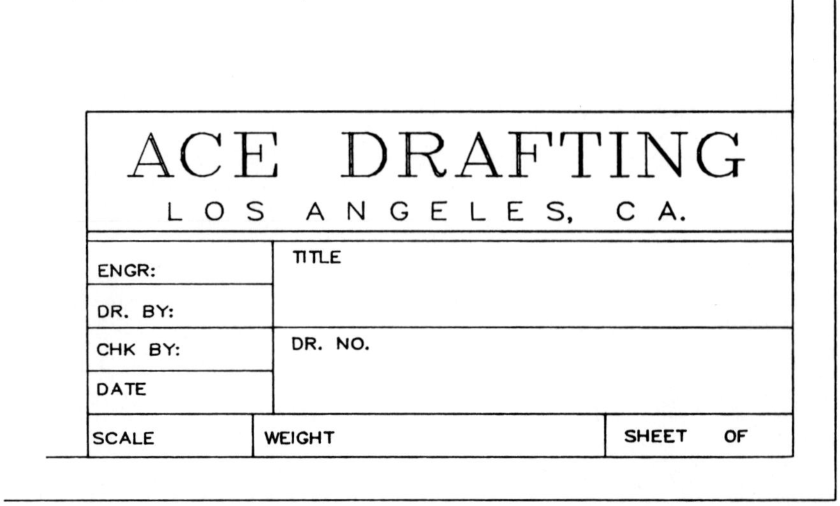

FIGURE 4–46 Various title blocks may be stored and called up with a CAD system.

EXERCISES

1. Sketch the isometric drawings as shown in **Fig. 4–47**.
2. Sketch the multiview drawings as shown in **Fig. 4–48**.
3. Sketch the items in **Fig. 4–49**.
4. Sketch a design for the following:
   - kitchen cabinet door pull
   - automobile's steering wheel with a computerized shift control
   - mailbox for a pole at the street's curb
5. Practice lettering on $1/8''$ grid paper.
6. Practice lettering on a blank sheet of paper with freehand guide lines.
7. Follow the six steps for sketching the pickup truck shown in **Fig. 4–50**.
8. Place lettering on a drawing with a CAD system.

FIGURE 4–47 Isometric sketching practice

60 Drafting in a Computer Age

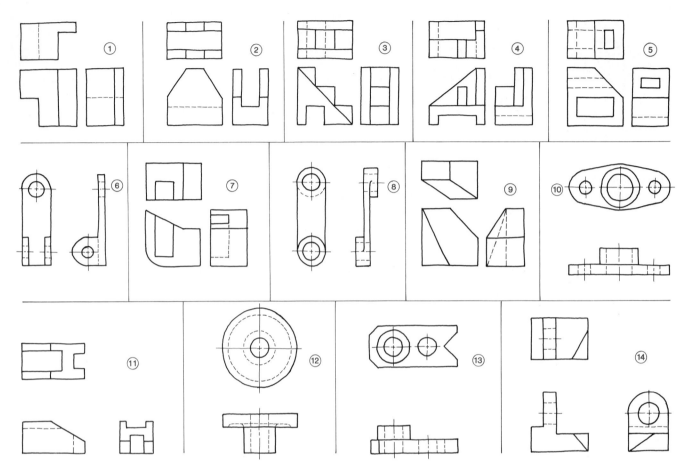

FIGURE 4–48 Orthographic multiview sketching practice

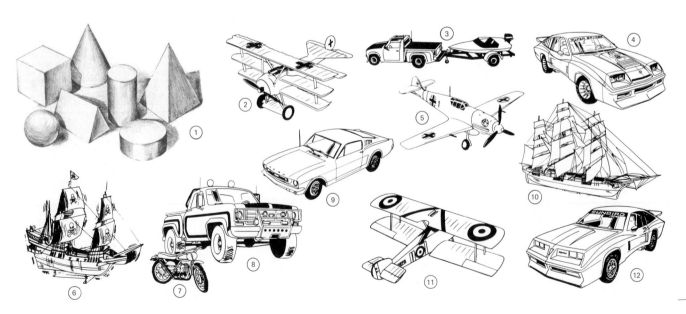

FIGURE 4–49 Sketching practice with real objects

Chapter 4 Sketching and Lettering **61**

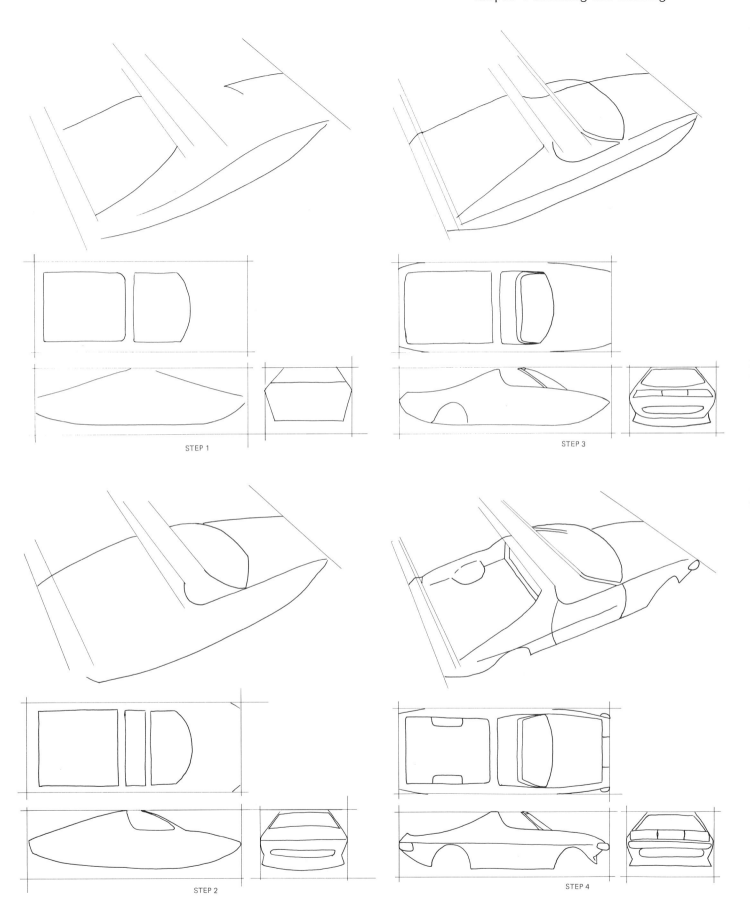

FIGURE 4–50 Follow the six progressive steps to draw a pick-up truck.

62 Drafting in a Computer Age

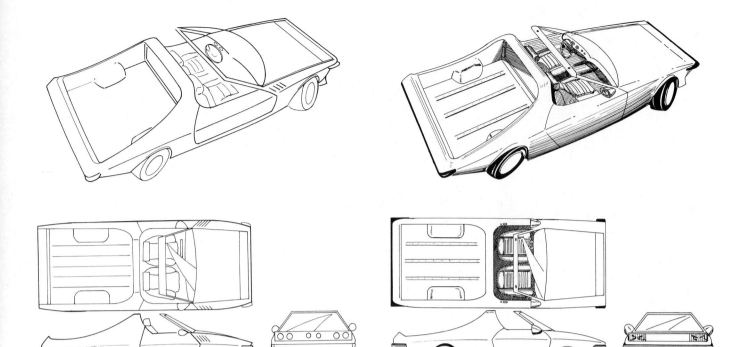

(FIGURE 4–50 continued)

 **KEY TERMS**

ANSI
Isometric
Line conventions
Location dimensions
Multiview

Oblique
One-point perspective
Orthographic
Overall dimensions
Pictorial

Shading
Size dimension
Three-point perspective
Two-point perspective

the world of CAD

Engineers at NASA's Marshall Space Flight Center in Huntsville, Alabama are using computer-aided design and computer-aided manufacturing (CAD/CAM) to design better spacecraft, payloads, and other projects for America's space agency.

The Marshall Center, named for the famous soldier and statesman, General of the Army George C. Marshall, manages the main propulsion system of NASA's Space Transportation System which includes the Space Shuttle solid rocket boosters, external tank, and main engines. The Center also manages the development of the Hubble Space Telescope and a large part of Space Station Freedom, which is scheduled to be in orbit in the mid-1990s, Fig. 1. Marshall Center also manages the Orbital Maneuvering Vehicle, the Advanced X-Ray Astrophysics Facility, the Tethered Satellite System, the Inertial Upper Stage and Orbital Transfer Vehicle, many future Spacelab missions, and much more. And engineers use CAD/CAM for work on many, if not all of these space projects.

The Marshall Center received its first CAD/CAM computer system in the early 1970s. Today, there are approximately 150 CAD/CAM workstations throughout the Center. These computers are used to do precision design for a wide variety of electrical, mechanical, engineering, and architectural work, as well as robotics, Fig. 2.

Engineers use CAD/CAM in the following six scientific and engineering laboratories at the Marshall Center: Structures and Dynamics; Propulsion; Systems Analysis and Integration; Materials and Processes; Test; and Information and Electronic Systems. A seventh laboratory, Space Science, has just started to use CAD/CAM.

Marshall's Program Development Directorate, where future spacecraft are designed, also uses

FIGURE 1 "SPACE STATION"— This is an artist's concept of the permanently manned international Space Station to be placed into a 300-mile-high equatorial orbit by NASA in the mid-1990s. The Space Station will serve as an orbiting research facility to enable scientific exploration, technology development, and private-sector research in space and will ultimately serve as the staging base for continued manned and unmanned exploration of the solar system. Acrylic by Vincent Di Fate.

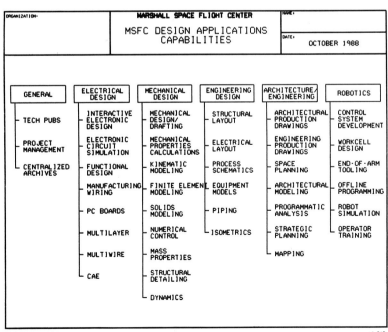

FIGURE 2 This chart shows in what areas the CAD/CAM computer system is being utilized by engineers at the Marshall Space Flight Center in Huntsville, Alabama.

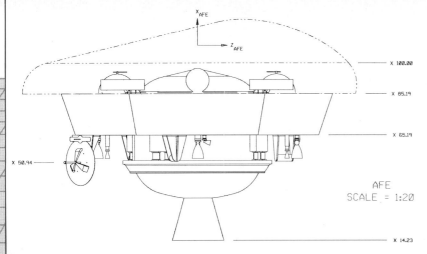

FIGURE 3 This is a CAD/CAM-generated sketch of one of the futuristic spacecraft that Marshall Space Flight Center engineers in Huntsville, Alabama are designing using the CAD/CAM system.

CAD/CAM extensively. These engineers look at the preliminary designs of projects that may be 2–10 years into the future. Projects they currently are working on include earth satellites, future space telescopes and spacecraft that may be used in a mission to the planet Mars, Fig. 3.

Very few mechanical drafting boards are used at the Marshall Space Flight Center in Huntsville. A 30-year NASA engineer who was trained on drafting boards now uses the CAD/CAM system exclusively to design a motor case for a future NASA rocket. Engineering college students, who work in Marshall's engineering laboratories, learn how to use CAD/CAM from the first day they start work. The training they receive on the computer system often supplements the training they receive at their technical college. All engineers agree that CAD/CAM is a very effective design tool and a great advance from the much older drafting board. Productivity is increased since a majority of projects can be completed in one-third the time that it would have taken if a drafting board was used to do the work.

When a NASA engineer sits at a CAD/CAM workstation, he or she can see up to eight different views of the drawing being worked on displayed on the computer screen, Fig. 4. The engineer also can rotate the drawing and get a three-dimensional view, thus providing a better insight into what he or she is designing.

One of the CAD/CAM features that NASA engineers use extensively is the function that allows dimensions to be automatically inserted into a drawing. Drafting time is saved when engineers do not have to actually measure the dimensions, and the computer can measure the radius of an object within 1/10,000 of an inch. The CAD/CAM system also can duplicate repetitious parts of a drawing, thus saving time an engineer would have to spend if he or she wanted to create the same sketch again. Another important aspect of CAD/CAM is that it enables engineers to reference components that other engineers have designed. Again, the engineer saves time by not having to recreate a component that was previously created. Engineers add that their computer drawings can be produced in black and white on paper, or in color as a transparency to be used in a formal presentation.

NASA engineers say that a new engineer can produce simple drawings after using the CAD/CAM system and a manual for about a week, and that within 3–6 months, that person will have become proficient with the computer. More than 90 percent of the engineers in the Structures Division within the Structures and Dynamics Laboratories are trained to use CAD/CAM. Whether they are designing Space Shuttle payloads or doing redesign work on the Shuttle's solid rocket motor, these engineers say that the CAD/CAM system aids in design, but does not replace the basic drafting techniques they learned while in high school or college. To effectively use this design aid, a designer first must have basic drafting skills.

(Courtesy of Jim Sahli, Public Affairs Office, NASA, Marshall Space Flight Center, Huntsville, AL)

FIGURE 4 This is a typical CAD/CAM system that is used at the Marshall Space Flight Center in Huntsville, Alabama. Engineers can view up to eight different parts of their drawing on the screen in front of them. The computer mouse shown is used to key in drafting symbols. The keyboard also is used to provide input to the drawings. Marshall engineers also have a desktop version of the CAD/CAM computer which they also use in designing spacecraft for NASA.

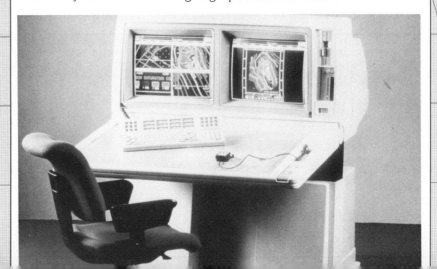

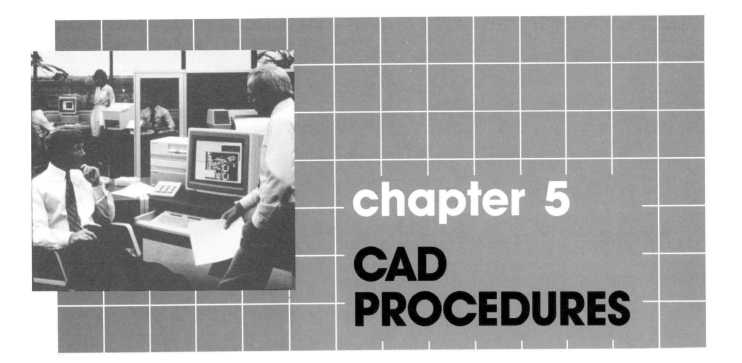

chapter 5
CAD PROCEDURES

OBJECTIVES

The student will be able to:

- identify all of the components used on the personal CAD system which are discussed in this chapter
- list the steps to turn on the system and boot the CAD program
- define the terms: limits, grid size, snap interval, units of measurement, and coordinates
- explain how to draw a line using the keyboard
- explain how to draw a line using the screen menu
- explain how to draw a line using the graphics tablet
- explain how to store and file a drawing

PERSONAL CAD SYSTEM

This chapter explains loading, setup, drawing, and saving procedures for personal CAD systems. Since there is a high degree of similarity among CAD systems, this information is transferred easily from one system to another. To provide maximum consistency, the system used throughout this text is based on AutoCAD software. The equipment used **(Fig. 5–1)** includes the following:

1. **Computer**—Personal computer with an internal **hard disk** and/or double **disk drives**. The AutoCAD software is loaded on the **hard disk**. (It is critical that the CAD system have the memory capacity to run the CAD software and one floppy disk drive for storing drawings.)
2. **Keyboard**—Standard computer keyboard with function keys
3. **Monitor**—Medium resolution (640 x 480 pixels) in color
4. **Graphics tablet**—11″ x 11″ graphics tablet with the AutoCAD menu affixed **(Fig. 5–2)**
5. **Puck**—Multibutton puck.

TURNING ON (BOOTING) THE SYSTEM

Switch on the computer, monitor, and graphics tablet, and **boot** the system. Boot (load) is a term used for starting up a software program on a computer. AutoCAD may automatically boot when the computer is turned on and the main menu will appear on the screen. This is normal if it is the primary system used on the machine. If it does not automatically boot, a **prompt** (instructions on the monitor) will appear on the screen. The operator will type ACAD

FIGURE 5–1 CAD system's hardware

64 Drafting in a Computer Age

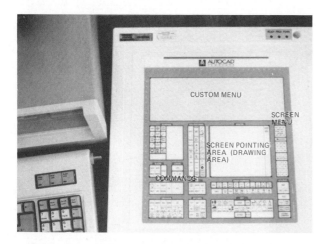

FIGURE 5-2 The graphics tablet menu

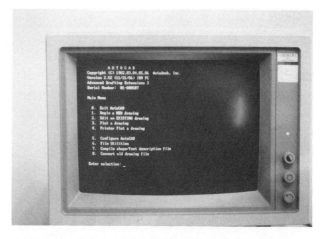

FIGURE 5-3 The main menu shown on the monitor

and press the RETURN key to boot the program.

The **main menu (Fig. 5-3)** appears on the screen as follows:

Main Menu

0. Exit AutoCAD
1. Begin a NEW drawing
2. Edit an EXISTING drawing
3. Plot a drawing
4. Printer Plot a drawing
5. Configure AutoCAD
6. File Utilities
7. Compile shape/font description file
8. Convert old drawing file

The two menu selections used for drawing are #1 and #2.

STARTING A DRAWING

Step 1. Type 1 and press RETURN to begin a new drawing. The prompt will ask for the name of the drawing.

Step 2. Give it any name up to eight letters, and press RETURN.

Example:
1, press RETURN
Name of drawing: TABLE, press RETURN

The name of the new drawing is TABLE. The screen goes blank and then displays the drawing for-

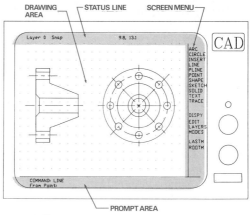

FIGURE 5-4 The drawing format shown on the monitor

mat **(Fig. 5-4)** which lists the following four areas:

1. **Drawing Area**—This is the area on which the drawing will appear. In the drawing area is the **cursor** (cross hairs on the screen). The cursor places the drawing entities.

2. **Screen Menu**—The screen menu gives prompts for drawing and editing. Selections are made with the puck by placing it in the screen menu area of the graphics tablet.

3. **Status Line**—The status line appears along the top of the screen. It displays the present drawing layer being used, the status of SNAP, ORTHO, and the **coordinates** (location of the cursor on the drawing area).

4. **Prompt Area**—The prompt area appears along the bottom of the screen. It displays the last three commands used by the CAD operator.

Setting Drawing Parameters

When starting a new drawing, the first thing to do is set up the **drawing parameters**. These are the **sheet size** (LIMITS), **grid size** (GRID), **snap interval** (SNAP), **units of measurement** (UNITS), and the ORTHO command (which makes sure the vertical and horizontal lines are at 90°). Then, turn on the moving **coordinates** (COORDS).

Limits. The **limits** determine the size of the drawing space.

To set limits, type LIMITS or pick it from the graphics tablet. The screen prompt will ask for the lower-left corner coordinates; they should be

0,0. The prompt then asks for the upper-right corner coordinates to be established. Type in the desired sheet size. Type the X axis first, then the Y axis.

Example:
If a C-size (17" x 22") sheet is desired, type in 22,17, press RETURN.

Pick ZOOM ALL with the puck, and the screen will adjust so the lower-left corner is 0,0 and the upper-right corner is 22,17.

Grid size. A grid is a series of light dots on the screen in a grid pattern. It is similar to the grid fade-out lines on a sheet of drafting paper. The grid is an aid to the CAD operator and does not print on the final drawing.

To set the grid size, type GRID or pick GRID with the puck on the menu. The screen prompt asks for the size desired. If .5 is entered, a .5" dotted grid pattern will appear on the screen.

Snap Interval. If snap is turned on, it controls the cursor movement. If snap is not on, the cursor floats across the monitor as the puck is moved and can be placed on any point. When snap is on, the cursor jumps from snap point to snap point. This gives the operator greater accuracy in placing the cursor on a desired point.

To set snap, type SNAP or pick SNAP with the puck. The screen prompt will ask for the snap interval; if .25 is typed, the cursor will now jump on a .25" interval.

Units of measurement. AutoCAD gives the following four choices for units:

1. Scientific notation—16.5" would look like 1.65E+01.
2. Decimal units are most commonly used in mechanical drawing—16.5" would look like 16.50".
3. Engineering units display feet and decimal inches—16.5" would look like 1'-4.50".
4. Architectural units display feet and fractional inches—16.5" would look like $1'-4\frac{1}{2}"$.

The unit command also allows the CAD operator to set the accuracy of the measurement. If decimal units are selected, the prompt asks how many decimals are desired (0 to 8 places). If architectural units are selected, the prompt asks what denominator of smallest fraction to display (2, 4, 8, 16, 32, or 64). The operator selects one.

To set the units, type UNITS. The screen prompt gives the operator a choice of the four units and the accuracy of the selection to be displayed. The prompt then asks how the angles are to be measured (decimal degrees, degrees/minutes/seconds, grads, or radians).

Coordinates. The coordinates of the cursor are displayed on the status line of the screen (top center). As previously mentioned, the type of units displayed and their accuracy are determined when the CAD operator selects the UNITS. The coordinates may be turned on or off. It is advantageous to keep the coordinates on while drawing to aid in drawing accuracy.

To turn the coordinates on, select MOVING COORDS on the menu tablet with the puck.

Cartesian Coordinate System

Before continuing with the CAD drawing, it is necessary to understand the cartesian coordinate system. The **cartesian coordinate system** determines the parameters for **digitizing** the coordinate points. A drawing is **digitized** into the memory of the computer by recording the coordinate points that will generate the drawing **(Fig. 5–5)**. The first number of the digitized point is the X axis. The second number is the Y axis. Once recorded, the computer may recall the coordinate on the monitor and plot a drawing.

A major consideration when digitizing is the placement of the **origin**, or zero grid point. **Figure 5–6**

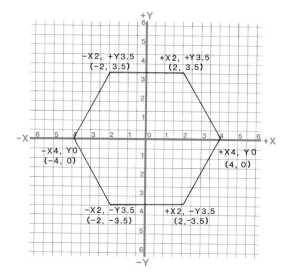

FIGURE 5–5 Digitizing the points of a drawing with the cartesian coordinate system

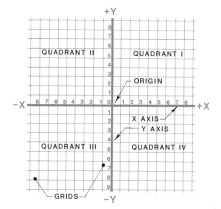

FIGURE 5–6 The cartesian coordinate system with the origin in the center

66 Drafting in a Computer Age

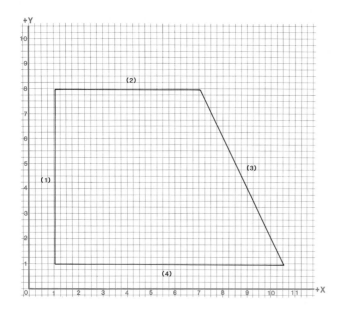

FIGURE 5–7 An example of a drawing on quadrant 1 of the cartesian coordinate system

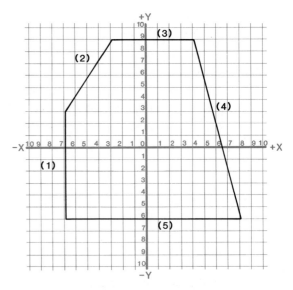

| COORDINATE DATA TABLE ||||||
LN	X	Y	go to	X	Y
1	1	1	→	1	8
2	8	1	→	7	8
3	7	8	→	1	10.5
4	1	10.5	→	1	1
			FINISH		

FIGURE 5–8 A coordinate data table with information for the drawing in Fig. 5–7

| COORDINATE DATA TABLE ||||||
LN	X	Y	go to	X	Y
1	-7	-6	→	-7	+3
2	-7	+3	→	-3	+9
3	-3	+9	→	+4	+9
4	+4	+9	→	+8	-6
5	+8	-6	→	-7	-6
			FINISH		

FIGURE 5–9 The coordinate points of a polygon are shown in the four quadrants and listed in a data table.

shows the four **quadrants** with the point of origin in the center. All coordinate points to the right and above the origin are positive (+). This area is quadrant 1. All coordinate points to the left and below the origin are negative (-). This area is quadrant 3. Quadrants 2 and 4 will have mixed positive (+) and negative (-) coordinates.

When the origin is placed at the lower-left corner of the drawing format, all coordinates are positive. Digitizing will be faster and simpler. **Figure 5–7** shows a simple drawing on a cartesian coordinate grid. The coordinate points are recorded in a data table **(Fig. 5–8)**. An example of a drawing with coordinate grids and the origin in the center is shown in **Fig. 5–9**. It is accompanied with its data table. A general practice is to record the negative symbols and leave off the positive symbols.

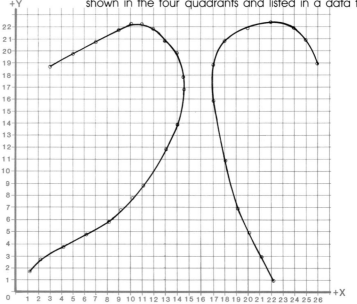

FIGURE 5–10 Data table coordinates for two curves. The first coordinate is the X axis, the following coordinate is the Y axis.

Curves may be digitized **(Fig. 5–10)**. The computer plots a straight line between all coordinate points, so the closer the placement of digitized points, the more accurately plotted the curve will be.

Entering Entities (Drawing Elements)

Entity is a term for a drawing element such as a line, circle, arc, or dimension. There are several ways of entering entities (drawing elements) on the screen. The following are three ways of drawing a 3″ line (from 3,2 to 3,5) on the screen:

Keyboard. Graphics tablets and menus are fast, convenient ways of drawing on a CAD system, but the keyboard can handle any of the commands. A CAD drawing can be generated with the keyboard using the cartesian coordinate system.

The steps to enter coordinates using a keyboard are as follows:

Step 1. TYPE: LINE, press RETURN
Prompt states: From point:

Step 2. TYPE: 3,2 (that is the start of the line), press RETURN
Prompt states: To point:

Step 3. TYPE: 3,5 (that is the end of the line), press RETURN.

A line will appear on the screen 3″ long with coordinates of 3,2 to 3,5 **(Fig. 5–11)**. Lines can be drawn anywhere on the screen and at any angle.

Screen Menu. The screen menu along the right edge of the screen has a list of **commands**. To select a command from the screen menu, move the cursor to the screen menu area. This is easiest on the graphics tablet screen menu area, but also may be done with the keyboard by using the arrow keys. As the cursor hits a screen menu item, the item lights up **(Fig. 5–12)**.

FIGURE 5–11 Line on monitor with the coordinates 3,2 to 3,5

FIGURE 5–12 Screen menu with the line command highlighted

The steps to enter coordinates using a screen menu are as follows:

Step 1. Move the cursor over the screen menu command MAIN, press RETURN or press the zero button on the puck.

Step 2. In the same way, select the LINE command from the screen menu.

Step 3. Move the cursor (with puck or arrows) to 3,2 on the screen (watch the coordinates on the status line for accurate placement). Press the zero button on the puck.

Step 4. Move the cursor to 3,5 and press the zero button on the puck.

The 3″ line is drawn. To draw more lines, continue in the same fashion.

Graphics Tablet. A graphics tablet is similar to a screen menu.

The steps to enter coordinates using a graphics tablet are as follows:

Step 1. Select LINE from the tablet by putting the puck's cross hairs over LINE and pressing button zero **(Fig. 5–13)**.

Step 2. Move the cursor to 3,2 on the screen and press button zero on the puck.

Step 3. Move to 3,5 and press zero on the puck.

The 3″ line is drawn. This is a rapid way to enter drawing commands.

Any of these methods, or a combination, may be used to enter the available drawing commands.

68 Drafting in a Computer Age

FIGURE 5-13 Selecting a menu command with a puck

CREATING A SIMPLE DRAWING

Figures 5-14A, **B**, and **C** show a series of commands used to create working drawings. Using these commands will permit the CAD operator to draw and manipulate all working drawings.

SAVING AND ENDING A DRAWING

After a work session, whether the drawing is finished or not, the drawing needs to be saved on one of the disks.

To save the drawing, type SAVE and press RETURN. The prompt will indicate the name of the drawing in brackets ⟨TABLE⟩. If that is the desired name, press RETURN. The drawing will be saved on the hard disk. If it is desired to save the drawing under a different name, type the new name before pressing the RETURN, and it will be saved under the new name. If the drawing is to be saved on a floppy disk, precede the name with A.

Example:
SAVE, press RETURN
⟨TABLE⟩:A:TABLE, press RETURN

This saves the drawing named TABLE on the floppy disk.

COMMAND	ICON DIAGRAMS			DESCRIPTION
LINE				DRAWS A LINE FROM ONE SELECTED POINT TO ANOTHER
ARC				DRAWS AN ARC THROUGH THREE SELECTED POINTS
CIRCLE				DRAWS A CIRCLE. THE CENTER IS SELECTED, THEN THE RADIUS. THE RADIUS MAY BE TYPED IN OR SPECIFIED WITH THE PUCK.
ELLIPSE				DRAWS AN ELLIPSE. THE MINOR AND MAJOR DIAMETERS ARE SPECIFIED AND THE ELLIPSE IS DRAWN. IT MAY THEN BE ROTATED AND MOVED.
POLYGON				DRAWS A POLYGON OF A SPECIFIED NUMBER OF SIDES AND SIZE. THE NUMBER OF SIDES IS TYPED IN, AND THE SIZE MAY BE SPECIFIED BY ITS INSIDE OR OUTSIDE DIAMETER.
POINT	SELECT POINT STYLE & SIZE			PLACES A POINT ON THE DRAWING. THE SIZE AND SHAPE OF THE POINT MAY BE SELECTED BY THE OPERATOR. THIS ONE IS A SIMPLE CROSS.
DONUT	SELECT ID & OD			DRAWS 2 CONCENTRIC CIRCLES WITH A SPECIFIED INSIDE AND OUTSIDE DIAMETER.
DIMENSION, DIAMETER				IF THE DIAMETER DIMENSION IS TOO SMALL TO APPEAR INSIDE THE CIRCLE, IT IS DISPLAYED ON THE OUTSIDE WITH A LEADER.
DIMENSION, DIAMETER				DIMENSION THE DIAMETER OF A CIRCLE. THE CIRCLE IS SELECTED, THE COMPUTER MEASURES IT AND DISPLAYS THE DIMENSION PRECEDED WITH THE Ø SYMBOL FOR DIAMETER.
DIMENSION, RADIUS				DIMENSION AN ARC. THE ARC IS SELECTED, THE COMPUTER MEASURES IT AND DISPLAYS THE DIMENSION PRECEDED BY AN R.
DIMENSION, ARC				DIMENSION AN ARC. THE TWO LINES OF THE ARC ARE SELECTED AND THE DIMENSION IS DISPLAYED WITH THE ANGLE SYMBOL.
DIMENSION, VERTICAL				A VERTICAL DIMENSION IS DISPLAYED BY SELECTING THE BEGINNING OF THE TWO EXTENSION LINES AND THE LOCATION OF THE DIMENSION.

A

FIGURES 5-14A, B and C These AutoCAD commands are standard on most CAD systems. In the following explanations, the terms **selected** and **specified** are used. A point may be selected with a mouse, puck, or by typing the coordinates on the keyboard. To specify is to give a measurement via the keyboard or show the distance with the puck. The command name is to the left and an explanation to the right of the diagrams.

COMMAND	ICON DIAGRAM			DESCRIPTION
DIMENSION, HORIZONTAL			0.6784	A HORIZONTAL DIMENSION IS DISPLAYED BY SELECTING THE BEGINNING OF THE TWO EXTENSION LINES AND THE LOCATION OF THE DIMENSION. NOTE THAT THE PROCESS IS THE SAME FOR VERTICAL AND HORIZONTAL DIMENSIONS. THE COMMAND SELECTION MAKES THE DIFFERENCE.
TEXT, LEFT			TEXT, LEFT JUSTIFIED	TEXT, LEFT JUSTIFIED ALIGNS THE TEXT TO ITS LEFT ALONG THE VERTICAL CURSOR.
TEXT, CENTER			TEXT, CENTER JUSTIFIED	TEXT, CENTER JUSTIFIED CENTERS THE TEXT ON THE VERTICAL CURSOR.
TEXT, RIGHT			TEXT, RIGHT JUSTIFIED	TEXT, RIGHT JUSTIFIED ALIGNS THE TEXT TO ITS RIGHT ALONG THE VERTICAL CURSOR.
SCALE				SCALE CHANGES THE SIZE OF SELECTED SHAPES. HERE A POLYGON HAS BEEN SCALED UP — ENLARGED. THE OPERATOR SPECIFIES THE AMOUNT OF SIZE CHANGE (2 = DOUBLE SIZE, .5 = ONE HALF SIZE, ETC.).
ROTATE				ROTATE REVOLVES A SELECTED OBJECT ABOUT A SELECTED POINT. THIS POLYGON HAS BEEN ROTATED 10 DEGREES ABOUT ITS CENTER.
OFFSET				OFFSET DRAWS A LINE OR LINES PARALLEL TO SELECTED OBJECTS. THE OPERATOR CHOOSES THE AMOUNT OF OFFSET AND ON WHICH SIDE IT WILL APPEAR.
MOVE				MOVE CHANGES THE LOCATION OF SELECTED OBJECTS. THIS ELLIPSE HAS BEEN MOVED DOWN.
ERASE, WINDOW				ERASE, WINDOW ELIMINATES THE SELECTED FEATURES THE OPERATOR INCLUDES IN THE WINDOW.
ERASE, OBJECT				ERASE, OBJECT ELIMINATES THE FEATURES THE OPERATOR PICKS WITH THE PUCK.
ERASE, LAST				ERASE, LAST ELIMINATES THE LAST FEATURE DRAWN BY THE OPERATOR.

B

(FIGURE 5–14 continued)

To end the drawing and return to the main menu (Fig. 5–3, page 64), type END, and press RETURN. This ends the drawing, the screen will go blank and the main menu will appear.

TO EDIT AN EXISTING DRAWING

A drawing called TABLE now exists in the computer's storage file. To recall the drawing to continue drawing on it or to revise it, select #2 on the main menu: Edit an existing drawing.

Example:
2, press RETURN
Name of drawing: TABLE
The screen will go blank and the drawing named TABLE will appear.

COMMAND	ICON DIAGRAMS			DESCRIPTION
ARRAY, POLAR				ARRAY, POLAR (CIRCULAR) COPIES A SELECTED FEATURE IN A CIRCULAR PATTERN ABOUT A SPECIFIED CENTER POINT. THE NUMBER OF TIMES THE FEATURE APPEARS IS SPECIFIED AND THEY ARE EQUALLY SPACED.
ARRAY, RECTANGULAR				ARRAY, RECTANGULAR COPIES FEATURES HORIZONTALLY AND VERTICALLY. THE SPACING AND THE NUMBER OF COLUMNS AND ROWS ARE SPECIFIED. THIS EXAMPLE HAS 3 COLUMNS AND 4 ROWS.
DIVIDE				DIVIDES A LINE INTO A SPECIFIED NUMBER OF EQUAL SPACES. THIS LINE HAS BEEN DIVIDED INTO 3 EQUAL PARTS.
MIRROR				MIRROR DRAWS A MIRRORED IMAGE OF THE SELECTED FEATURE. THE MIRROR LINE MAY BE AT ANY ANGLE. THIS EXAMPLE HAS BEEN MIRRORED VERTICALLY, THAT IS, THE MIRROR LINE IS A VERTICAL LINE.
BREAK				BREAK PUTS A GAP IN A LINE, CIRCLE OR ARC. THE LINE AND THE SIZE OF THE GAP IS SPECIFIED.
COPY				COPY REPEATS A SELECTED FEATURE ON THE DRAWING AS MANY TIMES AS DESIRED TO SELECTED LOCATIONS.
STRETCH				STRETCH ELONGATES OR SHORTENS FEATURES. THIS SLOT HAS BEEN ELONGATED HORIZONTALLY.
JOIN				JOIN MAKES A CLEAN CORNER AT THE JUNCTION OF TWO LINES. THE LINES ARE SELECTED AND AUTOMATICALLY JOIN AT THEIR INTERSECTION.
FILLET				FILLET MAKES A RADIUS AT THE JUNCTION OF TWO LINES. THE SIZE OF THE RADIUS IS SPECIFIED AND THE LINES ARE SELECTED.
CHAMFER				CHAMFER MAKES A SPECIFIED ANGLE BETWEEN TWO LINES AND TRIMS THE LINES.

(FIGURE 5–14 continued) c

EXERCISES

1. Sketch the drawings on ¼" grid paper with the data for each cartesian coordinate data table in **Fig. 5–15**.
2. Write a data table for the drawings in **Fig. 5–16**.
3. Write a data table for **Fig. 5–17**.
4. Write a data table for a hexagon that is 2" across flats.
5. Write a data table for a 3" equilateral triangle.
6. Write a data table for a 3" diameter circle.
7. Visit a CAD vendor for a demonstration.
8. Visit an industry with a CAD system and ask for a demonstration.
9. Visit a school with a CAD system and ask for a demonstration.
10. Learn how to turn on the hardware for a CAD system and boot up the CAD program.
11. Read the manual for your CAD system and become familiar with its vocabulary.

①

COORDINATE DATA TABLE					
LN	X	Y	go to	X	Y
1	1	1	→	1	4
2	1	4	→	3	6
3	3	6	→	6	6
4	6	6	→	8	4
5	8	4	→	8	1
6	8	1	→	1	1
			FINISH		

②

COORDINATE DATA TABLE					
LN	X	Y	go to	X	Y
1	2	2	→	9	12
2	9	12	→	8	4
3	8	4	→	2	2
			FINISH		

③

COORDINATE DATA TABLE					
LN	X	Y	go to	X	Y
1-2	11	13	→	7.5	2
2-3	7.5	2	→	16.5	9
3-4	16.5	9	→	5.5	9
4-5	5.5	9	→	14.5	2
5-1	14.5	2	→	11	13
			FINISH		

④

COORDINATE DATA TABLE					
LN	X	Y	go to	X	Y
1	11	1	→	2	5
2	2	5	→	2	7
3	2	7	→	4	11
4	4	11	→	10	14
5	10	14	→	17	11
6	17	11	→	17	4
7	17	4	→	11	1
8	11	1	→	11	4
9	11	4	→	15	6
10	15	6	→	10	8
11	10	8	→	6	6
12	6	6	→	11	4
			new start		
13	6	6	→	6	10
14	6	10	→	10	12
15	10	12	→	10	8
			new start		
16	10	12	→	15	10
17	15	10	→	15	6
			new start		
18	6	10	→	4	11
			new start		
19	15	10	→	17	11
			FINISH		

⑤

COORDINATE DATA TABLE					
LN	X	Y	go to	X	Y
1	1	1	→	1	6
2	1	6	→	3	6
3	3	6	→	8	3
4	8	3	→	8	1
5	8	1	→	1	1
			new start		
6	1	8	→	1	10
7	1	10	→	8	10
8	8	10	→	8	8
9	8	8	→	1	8
			new start		
10	3	8	→	3	10
			new start		
11	10	1	→	10	6
12	10	6	→	12	6
13	12	6	→	12	1
14	12	1	→	10	1
			new start		
15	10	3	→	12	3
			new start		
16	19	5	→	13	8
17	13	8	→	13	13
18	13	13	→	15	14
19	15	14	→	17	13
20	17	13	→	21	8
21	21	8	→	21	6
22	21	6	→	19	5
23	19	5	→	19	7
24	19	7	→	15	12
25	15	12	→	13	13
			new start		
26	15	12	→	17	13
			new start		
27	19	7	→	21	8
			FINISH		

⑥

COORDINATE DATA TABLE					
LN	X	Y	go to	X	Y
1	1	1	→	1	14
2	1	14	→	4	14
3	4	14	→	9	4
4	9	4	→	12	4
5	12	4	→	17	14
6	17	14	→	20	14
7	20	14	→	20	1
8	20	1	→	1	1
			new start		
9	17	2	→	4	2
10	4	2	→	4	3
11	4	3	→	17	3
12	17	3	→	17	2
13	19	5	→	15	5
14	15	5	→	19	12
15	19	12	→	19	5
			new start		
16	6	5	→	2	5
17	2	5	→	2	12
18	2	12	→	6	5
			FINISH		

FIGURE 5–15 Sketch the drawings on 1/4" grid paper from the data in each cartesian coordinate data table.

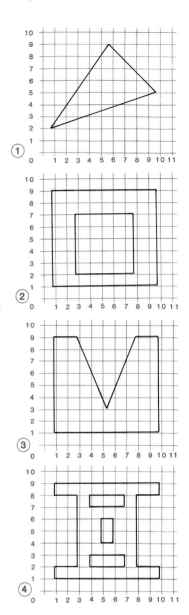

FIGURE 5–16 Write a data table for each figure.

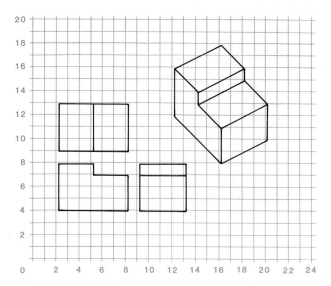

FIGURE 5–17 Write a data table for the multiview and pictorial drawings.

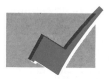

KEY TERMS

Boot
Cartesian coordinate system
Command
Coordinates
Cursor
Digitize
Disk drive
Drawing area
Drawing parameters
Entity

Graphics tablet
Grid size
Hard disk
Keyboard
Limits
Main menu
Monitor
Origin
Prompt

Prompt area
Puck
Quadrant
Screen menu
Selected
Snap interval
Specify
Status line
Units of measurement

the world of CAD

Computer-aided drafting and design (CADD) has had an enormous impact on the fields of architecture and engineering. Einhorn, Yaffee, Prescott, an architecture and engineering firm located in Albany, New York, uses CADD extensively for their projects. About 60% of both new construction and renovation are completed using CADD.

The Einhorn, Yaffee, Prescott (EYP) CAD System hardware is CALCOMP System 25, Series 600. Current software is run on the UNIX Operating System 5 with Berkley Enhancements MASSCOMP Version 3.0. EYP has six workstations linked together in a "building block concept." There is one CPU for each pair of workstations. One CPU acts as a system manager allowing all the workstations to communicate with one another via a local area network. Each pair of workstations is capable of operating independently should either of the other CPUs not be operating.

Presently, the workstations are in operation for two 6-hour shifts, five days a week, Figure 1. At times, they are in operation for 24 hours a day, seven days a week!

All of EYP's CADD renovation projects being with verifying and inputting of existing building conditions, plans, and elevations drawings. Once existing conditions drawings have been verified, they serve as base drawings for architectural/engineering design phases. Rather than sacrifice quality and correct inaccuracies as they are identified, EYP's CADD users input existing conditions drawings.

Drawings are produced on the CADD system very similar to the overlay drafting method. Each drawing has 256 available levels. Although not all 256 levels are used each time, it provides the luxury of breaking up the drawings in more than the conventional way. Levels can be turned "on or off" in any combination and at any time. In addition to the levels, each workstation has three picture planes available. Each picture plane holds one drawing. This allows us to view three different drawings at one time. For example, overlay the architecture background, electrical plan and mechanical plan for coordination. The drawings reside on a hard disk in one of the three CPUs.

What makes CADD so effective is the speed and accuracy with which a user can produce a set of drawings. On an average, electronic drawing is 2 to 3 times faster than by hand. Automatic dimensioning and precision is accurate down to 1/100th of an inch. As with any CADD system, objects were created once and copied, eliminating the need to redraw objects numerous times.

This improved drawings accuracy has resulted in a number of significant advantages for EYP. Specifically, the engineering staff can start design work sooner than under normal design/drafting processes. Given the level of accuracy required to begin mechanical and electrical engineering design, EYP's engineering staff usually began working at the construction documents phase—the point at which the architectural drawings were almost complete. Improved accuracy now gained at the architectural design phase allows EYP's engineering staff to begin working at the design development phase, giving them more time to complete their work.

There are other advantages to using CADD in addition to speed and accuracy. For example, the CADD system allow for simultaneous drawing overlays, making it possible to take each discipline's drawing and laying one upon another to be viewed on the screen. This ability to see the whole picture significantly reduces and/or virtually eliminates costly design errors, conflicts, and construction change orders. Also, the quality control (QC)/review phase is completed sooner. Using CADD allows

FIGURE 1 EYP has 6 workstations linked together in a "building block concept." There is a CPU for each pair of workstations. One CPU acts as a systems manager allowing all workstations to communicate with one another via a local area network.

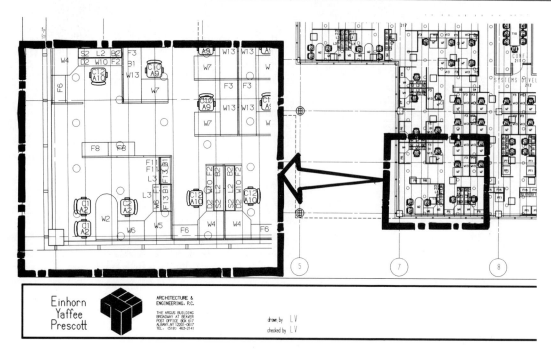

FIGURE 2 EYP's CADD system can enlarge a specified area of a drawing to better see the graphic elements representing FF & E.

TAG #	CATALOG NO.	FINISH CODE 1/COLOR	FINISH 2/COLOR	DESCRIPTION	HEIGHT	WIDTH	DEPTH	COUNT	TTL YDS	LIST PRICE	TOTAL PRICE
X1	TMKT-66	J GRAPHITE		T-MOUNT KIT	66"			1		$ 62.00	$ 62.00
E4	WC-6	J GRAPHITE		WIRE CHANNEL	72"	.5"	.5"	18		$ 10.00	$ 180.00
X2	WMK-35	J GRAPHITE		WALL MOUNT KIT	35"			1		$ 40.00	$ 40.00
W4	WRCS-324	H-1A DARK OAK	J GRAPHITE	90 D CORNER CANTILEVERED WORKSURFACE		36"	24"	26		$ 295.00	$ 7,670.00
W5	WRCS-324-W	RD MAHOGANY	J GRAPHITE	90 D CORNER CANTILEVERED CYGNIA WORKSURFACE		36"	24"	6		$ 574.00	$ 3,444.00
W6	WS-324-W	RD MAHOGANY	J GRAPHITE	CANTILEVERED CYGNIA WORKSURFACE		36"	24"	11		$ 368.00	$ 4,048.00
W7	WS-330	H-1A DARK OAK	J GRAPHITE	CANTILEVERED WORKSURFACE		36"	30"	52		$ 216.00	$ 11,232.00
W8	WS-4224	H-1A DARK OAK	J GRAPHITE	CANTILEVERED WORKSURFACE		42"	24"	12		$ 211.00	$ 2,532.00
W9	WS-430	H-1A DARK OAK	J GRAPHITE	CANTILEVERED WORKSURFACE		48"	30"	6		$ 252.00	$ 1,512.00
W10	WS-524	H-1A DARK OAK	J GRAPHITE	CANTILEVERED WORKSURFACE		60"	24"	26		$ 258.00	$ 6,708.00
W11	WS-524-W	RD MAHOGANY	J GRAPHITE	CANTILEVERED CYGNIA WORKSURFACE		60"	24"	1		$ 455.00	$ 455.00
W13	WS-624	H-1A DARK OAK	J GRAPHITE	CANTILEVERED WORKSURFACE		72"	24"	58		$ 352.00	$ 20,416.00
W15	WS-630	H-1A DARK OAK	J GRAPHITE	CANTILEVERED WORKSURFACE		72"	30"	12		$ 383.00	$ 4,596.00
EP1	WSSP-2427	V-1A DARK OAK		WORKSURFACE SUPPORT PANEL SPECIAL CUT	27"	24"	1.25"	34		$ 202.00	$ 6,868.00
EP2	WSSP-3027	V-1A DARK OAK		WORKSURFACE SUPPORT PANEL SPECIAL CUT	27"	30"	1.25"	64		$ 252.00	$ 16,128.00
X3	1220-1304	J GRAPHITE		TOP CAP				23		$ 2.00	$ 46.00

MANUFACTURERS TOTAL: $ 419,158.00

FIGURE 3 Each graphic element can have a database link, making it possible to accurately report the specifications of thousands of FF & E.

each user to complete QC checks while they are working, throughout the design and construction documents phases.

Changes are inevitable, and usually at the last minute. The computer's ability to move, copy, and delete 5 to 10 times faster enables EYP to continually respond to their client's changing facility needs. Architectural and engineering alternatives and modifications can be checked quickly to identify coordination conflicts and ensure accuracy. Furniture, furnishings, and equipment (FF & E) data can be updated as needed and inventory reports can be generated to reflect these revisions.

The EYP CADD system is capable of "intelligent" graphics processing. Each graphic element drawn (lines, geometrics, and polygons) can have a database link. This database link provides for a reporting capability of all data previously input. In 1986, EYP input the graphics and related data of nearly 500,000 square feet of office and health care facilities, FF & E. This has provided the ability to accurately report the specifications for thousands of FF & E items, sorted by type, area, manufacturer, cost, condition or other attributes, Figures 2, 3, and 4.

Finally, CADD offers consistent, good looking drawings. EYP has an electrostatic plotter that is not limited by printing medium or process. Anything can be instantly plotted without being bogged down by lengthy plotting processes. The fin-

FIGURE 4 A photograph of one of the work areas shown in Figure 2 with furnishings specified in Figure 3.

ished product is aesthetically pleasing, as well as very easy to read, Figures 5, 6, and 7. Consistency is maintained for all disciplines.

There are some common misconceptions about CADD and some reoccurring pitfalls that should be avoided. First, CADD cannot correct inefficiencies already present in a project. A computer is programmed and run by people. It will only do what it is told to do. The flow of a project, and information to be presented, is dictated by the user.

Second, CADD cannot do the design. CADD stands for computer-aided drafting and design. The application software performs many design calculations and shortens many design functions, but there is no CADD system made that incorporates artificial intelligence to make all design decisions, as yet. However, in the hands of a designer, CADD becomes a tool that speeds the design process and frees the designer for more creative time, more design options, and design refinement.

Third, CADD is not instantaneous, though sometimes it seems so. Changes on CADD take time—valuable time. Because changes seem effortless on CADD, the tendency is to make changes in the design and the drawing until perfected. In doing so, the CADD operator must be aware that what changes in one discipline may affect another discipline which requires additional time spent; coordinating with other dis-

FIGURE 5 A front elevation of a supermarket.

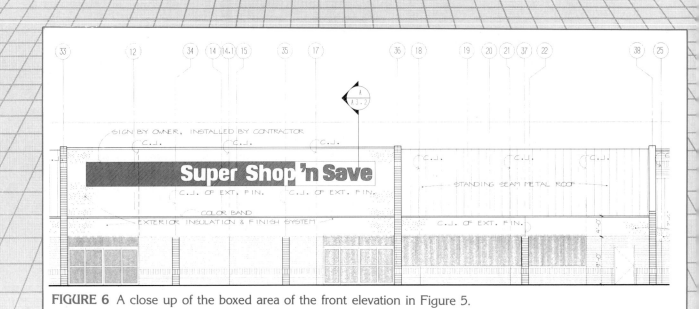

FIGURE 6 A close up of the boxed area of the front elevation in Figure 5.

ciplines; making the changes; plotting; etc.

Finally, CADD does not guarantee a complete and coordinated set of drawings. Typically when people look at a document produced by a computer, they assume that document to be correct and error-free. Though accuracy and coordination are greatly enhanced by CADD, quality control is still vital, requiring the human factor, and should not be compromised.

(Courtesy of Charles J. Sacco, AIA, Principal; Einhorn, Yaffee, Prescott, Architecture and Engineering, P.C., Albany, NY)

FIGURE 7 A photograph showing the completed supermarket front shown in Figure 6.

chapter 6
DRAFTING CONVENTIONS AND FORMATS

OBJECTIVES

The student will be able to:

- draw a basic working drawing using ANSI line conventions
- describe the types of working drawing conventions
- lay out and draw a school drafting format sheet
- lay out and draw an industrial company format sheet

STANDARDIZATION

Different industries and companies use unique processes and procedures for designing and manufacturing products. However, drawings used in the design and manufacture of products are prepared to identical sets of **standards**. Standardization of drawings enables all manufacturers and builders to interpret the drawings of many different designers, drafters, engineers, or architects. Before standardization, working drawings could only be interpreted by the designer or drafter who prepared the drawing **(Fig. 6–1)**.

National and World Standards

Dictionaries are used to standardize words in a written or spoken language. Likewise, the American National Standards Institute (**ANSI**) publications are used as the universal reference work for drafting standards in the United States. The International Drafting Standards (**ISO**) publication R–1000 describes metric standards used throughout the world. The consistent use of these two international standards insures a product can be designed anywhere in the world, and built or manufactured in any other part of the world. Each specialized standard for drafting is found in ANSI's Y–14 publication with the following identification numbers:

Y14.1 Drawing Sheet Size and Format
Y14.2 Line Conventions and Lettering
Y14.3 Multiview and Sectional-View Drawings
Y14.4 Pictorial Drawings
Y14.5 Dimensioning and Tolerancing
Y14.6 Screw-Thread Representation
Y14.7 Gears, Splines, and Serrations
Y14.7.1 Gear-Drawing Standards—Part 1
Y14.9 Forgings
Y14.10 Metal Stampings
Y14.11 Plastics
Y14.14 Mechanical Assemblies
Y14.15 Electrical and Electronics Diagrams, Including Supplements Y14.15a and Y14.15b
Y14.17 Fluid Power Diagrams
Y14.26 Dictionary of Terms for Computer-Aided Preparation of Product Definition Data

FIGURE 6–1 Early working drawings usually required interpretation because only the drafter or designer could read his or her drawings. (Courtesy of The Celotex Corporation)

73

LINE CONVENTIONS

Technical drawings contain many types of lines, each with a special meaning. Through the consistent use of the ANSI alphabet of lines, technical drawings become a language which can be read by anyone who understands the standard. The ANSI **line conventions** are shown in **Fig. 6–2**.

Line Widths

All lines are either thick (.032" or 0.7 mm average width) or thin (0.16" or 0.35 mm average width). Both thick and thin lines are always dark (**opaque**) to insure high reproduction quality. Lines used for layout and lettering guide lines (**Fig. 6–3**) are drawn extremely light and should be invisible on the finished drawing reproduction. Some drafters use non-reproducible blue pencils to insure invisibility.

Line Types

Visible **object lines** (**Fig. 6–4**) are solid, thick lines used to define the outline of objects. **Break lines** are used to interrupt the drawing if the object will not fit on a drawing sheet. Short break lines are thick, wavy lines, and long break lines are thin lines with zigzags spaced at approximately 1.5" intervals (**Fig. 6–5**). When objects are visually cut (sectioned) to show an interior view, a thick **cutting plane line** is used to show the location of the cut. Thin **section lines** are used to represent the material (**Fig. 6–6**) through which the cut is made. **Chain lines** (**Fig. 6–7**) are used to indicate specific areas. Object lines which fall behind the front plane of an object are drawn with dashed lines representing **hidden lines** (**Fig. 6–8A** and **B**). (Note that when a hidden line appears as the continuation of a solid object line, the drafter must leave a

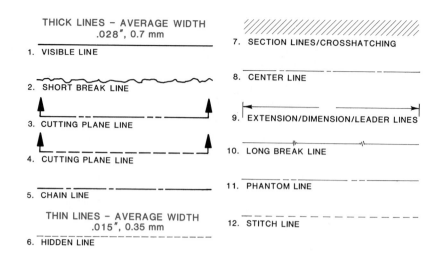

FIGURE 6–2 ANSI line conventions

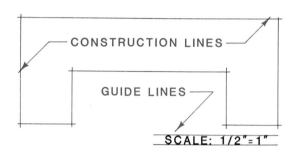

FIGURE 6–3 Construction lines and guide lines are drawn very lightly so they will not be seen in the finished working drawing or print.

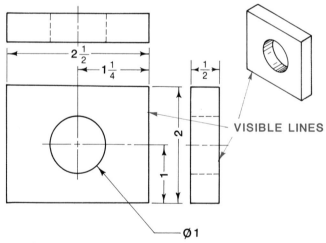

FIGURE 6–4 Note the width of the object lines and compare them to the widths of the other lines.

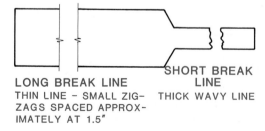

FIGURE 6–5 Long and short break lines

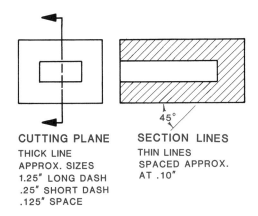

FIGURE 6–6 The cutting plane and the section lines (crosshatching)

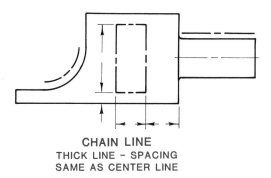

FIGURE 6–7 A chain line is used to indicate a specific area.

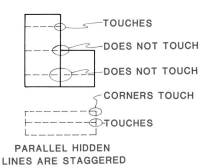

FIGURE 6–8A Drafting conventions for hidden lines in a working drawing. The hidden line is approximately 1/8″ long with a 1/32″ space.

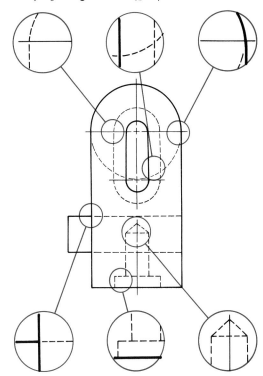

FIGURE 6–8B General rules for hidden lines

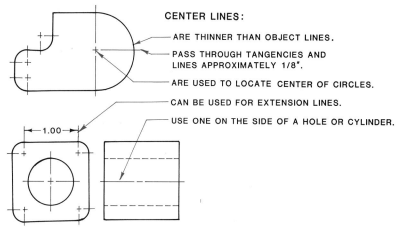

FIGURE 6–9 Drafting conventions for center lines on a working drawing. The approximate dimensions are: long dash 1 1/2″, short dash 1/8″, space 1/16″.

space.) When objects are symmetrical with a common center, the center is defined with a **centerline** constructed with thin long and short dash lines **(Fig. 6–9)**. The size of objects is described with the use of thin **leaders**, **extension lines** and **dimension lines (Fig. 6–10)**. When alternate positions of an object or matching part must be shown without interfering with the main drawing, thin **phantom lines** are used **(Fig. 6–11)**. To show the location of stitches in a material, **stitch lines** are used **(Fig. 6–12)**.

DRAFTING CONVENTIONS

The working drawings of a machine part shown in Fig. 6–13 through Fig. 6–24 show the application of many drafting conventions. **Figure 6–13** is a **multiview** drawing with inch-fraction dimensions and surface symbols.

Figure 6–14 shows the same multiview drawing with inch-decimal dimensions. In Fig. 6–14, a cutting plane line is positioned on the top view, while the corresponding full section is shown in the right-side view. Note the standard weld symbol and dual inch-millimeter chart in Fig. 6–14.

Metric (millimeter) dimensions and position tolerancing symbols are

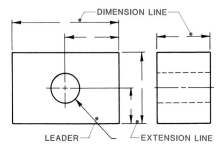

FIGURE 6–10 Leader, dimension, and extension lines

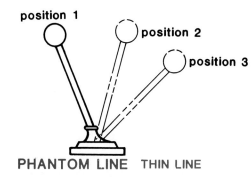

FIGURE 6–11 Phantom lines show alternate positions. Line sizes are the same as center lines.

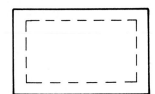

FIGURE 6–12 Stitch lines show various stitches sewn into materials.

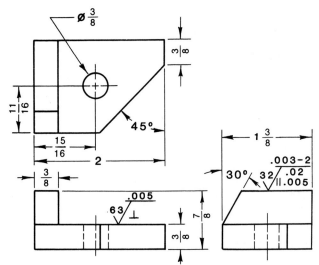

FIGURE 6–13 Multiview drawing with third angle projection, inch-fraction dimensions, and surface finish symbols

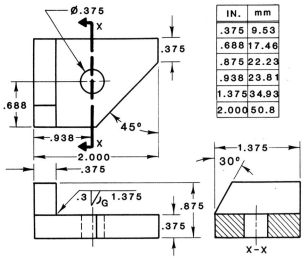

FIGURE 6–14 Multiview drawing with inch-decimal dimensions, full section, weld symbol, and dual-dimension table

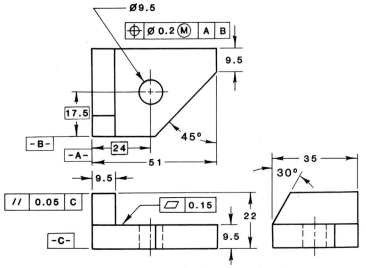

FIGURE 6–15 Multiview drawing with metric (millimeters) dimensioning and positional tolerancing symbols

added to the machine part drawing shown in **Fig. 6–15**. Tolerance dimensions are used to dimension the same drawing in **Fig. 6–16**.

Isometric drawings **(Fig. 6–17)** and isometric **section** drawings **(Fig. 6–18)** are all drawn with receding sides positioned 30° above the horizon, and extending to the right and left.

Pattern development drawings show the outline of all connected surfaces of an object when laid flat. Pattern development drawings are used to define the outline of an object when a continuous material is used on all surfaces. The pattern development drawing represents all the surfaces of an object **(Fig. 6–19)**.

A pictorial drawing is usually used to show the finished shape of the assembled object.

Oblique pictorial drawings show the front view of an object in true scale and form, with receding angles projected to only one side at 30° or 45°.

When receding sides are drawn at one-half the actual depth, the drawing is known as a **cabinet** drawing **(Fig. 6–20)**. When sides are drawn at full-scale depth, the drawing is known as a **cavalier** drawing **(Fig. 6–21)**. Since cavalier oblique drawings appear more distorted than cabinet drawings, using half-depth dimensions on the receding angle helps reduce the appearance of distortion.

Perspective drawings **(Fig. 6–22)** are classified into one-, two-, and three-point perspective drawings (see Chapter 14). **Dimetric** pictorial drawings **(Fig. 6–23)** are similar to isometric drawings, except that two axes have equal angles. In **trimetric** pictorial drawings **(Fig. 6–24)**, all three axes' angles are different.

DRAWING FORMATS

Drawing sheet sizes are standardized by ANSI. Letters A through E are used

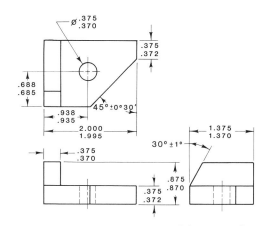

FIGURE 6–16 Multiview drawing with tolerance dimensions

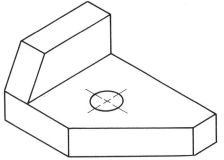

FIGURE 6–17 Isometric drawing

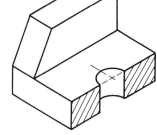

FIGURE 6–18 Isometric section drawing

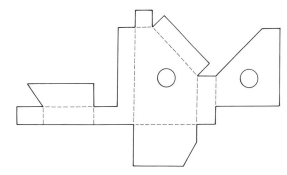

FIGURE 6–19 Pattern development drawing

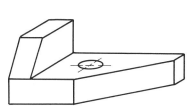

FIGURE 6–20 Cabinet drawing

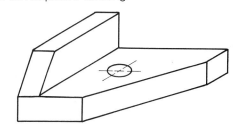

FIGURE 6–21 Cavalier drawing

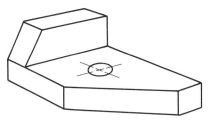

FIGURE 6–22 Perspective drawing

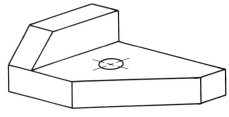

FIGURE 6–23 Dimetric drawing

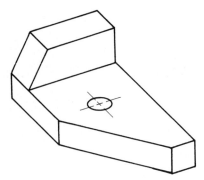

FIGURE 6-24 Trimetric drawing

to specify standard U.S. customary sheet sizes **(Fig. 6–25)**. In the same manner, metric sheet sizes are standardized by ISO. Designations of A0 through A10 are used to identify standard metric sheet sizes **(Fig. 6–26)**. In addition to sheet sizes, ANSI specifies **border** designations. **Figure 6–27** shows drawing sheets with standard alphanumeric zones which are used for quick referencing of a drawing's quadrant areas. Standard border sizes and layouts are shown also in **Fig. 6–28**.

Although the layout of drawing **title blocks** varies greatly among companies, title block information usually contains space for the following:

E = TOTAL SHEET SIZE

AMERICAN STANDARD SHEET SIZES (INCHES)		
DESG	WIDTHxLEN	WIDTHxLEN
A	9x12	8.5x11
B	12x18	11x17
C	18x24	17x22
D	24x36	22x34
E	36x48	34x44

FIGURE 6–25 ANSI standard paper sizes

- company name and division
- drawing title
- drawing number
- part identification number
- date
- tolerance limits
- surface quality
- designer's name
- drafter's name
- checker's name
- angle of projection
- material type
- scale

DES	WIDTHxLEN (mm)
A0	841x1189
A1	594x841
A2	420x594
A3	297x420
A4	210x297
A5	148x210
A6	105x148
A7	74x105
A8	52x74
A9	37x52
A10	26x37

A0 = TOTAL SHEET SIZE

FIGURE 6–26 ISO metric A-size paper series

- next assembly
- microfilming marks
- photoreduction marks
- revision record
- material lists

Typical industrial title blocks are shown in **Fig. 6–29**. Title blocks are abbreviated for educational use. They usually include the following:

- drawing title
- scale
- drawing number date
- student drafter's name
- school
- instructor's approval
- grade

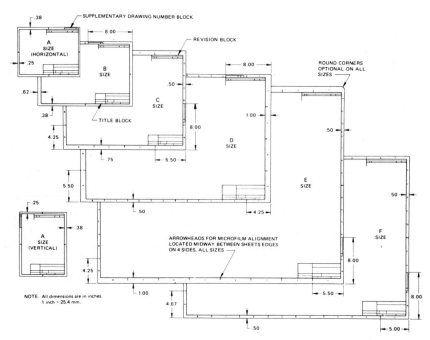

FIGURE 6–27 ANSI standard drawing formats Y14.1 © 1980 (Reprinted with permission of the American Society of Mechanical Engineers).

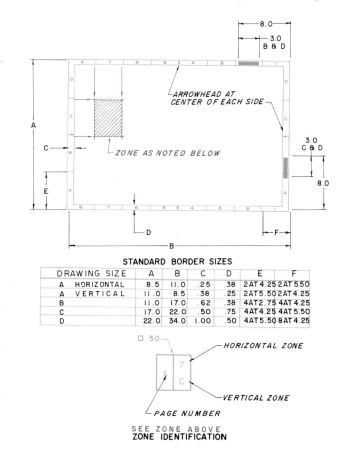

FIGURE 6-28 Border data and sizes for drawing formats (Reprinted from TECHNICAL DRAWING AND DESIGN by Goetsch and Nelson, © 1986 Delmar Publishers Inc.)

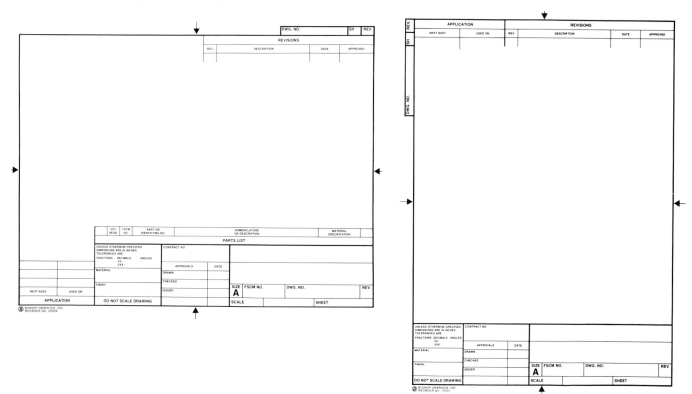

FIGURE 6-29 Examples of industrial horizontal and vertical title blocks (Courtesy of Bishop Graphics, Inc., Westlake Village, CA)

Figure 6–30 shows an A-size school format and **Fig. 6–31** shows a B-size school format.

Folding methods for prints are standardized for consistency and ease of use. **Figure 6–32** shows how standard sheet sizes fold to 8.5″ x 11″ or 9″ x 12″. Prints are always folded with the title block and drawing identification number on top for convenient recognition and filing.

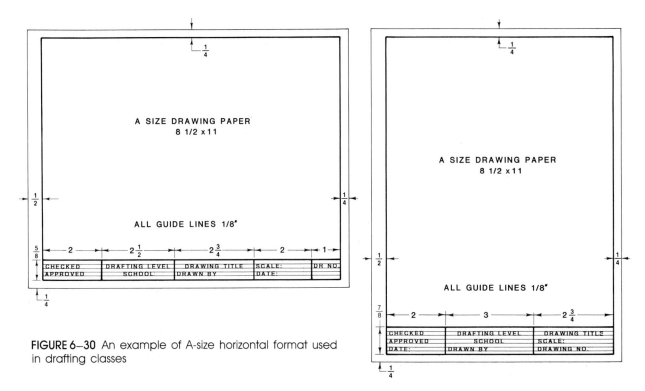

FIGURE 6–30 An example of A-size horizontal format used in drafting classes

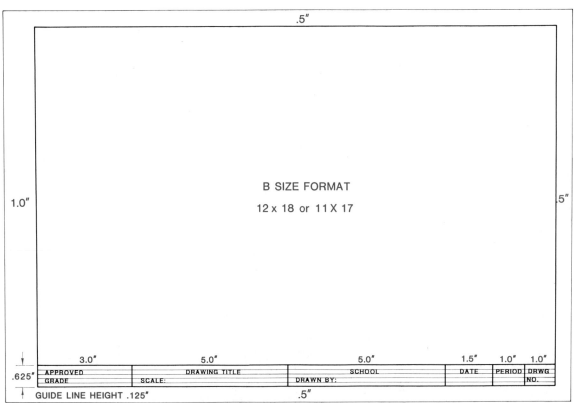

FIGURE 6–31 An example of a B-size format used in drafting classes

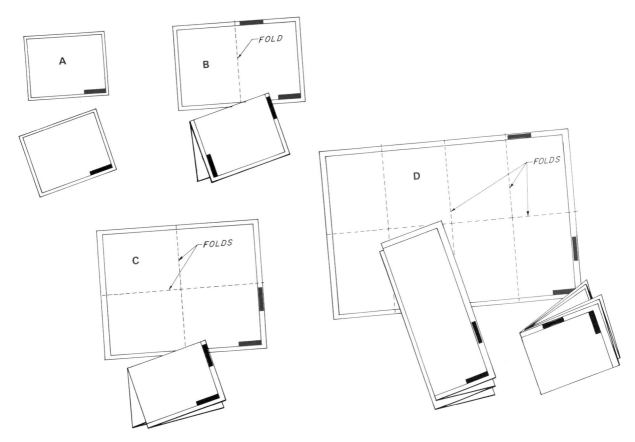

FIGURE 6–32 All standard formats will fold down to an A-size for easy storage (Reprinted from TECHNICAL DRAWING AND DESIGN by Goetsch and Nelson, © 1986 Delmar Publishers Inc.)

COMPUTER-AIDED DRAFTING & DESIGN

CAD LINE CONVENTIONS

CAD system multipen plotters are able to produce drawings with consistent and near-perfect line thicknesses. In order to use the CAD system properly, CAD operators must understand the correct use of ANSI drafting line widths and types.

Line Widths

When viewed on a monitor, line widths are identically thin and dense. Plotters create the difference in line widths. In multiple pen plotters, this is accomplished by using different size pen points. In single pen plotters, multiple pen strokes are used to widen lines to the desired thickness **(Fig. 6–33)**. Plotter pens are numbered, and each pen is assigned a layer color and line type. A blue line and a white line are the same width on a monitor, but may be plotted at different widths on the finished drawing.

Line Types

To produce proper line types, the LAYERING command is used. Each layer, assigned a name by the operator, represents a color and line type. Each layer should contain only one color and one line type. For example, a minimum of three layers is necessary to produce a drawing containing object lines, hidden lines, and centerlines. Most robust CAD systems have the capacity to produce 155 layers.

Figure 6–34 shows an example of the layers assigned to a drawing. In Fig. 6–34, the drawing contains an object line, hidden line, centerline, dimension line, phantom line, and construction line layer. The construction line layer is turned off in Fig. 6–34, which is common practice during the plotting of a drawing. In Fig. 6–34, five colors are

82 Drafting in a Computer Age

FIGURE 6–33 A single-pen plotter adjusts the proper line widths by multiple strokes. (Courtesy of Houston Instruments, a division of AMETEK)

used; white for object lines, yellow for hidden lines, green for centerlines or dimension lines, blue for construction lines, and cyan (blue-green hue) for phantom lines.

Continuous lines are the most common since they are used for all object and construction lines. The AutoCAD software used throughout this text has eight built-in line types **(Fig. 6–35)**.

CAD FORMATS

Formatting borders and title blocks on CAD systems is done by using either preprinted plotting paper, or preformatted data from the computer's memory.

Preformatting is one of the timesaving advantages of drawing on a CAD system. Preformatting is the same as drawing any repetitive detail. Borders and title block lines are drawn and are stored in the computer to be recalled at any time and inserted around any drawing. **Figure 6–36A** shows an example of a CAD-stored engineering drawing format, and **Fig. 6–36B** shows an architectural drawing format. **Figure 6–37** shows the insertion of a CAD-stored format around a newly plotted drawing.

FIGURE 6–34 Each drawing layer on the CAD system is given a name with the various line types and their color.

NAME	EXAMPLE	NAME	EXAMPLE
Continuous	———————————	Dashed	— — — — —
Hidden	— — — — — —	Dashdot	— · — · — · —
Center	—— — —— — ——	Divide	— · · — · · —
Phantom	—— — — —— — —	Border	— — · — — ·

FIGURE 6–35 CAD line types

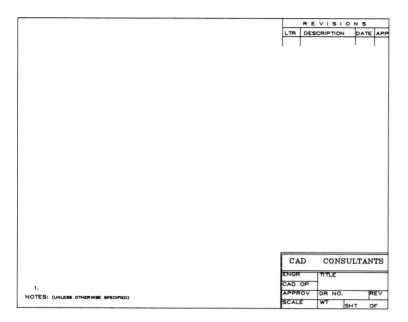

FIGURE 6–36A Example of a stored drawing format for an engineering firm

FIGURE 6–36B Example of a stored drawing format for an architectural firm

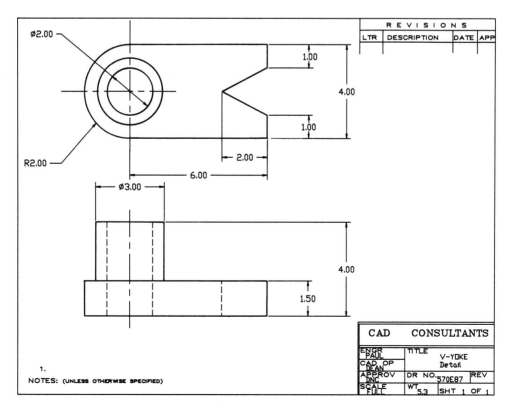

FIGURE 6–37 After the drawing is completed, a stored drawing format may be inserted.

EXERCISES

1. Draw **Figs. 6–38** and **6–39** on a CAD system and/or with instruments using the proper line weights.
2. Design a C-size drawing format for school use.
3. Design a C-size drawing format for the Ace Tool Design Company.
4. List each ANSI drafting standard used in this chapter with the ANSI reference page numbers.
5. Set up a standard drawing format for A-, B-, and C-size paper on a CAD system.
6. With a CAD system, draw the figures shown in **Fig. 6–40**. Adjust the line widths as needed for the final hard copy plot. Estimate all dimensions.

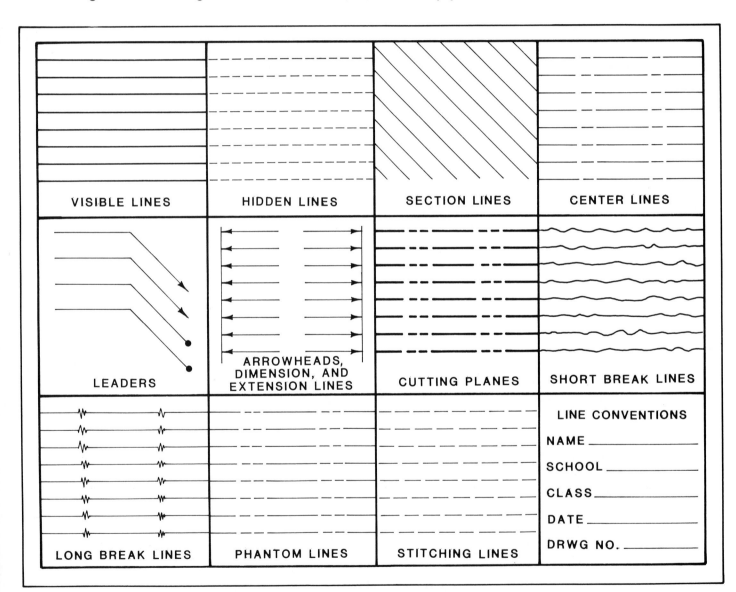

FIGURE 6–38 Practice drawing the line conventions with drawing instruments and/or a CAD system.

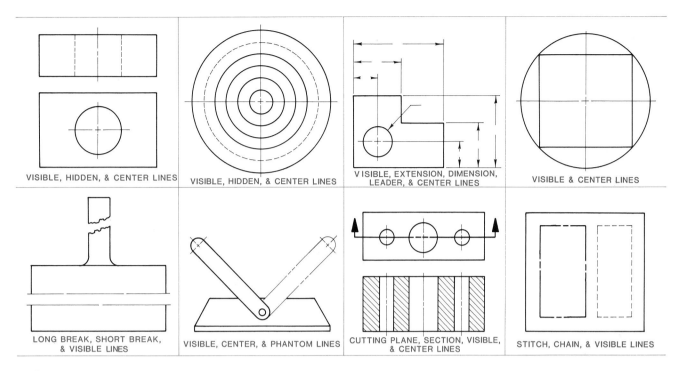

FIGURE 6-39 Redraw each figure two times as large using correct line widths.

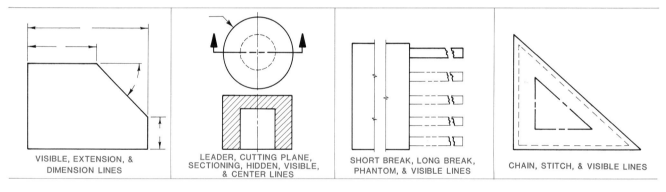

FIGURE 6-40 With a CAD system, redraw each figure. Use a separate layer for each line type.

KEY TERMS

- ANSI
- Border
- Break line
- Cabinet drawing
- Cavalier drawing
- Centerline
- Chain line
- Cutting plane
- Dimension line
- Dimetric drawing
- Extension line
- Hidden line
- ISO
- Isometric drawing
- Leader
- Line conventions
- Multiview drawing
- Object line
- Oblique drawing
- Opaque
- Pattern development drawing
- Perspective drawing
- Phantom line
- Section drawing
- Section line
- Standards
- Stitch line
- Title block
- Trimetric

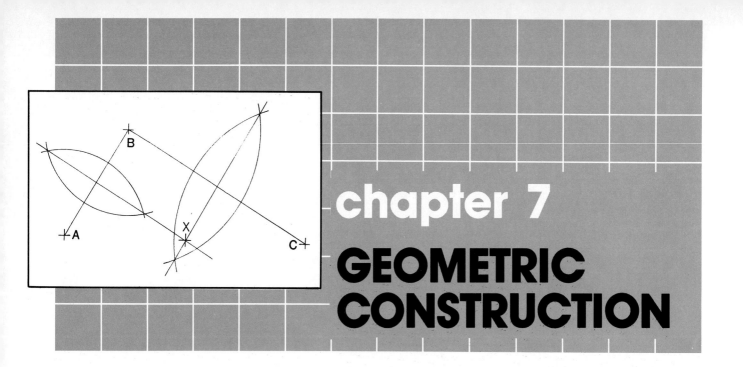

chapter 7
GEOMETRIC CONSTRUCTION

OBJECTIVES

The student will be able to:

- understand geometric form terminology
- recognize two-dimensional geometric forms
- recognize three-dimensional geometric forms
- construct basic geometric forms using drafting instruments
- construct basic geometric forms using a CAD system

CAD VERSUS DRAWING INSTRUMENTS

Construction of geometric forms with drawing instruments has always been a tedious but necessary drafting task **(Fig. 7–1)**. The use of geometric drafting templates **(Fig. 7–2)** is a significant time-saver for the contemporary drafter; however, drawing geometric forms on a CAD system is simpler and faster. A CAD drafter must be familiar with the basics of **geometric construction** in order to fully utilize the geometric drawing capacity of a CAD system.

GEOMETRIC FORMS

To communicate effectively, a CAD drafter must be able to use geometric terminology correctly. Geometric terminology commonly used in most fields of drafting and design is shown in **Fig. 7–3**. Geometric forms are either **two-dimensional** or **three-dimensional**.

Figure 7–4 shows basic two-dimensional forms that are variations of circles, triangles, squares, **polygons**, **ellipses**, and free forms. **Figure 7–5** shows a two-dimensional engineering drawing constructed with

FIGURE 7–1 Early Greek architecture refined the use of geometric construction. (Courtesy of The Celotex Corporation)

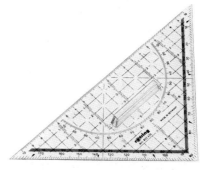

FIGURE 7–2 Drawing templates help with the drawing of geometric constructions. (Courtesy of Koh-I-Noor)

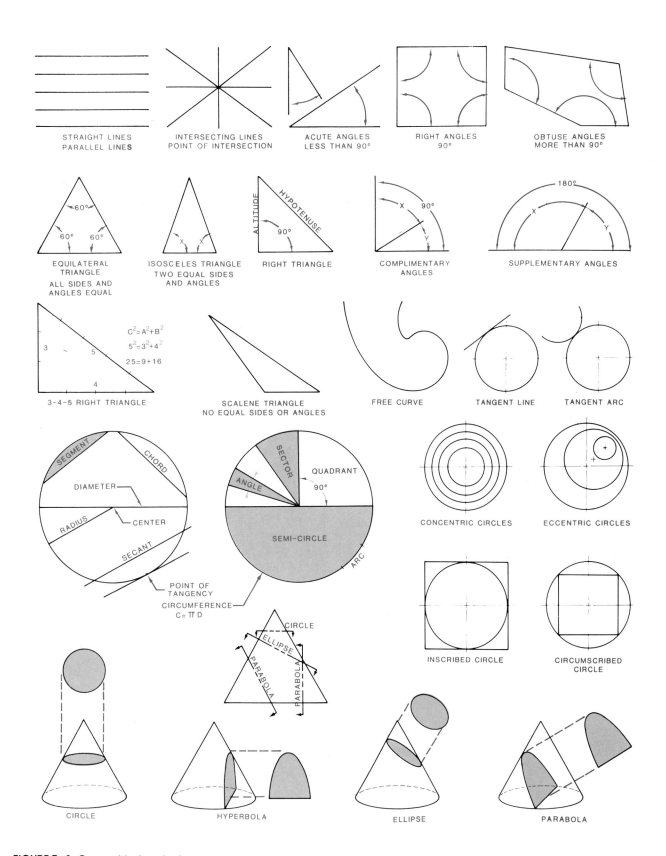

FIGURE 7-3 Geometric terminology

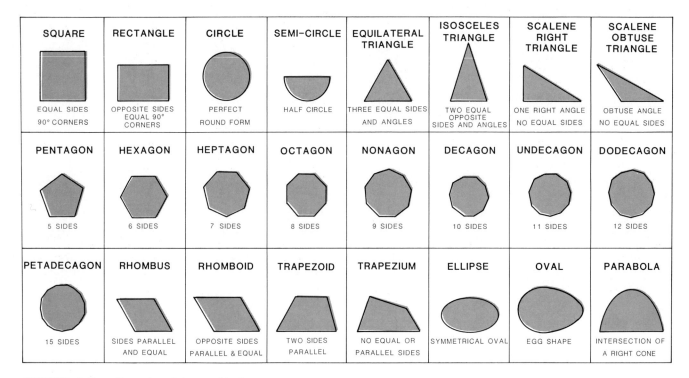

FIGURE 7–4 Two-dimensional geometric forms

a combination of straight lines, arcs, circles, and curves. Study Figs. 7–4 and 7–5, and identify the different types of geometric forms.

Three-dimensional geometric forms possess width, depth, and height. Although there are hundreds of different forms, all are variations of the basic forms found in **Fig. 7–6**.

CONVENTIONAL GEOMETRIC CONSTRUCTION

The procedures for manually constructing the most common geometric forms are shown in **Figs. 7–7** through **7–23**.

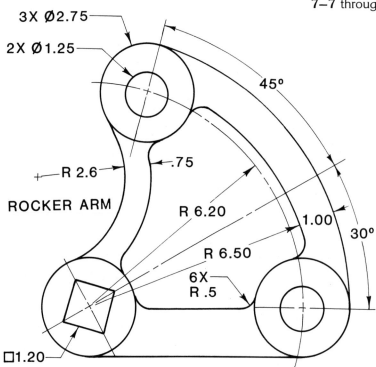

FIGURE 7–5 Any shape can be drawn with lines, circles, arcs, and curves.

Chapter 7 Geometric Construction 89

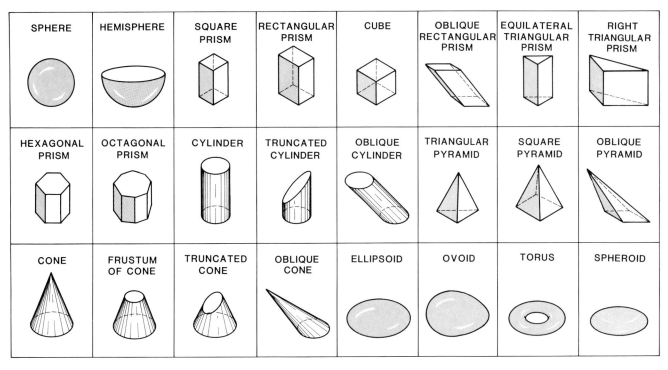

FIGURE 7–6 Three-dimensional geometric forms

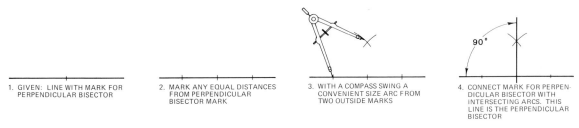

FIGURE 7–7 Perpendicular bisector

FIGURE 7–8 Parallel lines

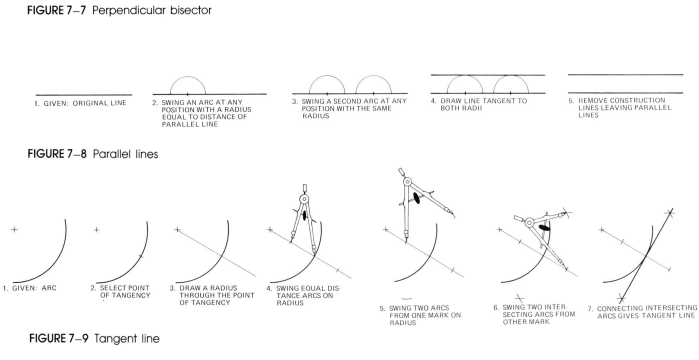

FIGURE 7–9 Tangent line

90 Drafting in a Computer Age

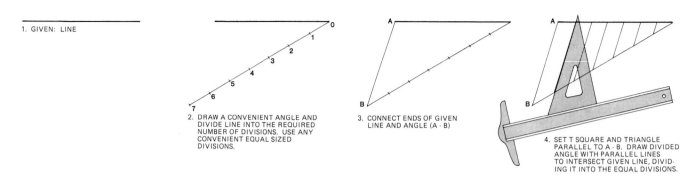

FIGURE 7-10 Divide a line into equal parts.

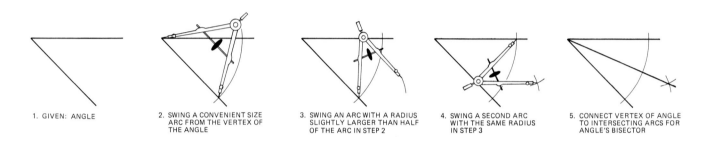

FIGURE 7-11 Bisect an angle.

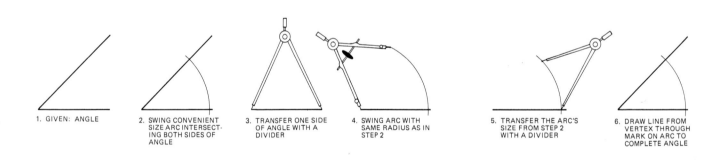

FIGURE 7-12 Duplicate an angle.

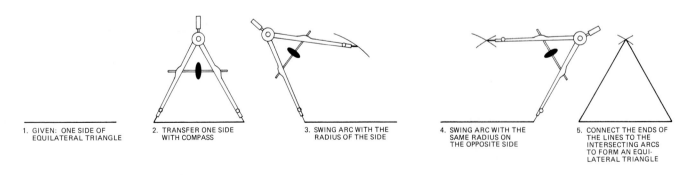

FIGURE 7-13 Constructing an equilateral triangle

Chapter 7 Geometric Construction 91

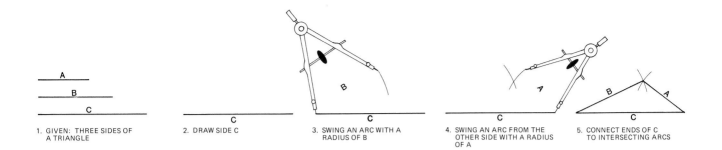

FIGURE 7-14 Constructing a triangle

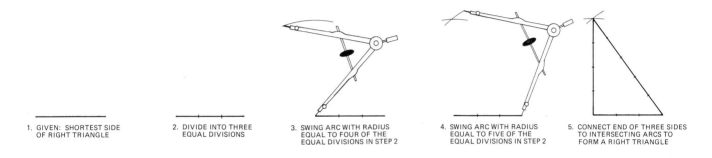

FIGURE 7-15 Constructing a right triangle

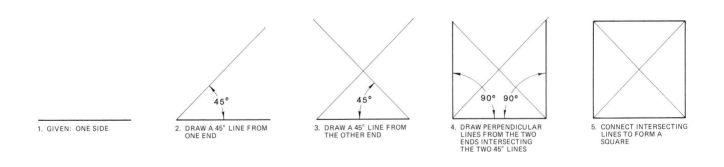

FIGURE 7-16 Constructing a square

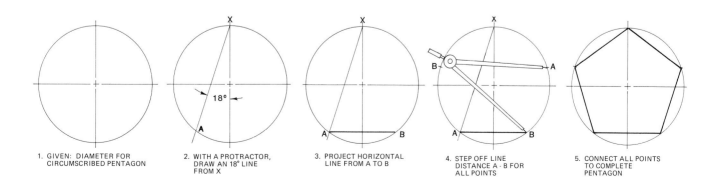

FIGURE 7-17 Constructing a pentagon (5 sides)

92 Drafting in a Computer Age

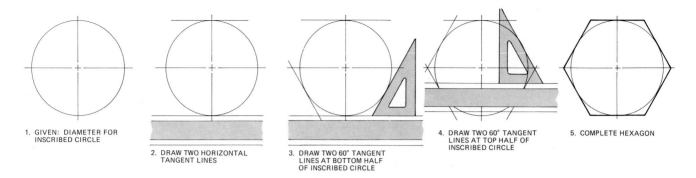

FIGURE 7–18 Constructing a hexagon (6 sides)

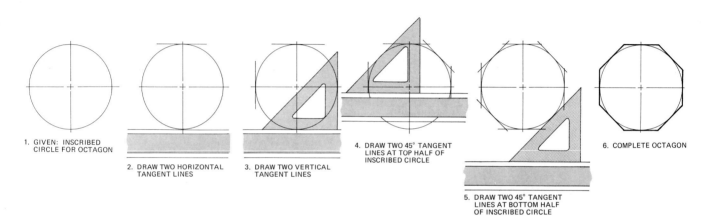

FIGURE 7–19 Constructing an octagon (8 sides)

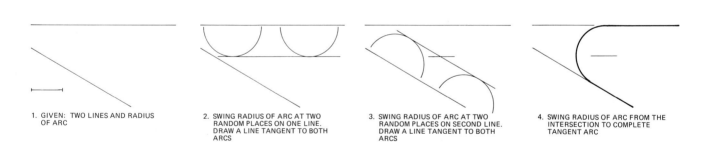

FIGURE 7–20 Drawing an arc tangent to two lines

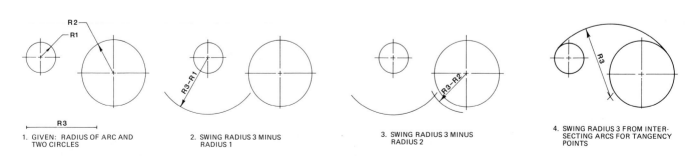

FIGURE 7–21 Drawing an arc tangent to two circles

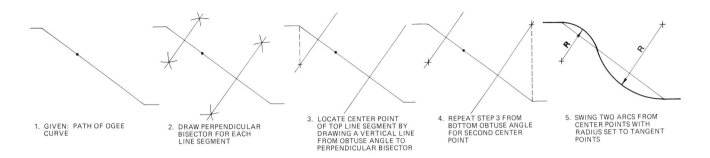

FIGURE 7–22 Drawing an **ogee curve** (also called a reverse curve). An ogee curve occurs in situations where a smooth contour is needed between two offset features.

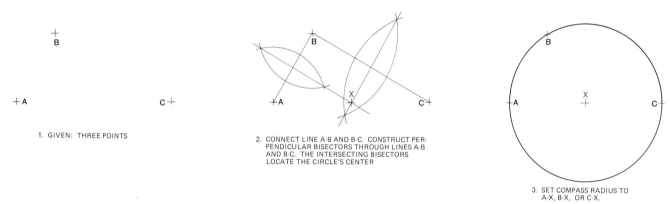

FIGURE 7–23 Drawing a circle through three points

COMPUTER-AIDED DRAFTING & DESIGN

COMPUTER-AIDED GEOMETRIC CONSTRUCTION

The construction of basic geometric forms is programmed into CAD-system software. To draw basic geometric forms on a CAD system, the drafter enters the data describing the type, size, and location of the form using a graphics tablet and menu **(Fig. 7–24)**.

Dividing a Line Into Equal Parts

The command used for this function is DIVIDE **(Fig. 7–25)**. First, the line is drawn. It may be any length and at any angle, it may even be an arc or irregular curve. Next, the DIVIDE command is selected. The line is then touched with the cursor and the required number of divisions is typed in. The line in Fig. 7–25 is divided into 5 equal parts.

Constructing Perpendicular Lines

The commands used here are LINE and PERPENDICULAR **(Fig. 7–26)**. Given the point and the line, the LINE command is selected and the line is started at the given point. Then, PERPENDICULAR is selected and the line is touched with the cursor anywhere. A new line automatically snaps **perpendicular** to the selected line. **Snaps** means a line or point appears at the closest grid coordinates located on the monitor.

Line Drawn at a Given Angle

The LINE command is used to perform this operation. When drawing a line, its length and angle are displayed on the screen. These distances and angles change as the puck is moved. The operator either can watch the **polar coordinate** (an-

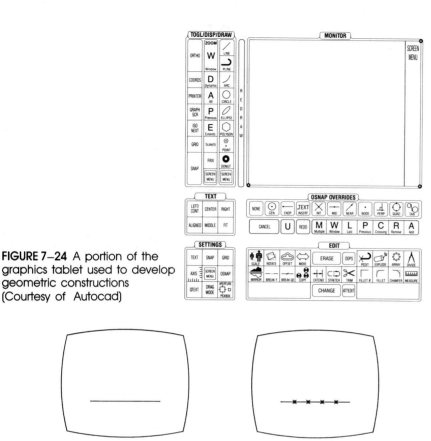

FIGURE 7-24 A portion of the graphics tablet used to develop geometric constructions (Courtesy of Autocad)

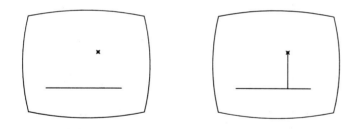

COMMAND: DIVIDE

FIGURE 7-25 Dividing a line into equal parts with a CAD system

COMMANDS: LINE, PERPENDICULAR

FIGURE 7-26 Constructing perpendicular lines with a CAD system

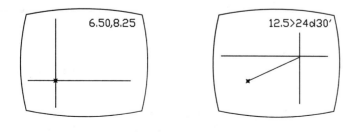

COMMAND: LINE

FIGURE 7-27 Drawing a line at a given angle with a CAD system

gular dimension) readout on the screen and position the line to the desired units; or, type the units to accurately draw the line. In **Fig. 7–27**, the line is 12.5 units long and is drawn at an angle of 24° and 30' from the horizontal (12.5 > 24d30').

Line Drawn From a Point Tangent to a Circle

To obtain this line, the commands LINE and TANGENT are used **(Fig. 7–28)**. Given the point and the circle, the LINE command is selected and the line is started at the given point. Then, TANGENT is selected and the circle is touched by the cursor anywhere on the side where the line is to be drawn. The new line automatically snaps **tangent** to the circle.

Line Drawn Tangent to Two Circles

Given two circles, LINE and then TANGENT are selected. Touch one of the circles on the side the line is to be placed, select TANGENT again and touch the other circle. The line automatically snaps tangent to both circles. This procedure is repeated for the lines in **Figs. 7–29** and **7–30**.

Construction of a Polygon

The POLYGON command constructs any **equilateral** polygon from a triangle (3 sides) to a 1024-sided polygon. The two examples in **Fig. 7–31** are a **hexagon** (6 sides) and an **octagon** (8 sides). The CAD method of construction for all polygons is the same. The POLYGON command is first selected. Then, the number of sides that are desired is typed in. Next, the center of the polygon is located using the puck. Type in the desired size and the polygon is drawn. The polygon may be erased, moved, or rotated the same as any other entity.

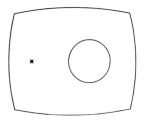
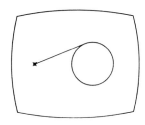

COMMANDS: LINE, TANGENT

FIGURE 7-28 Drawing a tangent line from a point to a circle with a CAD system

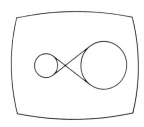

COMMAND: LINE, TAN, TAN
(DONE ONCE FOR EACH LINE)

FIGURE 7-30 Drawing tangent lines to circles with a CAD system

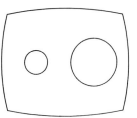
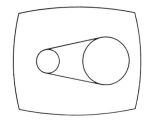

COMMAND: LINE, TAN, TAN
(DONE ONCE FOR EACH LINE)

FIGURE 7-29 Drawing tangent lines to circles with a CAD system

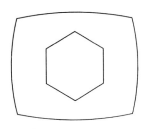
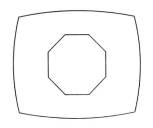

COMMAND: POLYGON, 6 SIDES, SIZE

COMMAND: POLYGON, 8 SIDES, SIZE

FIGURE 7-31 Drawing polygons with a CAD system

Construction of a Fillet

A **fillet** is a concave arc constructed tangent to two lines. The operator must set its radius by selecting the FILLET command, typing R (radius) and the desired fillet radius. To construct a fillet from two given lines, the operator simply selects the FILLET command and touches the two lines. In **Fig. 7-32**, this was done three times (one fillet and two rounds). Note that this command draws the desired arc tangent to the two lines and trims extraneous lines.

Figure 7-33 shows the application of geometric forms to a completed engineering CAD drawing. The student should study this drawing, find the geometric forms, and determine the sequence and procedures necessary to draw this object using a CAD system.

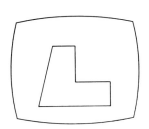
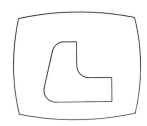

COMMAND: FILLET

FIGURE 7-32 Drawing a fillet with a CAD system

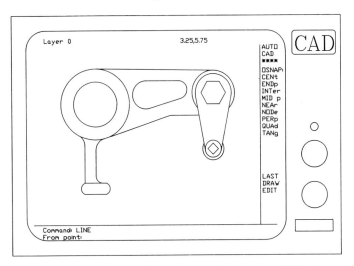

FIGURE 7-33 A CAD-generated drawing, using many of the geometric commands

EXERCISES

1. Construct a perpendicular bisector for a 4.75" line.
2. Construct a parallel line to a 3.5" line set at 30°.
3. Construct a tangent line to a 1.75" diameter circle.
4. Divide a 5" line into 8 equal parts.
5. Bisect a 42° angle.
6. Construct an equilateral triangle with 2.25" sides.
7. Construct a right triangle with the shortest side 1.5".
8. Construct a 1.375" square.
9. Construct a **pentagon** (5 sides) in a circumscribed circle with a 1.5" radius.
10. Construct a hexagon with a 2.0" radius inscribed circle.
11. Construct an octagon with a 3.75" diameter inscribed circle.
12. Construct a tangent arc within a 30° angle with a .5" radius.
13. Construct a right triangle with the short side 2.0" long.
14. Draw the basic geometric constructions shown in **Fig. 7–34**.
15. Draw the two-dimensional geometric forms shown in **Fig. 7–35**.
16. Draw the geometric forms shown in **Fig. 7–36** with a CAD system.
17. From the data table shown in **Fig. 7–37**, plot and draw the geometric form.
18. Create a data table for each of the geometric forms in **Fig. 7–38**.

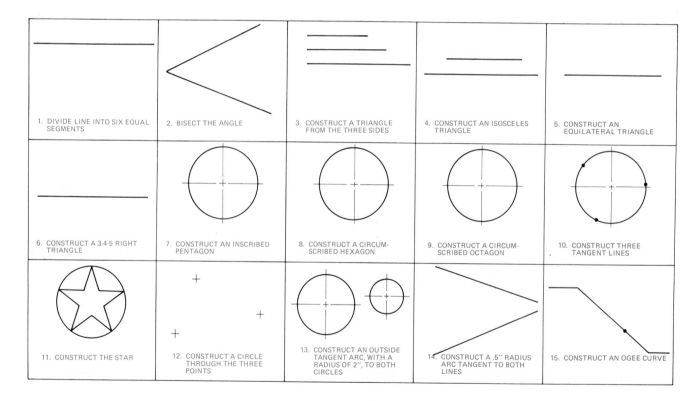

FIGURE 7–34 With drawing instruments or a CAD system, draw the geometric forms. Double the size of each exercise.

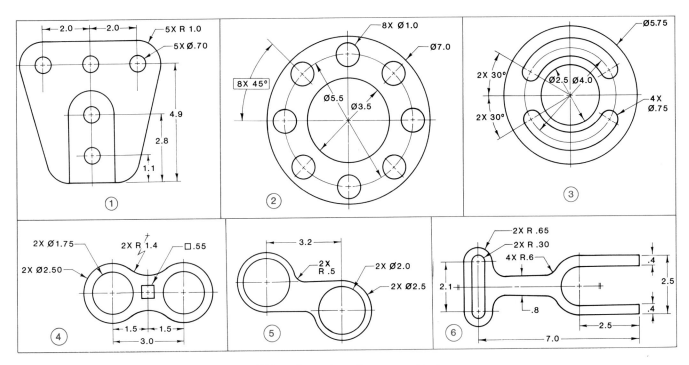

FIGURE 7–35 With drawing instruments or a CAD system, draw the geometric forms.

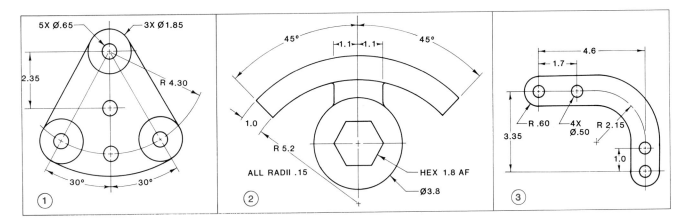

FIGURE 7–36 With drawing instruments or a CAD system, draw the geometric forms.

98 Drafting in a Computer Age

DATA TABLE					
line	X	Y	go to	X	Y
1-2	1	3	→	4.5	9
2-3	4.5	9	→	8	3
3-1	8	3	→	1	3
new start					
4-5	4.5	1	→	1	7
5-6	1	7	→	8	7
6-4	8	7	→	4.5	1
finish					

FIGURE 7-37 Lay out the drawing from the data table's coordinates on .25" grid paper.

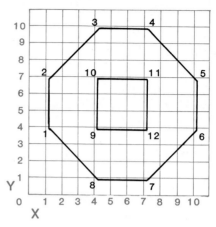

DATA TABLE					
line	X	Y	go to	X	Y
1-2	1	4	→	1	7
2-3	1	7	→	4	10
	4	10	→		

FIGURE 7-38 Complete the data table.

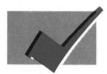

KEY TERMS

Ellipse
Equilateral
Fillet
Geometric construction
Hexagon
Octagon

Ogee curve
Parallel
Pentagon
Perpendicular
Polar coordinates

Polygon
Snaps
Tangent
Three-dimensional
Two-dimensional

the world of CAD

THE CAD OPERATOR

As the term "Operator" is commonly used, the closest possible parallel job description is that of "Draftsman" in traditional design practice. The similarities and differences from traditional drafting and design in modern CAD-based environments deserves investigation. This question is of great importance to those entering any aspect of the architectural, engineering, or general design fields.

The position, title, and work responsibilities of a "CAD Operator" are aspects of the modern design office which as yet, have not been universally defined. The term "Operator" is commonly used to refer to those persons who spend most of their working hours at a terminal or personal computer engaged in creating drawings on a CAD software package, or in editing and updating such drawings. The duties and qualifications of such people vary widely from one design firm to another. This is not surprising given that CAD for general applications is only now becoming widely available. We have reached an exciting era. The advent of powerful, fast, low-cost drawing microcomputers and software packages has redefined the role of CAD operators and designers:

- CAD operators and designers are more valuable than drafters
- CAD (hardware and software) are tools that extend human design capabilities
- Advanced CAD environments do not have draftsmen
- CAD will require higher skill than traditional manual drafting positions
- The prevailing notion of CAD "Operators" will be short-lived
- CAD systems will require talented well-trained people (in fact the human element will become recognized as the true measure and investment base in any CAD system)
- CAD will fully challenge the creativity and capability of all designers and builders
- CAD will lead to full integration of data management, and integration in the design and building trades

THE TRADITIONAL DESIGN OFFICE

It is useful to review the role of a person working in a conventional (i.e., manual) drawing environment. In the past, draftspersons have been responsible for the physical production and maintenance of working drawings and finish drawings in support of designers, engineers, and architects. In short, these individuals were and still remain as support personnel who translate and represent the ideas of others. Heavy emphasis is placed on artistic skills and dexterity to consistently develop drawings sufficient in clarity, detail, and accuracy to support legally binding construction contracts. Draftsman have always translated information developed by designers, engineers, and architects for presentation to the various trades so that designs could be constructed in the field or manufactured in the factory.

Manual drafting continues to employ the talents of skilled technicians with unique artistic abilities. However, CAD as a concept involves more than the production of drawings. Increasingly, design firms must adopt CAD equipment and methods to remain competitive. As a result of the CAD-associated realities of modern business, manual drafting personnel are being assigned and retrained to work on CAD systems. This conversion of labor skills is occurring at an overwhelming pace. Consider that as little as five years ago, CAD systems were almost unheard of in general practice.

THE DESIGN DRAFTSMAN OF THE FUTURE

CAD is a tool whose potential can only be tapped if properly staffed and managed. The role of those who will work on tomorrow's systems has thus become the center of management thinking. Likewise it should be foremost in the minds of those training for and seeking entry into the design professions. CAD is now drastically changing, not only the ways in which design work is performed, but also in the types and ways design services are provided to clients.

Most CAD systems represent a complex set of hardware, software, and peripherals equipment. However, the most critical element in any system is the quality and imagination of the people who manage and work on the system. Progressive modes for managing CAD will mean progressive modes for managing people and the design process they are engaged in. In turn, those individuals working on a system will need to maintain flexibility in the ways problems are approached. They will need to implement system routines and applications with a creative, open manner. All designers will need to recognize the inherent strengths and weaknesses of a given system. Effective operators will be those who recognize the constancy of change. They will adapt to it, and change while remembering that the reason for design will remain fixed: conceptualizing and detailing solutions which are practical, affordable, and address the problem at hand. CAD-system operators and designers are a vital part of the "design-build" team. The importance of this role is growing along with the work responsibilities associated with the position. Figure 1 provides a summary and comparison of the duties associated with drafting in the manual environment

COMPARATIVE ROLES AND DUTIES

TRADITIONAL (Manual)

- Inputs: Paper, Old Drawings, Calculations, Old Jobs
- Research and calculations
- Concept Development
- Engineering/Arch. (Preliminary Design)
- Manual Drafting: Cycles/Iterations of Changes, Check Reviews and Mark-ups
- Final; Prints

CAD

- Inputs: Electronic Surveys, Scanned Drawings of Existing Information, Electronic Media Research, Product Information, Standard Specification/Contract Documents
- Design Team: Conceptualization, Calculations, Research
- Create Drawing and Project Database
  - Engineering/Arch. Design
  - Design/Draft Concurrent Time Frame
  - Plans and Specs.
  - Interface of Contractors and Vendors (Incorporation of Field Modifications)
- Data Provided to Owner
- Facilities Maintenance and Adaptation
- Process May Be Repeated

(TIME axis)

---

with those emerging as the CAD draftsman role now and in the future. The following list shows the critical nature of this function and some of the duties which will move the CAD professional to the hub of the design process:

- Production record organization
- Production organization scheduling
- Field data (electronic) compilation
- Digitizing/Scanning of existing on-CAD drawings
- Products research
- Calculation organization
- Standard drawing development
- Standard detail development
- Bill of material calculations
- Specification checklists
- Production of job specific drawings
- Edit functions (design refinement)
- Drawings assembly
- Production coordination—drawing output

In short, CAD is a system. In the ideal case, the system is properly implemented and managed, creative design time is optimized and designers are limited only by their imaginations. We have not yet reached such a state. However, real experiences with CAD demonstrate that committed people, adapting to changes and working together, have used CAD to support the design process in efficient, creative ways. In the future, the true test of systems will be the performance of the people working with it.

TOMORROW'S OPERATOR—THE DESIGNER

Today's progressive design firms (and certainly all firms in the future) employ CAD technologies and design approaches. Young people entering the field need to be aware of the enormous magnitude of the changes now taking place and how they will be effected.

It is now clear that hardware and software are and will continue to change rapidly and significantly. Designers will work with engineers and architects in a fast-paced and exciting process that will center on the management and manipulation of project specific data bases. This means traditional drafting will be transformed from mere drawing production to the processing of existing data to electronic media, participating more and more in the actual design and component selection process, assisting directly in calculation and parametric routines for component sizing, product specification, project scheduling, project cost estimating. In association with all of these task areas will be the aspect of continuing editing and updating of the project information (throughout the design and building phases). Ultimately, the skills of any individual in the operation of a system will be paramount. Of equal importance will be the ability of the "new designers" to find and fully employ the ever expanding tool base of CAD.

As these applications expand, the draftsman has become the designer. Today's designer has become a much more valuable part of the design process.

(Courtesy of Besha Associates Engineering Corporation, Albany, New York)

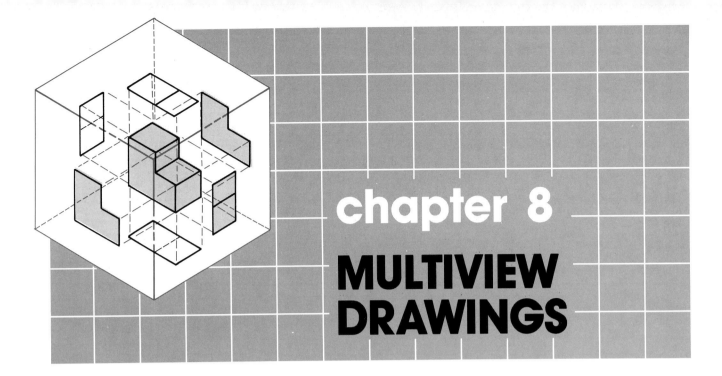

chapter 8
MULTIVIEW DRAWINGS

OBJECTIVES

The student will be able to:

- visualize the planes of projection
- select the front view for a multiview drawing
- draw third-angle projection multiview drawings
- draw first-angle projection multiview drawings
- center a multiview drawing
- draw runouts on a multiview drawing
- project an incline surface on a multiview drawing
- project an ellipse on a multiview drawing
- project an oblique surface on a multiview drawing
- generate multiview drawings with a CAD system

MULTIVIEW ORTHOGRAPHIC PROJECTIONS

Orthographic projection is a projection of a single view of an object on a drawing surface that is perpendicular to both the view and the lines of projection. Horizontal orthographic projection lines are used to align views with the same height dimensions. Vertical orthographic projection lines are used to align views with the same width dimensions **(Fig. 8–1)**. The technique of orthographic projection is used to create **multiview** drawings of a single object. The related views of the object appear as if they were all in the same plane.

Multiview drawings describe the exact size and shape of an object. **Figure 8–2** shows a CAD-generated multiview drawing of a machine part.

SELECTION OF VIEWS

Describing the shape of an object with a multiview drawing should always be accomplished with the minimum number of views. The first view chosen is the **front view** because it best shows the form of an object. **Figure 8–3** shows the selection of the front view of several objects. In each case, the front view reveals the most unique and distinguishing shape feature of the object.

Simple flat objects, such as a gasket **(Fig. 8–4)**, require only one view, since the thickness is uniform and can be described in a note.

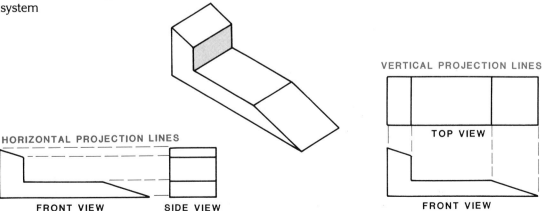

FIGURE 8–1 An example of the orthographic projection of the side and top views from the front view

100 Drafting in a Computer Age

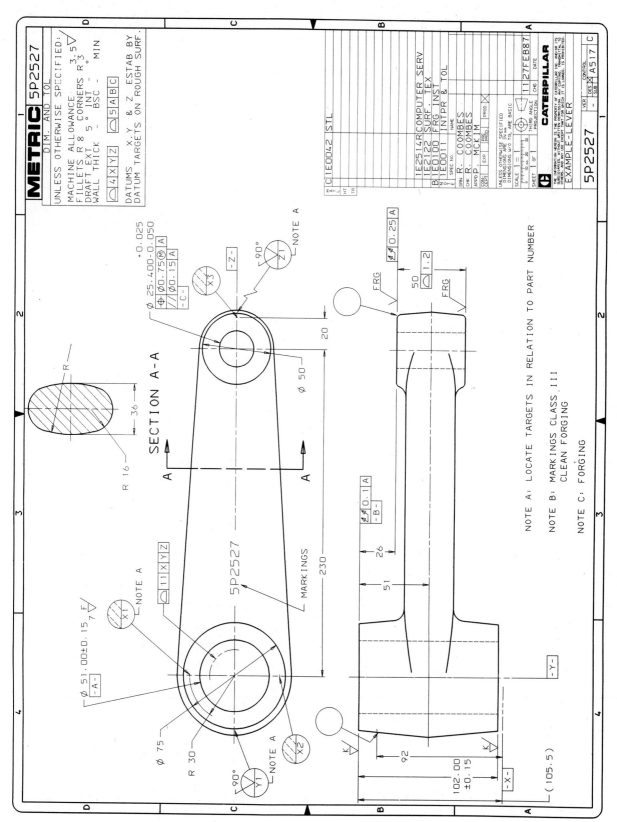

FIGURE 8-2 An example of a CAD-generated multiview working drawing (Courtesy of Caterpillar, Inc.)

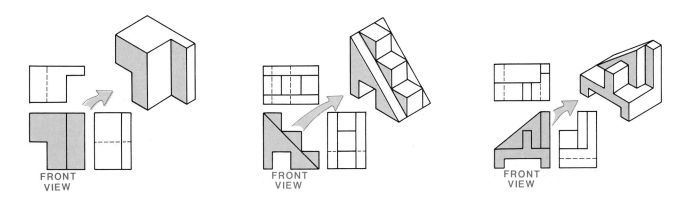

FIGURE 8–3 The front view selection depends on which view best describes the shape of the item being drawn.

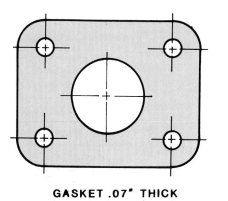

FIGURE 8–4 Simple flat objects can be defined with a single view.

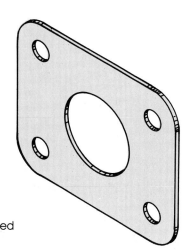

Symmetrical objects may need only one or two views, since a second or third view is identical to the one of the other views. In a multiview drawing of a solid sphere (ball), all views are identical so only one view is needed. Only two views are required to describe the object in **Fig. 8–5**, since the front and side views are the same.

Three views are used to describe a clock radio. The front view **(Fig. 8–6)** shows the height and width of the clock radio; the **top view**

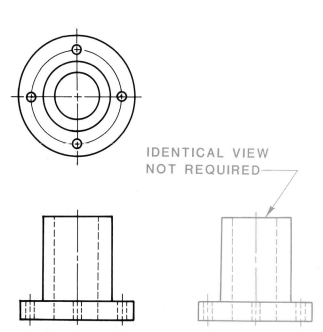

FIGURE 8–5 A symmetrical object requires only two views for a full description.

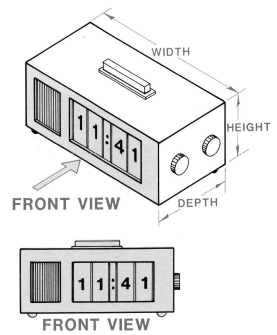

FIGURE 8–6 A multiview drawing's front view of a digital clock

(Fig. 8–7) shows the depth and width; and the **right-side view (Fig. 8–8)** shows the height and depth. The size and shape of most objects can usually be described in three views **(Fig. 8–9)** when the bottom, left side, and rear views would not help to further clarify any aspect of the object.

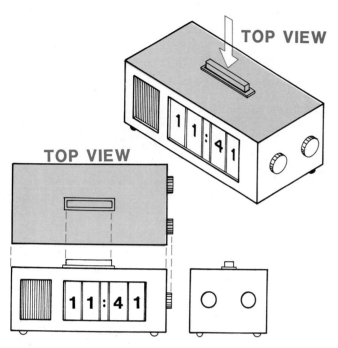

FIGURE 8–7 The digital clock's top view

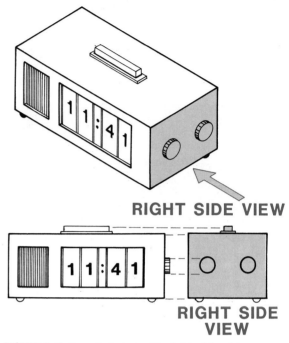

FIGURE 8–8 The digital clock's right-side view

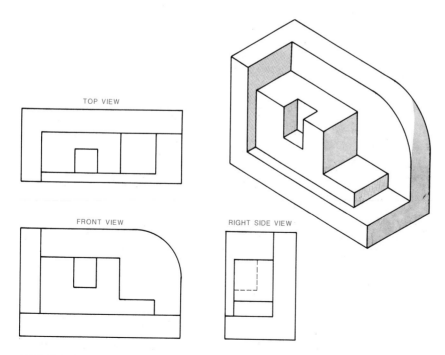

FIGURE 8–9 Most objects can be fully described with three views.

Other objects may require up to six views to show the exact size and shape of each side **(Fig. 8–10)**. **Figure 8–11** shows a six-view drawing of the same clock radio shown in the three views of Figs. 8–6 through 8–8. Although the three views have identical dimensions, the details of each of the six views shown in Fig. 8–11 are different. Six views, therefore, are needed to show all the details necessary for manufacturing the clock radio case. Many additional multiview working drawings are needed for the manufacture of the interior of the radio.

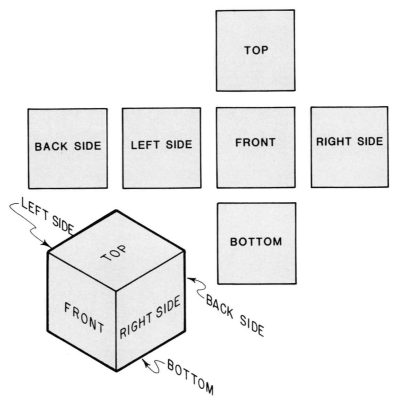

FIGURE 8–10 Orthographic projection of six views using third-angle projection for the placement of views

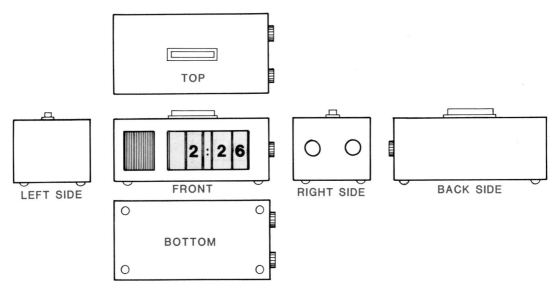

FIGURE 8–11 A multiview drawing may have a maximum of six orthographic views.

PLANES OF PROJECTION

Each view in a multiview drawing is drawn as it appears on an imaginary **plane of projection**. Imagine a sheet of glass (plane of projection) placed between a viewer and an object **(Fig. 8–12)**. Now imagine the viewer's lines of sight as parallel and connected through the glass to each line and corner of the object. The shape of the object can be visualized where the lines of sight intersect the plane of projection.

To visualize all six planes of projection, imagine a glass box surrounding an object **(Fig. 8–13)** with each view of the object drawn on each glass plane as viewed from the outside. When the sides of the glass box are hinged open **(Fig. 8–14)**, the relative position of each view in a multiview drawing is revealed. The hinge lines between the planes are also called **fold lines** or **reference lines**, and are used to locate the position of each view **(Fig. 8–15)**.

Views can be located also by drawing a 45° line which is used to project a third view once two base views are established. **Figure 8–16** shows the use of this method in projecting a side view from a front and top view, or a top view from a front and side view.

Another method of drawing a third view from two established views is the use of dividers to transfer distances from completed views. For example, depth dimensions from the top view can be transferred to the side view with dividers.

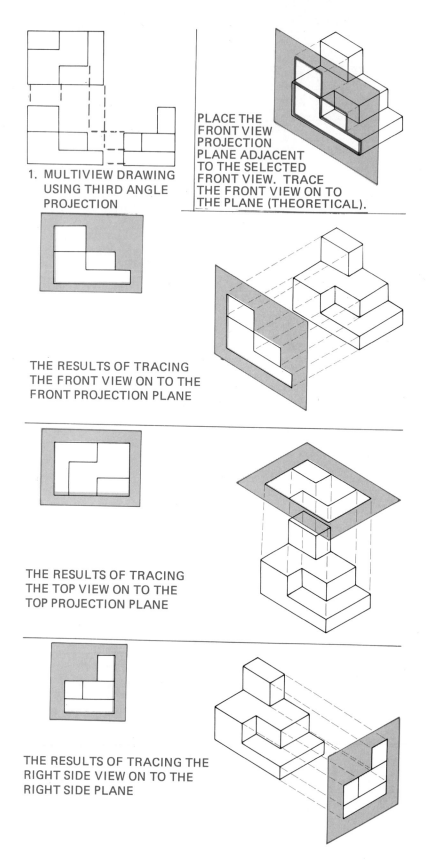

1. MULTIVIEW DRAWING USING THIRD ANGLE PROJECTION

PLACE THE FRONT VIEW PROJECTION PLANE ADJACENT TO THE SELECTED FRONT VIEW. TRACE THE FRONT VIEW ON TO THE PLANE (THEORETICAL).

THE RESULTS OF TRACING THE FRONT VIEW ON TO THE FRONT PROJECTION PLANE

THE RESULTS OF TRACING THE TOP VIEW ON TO THE TOP PROJECTION PLANE

THE RESULTS OF TRACING THE RIGHT SIDE VIEW ON TO THE RIGHT SIDE PLANE

FIGURE 8–12 Planes of projection

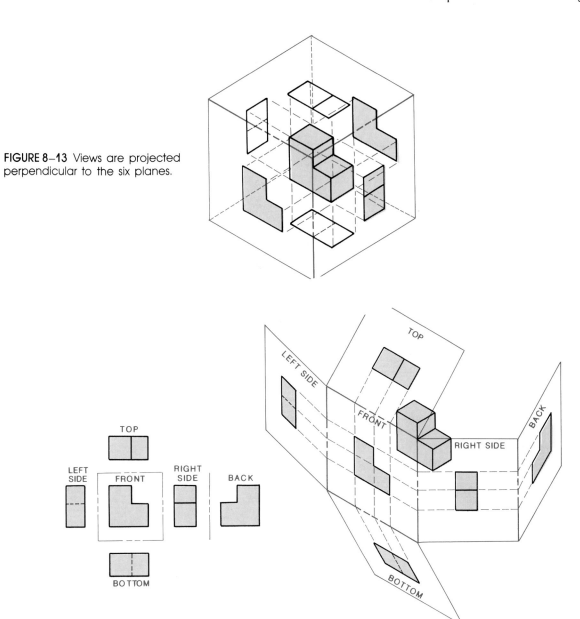

FIGURE 8–13 Views are projected perpendicular to the six planes.

FIGURE 8–14 Opening the projection box on its fold lines (reference lines)

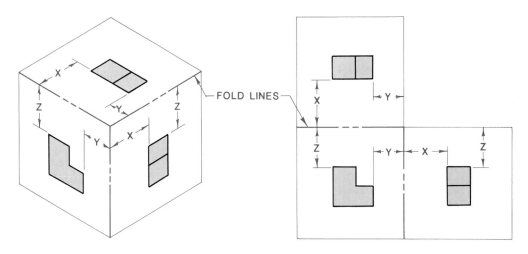

FIGURE 8–15 Locating views from the fold lines (reference lines)

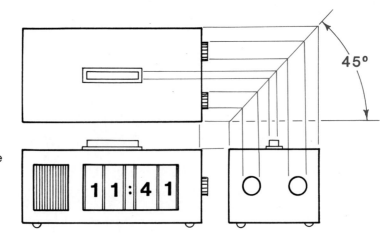

FIGURE 8-16 A third view can be projected from two established views.

ANGLES OF PROJECTION

The three basic views used in multiview drawings in the United States are the front view, right-side view, and top view. This is known as **third-angle projection (Fig. 8–17)**. In Europe, the three basic views are the front view, left-side view, and bottom view. This is known as **first-angle projection (Fig. 8–18)**. Figure 8–19 shows a comparison of a multiview drawing of the same object using first-angle and third-angle projection.

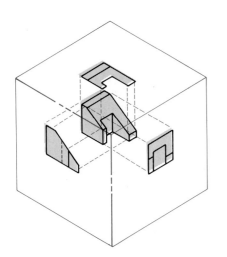
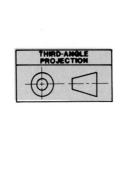
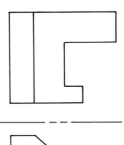
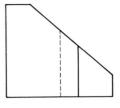
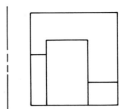

FIGURE 8-17 Third-angle projection

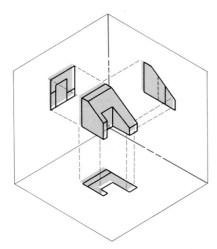
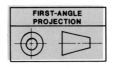
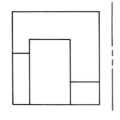
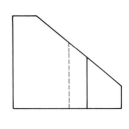
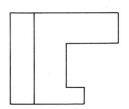

FIGURE 8-18 First-angle projection

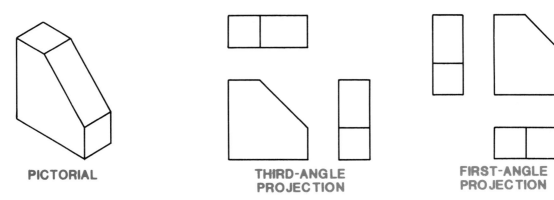

FIGURE 8-19 A comparison of first-angle and third-angle projection

Line Precedence

When objects are viewed through a plane of projection, some lines fall on an identical perpendicular plane although one line may be further from the viewer than the other. When lines overlap in this manner, the darkest line takes **precedence** on the drawing **(Fig. 8-20)** in the following order:

1. Cutting plane line
2. Object line
3. Hidden line
4. Centerline

VISUALIZATION

All flat surfaces in multiview drawings are either normal, inclined, or oblique **(Fig. 8-21)**. **Normal surfaces** are true size, and are parallel or perpendicular to the plane of projection. **Inclined surfaces** are angu-

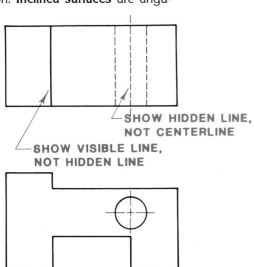

FIGURE 8-20 Line precedence for overlapping lines

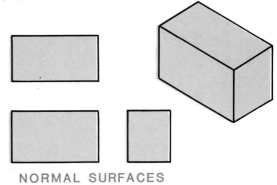

FIGURE 8-21
All flat surfaces in a multiview may be classified as normal, inclined, or oblique.

NORMAL SURFACES

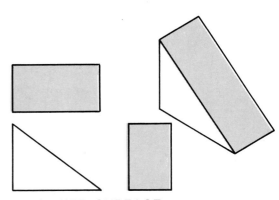

INCLINED SURFACE

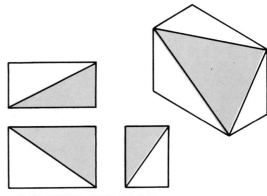

OBLIQUE SURFACE

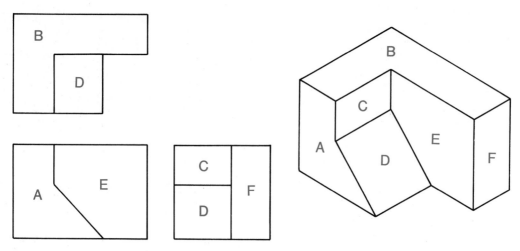

FIGURE 8–22 Identifying normal and inclined surfaces

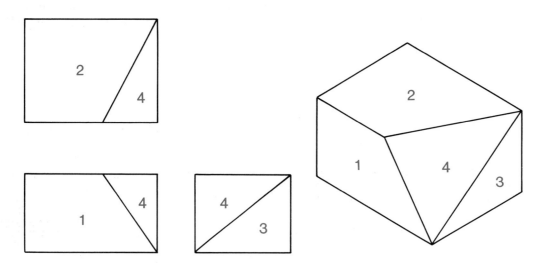

FIGURE 8–23 Identifying normal and oblique surfaces

lar; their planes are not parallel or perpendicular to the plane of projection. Inclined surfaces are slanted to normal surfaces. **Figure 8–22** shows the relationship between normal and inclined surfaces. **Oblique surfaces** are not parallel or perpendicular to any normal surface. **Figure 8–23** shows the relationship between normal and oblique surfaces. Numbers and letters are used on Figs. 8–22 and 8–23 to identify identical surfaces on different views.

INCLINED AND OBLIQUE SURFACE PROJECTIONS

Surfaces perpendicular or parallel to the plane of projection always appear as a true size surface or edge on the orthographic plane. Surfaces that lie on any angle to the plane of projection appear foreshortened (not true size surfaces) on the orthographic plane. **Figure 8–24** shows a comparison of true size surfaces and foreshortened surfaces on an orthographic view.

Holes and cylinders appear as true circles on normal surfaces but appear as **ellipses** on inclined surfaces **(Fig. 8–25)**. Since inclined surfaces recede from the plane of projection at right angles, the ellipse's major diameter parallel to the plane of projection is true size. The ellipse dimension perpendicular to the plane of projection is foreshortened.

Chapter 8 Multiview Drawings 109

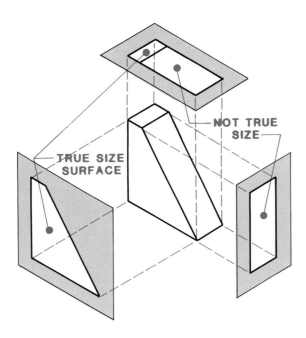

FIGURE 8-24 Projection of perpendicular and inclined surfaces

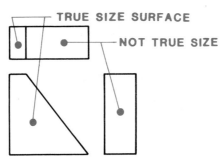

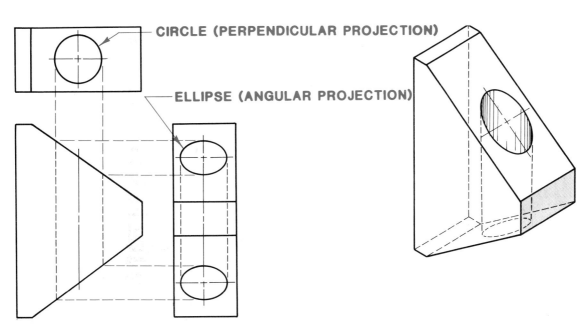

FIGURE 8-25 A hole will appear as an ellipse on an inclined plane.

The amount of foreshortening is related to the angle of incline. Small angles produce narrow ellipses and large angles produce wide ellipses **(Fig. 8–26)**. **Figure 8–27** shows ellipse widths derived from common inclined angles.

The use of standard ellipse templates eliminates plotting ellipse layouts on inclined surfaces. The steps shown in **Fig. 8–28** must be followed to insure ellipse accuracy.

Since **oblique** surfaces **(Fig. 8–29)** are not parallel to any normal plane, oblique surfaces, including circles, must be plotted from known true distances on other views.

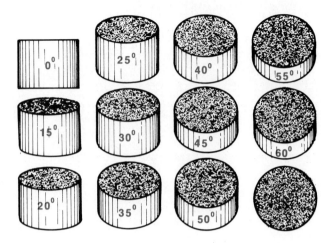

FIGURE 8–26 Examples of ellipses

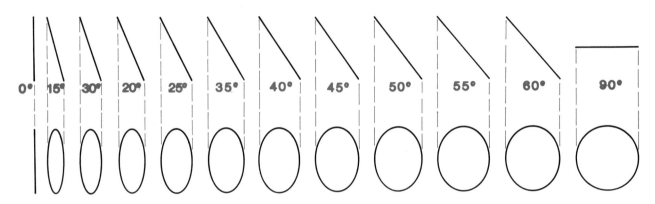

FIGURE 8–27 The angle of projection determines the minor diameter of the ellipse.

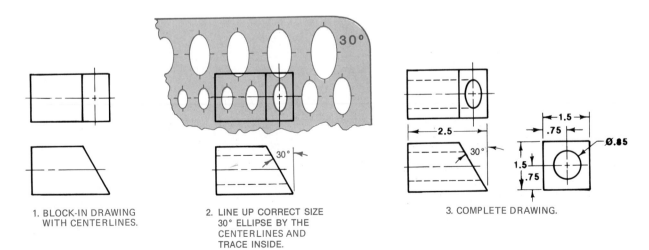

1. BLOCK-IN DRAWING WITH CENTERLINES.
2. LINE UP CORRECT SIZE 30° ELLIPSE BY THE CENTERLINES AND TRACE INSIDE.
3. COMPLETE DRAWING.

FIGURE 8–28 Drawing an ellipse with an ellipse template

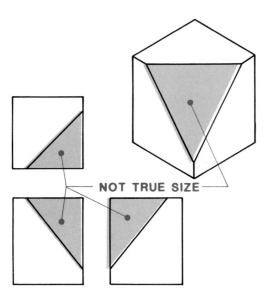

FIGURE 8–29 An oblique surface will not appear true size in any view.

Runouts

Surfaces with round or smooth contours do not intersect with straight lines. When surfaces blend together in this manner, the merging surface is called a **runout**. Square runouts and round runouts are shown in **Fig. 8–30**.

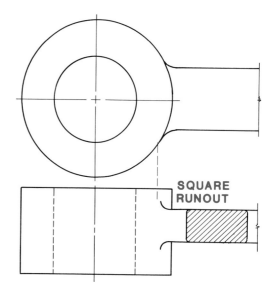

FIGURE 8–30 Examples of runouts on a multiview drawing

DIMENSIONING

Conventions for dimensioning are covered in Chapter 9. Some consideration of dimensioning needs is necessary, however, when planning and preparing multiview drawings. Use the following guidelines in planning for the dimensioning of multiview drawings:

1. Spacing between views depends on the number of dimensions to be placed between the views **(Fig. 8–31)**.
2. Place the closest dimension to the view at .375" clearance. Every additional dimension is spaced .25" apart **(Fig. 8–32)**.
3. Place as many dimensions as possible between views **(Fig. 8–33)**, rather than on the outside.

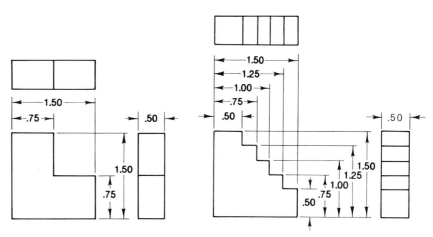

FIGURE 8–31 The numbers of dimensions between views will determine the space between the views.

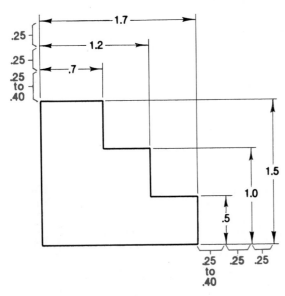

FIGURE 8-32 Typical dimension spacing

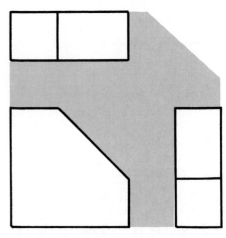

FIGURE 8-33 Place as many dimensions as possible between the views.

4. Place dimensions on the part of the view that best describes the form **(Fig. 8-34)**.
5. A working drawing must be completely dimensioned with enough information to manufacture the item without guesswork or fabrication redesign.

Most companies position the first drawing on a sheet in a corner to allow space for additional details. Nevertheless a drafter must know how to center a drawing in a given space to be certain the finished drawing will fit into the available space. To center a drawing, follow the steps in **Fig. 8-35**.

DRAWING PROCEDURES

When drawing an object, first determine the number of views needed, then decide which surface will become the front view. Next, establish the space needed between views for dimensioning and calculate the center of the drawing. Once these preliminary decisions have been made, follow the steps in **Fig. 8-36**.

Follow the steps in **Fig. 8-37** to draw arcs and circles on inclined surfaces. Refer to **Fig. 8-38** to prepare multiview drawings which contain oblique surfaces.

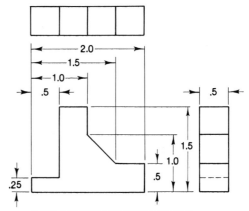

AN EXAMPLE OF CORRECT DIMENSION PLACEMENT. THE DIMENSION IS PLACED ON THE SEGMENT THAT BEST DESCRIBES ITS FORM.

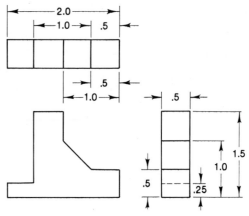

AN EXAMPLE OF INCORRECT DIMENSION PLACEMENT. EXPLAIN THE ERRORS OF THE IMPROPERLY PLACED DIMENSIONS.

FIGURE 8-34 Comparison of correct and incorrect dimension placements on a multiview drawing

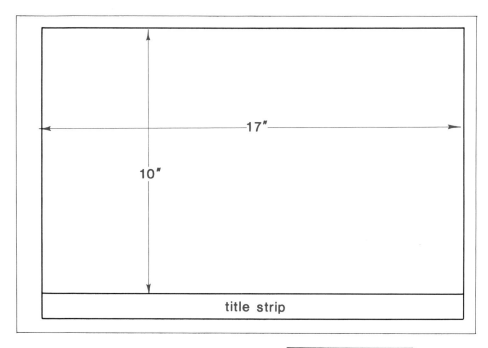

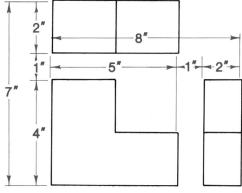

1. DEFINE THE DRAWING AREA.
2. ADD THE VERTICAL AND HORIZONTAL DISTANCES OF THE DRAWING AND THE SPACES BETWEEN THE VIEWS.
3. SUBTRACT THE FULL DRAWING'S HEIGHT FROM THE DRAWING AREA'S HEIGHT (10"−7"=3"). DIVIDE 3" IN HALF AND START 1.5" FROM TOP OR BOTTOM.
4. SUBTRACT THE FULL DRAWING'S LENGTH FROM THE DRAWING AREA'S LENGTH (17"−8"=9"). DIVIDE 9" IN HALF AND START 4.5" FROM EITHER SIDE.
5. COMPLETE CENTERED DRAWING.

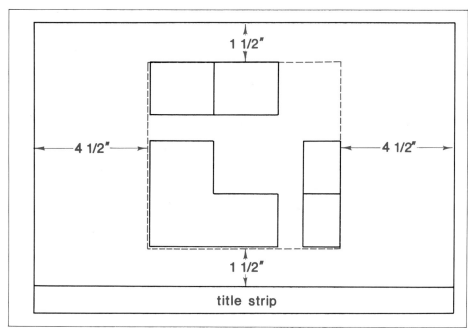

FIGURE 8–35 Centering a multiview drawing

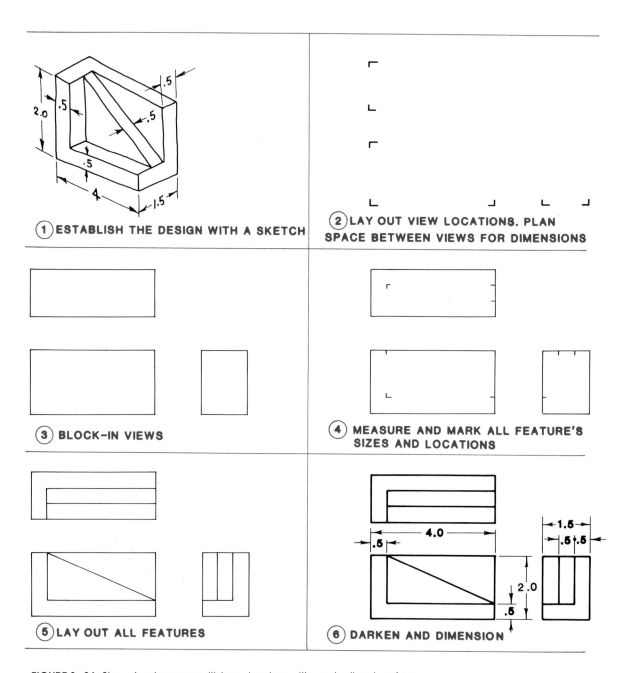

FIGURE 8-36 Steps to draw a multiview drawing with an inclined surface

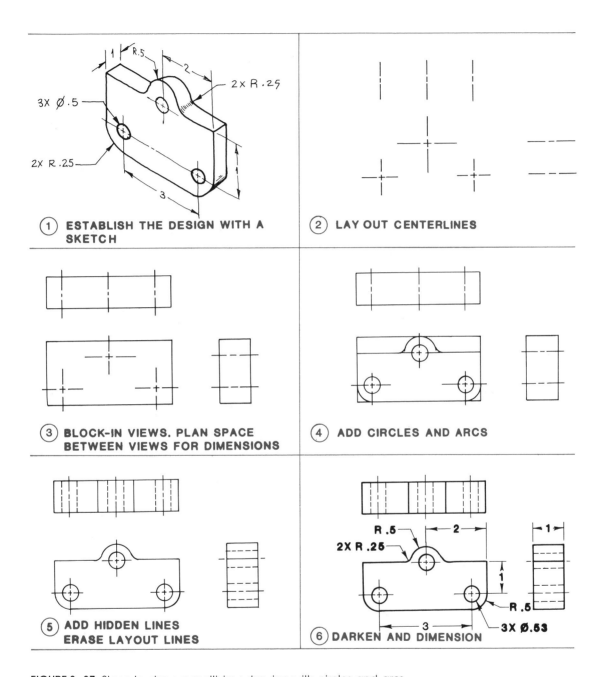

FIGURE 8-37 Steps to draw a multiview drawing with circles and arcs

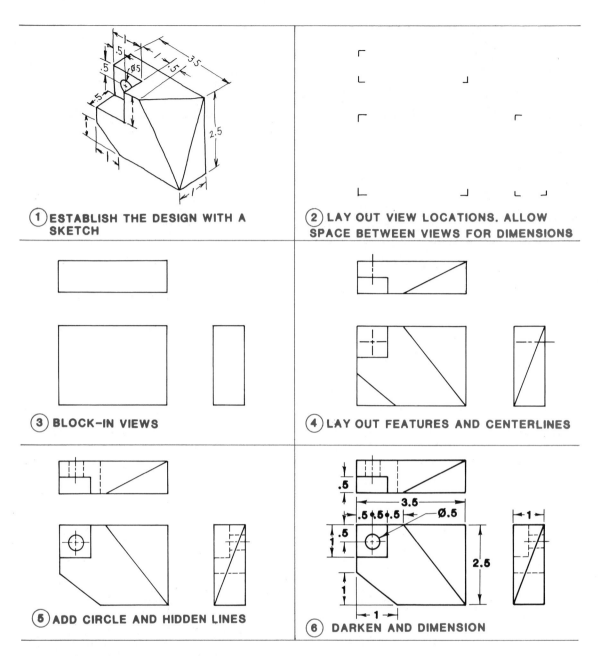

FIGURE 8-38 Steps to draw a multiview drawing with an oblique surface

COMPUTER-AIDED DRAFTING & DESIGN

COMPUTER-AIDED MULTIVIEW DRAWINGS

Preparing multiview drawings on a CAD system is similar to drawing manually, but the sequences differ greatly. Multiview drawings are combinations of lines, circles, and arcs. The use of grid marks, orthographic projection lines, and different layers for object, hidden, and construction lines, simplifies and increases the speed of preparing multiview drawings on a CAD system. Figures 8–39 through 8–42 illustrate CAD procedures used in completing three views of a cup. The steps in drawing this cup on a CAD system are as follows:

Step 1. Accurately draw the front view. Use the GRID and SNAP commands during construction. Lines, circles, and arcs may be entered via the keyboard to insure accuracy **(Fig. 8–39)**.

Step 2. Draw projection lines to align the top and right-side views. Use a construction layer that can be turned off at the end of the process. To guarantee the straightness of the projection lines, use the ORTHO command **(Fig. 8–40)**.

Step 3. Complete the necessary views using the procedures from Step 1 **(Fig. 8–41)**.

Step 4. Using a hidden line layer, draw all hidden lines. When all views are complete, turn off the construction layer **(Fig. 8–42)**.

These CAD procedures apply to the completion of most multiview drawings, such as the doorstop in **Figs. 8–43** through **8–46**.

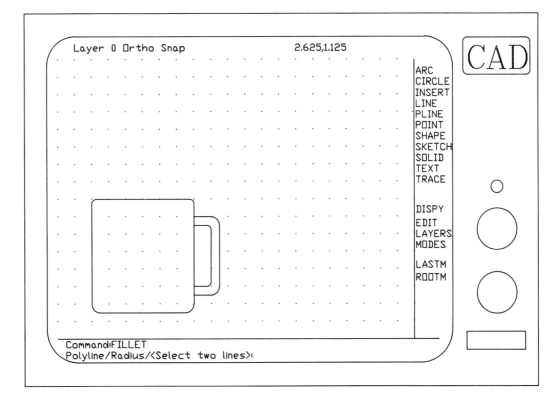

FIGURE 8–39 Front-view layout with a CAD system

118 Drafting in a Computer Age

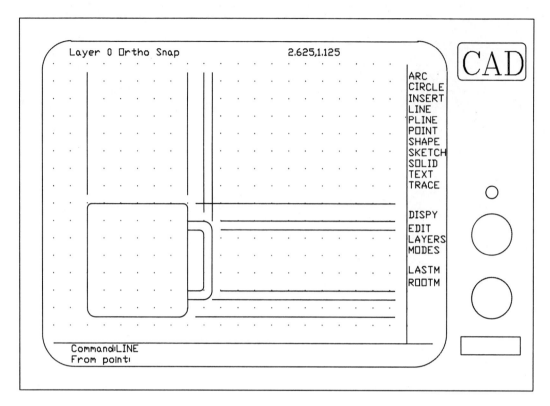

FIGURE 8–40 Draw the projection lines for alignment.

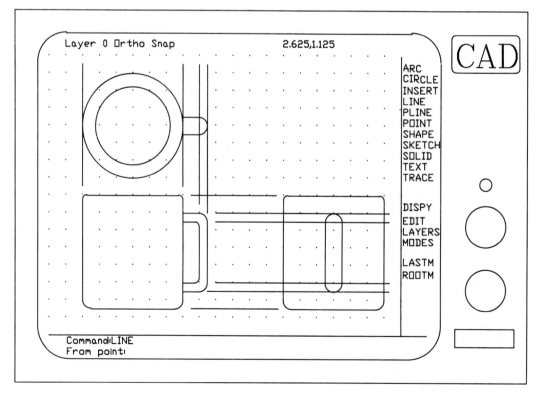

FIGURE 8–41 Complete the top and side views.

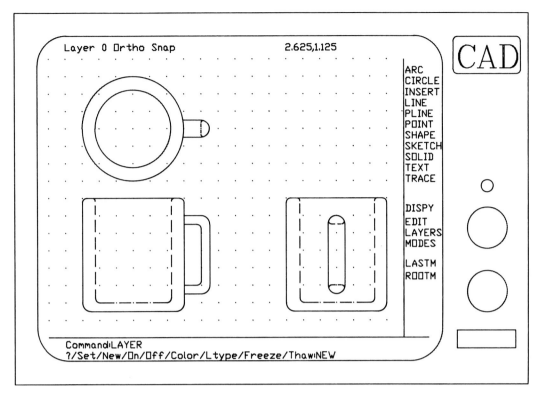

FIGURE 8–42 Add the hidden lines and remove the projection lines.

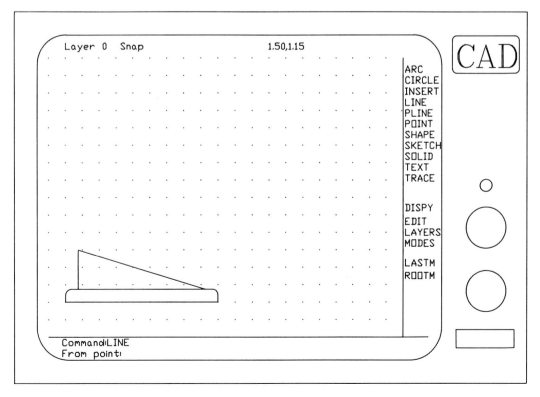

FIGURE 8–43 Lay out the front view.

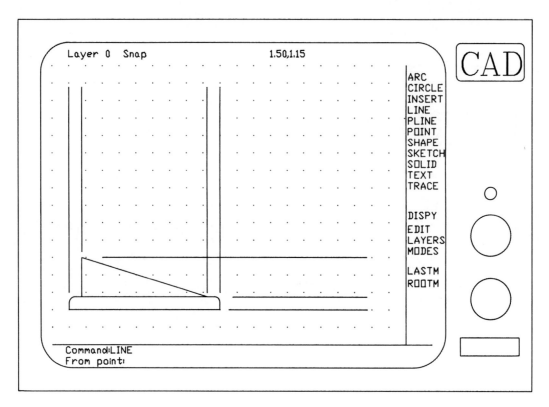

FIGURE 8–44 Place the projection lines for alignment.

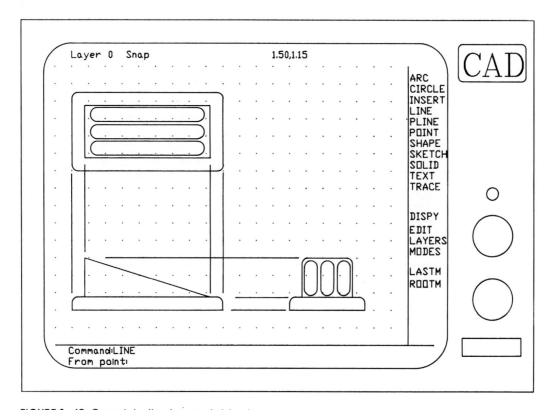

FIGURE 8–45 Complete the top and side views.

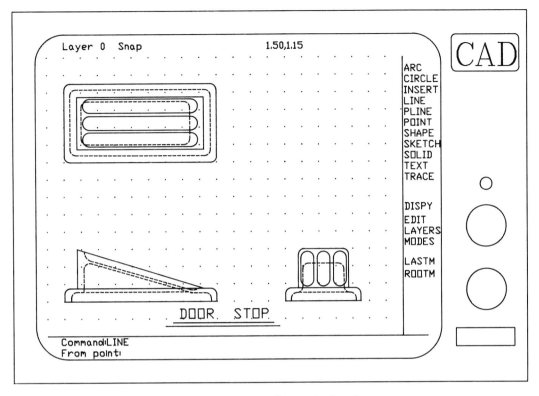

FIGURE 8-46 Add the hidden lines and remove the projection lines.

EXERCISES

1. Sketch the multiview drawings in **Fig. 8–47**.
2. Sketch the necessary multiview drawings for each isometric drawing in **Fig. 8–48**.
3. With drafting instruments or a CAD system, draw multiview drawings for each pictorial drawing in **Figs. 8–49** through **8–57**.

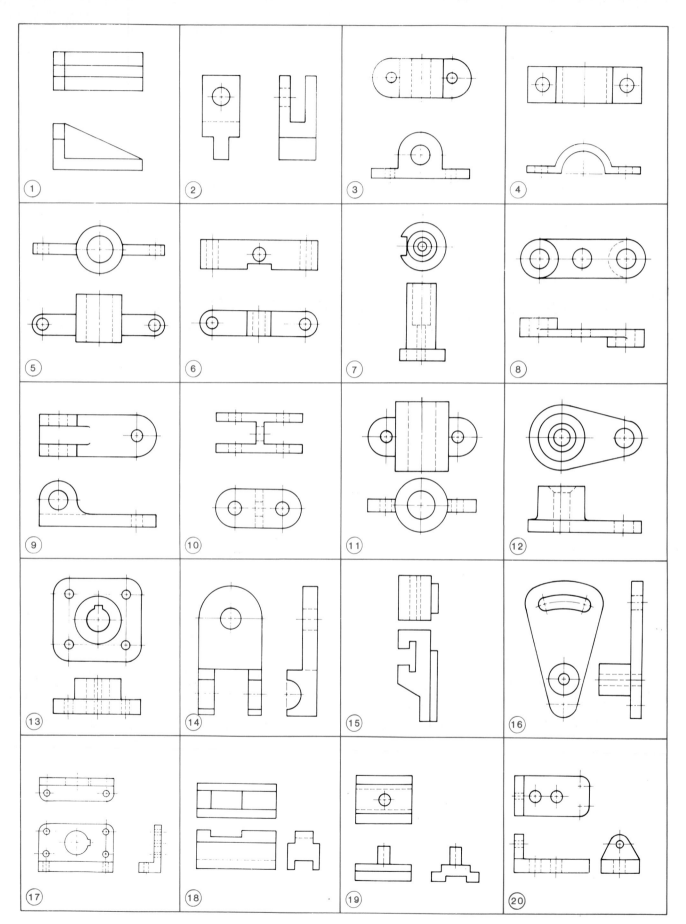

FIGURE 8–47 Sketch the multiview drawings.

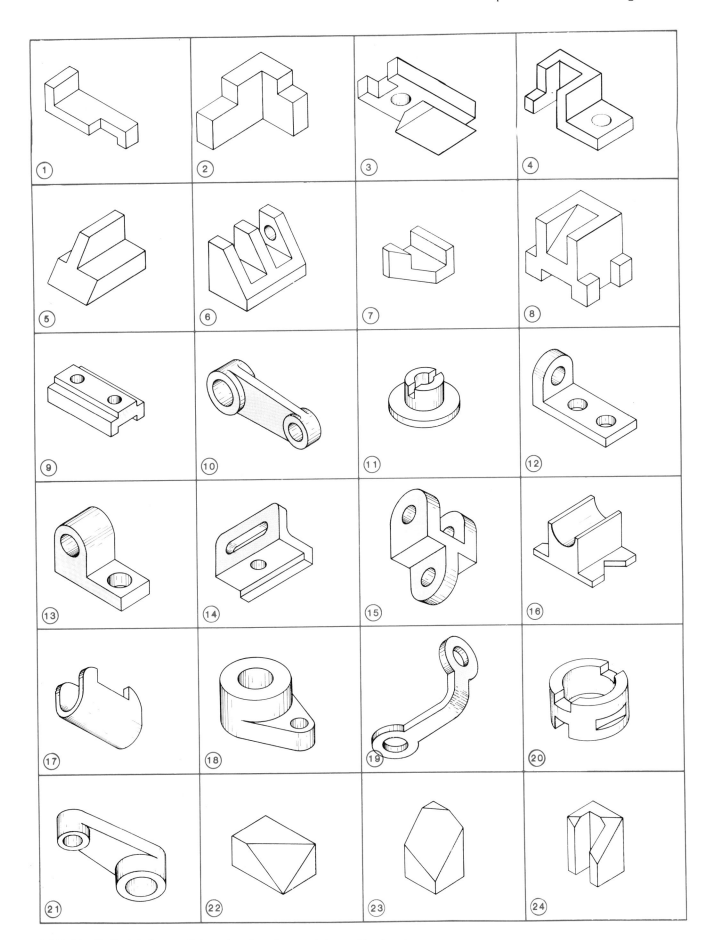

FIGURE 8–48 Sketch the multiview drawings for each isometric drawing.

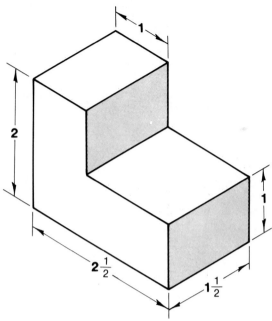

FIGURE 8–49 Draw the multiview drawing of the step block.

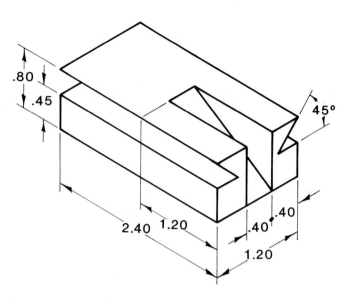

FIGURE 8–51 Draw the multiview drawing of the wedge support.

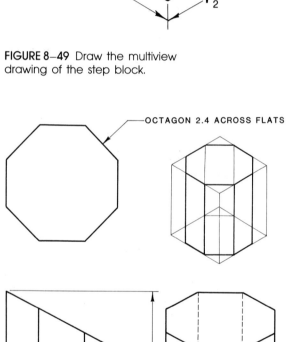

FIGURE 8–50 Draw the multiview drawing of the truncated octagonal prism and complete the isometric drawing.

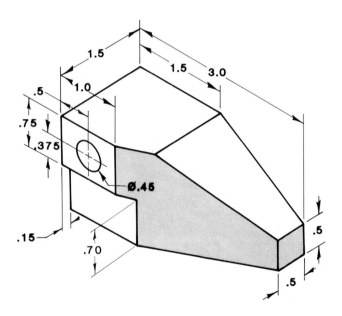

FIGURE 8–52 Draw the multiview drawing of the wedged bearing guide.

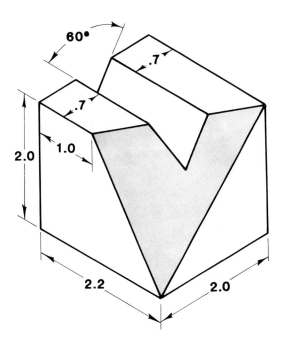

FIGURE 8–53 Draw the multiview drawing of the V-groove corner block.

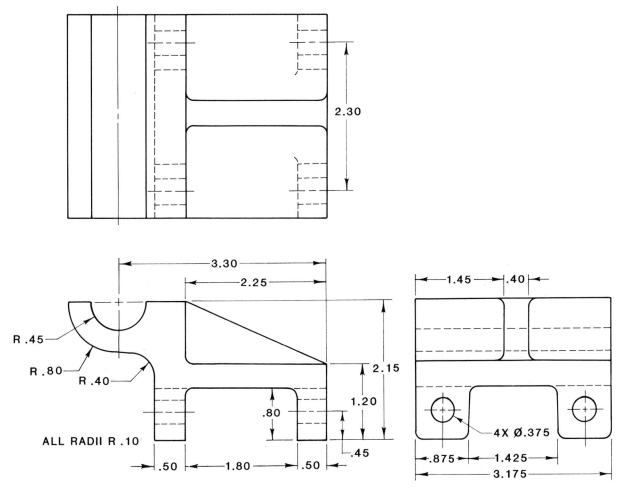

FIGURE 8–54 Draw the multiview drawing of the pipe support bracket.

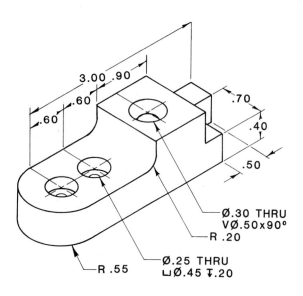

FIGURE 8–55 Draw the multiview drawing of the locator bracket.

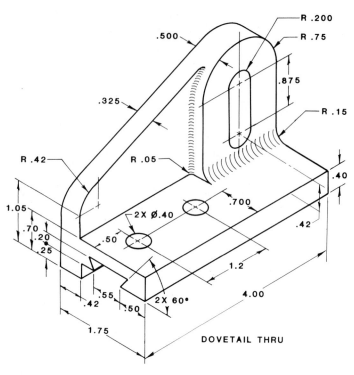

FIGURE 8–56 Draw the multiview drawing of the dovetail slide guide.

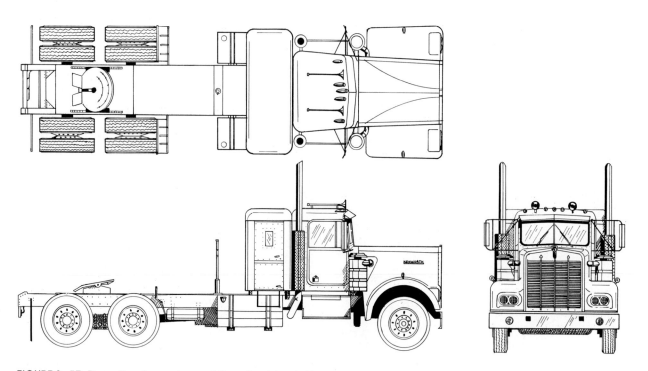

FIGURE 8–57 Draw the three views of the diesel truck. Use dividers to transfer dimensions.

4. Draw the multiview drawing in **Fig. 8–58**. Referring to Chapter 17 about ECOs, make the following engineering change orders:

- Change the slot width to .55" (keep the left side at .65").
- Change the slot depth to .55".
- Change the base thickness to .50".

5. Draw the multiview drawing in **Fig. 8–59**. Make the following ECOs:

- Change the Woodruff key slot to a No. 506 Woodruff key. Check the table in the index.
- Change the small bearing shaft hole to fit a .40" diameter shaft.
- Change the side slot width from 1.05" to 1.00". Shorten the right side of the slot.
- Change web thickness from .35" to .40". Be sure it is centered.

6. Design and draw a multiview drawing for a doorstop for a factory's 10' x 8' steel door. The space between the bottom of the door and the concrete floor is .5". The stop should not pro-

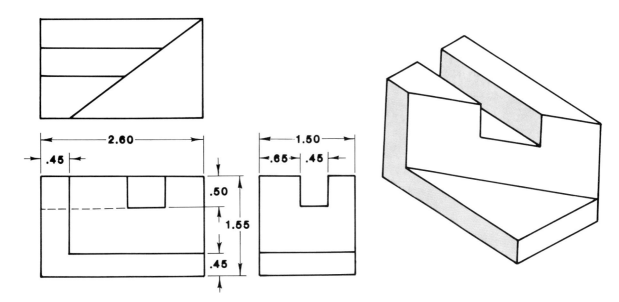

FIGURE 8–58 Draw the multiview drawing of the base guide. Make the required engineering change orders (ECOs) given in Exercise No. 4.

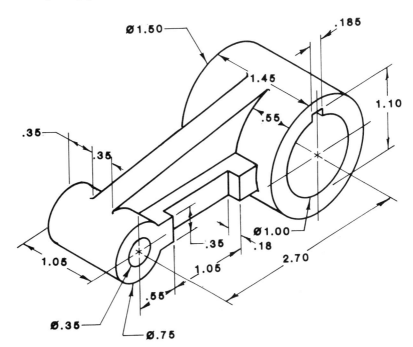

FIGURE 8–59 Draw the multiview drawing of the bearing bracket making the required ECOs given in Exercise 5.

ject more than 2" from the door into the passageway.

7. A heavy wood 10' x 3' worktable, with four 3" x 3" legs, is 30" in height. Design and draw a multiview drawing of an adjustable leg lift capable of changing the height to 34". Changes in the table's height will be made several times a week.

8. Using the cartesian coordinate system, draw the figure from the data table of coordinate points in **Fig. 8-60**. Use $\frac{1}{4}$" grid paper.

9. Using the cartesian coordinate system, complete the data table for the drawing in **Fig. 8-61**.

10. With a CAD system or instruments, draw the exercises in **Fig. 8-62**.

11. With a CAD system or instruments, draw the metric multiview drawings in **Figs. 8-63** through **8-70**.

12. Complete the multiview assembly drawings of the vehicles in **Figs. 8-71** through **8-75**, using the metric or U.S. customary measuring system.

DATA TABLE					
LINE	X	Y	go to	X	Y
1-2	1	1	→	1	7
2-3	1	7	→	9	7
3-4	9	7	→	9	1
4-1	9	1	→	1	1
new	start				
5-6	4	7	→	4	3
6-7	4	3	→	9	3
new	start				
8-9	11	1	→	15	1
9-10	15	1	→	15	7
10-11	15	7	→	11	7
11-8	11	7	→	11	1
new	start				
12-13	11	3	→	13	3
13-14	13	3	→	13	7
new	start				
15-16	1	9	→	1	13
16-17	1	13	→	9	13
17-18	9	13	→	9	9
18-15	9	9	→	1	9
new	start				
19-20	4	9	→	4	11
20-21	4		→	9	11
end					

FIGURE 8-60 Place the coordinates from the data table on $\frac{1}{4}$" grid paper to lay out the drawing.

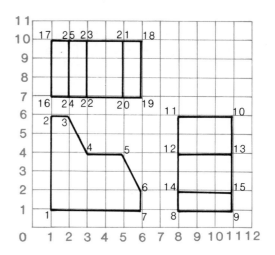

DATA TABLE					
LINE	X	Y	go to	X	Y
1-2	1	1	→	1	6
2-3	1	6	→	2	6
3-4	2		→		
4-5			→		
5-6					
6-7					

FIGURE 8-61 Complete the data table from the drawing.

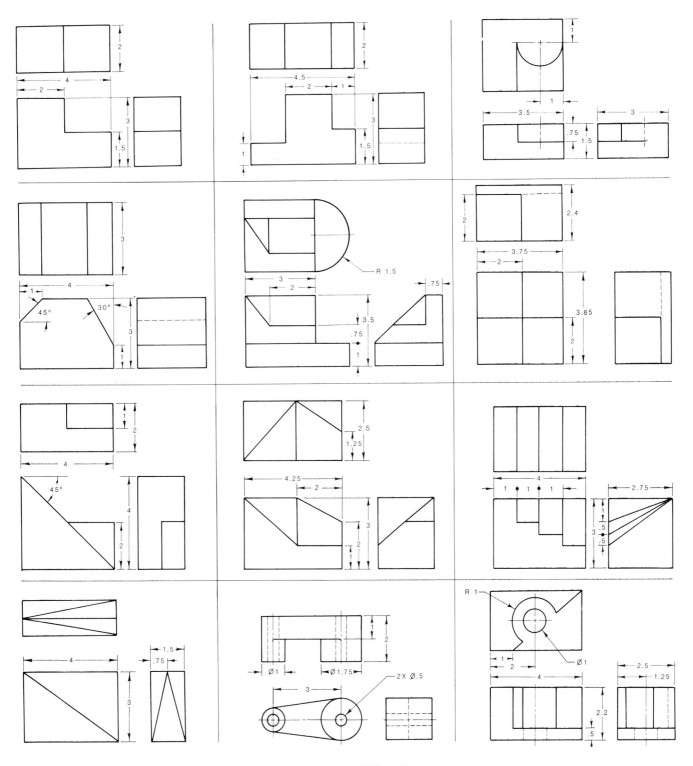

FIGURE 8-62 Draw the exercise with drawing instruments or a CAD system.

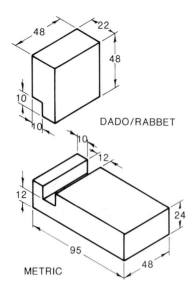

FIGURE 8–63 Draw the multiview drawing (metric) with instruments or a CAD system.

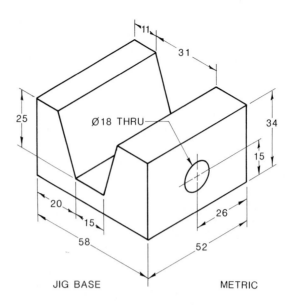

FIGURE 8–65 Draw the multiview drawing (metric) with instruments or a CAD system.

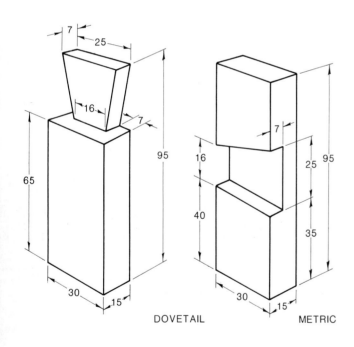

FIGURE 8–64 Draw the multiview drawing (metric) with instruments or a CAD system.

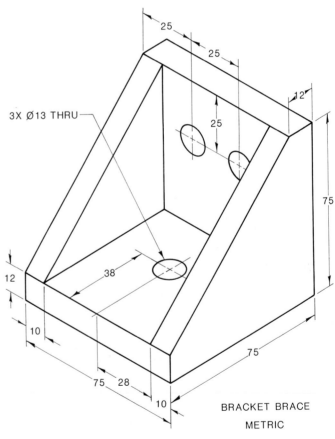

FIGURE 8–66 Draw the multiview drawing (metric) with instruments or a CAD system.

Chapter 8 Multiview Drawings **131**

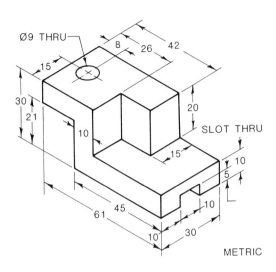

FIGURE 8–67 Draw the multiview drawing (metric) with instruments or a CAD system.

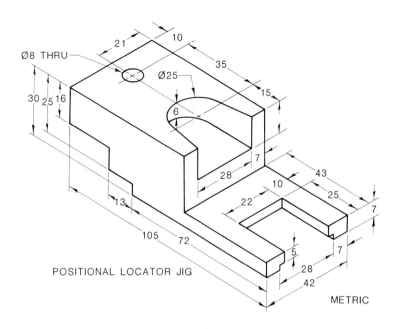

FIGURE 8–69 Complete the multiview drawing (metric) with instruments or a CAD system.

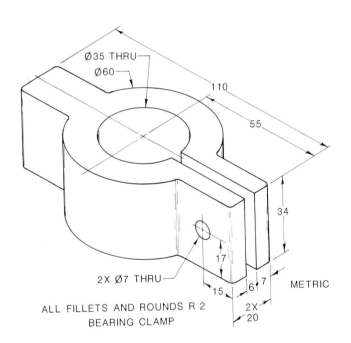

FIGURE 8–68 Draw the multiview drawing (metric) with instruments or a CAD system.

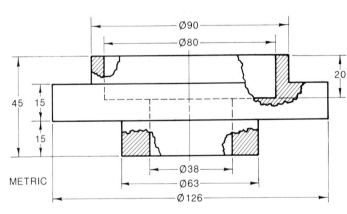

FIGURE 8–70 Complete the multiview drawing (metric) with instruments or a CAD system.

132 Drafting in a Computer Age

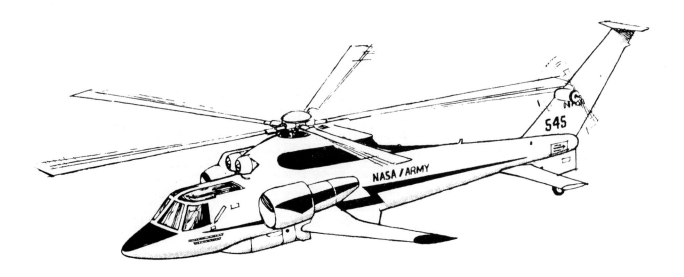

FIGURE 8–71 Draw the multiview drawing of a helicopter. (Courtesy of NASA)

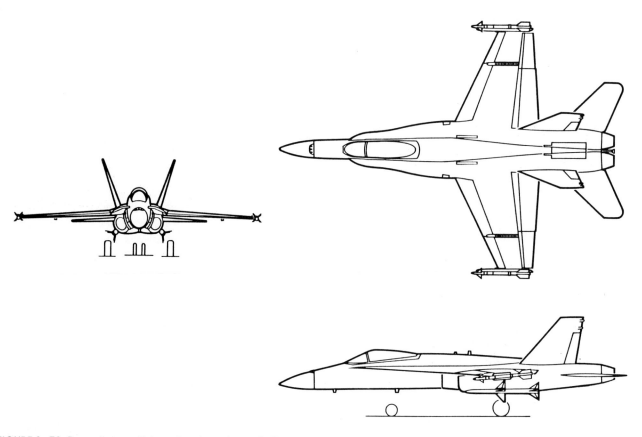

FIGURE 8–72 Draw the multiview drawing of an airplane.

Chapter 8 Multiview Drawings   **133**

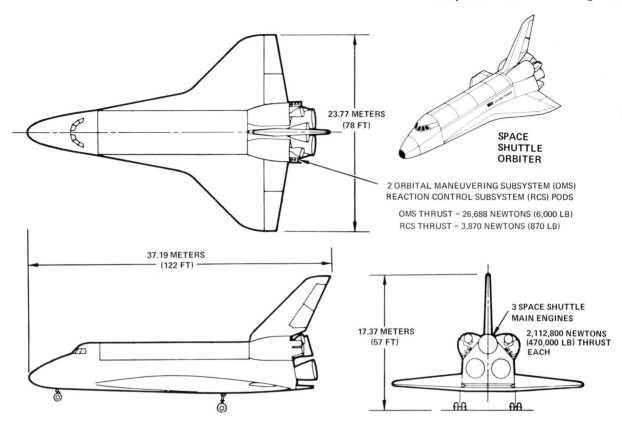

FIGURE 8-73 Draw the multiview drawing of the space shuttle. (Courtesy of NASA)

SPECIFICATIONS

Length	48 feet
Wing SPAN	27 feet
Height	14 feet
Empty weight	13,600 lbs.
Takeoff weight	17,600 lbs.

Engine: G.E. F404-GE-400, 16,000 lb. thrust class

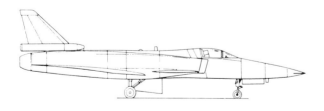

FIGURE 8-74 Draw the multiview drawing of the X-29 aircraft. Reposition the views so they can be drawn with orthographic projection. (Courtesy of NASA)

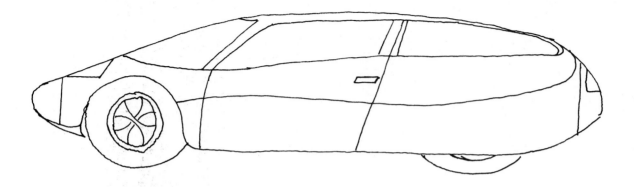

FIGURE 8–75 Draw the futuristic auto and complete the top, left-side, and right-side views.

 **KEY TERMS**

Ellipse
First-angle projection
Fold line
Front view
Inclined surface
Line precedence

Multiview
Normal surface
Oblique surface
Orthographic projection
Plane of projection
Reference line

Runout
Side view
Symmetrical
Third-angle projection
Top view
True size surface

the world of CAD

WHAT IS 3-D COMPUTER MODELING?

Designers often create models of proposed buildings in clay, cardboard, and other materials. Complex and accurate replicas can now be built using only microcomputers and inexpensive programs. Most often, these likenesses are constructed in an imaginary X–Y–Z coordinate space. Many programs use a collection of regular elements, such as rectangular blocks, cylinders, free-form shapes and walls. These objects can be combined and modified to form more elaborate components and even complete buildings or cities.

Within this category of CAD programs, there are many different types.

Wireframe—These programs work only with straight lines and a few other elements (such as arcs) which may be arbitrarily oriented in space. They do NOT include surfaces, so hidden line removal is NOT possible, Figure 1. Working with one of these programs is much like building a model from toothpicks and popsicle sticks.

Surface—Flat surfaces (bounded by straight lines) may be defined, arbitrarily oriented in space. In some cases, the total number of edges in a model may be limited. Hidden line removal is possible, but curved surfaces are often approximated with a series of many narrow facets, Figure 2. Some programs include special forms of curved surfaces, such as Bezier patches, surfaces of revolution or translation, and so forth. These programs recognize surfaces, so drawings can be made without lines that should be hidden, but they do NOT allow new surfaces to be created by removing part of a volume. In a way, using one of these programs is like building a model with cardboard or wooden blocks, where you cut individual shapes and add them together.

Solid—These programs usually work with one of two types of information: Boundary representation (B-rep) or Constructive Solid Geometry (CSG). The database for a program that uses B-rep is similar to one for a surface modeling program, but objects can be modified by cutting holes, notches, or depressions, and the generation of new surfaces is completely automatic. Programs that use CSG store the individual elements (rectangular blocks, cylinders, cones, etc.) and the rules (such as addition or subtraction of volumes), that were used to build the model. Manipulating one of these models is similar to forming a clay figure, from which pieces are added or from which other shapes are carved away, Figure 3.

HOW CAN 3-D MODELS BE DISPLAYED?

Typical drafting programs allow you to look at any section of a drawing, from the whole to a small part. In

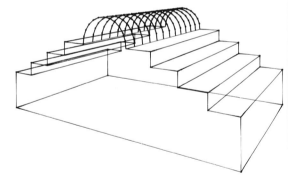
FIGURE 1 Example of a wire-frame model

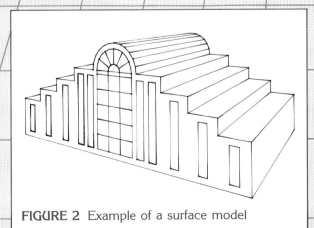
FIGURE 2 Example of a surface model

FIGURE 3 Example of a solid model

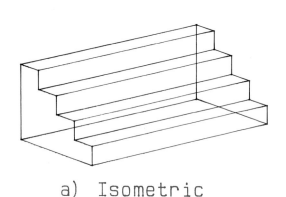 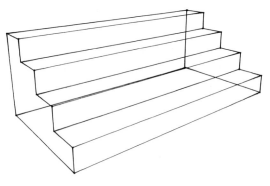

FIGURE 4 Wireframe models: A) isometric, B) perspective

addition, 3-D computer models can be viewed in a number of ways. It is very important that a program allow users to see and work in more than one "window" at a time, because what looks correct from one direction may be completely incorrect in other views. Display techniques include orthographic projections and perspective projections.

Wireframe drawings show all edges of every object within the view. With some practice, it is easy to understand the spatial relationship between objects in a wireframe drawing if it is also shows perspective. Wireframe orthographic drawing are more difficult to interpret because our brain tends to "flip" the image, which reverses near and far objects, Figure 4. Most 3-D drawing systems show the model as a wireframe, unless you give a command to remove the hidden lines.

Hidden line drawings include the following techniques that eliminate edges that should not be seen behind other surfaces:

Back surface removal—Calculates which surfaces are facing away from the viewer. This is a very fast algorithm, but leaves many lines that must still be removed, such as the front of buildings behind other objects, Figure 5A.

Discrete objects—The program assumes that NO surfaces intersect any other surfaces, which makes the resulting calculation reasonably fast, but places a burden on the user to build the model consistently. Most programs for microcomputers are limited to this type of hidden line removal, Figure 5B.

Full hidden line—These programs generate a new edge where two surfaces intersect. Edges should be calculated as vectors, which can be sent to a plotter or 2-D drafting program. For example, a chimney through the roof of a house would show the line of intersection, even though the chimney started on the first floor, Figure 5C.

Surface shading is another term that covers a variety of techniques. Each defined surface in a model will be filled with color. If the display can show a wide range of colors, these drawings can represent illumination from one or more light sources, so that the various sides of an object can be distinguished.

Lambert shading—The most common technique used on microcomputers. For each surface, a single angle to a light source is calculated, and the entire surface will be displayed with a single shade. This is appropriate for an isolated object that is illuminated by a single light source that is very far away. In most real situations, the shade will vary over a surface as the angle to a nearby light source differs.

Gourad and Phong shading—Gives a reasonably realistic appearance to a curved surface. Phong shading calculates highlights from specular (shiny) surfaces. The calculations involved are too slow for all but the

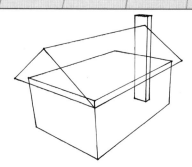 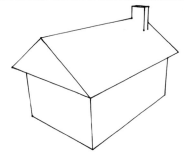 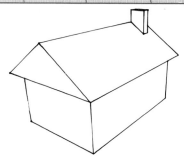

A – Back surface   B – Discrete objects   C – Full hidden line

FIGURE 5 Three techniques of hidden-line removal: A) back surface, B) discrete objects, C) full hidden line

fastest microcomputers, and the results look somewhat plastic.

Painter's Algorithm—A technique that produces shaded surface screen images very quickly. The order in which the surfaces are drawn is from the most distant to the closest, so that the final result will be a correct image. However, there are many situations in which this method will fail, because there might NOT be a correct order by distance.

Shadow projection—given one or more light sources, it is possible for a program to calculate where the shadows of individual objects will fall on the ground plane or other objects in the model. This is an excellent capability for designers, Figure 6.

Ray-tracing rendering—Calculates a beam of light back from the viewer's eye through every pixel in the image. When the ray strikes an object, it may be reflected, or partially refracted, or reach a light source. The combination of effects will determine the color of that pixel. These images can be quite realistic, but they are very slow to produce, and have hard edges of shadows. Any change in viewer position, surface color or light source intensity requires a complete recalculation. Even a very fast microcomputer can take several hours to construct a simple image.

View-factor rendering—First calculates how much each surface can "see" every other surface in the model. After this process, any particular view can be displayed relatively quickly, as long as there are no changes in the geometric configuration.

WHY ARE 3-D COMPUTER MODELS USED?

The following are a few reasons why 3-D computer models are used:

Design understanding—Most clients cannot interpret plans, elevations, and sections to appreciate the future experience of a design. A good 3-D program can allow designers to show a realistic view from anywhere within or outside a model. With sufficient experience, changes can be made immediately, eliminating the need for additional meetings before continuation of the design is authorized, Figure 7.

Zoning review—In many cases, the approval of a proposal depends on its visual impact. An array of perspectives from a computer program is more compelling than a single rendering and less expensive than a physical model.

Promotion—Innovative computer display techniques can be part of an advertising and sales campaign for commercial projects. Video recordings and live computer demonstrations can impress potential customers and clients.

Coordination of drawings—Because all plans, elevations, and sections are derived from a single 3-D database, all drawings are automatically consistent. For example, correct roof

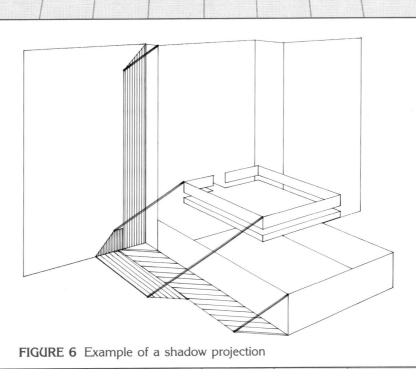

FIGURE 6 Example of a shadow projection

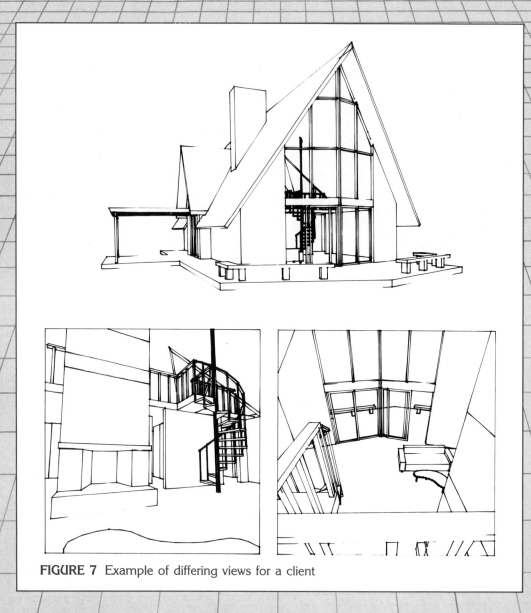

FIGURE 7 Example of differing views for a client

plans and elevations are tedious to construct and occasionally inaccurate. When the roof has been assembled from correctly sloping planes, all of the orthographic views will also be correct.

Contractor understanding—Often a simple 3-D drawing can clarify the construction requirements of a design. This reduction in uncertainty can have the effect of making the bid more precise.

(Courtesy of William Glennie, School of Architecture, Rensselaer Polytechnic Institute, Troy, NY. Drawings created with MegaMODEL and MegaDRAFT programs from MegaCADD, Seattle, WA.)

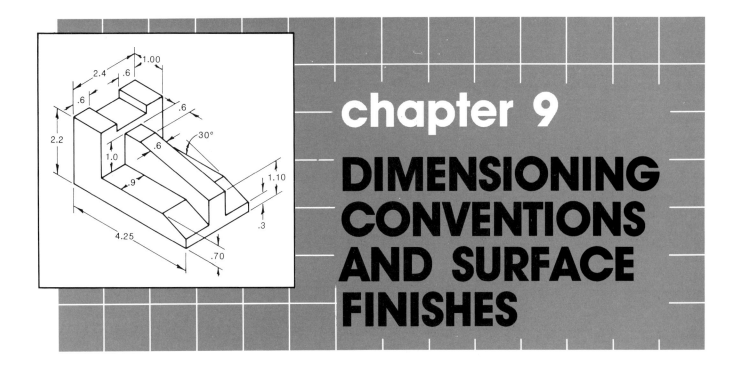

chapter 9
DIMENSIONING CONVENTIONS AND SURFACE FINISHES

OBJECTIVES

The student will be able to:

- recognize standardized dimensioning symbols
- apply correct dimensioning techniques to a working drawing
- dimension all geometric forms
- dimension machine operations
- dimension an isometric drawing
- add surface control symbols to a working drawing

INTRODUCTION

Engineering working drawings must contain all information necessary for manufacturing. This includes exact descriptions of the size, shape, and material specifications of every part of a product. These descriptions are accomplished through standardized **dimensioning** conventions **(Fig. 9–1)**. The machine part in Fig. 9–1 can be manufactured in any plant because the drawing contains universally accepted ANSI dimensional

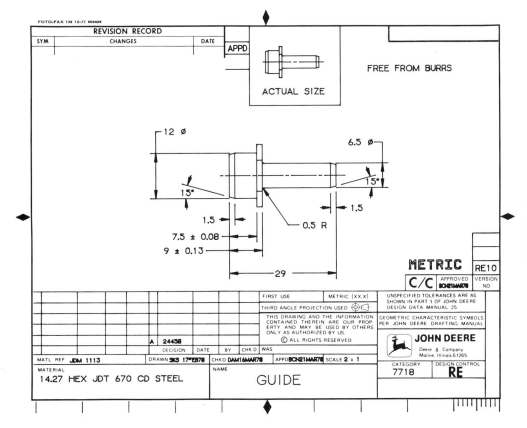

FIGURE 9–1 Dimensions and notations will assure accurate manufacturing. (Courtesy of Deere & Company)

135

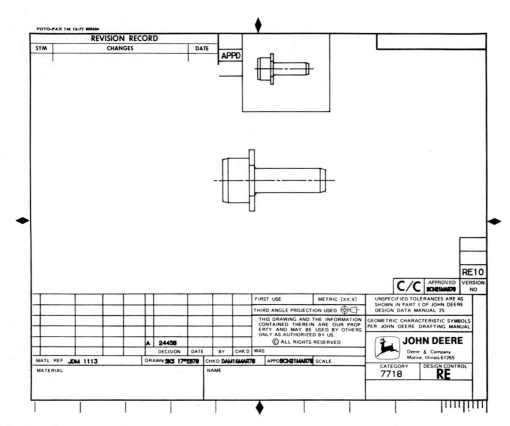

FIGURE 9-2 Without dimensions and notations, manufacturing of this part is impossible.

symbols and notations. Without these symbols and notations, the drawing **(Fig. 9-2)** is meaningless and the part could not be manufactured without extensive guesswork. Product quality and accuracy would depend on the interpretation of the manufacturer. Adhering to established standards is important in preparing working drawings manually or with a CAD system.

Working drawing dimensions are of two types; **size dimensions** and **location dimensions.** Size dimensions describe the size of each geometric form **(Fig. 9-3)**. Location dimensions provide the exact location of each geometric part of a drawing **(Fig. 9-4)**. In addition to size and location dimensions, **notations** are needed to specify materials **(Fig. 9-5)** and any other information necessary for manufacturing. If any dimension or notation is omitted, costly production errors and delays could result.

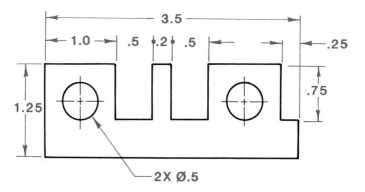

FIGURE 9-3 Size dimensions

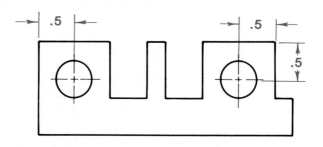

FIGURE 9-4 Location dimensions

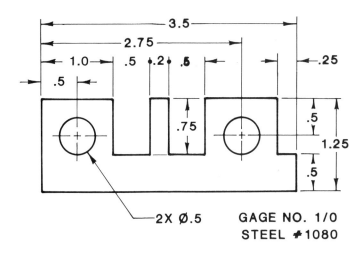

FIGURE 9–6 ANSI conventions for U.S. customary dimensions

1. Use the decimal inch for dimension values (5.75).
2. Do not use a zero before the decimal point for values less than one inch (.50).
3. The dimension value is expressed to the same number of decimal places as its tolerance (5.500 ± .002).
4. The number of decimal places may represent the tolerance:
   - 5.5 = large tolerance
   - 5.50 = average tolerance
   - 5.500 = small tolerance

FIGURE 9–5 A fully dimensioned drawing

SYSTEMS OF DIMENSIONING

The principles and rules for dimensioning, including dimensioning systems, are standardized by the American National Standards Institute (ANSI).

U.S. Customary Dimensioning

ANSI standards for **U.S. customary dimensioning,** with the use of decimal-inch values are shown in **Fig. 9–6**. In using decimal-inch dimensions, a zero is not used to the left of the decimal point for values less than one inch. The same number of decimal places should always be used for both dimensions in a tolerance dimension range. **Tolerance** is the total acceptable variation within a dimension (see to Chapter 10). Establishing the number of decimal places represents the level of tolerance desired. Three decimal places implies a higher level of tolerance than two.

Metric Dimensioning

ANSI standards for **metric dimensioning (Fig. 9–7)** requires all dimensions to be expressed in millimeters. Therefore, the millimeter symbol (mm) is not needed on each dimension, but it is used when a dimension is used in a notation. In this case, a single space separates the numeral and the millimeter symbol. Zeros precede the decimal point when the value is less than one millimeter; a zero is not used when the dimension is a whole number. Commas are not used between thousand units. A metric dimensional note should be displayed in or near the title block.

Dual Dimensioning

Working drawings are normally prepared with all U.S. customary or all metric dimensions. However, when the object is to be manufactured using both metric and U.S. customary measuring systems, **dual dimensions** may be necessary. Dual dimensioning may also be necessary when converting a metric drawing to a U.S. customary drawing and vice versa. When dual dimensions are used, the metric dimensions are placed under the dimension line in parenthesis. Sometimes a metric conversion chart is used **(Fig. 9–8)**.

DIMENSIONING ELEMENTS

ANSI has also standardized the use of lines, numerical values, symbols, notations, and their placement on

1. All dimensions will be in millimeters.
2. Do not use millimeter symbol mm with dimensions (25 not 25 mm).
3. When a dimension value is less than one millimeter, a zero will precede the decimal point (0.55 not .55).
4. When the dimension is a whole number, do not use a zero or a decimal point (150 not 150.0).
5. When the dimension uses a decimal value, do not end it with a zero (75.5 not 75.50).
6. An exception to convention No. 5 is: if the dimension value is followed by a tolerance, the number of decimal places must be equal (75.50 ± 0.05 not 75.5 ± 0.05).
7. Do not use commas or spaces (1575 not 1,575 or 1 575).
8. When specifying millimeters off the drawing, as in a document, leave a space between the number and the symbol (2585 mm not 2585mm).
9. Display the word "METRIC" with large letters near the title block.
10. Display note: "UNLESS OTHERWISE SPECIFIED, ALL DIMENSIONS ARE MILLIMETERS."

FIGURE 9–7 ANSI conventions for metric dimensions

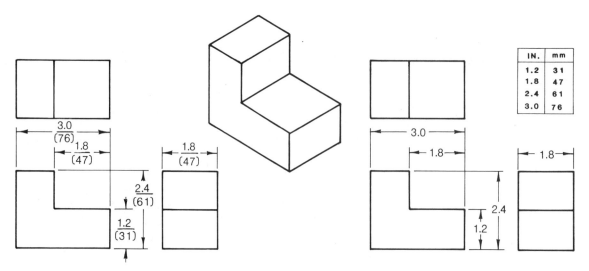

FIGURE 9-8 Two procedures used for dual dimensioning

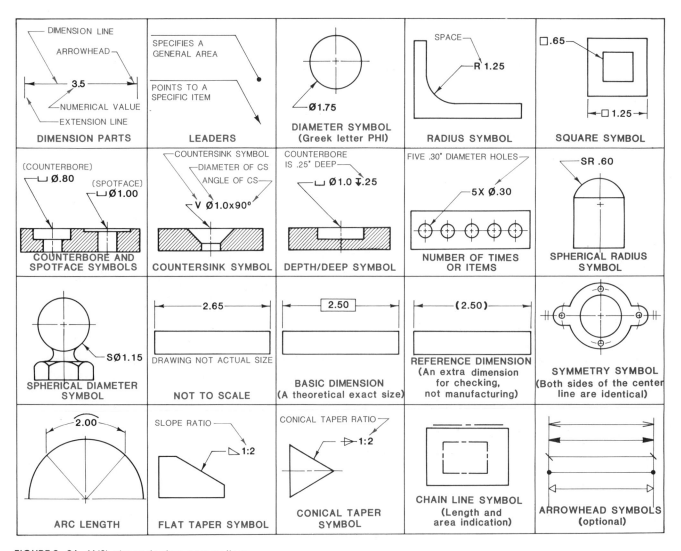

FIGURE 9-9A ANSI dimensioning conventions

Chapter 9 Dimensioning Conventions and Surface Finishes 139

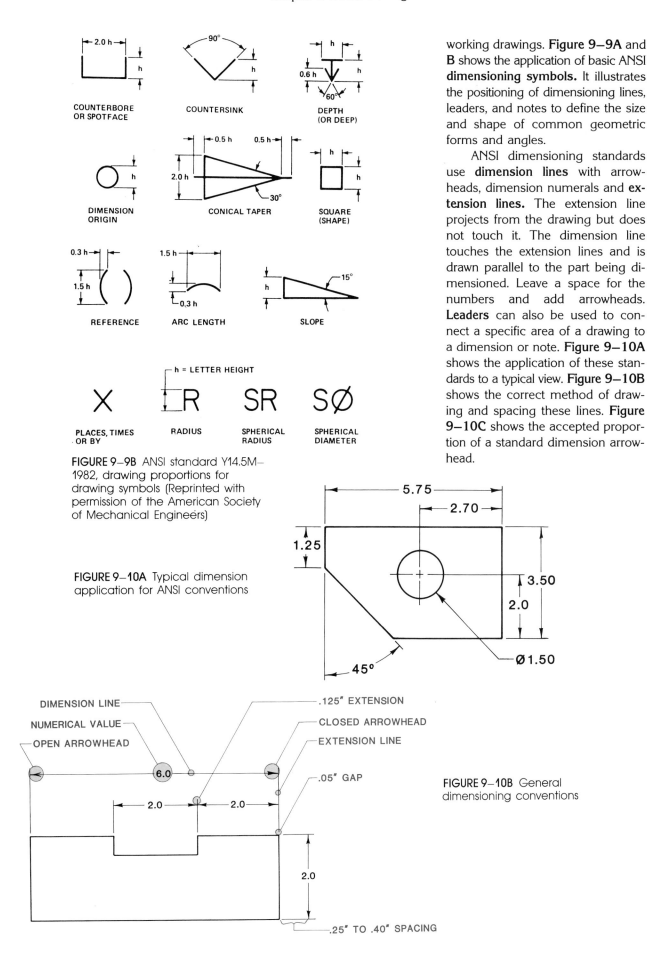

working drawings. **Figure 9–9A** and **B** shows the application of basic ANSI **dimensioning symbols.** It illustrates the positioning of dimensioning lines, leaders, and notes to define the size and shape of common geometric forms and angles.

ANSI dimensioning standards use **dimension lines** with arrowheads, dimension numerals and **extension lines.** The extension line projects from the drawing but does not touch it. The dimension line touches the extension lines and is drawn parallel to the part being dimensioned. Leave a space for the numbers and add arrowheads. **Leaders** can also be used to connect a specific area of a drawing to a dimension or note. **Figure 9–10A** shows the application of these standards to a typical view. **Figure 9–10B** shows the correct method of drawing and spacing these lines. **Figure 9–10C** shows the accepted proportion of a standard dimension arrowhead.

FIGURE 9–9B ANSI standard Y14.5M–1982, drawing proportions for drawing symbols (Reprinted with permission of the American Society of Mechanical Engineers)

FIGURE 9–10A Typical dimension application for ANSI conventions

FIGURE 9–10B General dimensioning conventions

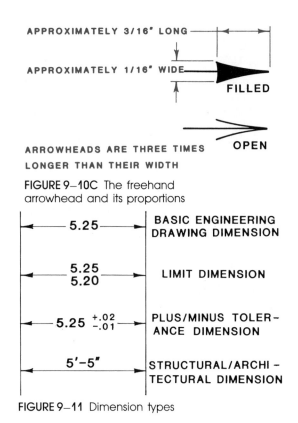

FIGURE 9-10C The freehand arrowhead and its proportions

FIGURE 9-11 Dimension types

DIMENSIONING GUIDELINES

Figure 9-11 illustrates basic dimensioning types. **Fig. 9-12** shows the two standard dimension positioning systems; unidirectional and aligned. **Unidirectional dimensions** all read from the bottom and are easier to read. **Aligned dimensions** read from the bottom or right of the drawing. Aligned dimensions conserve space on a drawing since they are aligned with dimension lines. Either system is acceptable if used consistently throughout a drawing.

If dimensions are not spaced properly, the drawing surface area will be wasted or dimensions may be too close to be read without confusion. Closely spaced dimensions can also inhibit the easy reading of object lines. For these reasons, spacing of dimensions **(Fig. 9-13)** is recommended.

Many manufactured objects contain geometric forms or parts that are identical. When this occurs, only one dimension is needed **(Fig. 9-14)**. A dimension which covers many identical (repetitive) parts is always accompanied by a label indicating the number of parts of the same size.

Regardless of the object, always dimension to visible lines, not to hidden lines **(Fig. 9-15)**. If it is necessary to dimension to a hidden line, another view or sectional drawing is required.

Polar Coordinate Dimensioning

When parts of a drawing align on a common **radial centerline**, polar coordinate dimensions are used **(Fig. 9-16)**. In this case, all radial dimension lines originate from a baseline with the largest dimension

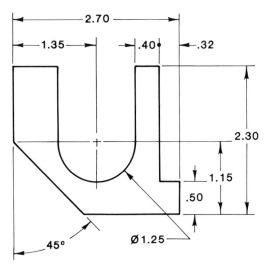

UNIDIRECTIONAL DIMENSIONS
ANSI STANDARD

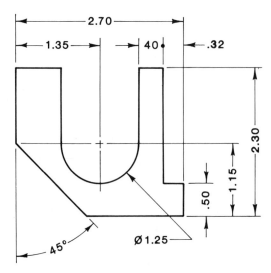

ALIGNED DIMENSIONS

FIGURE 9-12 Positioning systems for reading dimensions

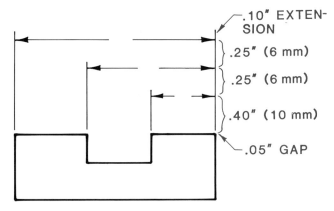

FIGURE 9–13 Dimension line spacing

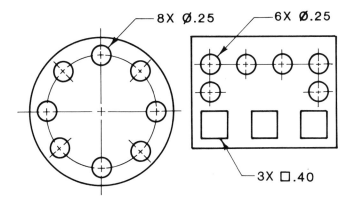

FIGURE 9–14 Repetitive dimensions

on the outside. All other dimensions progress inward to the smallest dimension.

Rectangular Coordinate Dimensioning

When a large number of parts, projections, holes, or complex contours are contained in a small area, **rectangular coordinate** dimensions may be used in lieu of conventional dimension practices. Rectangular coordinate systems eliminate the need for dimension lines by establishing distances from base intersections, usually a corner **(Fig. 9–17A)**. The baseline is assigned a value of zero. The dimension to each extension line represents the distance from zero. To eliminate size dimensions, each standard hole or projection is assigned a letter which relates to the size specified in an accompanying chart.

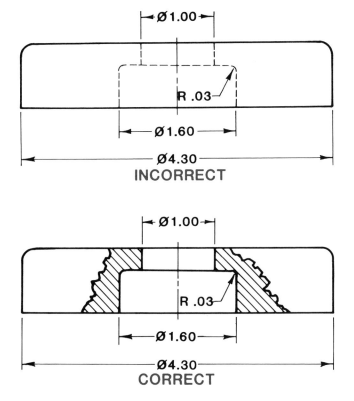

FIGURE 9–15 Always dimension to visible lines.

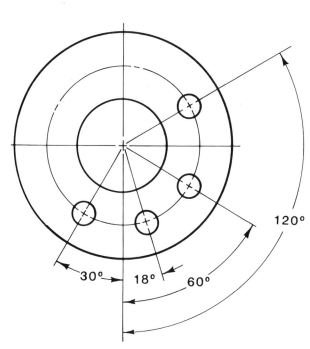

FIGURE 9–16 Example of polar coordinate dimensioning

Rectangular coordinate grid systems are used extensively to define and locate irregular contour lines (**Fig. 9–17B**). In this application, the grid lines are spaced evenly so each square can be increased (or decreased) in size to duplicate the outline of the design. Some rectangular coordinate systems completely eliminate all detail dimensions with a chart which contains distances from established baselines. In **Fig. 9–18**, the location and depth of holes on X, Y, and Z surfaces is shown in the top chart. The bottom chart shows the number and diameter of each hole.

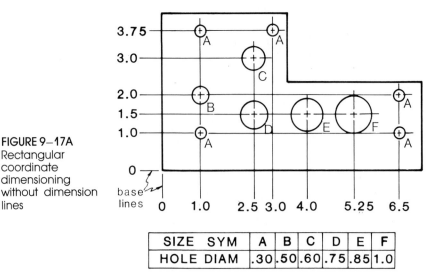

FIGURE 9–17A Rectangular coordinate dimensioning without dimension lines

Tabulated Outline Dimensioning

Another type of rectangular coordinate dimensioning is the **tabulated outline** method (**Fig. 9–19**). In this example, the X–Y distance of each plotting point on a contour line is tabulated on an X–Y coordinate chart. The conventional method of dimensioning a curve of this type is shown in **Fig. 9–20**. Notice all dimensions originate from baselines.

SPECIAL DIMENSIONING PROCEDURES

Since some object configurations or areas are difficult to dimension, special dimensioning procedures are standardized.

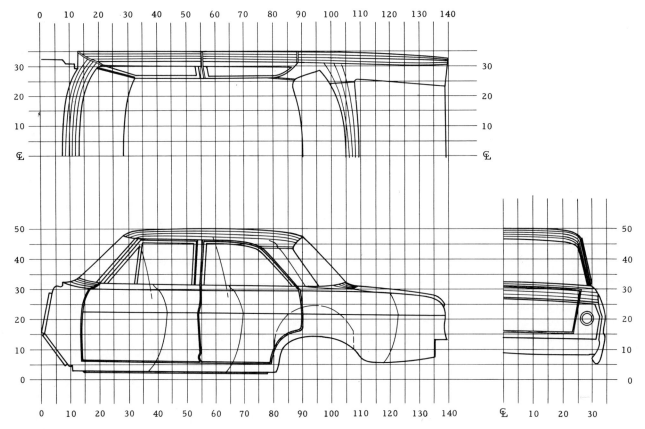

FIGURE 9–17B Rectangular coordinates form zones for area locations. (Courtesy of General Motors Corporation Engineering Standards)

Chapter 9 Dimensioning Conventions and Surface Finishes 143

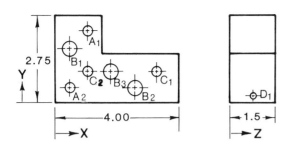

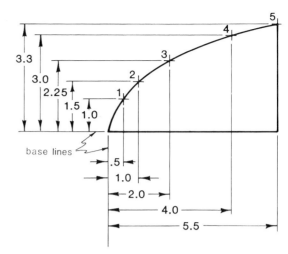

FIGURE 9–20 Coordinate or offset outline dimensioning

HOLE	FROM SURF	X	Y	Z
A_1	X,Y	1.0	2.25	THRU
A_2	X,Y	.5	.5	THRU
B_1	X,Y	.5	1.7	.50
B_2	X,Y	2.6	.5	.50
B_3	X,Y	1.8	1.0	.50
C_2	X,Y	1.0	1.0	THRU
C_1	X,Y	3.3	1.0	THRU
D_1	Z,Y	THRU	.1	.75

HOLE	DESCRIPTION	QUANTITY
A	Ø.30	2
B	Ø.45	3
C	Ø.28	2
D	Ø.20	1

FIGURE 9–18 Rectangular coordinate dimensioning in tabular form

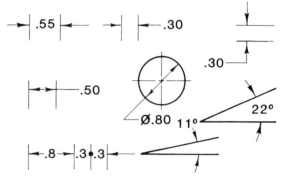

FIGURE 9–21 Dimensioning conventions for crowded areas

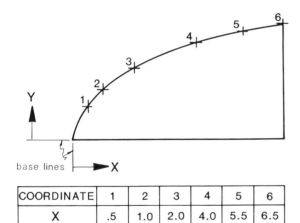

COORDINATE	1	2	3	4	5	6
X	.5	1.0	2.0	4.0	5.5	6.5
Y	1.0	1.5	2.25	3.0	3.3	3.5

FIGURE 9–19 Tabulated outline

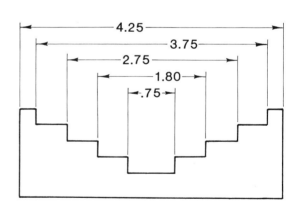

FIGURE 9–22 Examples of staggered dimensions

Dimensioning Crowded Areas

Figure 9–21 shows examples of dimensioning practices for crowded areas. Figure 9–22 shows how dimensions should be staggered for easier reading. In rare cases, oblique dimensioning (Fig. 9–23), is used to remove a dimension from a crowded area. Often a corner or area to be dimensioned is outside the object outline or may be eliminated during manufacturing. In this case, extension lines are drawn to imaginary intersections. Extension and dimension lines are then added (Fig. 9–24).

Since a wide variety of machined holes are used in manufactured products, dimensioning standards include ANSI symbols (Fig. 9–25). From these notation symbols, the print reader can visualize the information shown in the accompanying sectional view.

144 Drafting in a Computer Age

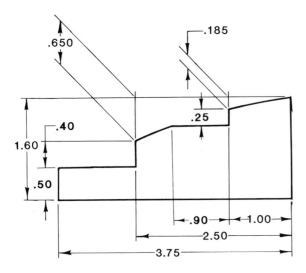

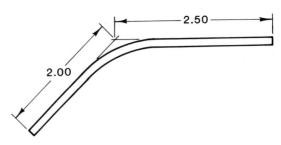

FIGURE 9–24 Locating a point for dimensioning

FIGURE 9–23 Examples of oblique dimensions

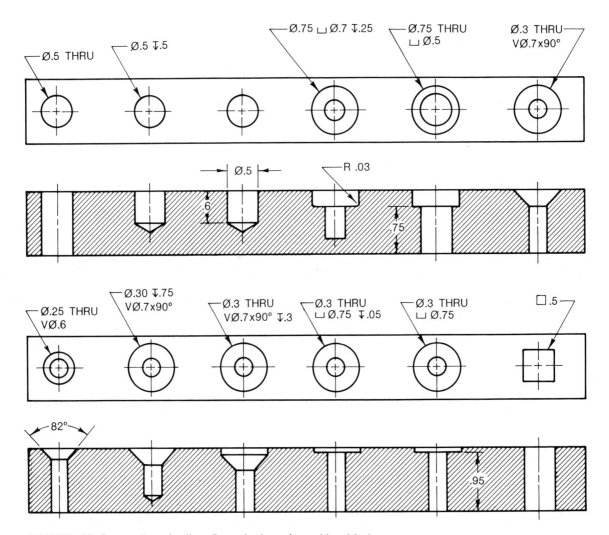

FIGURE 9–25 Conventions for the dimensioning of machined holes

Dimensioning Slots, Radii, Fillets, Rounds, and Chamfers

Often the dimensioning of rounded ends (Fig. 9–26), slots, radii, fillets, and chamfers offers an acceptable opportunity to shortcut normal dimension procedures. Figure 9–27 shows three methods of dimensioning slots.

Radii dimensioning shows the distance from apex to arc. If the apex of the radii is not located near the drawing, an offset dimension line can be used. Very small radii are dimensioned without the use of an apex (Fig. 9–28).

Fillets are inside rounded corners. Rounds are outside rounded corners. Fillets and rounds are dimensioned in a similar manner (Fig. 9–29).

Chamfers are flat corners and are either external, internal, or oblique (Fig. 9–30).

Knurling is another common manufacturing process which is dimensioned with a symbolic notation (Fig. 9–31).

Dimensioning Chords, Arcs, and Angles

Chords, arcs, and angles may appear similar in many drawings, but should be dimensioned differently (Fig. 9–32). Chords are dimensioned numerically with a straight line, and arcs are dimensioned numerically with a curved line. In dimensioning angles, extend straight extension lines, and connect them with curved dimension lines (Fig. 9–33).

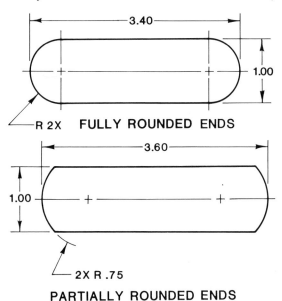

FIGURE 9–26 Conventions for the dimensioning of rounded ends

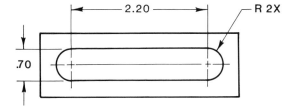

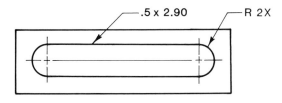

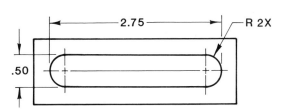

FIGURE 9–27 Conventions for the dimensioning of slotted holes

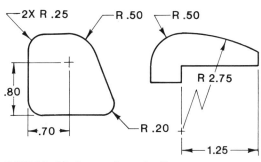

FIGURE 9–28 Conventions for the dimensioning of radii

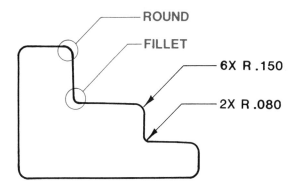

FIGURE 9–29 Conventions for the dimensioning of fillets and rounds

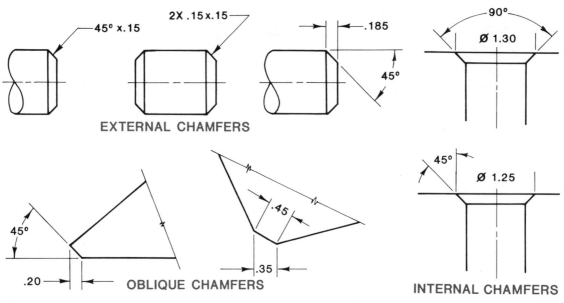

FIGURE 9-30 Conventions for the dimensioning of chamfers

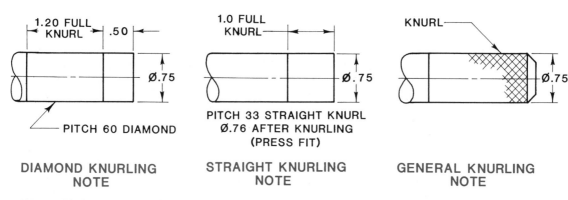

FIGURE 9-31 Conventions for knurling dimensions and notations

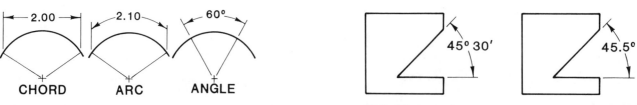

FIGURE 9-32 Conventions for the dimensioning of chords, arcs, and angles

FIGURE 9-33 Conventions for dimensioning angles

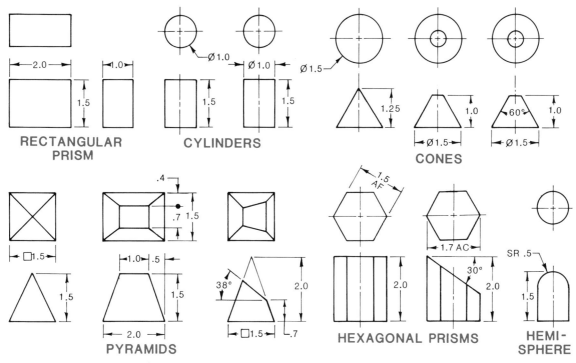

FIGURE 9–34 Examples of dimension placement on basic forms

Dimensioning Prisms, Cylinders, Cones, Pyramids, and Spheres

Other basic forms, such as prisms, cylinders, cones, pyramids, and spheres, are dimensioned by ANSI standards **(Fig. 9–34)**.

Pictorial Dimensioning

Pictorial drawings are rarely used as working drawings and are usually accompanied by a fully dimensioned multiview drawing. If a pictorial drawing must be used as a working drawing, it should be dimensioned as shown on the isometric drawing in **Fig. 9–35**. (See Chapter 15 for pictorial drawing details.)

SURFACE CONTROL

The surface texture for manufacturing processes is specified on working drawings through **finish symbols** **(Fig. 9–36)**. The **surface texture**

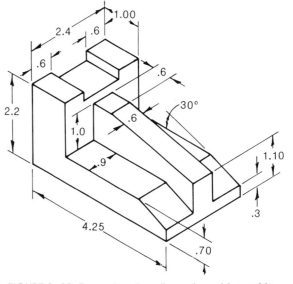

FIGURE 9–35 Example of a dimensioned isometric drawing

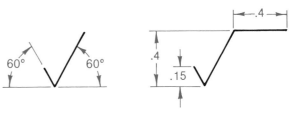

FIGURE 9–36 Typical size dimensions for the surface finish symbol

value is controlled with a symbol or notation on the working drawing **(Fig. 9–37)**. The point of the ✓ symbol is placed on the surface for which a surface finish is specified **(Fig. 9–38)**. Different forms of the surface finish symbol are used to designate different surface finishes **(Fig. 9–39)**.

Surface **smoothness** is a relative term. No surface appears smooth under a microscope. Some surfaces require a higher degree of smoothness than others. Bearings, seals, gears, pistons, and most matching and moving parts require a high degree of smoothness. Other surfaces, such as outside walls of machine parts, can be as rough as the final castings or forgings permit. It is, therefore, important to correctly specify the required degree of smoothness or **roughness** on working drawings. Roughness is the opposite of smoothness. **Figure 9–40** shows the basic characteristics of surface textures which are used to define degrees of roughness.

In specifying degrees of roughness, the drafter must specify values for roughness height **(Fig. 9–41)**, roughness width **(Fig. 9–42)**, waviness height **(Fig. 9–43)**, and waviness width **(Fig. 9–44)**.

Lay is the primary direction of the surface pattern made by machine tool marks. Seven symbols are given for the lay patterns produced by different manufacturing processes **(Fig. 9–45)**.

Roughness-width cutoff is the greatest spacing of repetitive surface irregularities included in the average roughness height. It is specified in inches on working drawings **(Fig. 9–46)**.

Figure 9–47 shows a surface control symbol with all its controlling features. **Figure 9–48** shows the method used for specifying minimum **contact area** for mating parts.

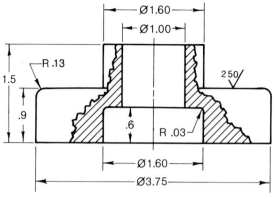

FIGURE 9–37 The surface texture value may be controlled with symbols or a notation placed on the drawing.

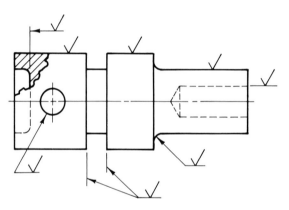

FIGURE 9–38 Examples for various positions of the surface finish symbols

✓	BASIC SURFACE FINISH SYMBOL
✓	SYMBOL FOR PLACEMENT OF ADDITIONAL SURFACE FINISH DATA
✓	SYMBOL FOR MATERIAL REMOVAL
.01 ✓	SYMBOL FOR MATERIAL REMOVAL ALLOWANCE
✓	SYMBOL FOR PROHIBITING MATERIAL REMOVAL

FIGURE 9–39 Various types of surface finish symbols

Chapter 9 Dimensioning Conventions and Surface Finishes

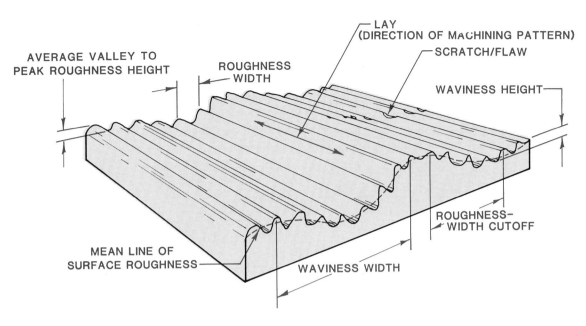

FIGURE 9–40 Characteristics of surface texture

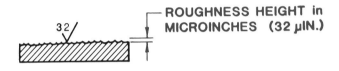

Recommended Roughness Average Rating Values
Microinches, μin. [Micrometers, μm]*

μin.	μm	μin.	μm	μin.	μm	μin.	μm
1	[0.025]	10	[0.25]	50	[1.25]	250	[6.3]
2	[0.050]	13	[0.32]	63	[1.6]	320	[8.0]
3	[0.075]	16	[0.40]	80	[2.0]	400	[10.0]
4	[0.100]	20	[0.50]	100	[2.5]	500	[12.5]
5	[0.125]	25	[0.63]	125	[3.2]	600	[15.0]
6	[0.15]	32	[0.80]	160	[4.0]	800	[20.0]
8	[0.20]	40	[1.00]	200	[5.0]	1000	[25.0]

*Boldface values preferred

FIGURE 9–41 The roughness height values noted on the finish symbol. The prefix micro is one-millionth (.000 001 or 10^{-6}).

FIGURE 9–42 The callout for the roughness width noted on the finish symbol

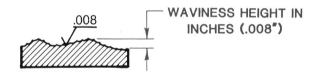

Recommended Waviness Height Values Inches, in. Millimeters, mm*

in.	mm	in.	mm	in.	mm
.00002	[0.0005]	.00003	[0.008]	.005	[0.12]
.00003	[0.0008]	.00005	[0.021]	.008	[0.20]
.00005	[0.0012]	.00008	[0.020]	.010	[0.25]
.00008	[0.0020]	.00010	[0.025]	.015	[0.38]
.00010	[0.0025]	.002	[0.05]	.020	[0.50]
.0002	[0.005]	.003	[0.08]	.030	[0.80]

*Boldface values preferred

FIGURE 9–43 The waviness height values noted on the finish symbol

FIGURE 9–44 The waviness width noted on the finish symbol

Roughness Tolerance

Different manufacturing processes, such as depth of cut, amount of feed, and machining repetition, result in varying degrees of surface smoothness. Engineers must use design processes that produce the proper degree of smoothness with the least amount of manufacturing processing. Producing surfaces that are smoother than necessary may drastically increase production time and costs. Producing surfaces that are too rough may result in an inferior quality product. In specifying degrees of smoothness, that is, tolerance for roughness, refer to the charts shown in **Fig. 9–49A** and **9–49B**. These charts show the degree of roughness height produced by various manufacturing and machining processes.

STANDARD ROUGHNESS-WIDTH CUTOFF VALUES*						
mm	0.075	0.250	0.750	2.500	7.500	25.000
in	0.003	0.010	0.030	0.100	0.300	1.000

*When no value is specified, the value .030 in. (0.750 mm) is assumed.

FIGURE 9–46 The roughness-width cutoff noted on the finish symbol. (The greatest spacing of repetitive surface irregularities, due to machining, that is included in the average roughness height.)

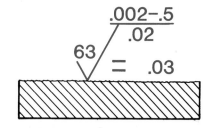

FIGURE 9–47 Surface control symbol in full use

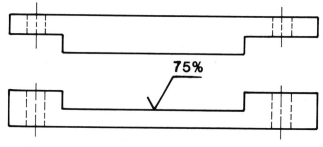

FIGURE 9–48 Specifying the minimum contact area

SYM	DESCRIPTION OF LAY	TOOL MARKS AND FINISH SYMBOLS
=	Lay parallel to the line representing the surface to which the symbol is applied.	
⊥	Lay perpendicular to the line representing the surface to which the symbol is applied.	
X	Lay angular in both directions to line representing the surface to which symbol is applied.	
M	Lay multidirectional/random pattern.	
C	Lay approximately circular relative to the center of the surface to which the symbol is applied.	
R	Lay approximately radial relative to the center of the surface to which the symbol is applied.	
P	Lay particulate, nondirectional, or protuberant.	

FIGURE 9–45 The lay symbols and their characteristics

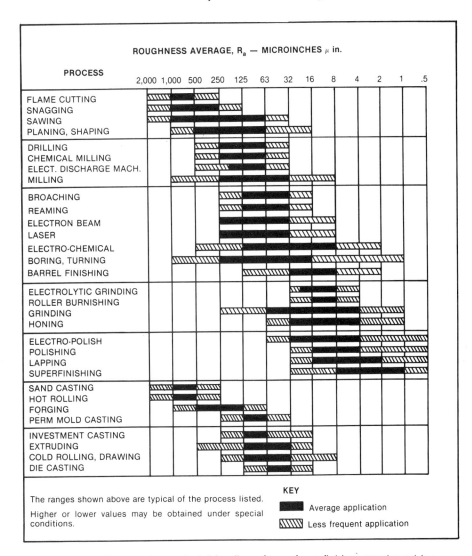

FIGURE 9-49A The roughness height ratings for surface finishes produced by common production procedures (Reprinted from BLUEPRINT READING FOR MANUFACTURING by Hoffman and Wallach © 1988, Delmar Publishers Inc.)

ROUGHNESS HEIGHT MICROINCHES (μ INCH)	SURFACE DESCRIPTION	MACHINING PROCESSES
1000 ∇	Very rough	Saw or torch cutting, forging, or sand casting
5000 ∇	Rough machining	Coarse feeds and heavy cuts in machining
250 ∇	Coarse	Coarse surface grind, medium feeds, and average cuts in machining
125 ∇	Medium	Sharp tools, light cuts, fine feeds, high speeds with machining
63 ∇	Good finish	Sharp tools, light cuts, extra fine feeds, light cuts with machining
32 ∇	High grade finish	Very sharp tools, very fine feeds, and cuts
16 ∇	Higher grade finish	Surface grinding, coarse honing, coarse lapping
8 ∇	Very fine machine finish	Fine honing and fine lapping
3 ∇	Extremely smooth machine finish	Extra fine honing and lapping

FIGURE 9-49B Typical roughness height values

COMPUTER-AIDED DRAFTING & DESIGN

DIMENSIONING WITH CAD

CAD systems contain automatic dimensioning capabilities that can dimension horizontally, vertically, and on any angle. Oblique, aligned, and unidirectional dimensioning is possible, and most systems can also automatically dimension diameters and radii. CAD systems that require an operator to individually input dimension lines, extension lines, arrowheads, and numerals are not considered automatic. In fully automatic systems, the computer calculates each dimension from the established graphics data. For this reason, objects must be accurately layed out and drawn. Drawing geometrically accurate forms is absolutely necessary for the effective functioning of the automatic dimensioning features of a CAD system.

The procedure of dimensioning on a CAD system differs from manual dimensioning practices. Although relatively simple, the correct procedure is necessary to insure dimensional accuracy. The dimensioning is controlled with a graphics tablet **(Fig. 9–50)**.

The steps used for horizontal and vertical dimensioning **(Fig. 9–51)** are as follows:

Step 1. Select DIM1 and then HORIZ or VERT from the graphics tablet.

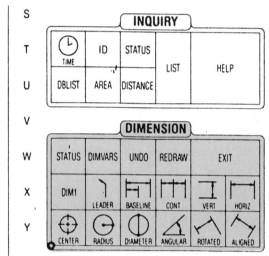

FIGURE 9–50 The section of the graphics tablet used for dimensioning

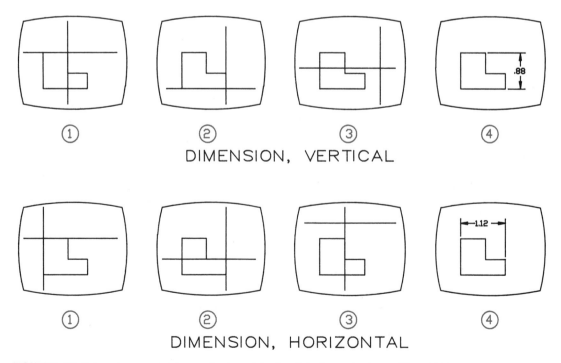

FIGURE 9–51 Procedures used for vertical and horizontal dimensioning with a CAD system

Select the origin of the first extension line by locating the cross hairs with the puck.

Step 2. Select the origin of the second dimension line by positioning the cross hairs with the puck.

Step 3. Select the location of the dimension. This determines how far the dimension will be from the object. At this point, the computer measures the size of the dimension and displays it in the screen prompt area. By typing in a different size at this point, the operator may change the dimension's value.

Step 4. Press the RETURN key and the dimension automatically is displayed, including the extension lines, dimension lines, and arrows.

The steps used for angular dimensioning **(Fig. 9–52)** are as follows:

Step 1. Select DIM1 and ANGULAR from the graphics tablet. Select one of the lines on the drawing that makes up the angle by locating the cross hairs with the puck on the line.

Step 2. Select the other line on the drawing that makes up the angle.

Step 3. Select the location for the numerical value of the angle. At this point, the measured angle will be displayed in the screen prompt area of the monitor.

Step 4. Press the RETURN key and the dimension automatically is displayed, including the extension lines, dimension arcs, arrows, and the angular dimension.

The steps used for diameter and radius dimensioning **(Fig. 9–53)** are as follows:

Step 1. Select DIM1 and DIAMETER or RADIUS, as required, from

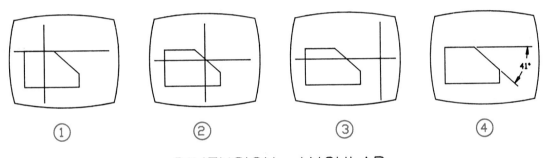

DIMENSION, ANGULAR

FIGURE 9–52 Procedure used for angular dimensioning with a CAD system

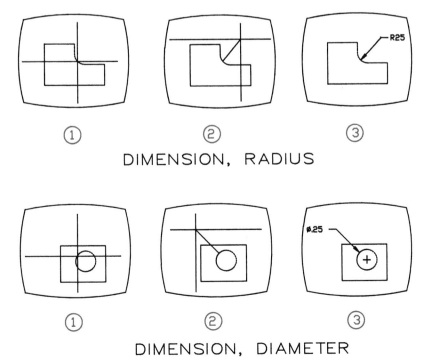

DIMENSION, RADIUS

DIMENSION, DIAMETER

FIGURE 9–53 Procedures used for dimensioning diameters and radii with a CAD system

the graphics tablet. Place the cross hairs on the circle or arc to be dimensioned.

Step 2. Place the cross hairs in the desired location of the dimension.

Step 3. Press the RETURN key and the dimension is automatically displayed, including the leader, arrow, dimension, and the proper symbol (∅ or R) before the dimension.

Other CAD dimensioning tasks are performed in a manner similar to horizontal, vertical, angular, diameter, and radii dimensioning. Specific procedures are found in each system's documentation manual.

Figure 9–54 shows an example of the application of all CAD dimensioning practices covered in this chapter. Note, the lines in **Fig. 9–54** are all the same width, which is how the drawing appears on the monitor before plotting. The correct line widths will be created by the plotter.

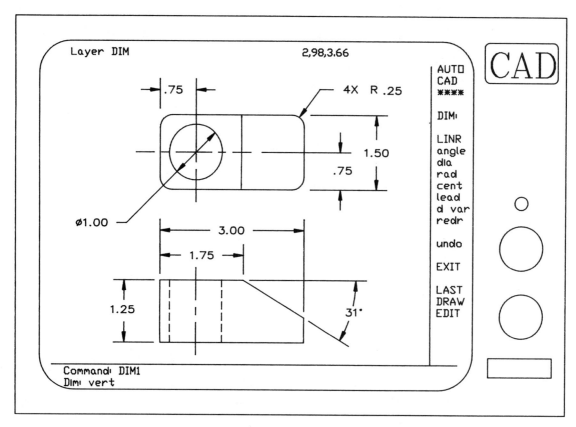

FIGURE 9–54 A fully dimensioned multiview drawing generated with a CAD system

EXERCISES

1. Sketch the single view drawings in **Fig. 9–55** on grid paper and add dimensions.
2. With drafting instruments, draw and dimension the drawings in **Figs. 9–56** and **9–57**.
3. With a CAD system, draw and dimension the multiview drawings in **Fig. 9–57**.
4. List the dimensioning errors in **Fig. 9–58**. Redraw and dimension it correctly.

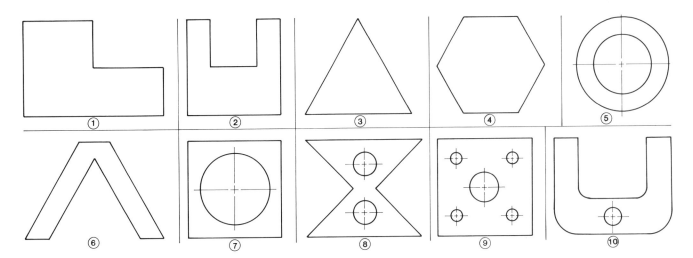

FIGURE 9–55 Sketch and dimension each example. Double the size of each drawing with a scale or dividers.

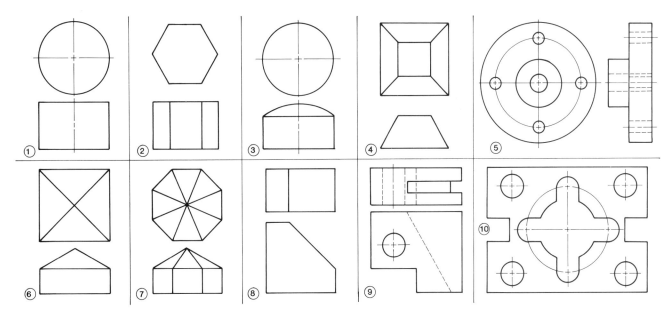

FIGURE 9–56 With drafting instruments, draw and dimension each example. Double the size of each multiview drawing with a scale or dividers, and dimension. Allow additional space between views as needed for dimensioning.

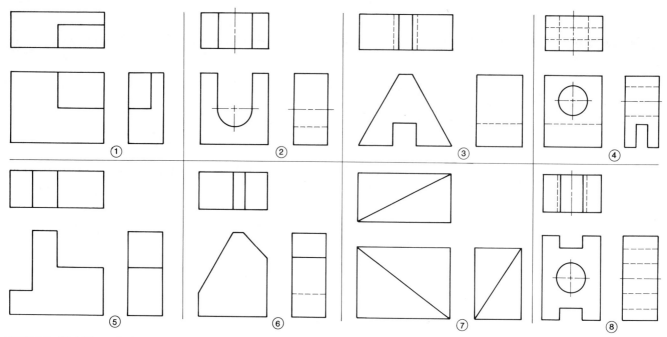

FIGURE 9–57 With a CAD system, draw and dimension each example. Measure dimensions directly from the page. Allow additional space between views as needed for dimensioning.

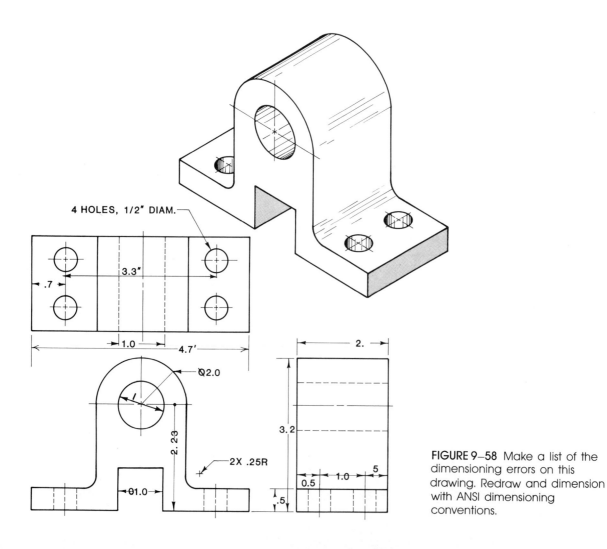

FIGURE 9–58 Make a list of the dimensioning errors on this drawing. Redraw and dimension with ANSI dimensioning conventions.

 ## KEY TERMS

Aligned dimensions
Chamfer
Contact area
Dimensioning
Dimension line
Dimensioning symbols
Dual dimensioning
Extension line
Fillets
Finish symbol
Knurling

Lay
Leaders
Location dimensions
Metric dimensioning
Notations
Polar coordinates
Radial centerline
Radii
Rectangular coordinates
Roughness

Roughness-width cutoff
Rounds
Size dimensions
Slots
Smoothness
Surface texture
Tabulated outline dimensioning
Tolerance
Unidirectional dimensions
U.S. customary dimensioning

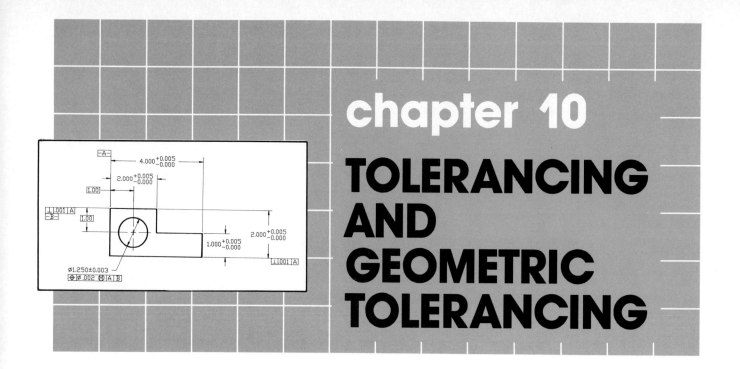

chapter 10
TOLERANCING AND GEOMETRIC TOLERANCING

OBJECTIVES

The student will be able to:

- place limit dimensions on a working drawing
- place plus and minus tolerance dimensions on a working drawing
- recognize geometric tolerancing symbols
- design various fits for mating parts
- dimension a working drawing with form tolerance symbols
- understand the concepts for each form tolerancing symbol

INTRODUCTION

Absolute accuracy is not completely attainable in any field. An expert marksman can't hit the bulls-eye dead center every time. But, is hitting the edge of the bulls-eye tolerable? Or is hitting the second ring acceptable occasionally? And is hitting the third ring never acceptable? Whichever goal is tolerated, that becomes the limit of accuracy. There are limits of accuracy in the manufacture of products as well. In fact, each dimension must contain a working limit of accuracy. This limit is called a **tolerance.** A tolerance is the permissible amount a dimension may vary in size during manufacturing or construction. For example, the width of a highway can vary $1/4''$ from the dimension shown on the drawing without serious consequences. However, if a watch part varies only a thousandth of an inch, the part will not fit or function. To control these variances (tolerances), manufacturers are assigned degrees of accuracy required in each product. This is done through a tolerance range to each dimension, or through a general tolerance notation.

TOLERANCE DIMENSIONS

The use of **basic (simple) dimensions** which do not include tolerance limits, allows manufacturers to establish their own acceptable limits which may not produce the desired result. **Figure 10–1** shows a comparison of a drawing dimensioned with basic dimensions and with tolerance dimensions of .05″. The tolerance dimensions shown on the right assures the shaft will fit into the hole with the clearance desired. Basic dimensions **(Fig. 10–2)** show only the general size of a part without any tolerance limitations.

Tolerance is the difference between two dimensions. If a 190′-6″

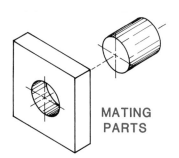
MATING PARTS

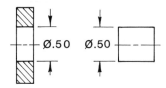
SIMPLE DIMENSIONS (BASIC)

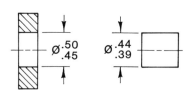
TOLERANCE DIMENSIONS WITH TOLERANCE OF .05″

FIGURE 10–1 Comparison of basic dimensions and tolerance dimensions

158

Chapter 10 Tolerancing and Geometric Tolerancing   159

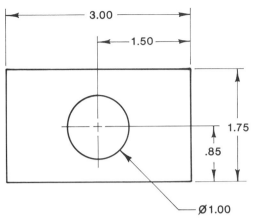

FIGURE 10–2 The basic dimension is the theoretical exact size.

sidewalk is acceptable if constructed at 190'-5" or 190'-7", that's one inch over and one inch under the basic dimension. But the tolerance range is actually two inches (190'-7" minus 190'-5" = 2").

TOLERANCE PRACTICES

A **limit tolerance** is a statement of the variations that can be permitted from a given dimension. When tolerance limits are aligned vertically, the larger dimension is placed on top. When aligned horizontally, the smaller dimension is placed to the left of the larger dimension. The number of decimal places in the basic dimension and in the tolerancing dimension must be equal **Fig. 10–3**.

Plus and **minus tolerancing dimensions (Fig. 10–4)**, are used to indicate the tolerance range above and below the basic dimension. The tolerance level higher than the basic dimension is marked with a plus sign (+). The tolerance level less than the basic dimension is marked with a minus sign (−). Plus and minus tolerance dimensions are placed to the right of the basic dimension, with the plus dimension on top of the minus dimension. If the plus tolerance dimension is the same as the minus tolerance dimension, a single number may be used preceded by a combined ± sign. The tolerance of 3.00" ± .03 **(Fig. 10–4)** indicates

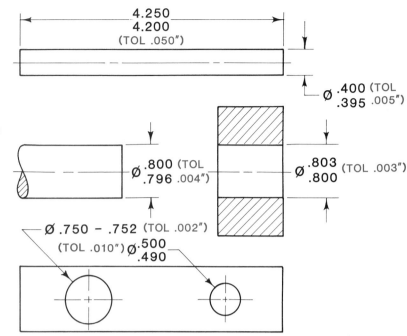

EXAMPLES OF LIMIT DIMENSIONING

FIGURE 10–3 A limit tolerance is a precise statement of the variations that can be permitted within a dimension. Note the large dimension is always on the top of the vertical tolerances, and to the right of the horizontal tolerances.

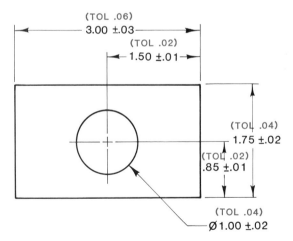

FIGURE 10–4 Plus and minus tolerance dimensions

the actual dimension may be as large as 3.03" or as small as 2.97". This represents a tolerance range of .6".

Bilateral and Unilateral Tolerancing

Tolerance dimensions that allow variation in both directions from a basic dimension (+ and −) are **bilateral tolerance dimensions (Fig. 10–5)**. Tolerances that allow variances in only one direction are **unilateral tolerance dimensions (Fig. 10–6)**. In the top of Fig. 10–5, the dimension may fall between 5.19" and 5.05". In the bottom of Fig. 10–6, the dimension may fall between 1.000" and 1.003".

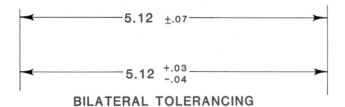

BILATERAL TOLERANCING

FIGURE 10–5 A bilateral tolerance has variations of size of larger and smaller (+ and −)

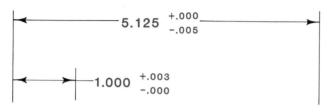

UNILATERAL TOLERANCING

FIGURE 10–6 A unilateral tolerance only has a variation in one direction—larger or smaller from the basic dimension

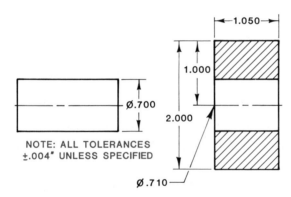

FIGURE 10–7A A general tolerance notation may be placed near the drawing or in the title block.

```
TOLERANCES UNLESS OTHERWISE SPECIFIED:
      0 TO 1.00 ARE ± .001
      1.01 TO 6.00 ARE ± .004
      OVER 6.00 ARE ± .008
```

FIGURE 10–7B General tolerance note

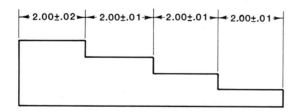

FIGURE 10–8 Chain dimensions accumulate tolerances. Accumulation in this drawing can be plus or minus .05″.

General Tolerancing

When all tolerances are identical, a **general tolerance note** (Fig. 10–7) may be used. This saves repeating tolerance limits on each dimension. A note can apply to an entire product or an individual part. Another type of general tolerancing note is shown in **Fig. 10–7B**. The dimensions and tolerances may vary with the design of different items. The tolerance levels in Fig. 10–7B are expressed as follows:

- All dimensions below one inch may have a tolerance of ±.001″.
- All dimensions between one and six inches may have a tolerance of .004″.
- All dimensions above six inches may have a tolerance of .008″.

Chain Dimensioning

Tolerance dimensions on **chain dimensions** are not recommended. Chain dimensioning accumulates tolerances. For example, the cumulative tolerance for the length of the part in **Fig. 10–8** can be plus .05″ (.02 + .01 + .01 + .01), or minus .05″, a total tolerance range of .10″, or anywhere in between.

Datum Dimensioning

In assigning tolerances to chain dimensions as in Fig. 10–8, a dimensioned 8″ length could be as short as 7.95″ or as long as 8.05″. For this reason **datum dimensioning** is recommended where tolerance limits are critical. In datum dimensioning, all dimensions originate from a single point and do not allow tolerances to accumulate **(Fig. 10–9)**. The total length in Fig. 10–9 cannot deviate more than ±.03″.

When the distance between any two points is critical, a **direct dimension** should be used to insure the least tolerance buildup. Points A and

Chapter 10 Tolerancing and Geometric Tolerancing **161**

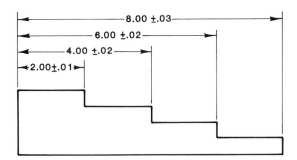

FIGURE 10-9 Datum dimensioning will not accumulate tolerances. The maximum tolerance in this drawing is plus or minus .03".

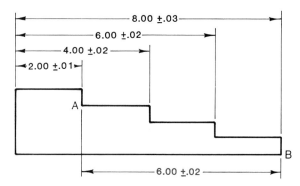

FIGURE 10-10 Direct dimensioning for a specific dimension will have the most accuracy.

B in **Fig. 10–10** should be within .04 of 6.00". Therefore, a tolerance dimension is used to connect these two points even though the remainder of the drawing is datum dimensioned.

Angle Tolerancing

While most tolerances relate to linear measurements, angle tolerances are also needed on many working drawings. **Figure 10–11** shows the correct method of specifying angle tolerances in degrees, minutes and seconds, and with degrees and a decimal part of a degree.

MATING PARTS

Appropriate tolerance levels for **mating parts** are very critical. If too much space is allowed between parts, the fit may wobble or fall apart. If too little space is allowed, parts may not fit or excessive friction may occur when moved. Very close tolerances also require more precise manufacturing methods and are expensive to produce. For these reasons, it is important that the appropriate fit be designed, dimensioned, and manufactured.

Figure 10–12 is a table showing the appropriate tolerances for various industrial applications at different sizes. The tolerance grade range extends from Grade 4 to Grade 13 with the lower number representing the closest tolerance limits.

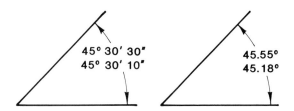

FIGURE 10-11 Specifying angle tolerances with degrees, minutes, and seconds, and with degrees and a decimal part of a degree.

	STANDARD TOLERANCES									
Nominal Size Range in Inches	Slip Gauges, Production and Wear Tolerance of Gauges and Measuring Instruments		Parts Subject to Very Tight Tolerances, Precision Bearings, Precision Assemblies		Precision Engineered Designs (General)		Engineering Work in General, Giving Scope for Wider Tolerances		Rough Work, Steel Structures, Castings, Agricultural Machinery	
Over–To	Grade 4	Grade 5	Grade 6	Grade 7	Grade 8	Grade 9	Grade 10	Grade 11	Grade 12	Grade 13
0.04–0.12	.00015	.00020	.00025	.0004	.0006	.0010	.0016	.0025	.004	.006
0.12–0.24	.00015	.00020	.0003	.0005	.0007	.0012	.0018	.0030	.005	.007
0.24–0.40	.00015	.00025	.0004	.0006	.0009	.0014	.0022	.0035	.006	.009
0.40–0.71	.0002	.0003	.0004	.0007	.0010	.0016	.0028	.0040	.007	.010
0.71–1.19	.00025	.0004	.0005	.0008	.0012	.0020	.0035	.0050	.008	.012
1.19–1.97	.0003	.0004	.0006	.0010	.0016	.0025	.0040	.006	.010	.016
1.97–3.15	.0003	.0005	.0007	.0012	.0018	.0030	.0045	.007	.012	.018
3.15–4.73	.0004	.0006	.0009	.0014	.0022	.0035	.005	.009	.014	.022
4.73–7.09	.0005	.0007	.0010	.0016	.0025	.0040	.006	.010	.016	.025
7.09–9.85	.0006	.0008	.0012	.0018	.0028	.0045	.007	.012	.018	.028
9.85–12.41	.0006	.0009	.0012	.0020	.0030	.0050	.008	.012	.020	.030
12.41–15.75	.0007	.0010	.0014	.0022	.0035	.006	.009	.014	.022	.035
15.75–19.69	.0008	.0010	.0016	.0025	.004	.006	.010	.016	.025	.040
19.69–30.09	.0009	.0012	.0020	.003	.005	.008	.012	.020	.030	.050
30.09–41.49	.0010	.0016	.0025	.004	.006	.010	.016	.025	.040	.060
41.49–56.19	.0012	.0020	.003	.005	.008	.012	.020	.030	.050	.080
56.19–76.39	.0016	.0025	.004	.006	.010	.016	.025	.040	.060	.100
76.39–100.9	.0020	.003	.005	.008	.012	.020	.030	.050	.080	.125
100.9–131.9	.0025	.004	.006	.010	.016	.025	.040	.060	.100	.160
131–171.9	.003	.005	.008	.012	.020	.030	.050	.080	.125	.200
171.9–200	.004	.006	.010	.016	.025	.040	.060	.100	.160	.250

FIGURE 10-12 Standardized grades of tolerances

Gauge instruments .04" to .12", Grade 4, have the smallest acceptable tolerance, .00015". Castings, 171.9" to 200", Grade 13, have the largest acceptable tolerance, .250".

Clearance Fits

Tolerancing of mating parts relies on the control of clearance desired between parts. Clearance (allowance) is the **tolerance zone** between mating parts. **Clearance fits** require separate tolerancing limits for each part.

The shaft's dimensions in **Fig. 10–13** are .50"/.45" with a tolerance of .05". The hole's dimensions are .53"/.51" with a tolerance of .02". When the largest shaft diameter of .50" mates with the smallest hole with a diameter of .51", the difference or allowance will be .01". When the smallest shaft (.45" diameter) mates with the largest hole (.53" diameter), the allowance is .08". With mass production, the mating of parts may have any allowance in the range of .01" to .08".

Interference Fits

When mating parts are designed to be permanently attached and not movable, an **interference fit** may be specified. Interference fits (force fits) have negative clearances so that the parts must be forced together **(Fig. 10–14)**. The amount of interference tolerance depends on the flexibility and hardness of the material. Soft materials force together more easily than hard materials.

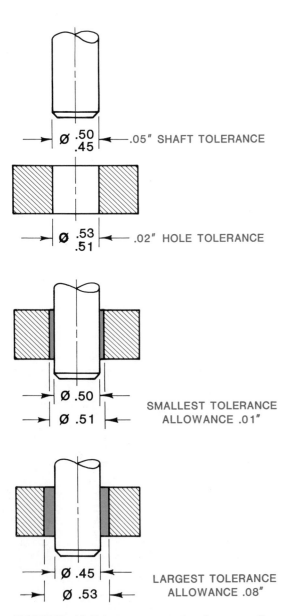

FIGURE 10–13 Tolerance zones for clearance fits

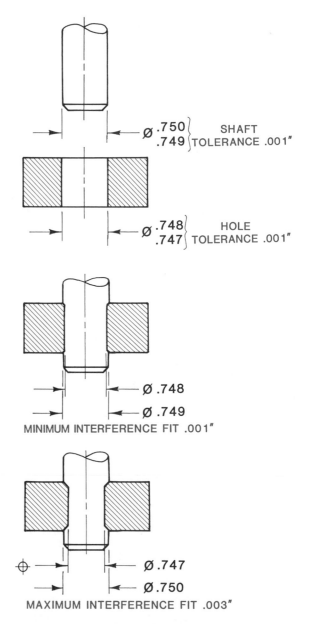

FIGURE 10–14 Interference fits (force fits)

Transition Fits

A **transition fit** is one in which parts overlap in size **(Fig. 10–15)**.

Basic Sizes

The appropriate matching of mating parts can be accomplished with custom machining; however, standard size parts should be specified when possible. For example, in the mating of a hole and shaft, the shaft size can be adjusted to match a standard hole size or a hole size can be adjusted to match a standard shaft size. Matching a shaft size to a **basic hole size** is shown in **Fig. 10–16**. Matching a hole size to a **basic shaft size** is shown in **Fig. 10–17**. The basic size of a shaft refers to the standard manufactured size. The basic size of a hole refers to the standard drill bit size.

Material Condition

When establishing hole tolerance sizes of mating parts, material conditions such as **maximum material condition (MMC)** and **least material condition (LMC)** must be considered. The MMC for a hole is the smallest allowable diameter which leaves the maximum amount of material. The MMC for a shaft is the largest allowable diameter which leaves the maximum amount of material. The LMC for a shaft is the smallest allowable diameter which leaves the minimum amount of material. The MMC and LMC for a shaft is shown in **Fig. 10–18A**, and the MMC and LMC for a hole is shown in **Fig. 10–18B**.

RFS means **regardless of feature size**. This indicates the tolerance must be met regardless of where the feature lies within its size tolerance.

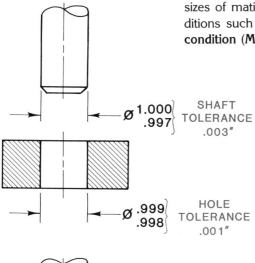

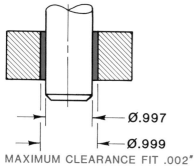

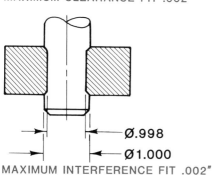

FIGURE 10–15 Transitional fits may be interference and/or clearance fits

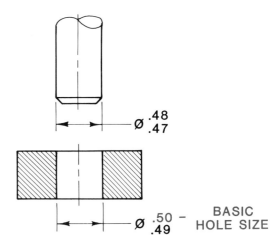

FIGURE 10–16 System of basic hole tolerances

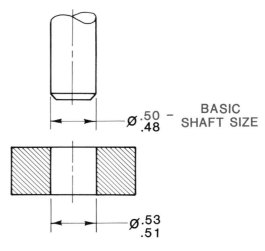

FIGURE 10–17 System of basic shaft tolerances

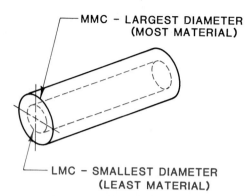

FIGURE 10–18A Maximum and minimum material conditions for a shaft

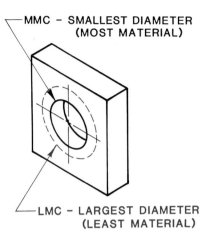

FIGURE 10–18B Maximum and minimum material conditions for a hole

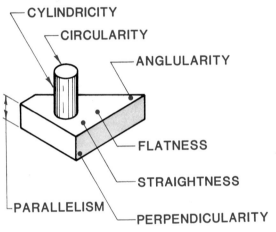

FIGURE 10–19 Form tolerancing controls

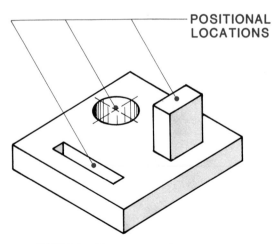

FIGURE 10–20 Positional tolerancing controls

Basic dimensioning controls only the shape and size of an item on a technical drawing. **Geometric tolerancing** is an extension of the dimensioning process that describes more precisely how an item is to be manufactured. Geometric tolerancing includes form tolerancing and positional tolerancing. **Form tolerancing** controls the geometric shape of objects **(Fig. 10–19)**. **Positional tolerancing** controls the exact location of product features such as holes, depressions, and projections **(Fig. 10–20)**.

Geometric characteristic symbols are used to control the tolerance limits of form, profile, orientation, location, and runouts **(Fig. 10–21)**. These symbols are a form of dimensioning shorthand, each representing a geometric feature to be controlled. Other modifying symbols used for geometric tolerancing are summarized in **Fig. 10–22**. See pages 165–168 for further explanation of geometric characteristic symbols.

TYPE OF TOLERANCE		CHARACTERISTIC	SYMBOL
FOR INDIVIDUAL FEATURES	FORM	STRAIGHTNESS	—
		FLATNESS	▱
		CIRCULARITY (ROUNDNESS)	○
		CYLINDRICITY	⌭
FOR INDIVIDUAL OR RELATED FEATURES	PROFILE	PROFILE OF A LINE	⌒
		PROFILE OF A SURFACE	⌓
FOR RELATED FEATURES	ORIENTATION	ANGULARITY	∠
		PERPENDICULARITY	⊥
		PARALLELISM	//
	LOCATION	POSITION	⌖
		CONCENTRICITY	◎
	RUNOUT	CIRCULAR RUNOUT	↗*
		TOTAL RUNOUT	↗↗*

*Arrowhead(s) may be filled in.

FIGURE 10–21 ANSI standard Y14.5M–1982, geometric characteristic symbols (Reprinted with permission of the American Society of Mechanical Engineers)

MODIFYING SYMBOLS	
TERM	SYMBOL
MMC—AT MAXIMUM MATERIAL CONDITION	Ⓜ
RFS—REGARDLESS OF FEATURE SIZE	Ⓢ
LMC—AT LEAST MATERIAL CONDITION	Ⓛ
PROJECTED TOLERANCE ZONE	Ⓟ
DIAMETER	⌀
SPHERICAL DIAMETER	S⌀
RADIUS	R
SPHERICAL RADIUS	SR
REFERENCE	( )
ARC LENGTH	⌒

FIGURE 10–22 ANSI standard Y14.5M–1982, modifying symbols used in specifying geometric tolerancing (Reprinted with permission of the American Society of Mechanical Engineers)

Datums

All geometric tolerancing begins with the establishment of a datum. A **datum** is a reference location easy to locate and measure. Theoretically, it is a perfect plane, point, line, or axis. The location of all geometric characteristics originate from a datum. On a technical drawing, the **datum identification box** or **datum target** is used to locate specific surfaces for reference **(Fig. 10–23)**.

Feature Control Frame

The **feature control frame** contains data for each controlled specification and includes the characteristic symbol, tolerance value, and datum reference letter. In **Fig. 10–24**, the marked surface is to be parallel within .003" of datum A.

Geometric Characteristic Symbols

As previously mentioned, each geometric characteristic symbol represents a geometric feature that must be controlled to the level specified. Most symbols are referenced to a datum; however, straightness, flatness, circularity (roundness), and cylindricity may stand alone. **Figures 10–25** through **10–37** show characteristic symbols accompanied with an explanation and a working drawing example.

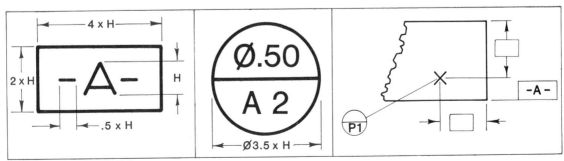

SYMBOL WITH DIMENSIONS FOR DATUM IDENTIFICATION BOX

SYMBOL AND DIMENSION FOR DATUM TARGET

USE OF DATUM SYMBOLS ON DRAWING

FIGURE 10–23 Use of reference datum symbols and their drawing sizes

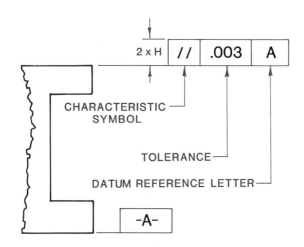

FIGURE 10-24 The feature control frame's drawing size. It states "The surface is to be parallel within .003" of surface –A–."

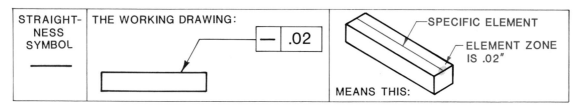

FIGURE 10-25 STRAIGHTNESS refers to an element of a surface or an axis which is a straight line. The straightness tolerance specifies a tolerance zone where the element or axis lies.

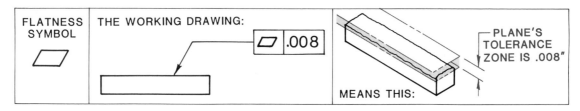

FIGURE 10-26 FLATNESS is the condition of a plane. The flatness tolerance specifies a tolerance zone where the surface must lie.

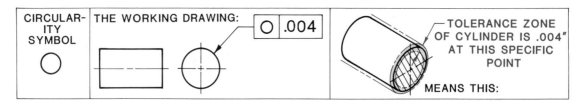

FIGURE 10-27 CIRCULARITY (ROUNDNESS) specifies a tolerance zone bounded by two concentric circles at a single specific point.

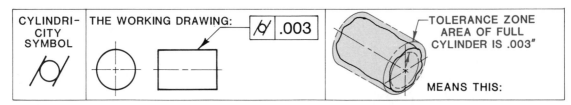

FIGURE 10-28 CYLINDRICITY is the combination of roundness and straightness for the whole cylinder. The tolerance zone is bounded by two concentric circles.

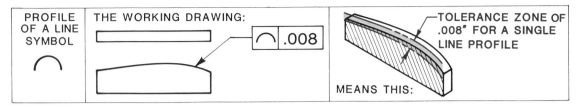

FIGURE 10-29 LINE PROFILE controls a single line profile (section) of an irregular surface. The tolerance zone extends on both sides of the profile along its full length.

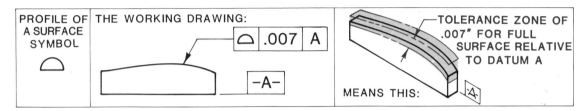

FIGURE 10-30 SURFACE PROFILE controls the entire surface for an irregular area. The tolerance zone extends the length and width of the surface.

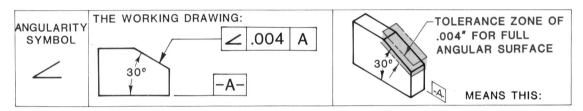

FIGURE 10-31 ANGULARITY is the condition of a surface or axis at an angle, other than 90°, from a datum plane or axis. The tolerance zone is defined by two parallel lines at the specific angle from the datum.

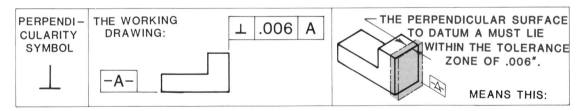

FIGURE 10-32 PERPENDICULARITY is the condition of a surface, plane, or axis, at a right angle to a datum plane or axis.

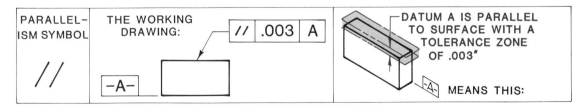

FIGURE 10-33 PARALLELISM is the condition of a surface equidistant at all points from a datum plane or axis. The tolerance zone is defined by two planes or lines parallel to the datum.

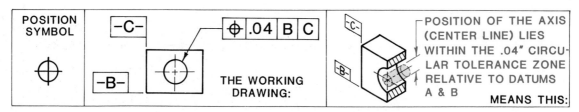

FIGURE 10-34 FEATURE POSITION is the theoretical exact location of its axis or center plane from the features from which it is dimensioned. The positional tolerance is the permissable error for the location of a feature relative to its other features.

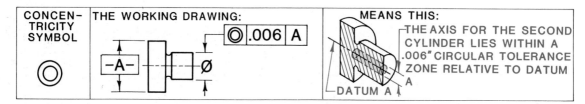

FIGURE 10-35 CONCENTRICITY is the condition where the axis of a cylinder aligns to a common axis datum feature. The tolerance zone forms a cylinder where the axis must lie.

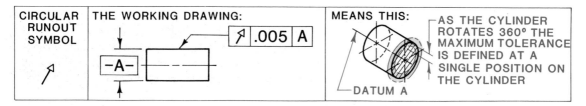

FIGURE 10-36 CIRCULAR RUNOUT controls a circular element at one position on a surface. The tolerance zone lies within two circles as the surface rotates one revolution.

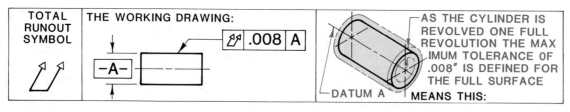

FIGURE 10-37 TOTAL RUNOUT controls the entire circular surface as the part is rotated 360°.

An example of a fully dimensioned metric working drawing, complete with dimensional and geometric tolerancing notations, is shown in **Fig. 10-38**. Interpret each dimension and feature control frame in this illustration.

Chapter 10 Tolerancing and Geometric Tolerancing 169

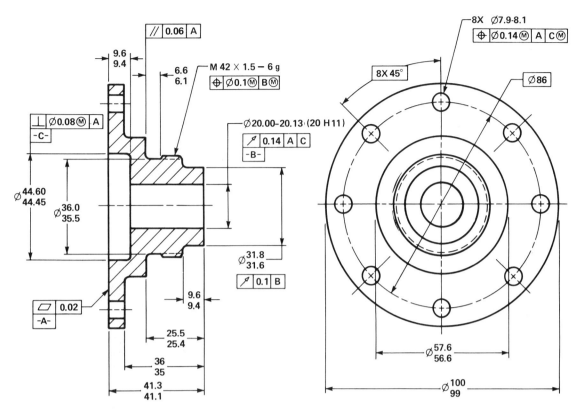

FIGURE 10-38 An example of a fully dimensioned metric working drawing with tolerances and geometric tolerancing, ANSI Y14.5M-1982 (Reprinted with permission of the American Society of Mechanical Engineers)

COMPUTER-AIDED DRAFTING & DESIGN

TOLERANCING WITH CAD

On a CAD system, tolerancing is one of the dimensioning functions. With dimensioning (see Chapter 9), a tolerance can automatically be applied by changing some of the dimensioning variables in the CAD software. Since tolerancing is a common drafting function, changing the dimensioning variables has been made easy for the CAD operator.

Figure 10-39 illustrates five methods that can be used to show the same 4.000" dimension on a CAD drawing using different dimensioning variables. Figure 10-39 states the dimension as

A—no specified tolerance,
B—bilateral plus and minus tolerance,

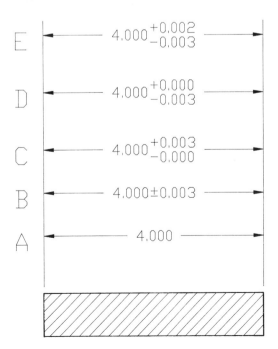

FIGURE 10-39 Examples of tolerances generated with a CAD system

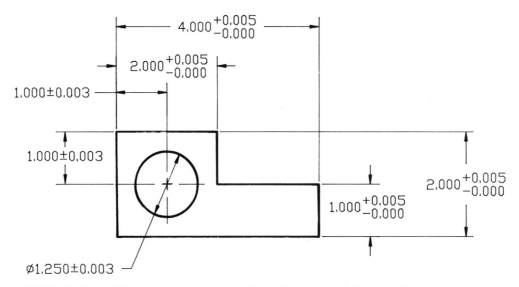

FIGURE 10–40 A CAD-generated drawing with unilateral and bilateral tolerances

C—unilateral with all the tolerance on the plus side,
D—unilateral with all the tolerance on the minus side, and
E—bilateral with different plus and minus tolerances.

Figure 10–40 shows a simple shape with specified tolerances. Note that the size tolerances for this part are stated unilaterally, and the location and size of the hole are stated bilaterally, with ± (plus and minus tolerances).

Applying geometric and positional tolerances on a CAD drawing is more difficult than simple tolerancing. ANSI Y14.5M is the national standard for tolerancing. The symbols have been standardized and tablet menu overlays have been developed. The tablet menu overlay **(Fig. 10–41)** includes all of the standard geometric symbols, as well as the reference blocks and other dimensioning aids. To use the overlay, the CAD operator places a stylus or a puck over the desired symbol, presses a button, and inserts the symbol where desired on the drawing **(Fig. 10–42)**.

Geometric symbols have been added to the drawing in **Fig. 10–43**.

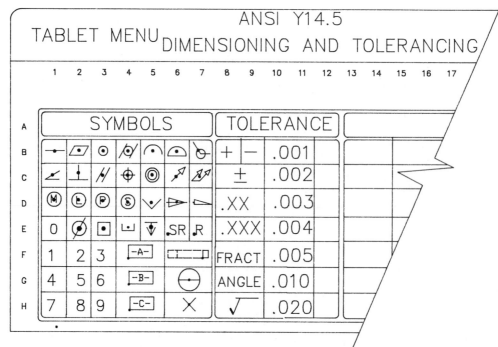

FIGURE 10–41 A segment of a graphics tablet menu overlay used for basic dimensioning, tolerancing, and geometric tolerancing.

FIGURE 10-42 Selecting a symbol from a graphics tablet menu overlay (Courtesy of Cascade Graphics Systems)

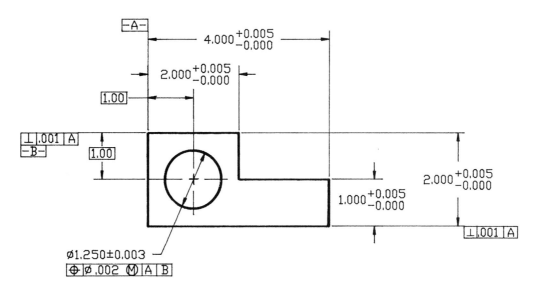

FIGURE 10-43 A CAD-generated drawing with tolerancing and geometric tolerancing symbols

This is the same drawing as is in Fig. 10-40. The following have been added:

1. datums A and B
2. form tolerances of perpendicularity and parallelism
3. hole locational dimensions have been changed to basic sizes
4. the positional tolerance of true position

All of these were added using the graphics tablet overlay. If an overlay is not available, the CAD operator could construct each symbol, and block and insert them as specified.

Figures 10-44 and 10-45 include both tolerancing and geometric tolerancing.

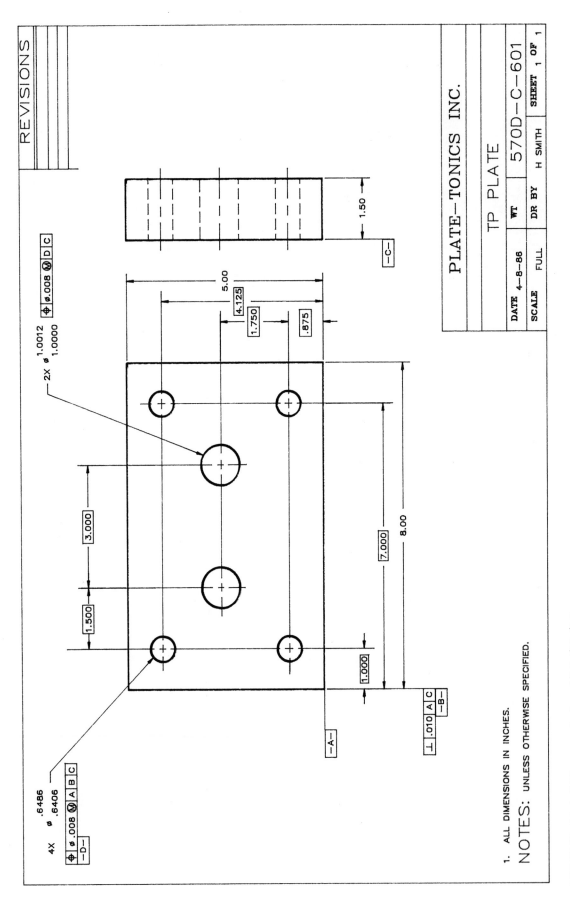

FIGURE 10-44 A CAD-generated working drawing

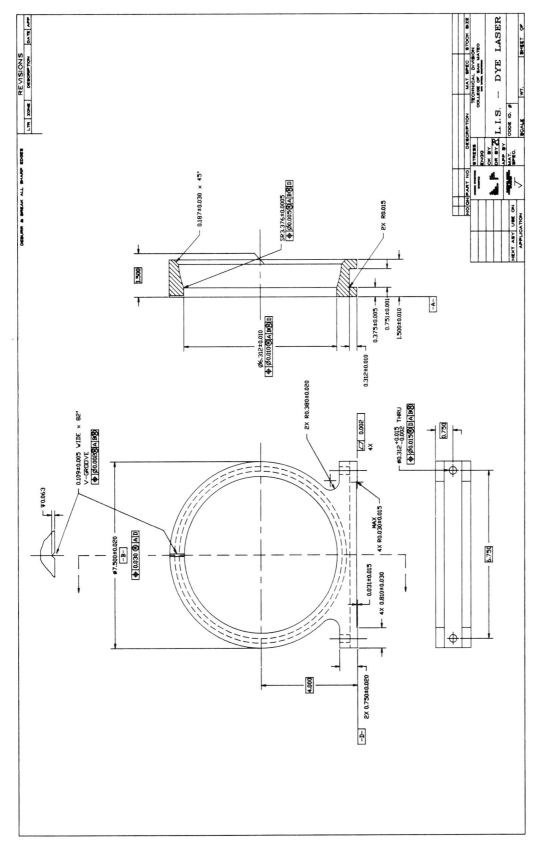

FIGURE 10–45 A CAD-generated working drawing (Courtesy of the College of San Mateo)

EXERCISES

1. With drawing instruments or a CAD system, draw the dimensioning and tolerancing symbols in **Fig. 10–46**.
2. What is the maximum and minimum tolerances and clearances for the mating parts in **Figs. 10–47 and 10–48**?
3. With instruments or a CAD system, draw the working drawings and complete the dimensioning, with tolerances and geometric tolerancing, as instructed in the captions of **Figs. 10–49 through 10–52**.
4. Draw the working drawing and make the ECOs (Engineering Change Orders—see Chapter 17) given in the caption of **Fig. 10–53**.

FIGURE 10–46 Draw each symbol with correct proportions. Label each symbol.

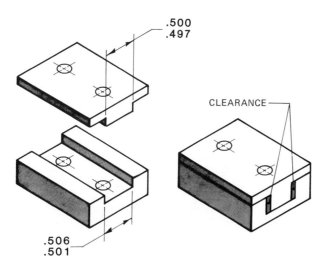

FIGURE 10–47 What is the maximum and minimum tolerances and clearances for the mating parts?

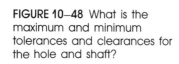

FIGURE 10–48 What is the maximum and minimum tolerances and clearances for the hole and shaft?

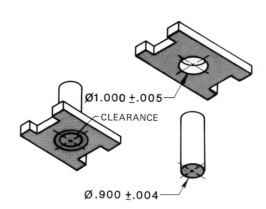

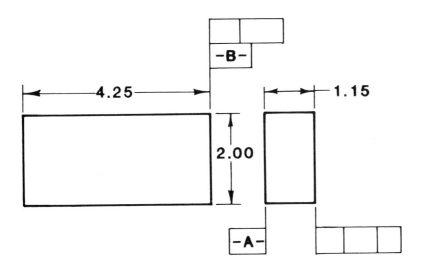

FIGURE 10–49 Draw the multiview drawing and dimension with Grade 12 tolerances. Add geometric tolerancing symbols for the following:
1. Datum A is parallel to its opposite side within .01".
2. Datum B is flat within .005".
3. Datum B is perpendicular to the bottom surface within .02".

FIGURE 10–50 Draw the multiview drawing and dimension with Grade 11 tolerances. Add geometric tolerancing symbols for the following:
1. Datum A is parallel to the opposite end within .004".
2. Roundess is within .008".

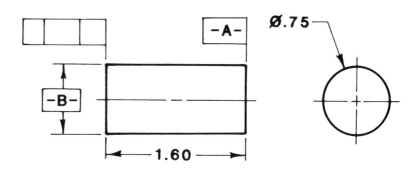

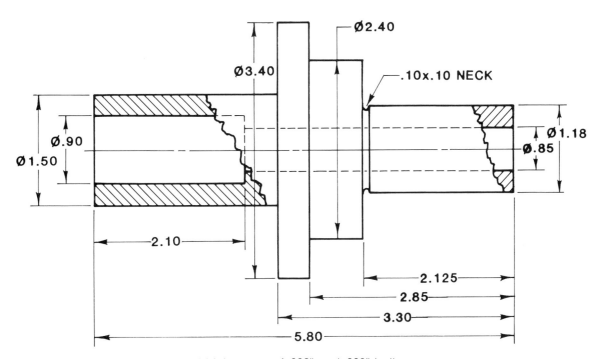

FIGURE 10–51 As a designer, add tolerances of .002" and .008" to the appropriate dimensions. Add geometric characteristic symbols of flatness, straightness, perpendicularity, and circularity where they apply.

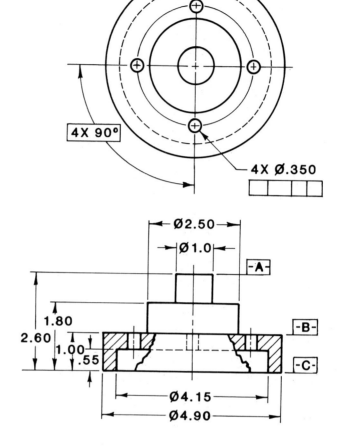

FIGURE 10-52 Draw the working drawing and dimension with Grade 8 tolerances. Add geometric tolerancing symbols for true position within .005" for the holes, and flatness within .003" for datums A, B, and C.

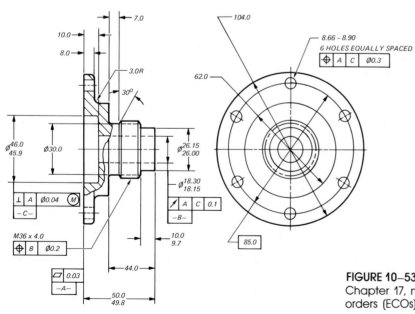

FIGURE 10-53 Redraw the bearing hub and, referring to Chapter 17, make the following engineering change orders (ECOs):

1. Change the feature control frame to conform to ANSI's new standards.
2. Convert all metric dimensions to inch/decimal.
3. Use Grade 7 tolerances.

KEY TERMS

Angularity
Basic (simple) dimension
Basic hole size
Basic shaft size
Bilateral tolerance dimensions
Chain dimension
Circular runout
Circularity (roundness)
Clearance fit
Concentricity
Cylindricity
Datum
Datum dimensioning
Datum identification box

Datum target
Direct dimension
Feature control frame
Feature position
Flatness
Form tolerancing
General tolerance note
Geometric characteristic symbols
Geometric tolerancing
Interference fit
Least material condition (LMC)
Limit tolerancing
Line profile
Mating parts

Maximum material condition (MMC)
Minus tolerancing dimensions
Parallelism
Perpendicularity
Plus tolerancing dimensions
Positional tolerancing
Regardless of feature size (RFS)
Straightness
Surface profile
Tolerance
Tolerance zone
Total runout
Transition fit
Unilateral tolerance dimensions

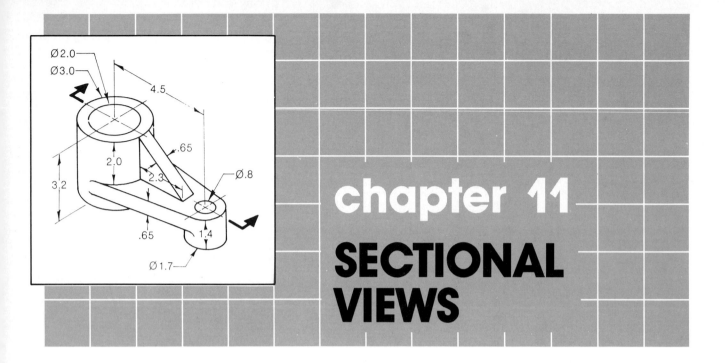

chapter 11
SECTIONAL VIEWS

OBJECTIVES

The student will be able to:

- draw a full section drawing
- draw a half section drawing
- draw a broken-out section drawing
- draw an offset section drawing
- draw an assembly section drawing
- draw a removed section drawing
- draw a revolved section drawing
- draw a thin material section drawing
- draw a pictorial section drawing
- rotate and align features for a section drawing

INTRODUCTION

Multiview drawings are adequate for describing the size and shape of an object when all important features can be seen on the normal orthographic views. However, if the internal surfaces require the extensive use of hidden lines, the drawing may be difficult to understand. When this occurs, a sectional view is usually prepared to show the size and shape of the interior more clearly **(Fig. 11–1)**. A **sectional view,** or section, is a drawing of an object as it would appear if cut in half or quartered. Sectional views do not contain hidden lines because dashed lines can confuse the clarity of the sectional view **(Fig. 11–2)**.

CUTTING PLANE

To understand sectional views, imagine a saw cutting an object in half. Replace the position of the saw

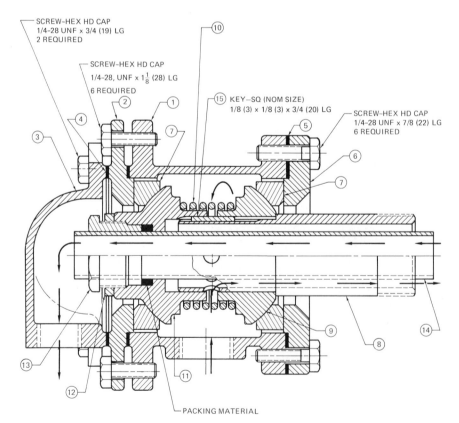

FIGURE 11–1 This working drawing view would be difficult to read if not drawn as a section. (Reprinted from TECHNICAL DRAWING AND DESIGN by Goetsch and Nelson, © 1986 Delmar Publishers Inc.)

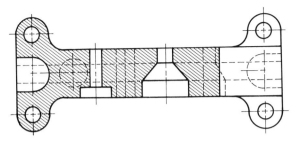

FIGURE 11-2 Hidden lines will confuse the clarity of a sectional view.

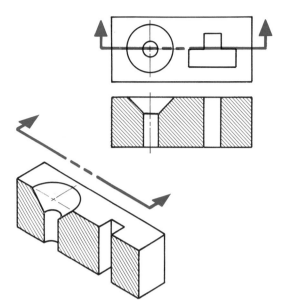

FIGURE 11-3 The arrows on the cutting plane indicate the direction of the line of sight for the sectional view.

with an imaginary plane and remove the nearest part of the object. What remains is a sectional view.

To define the location of the imaginary sectional plane, a **cutting plane line** is drawn on the view where the cut was made. Cutting plane lines are very wide and dense, with arrows placed on the ends to indicate the line of sight direction for the sectional drawing **(Fig. 11-3)**. Several types of cutting plane lines are used, depending on the type of drawing, size, and location of the section **(Fig. 11-4)**. **Figure 11-5** shows the relationship between the cutting plane, the cutting plane line, and the resulting sectional view.

SECTION LINING

Where the cutting plane passes through solid material, **section lining** lines are drawn. **Crosshatching** is the universal symbol for section lining (Fig. 11-5); however, section lining symbols are standardized for most manufacturing and construction materials (Fig. 11-25, page 186). When the universal crosshatching symbol is used rather than the standard material section lining symbol, a note near the section is used to indicate the type of material. The

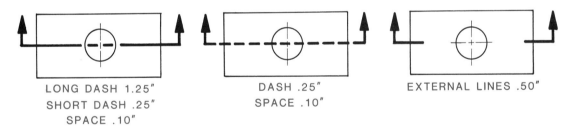

FIGURE 11-4 Average size and spacing recommended for the cutting plane lines

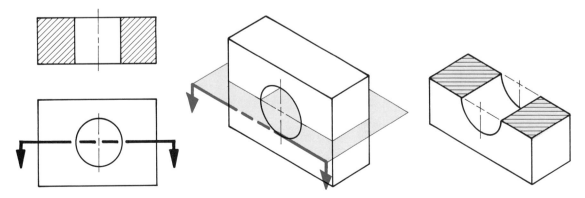

FIGURE 11-5 The cutting plane defines the surface to be sectioned.

spacing of the crosshatching will vary depending on the size of the section **(Fig. 11–6)**. Do not take the time to measure the spacing. Estimate with your eye when drawing. When adjacent parts are crosshatched, the direction of the lines is reversed **(Fig. 11–7)**.

Figure 11–8 shows the method of interrupting the section lining to place dimensions and notes. This should only be done in large sectioned areas. It is preferable to place notes and dimensions on the outside of the view.

Section Lining Exceptions

Although section lining is added to areas intersected by the cutting plane, there are some exceptions to this principle.

Ribs, webs, spokes, gear teeth, lugs, and ball bearings are not section lined, because they are either comparatively thinner than the remainder of the object or they appear intermittently **(Fig. 11–9)**.

Shafts and fasteners, such as nuts, bolts, rivets, pins, and screws are also not section lined **(Fig. 11–10)**.

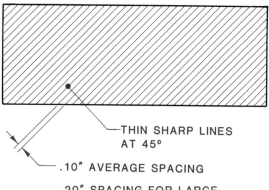

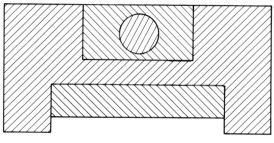

FIGURE 11–7 Reverse the direction of the crosshatching for adjacent parts.

FIGURE 11–6 Crosshatch spacing will vary according to the drawing size. Space the crosshatching by "eye" when drawing.

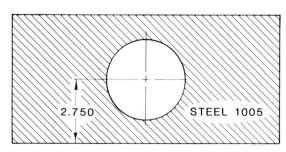

FIGURE 11–8 If necessary, notes and dimensions may be placed on the sectional view as shown.

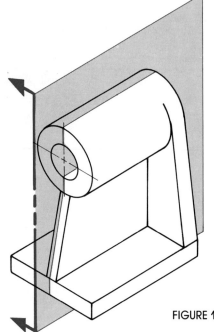

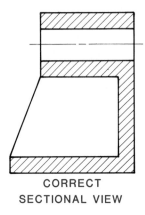

CORRECT SECTIONAL VIEW

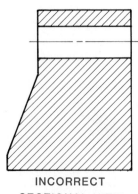

INCORRECT SECTIONAL VIEW (TRUE PROJECTION)

FIGURE 11–9 Do not crosshatch thin supporting materials.

Chapter 11 Sectional Views **181**

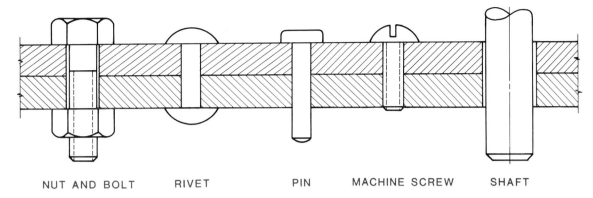

FIGURE 11-10 Do not crosshatch fasteners or shafts.

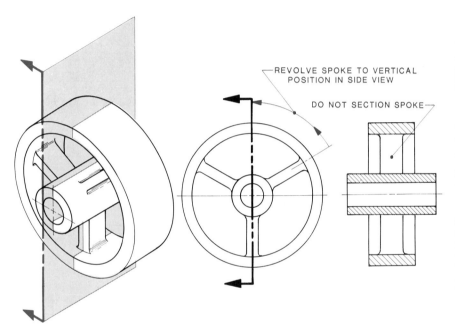

FIGURE 11-11 Revolve a spoke to a vertical position in a sectional view.

Objects with webs and spokes that do not align with the cutting plane line are rotated **(Fig. 11-11)** before projecting the sectional view. The spoke is not section lined.

Holes that do not intersect the cutting plane are also rotated to reveal a hole through the section view **(Fig. 11-12)**. Objects with protrusions are rotated and sectioned in the same manner **(Fig. 11-13)**.

When objects containing curved surfaces are sectioned, the true projection may reveal curved lines outlining the section lined areas. When this occurs, the section is drawn using straight lines separations **(Fig. 11-14)**.

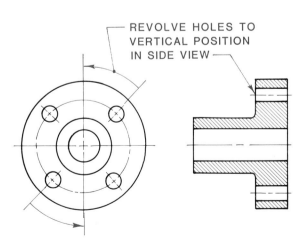

FIGURE 11-12 Revolve holes to a vertical position in a sectional view.

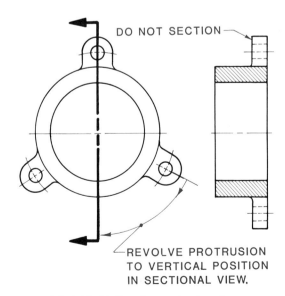

FIGURE 11-13 Revolve protrusions to a vertical position in a sectional view.

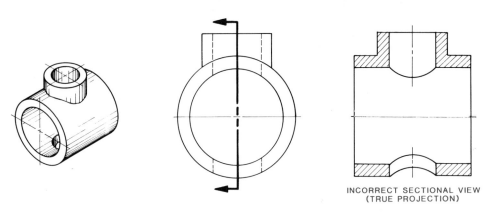

FIGURE 11-14 Drafting conventions simplify intersections for curved surfaces.

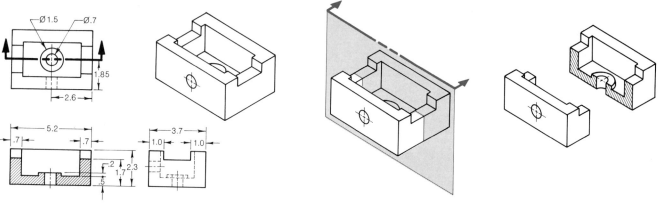

FIGURE 11-15 Full-section working drawings for a jig base

SECTIONAL VIEW TYPES

The type of sectional view selected depends on the size, scale, and complexity of the area to be sectioned.

Full Sections

When the cutting plane extends entirely through an object, the resulting section is a **full section**. Figure 11-15 shows a multiview drawing with a cutting plane on the top view, and the corresponding full sectional view in the front view position. Full sections of a front view will place the cutting plane on either horizontal (top) views or vertical (profile) views. In architectural drawings, a full section top view is called a plan view or floor plan. A full section for a profile view is an elevation section.

Half Sections

A **half section** is one-half of a full section. Half sections are prepared for symmetrical objects where the details on both sides of a center line are identical. Two perpendicular cutting plane lines, one on the vertical plane and one on the horizontal plane, define the section area. These align with center lines and remove an imaginary quarter of the object (Fig. 11-16).

Broken-Out Sections

When only a small part of an object needs to be sectioned, a **broken-out section** of that part is prepared. An irregular short break line separates the sectioned area from the remainder of the drawing. Figure 11-17 shows a broken-out section in both pictorial and orthographic views.

Offset Sections

A cutting plane normally extends through an object in a straight line. However, a straight cutting plane may miss key features which should be shown for maximum clarity. When this occurs, the cutting plane is **offset** to align with special features. The cutting plane line in **Fig. 11-18** is offset to align with a countersunk hole, a counterbored hole, and a slot, none of which fall on a straight line. When offsetting cutting plane lines, always bend the line at right angles to the cutting plane and perpendicular to the plane of projection. Note that the bends in the cutting plane are never shown in the sectioned view.

Chapter 11 Sectional Views **183**

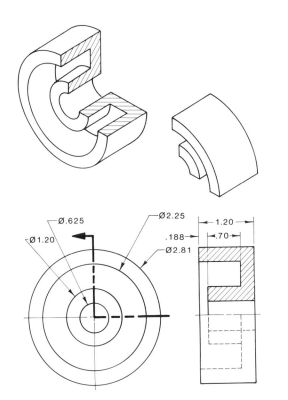

FIGURE 11-16 Half-section working drawing of a belt pulley

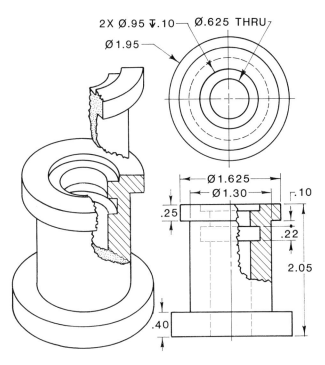

FIGURE 11-17 Broken-out section working drawing of a sleeve bearing

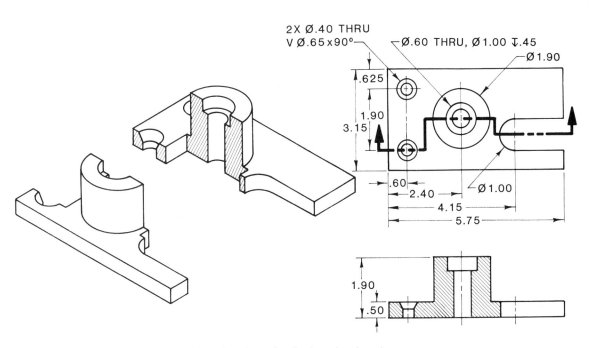

FIGURE 11-18 Offset-section working drawing of a jig bearing locator

Removed Sections

When sections are drawn in a location other than in one of the six multiview positions, they are **removed sections.** Sections are often removed to be drawn at a different scale or to dimension some portion in detail. When sections are removed, they must be indexed to a cutting plane line containing identifying letters **(Fig. 11-19).**

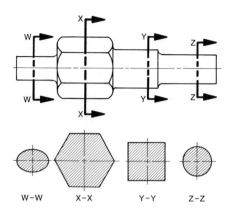

FIGURE 11-19 Removed sections in a drawing of a spindle arbor

Revolved Sections

Sectional views are often revolved 90° to reveal a section of an object perpendicular to the plane of projection. In visualizing a revolved section, imagine a cutting plane passing through an object creating a slice of the object. Now, imagine rotating the slice 90°. The rotated slice is a **revolved section.** Some revolved sections break the object to show the section, and others show the section directly on the view **(Fig. 11-20).**

Assembly Section

One common use for sectional drawings is to show the assembly of parts. When multiple-part assembly drawings are sectioned, the adjoining parts are section-lined at different angles. Figure 11-21 shows an example of an **assembly section** with multiple parts sectioned.

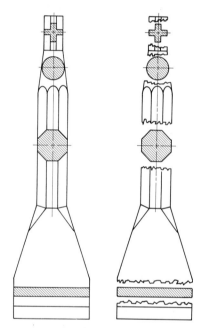

FIGURE 11-20 Two methods for drawing revolved sectional views for a jack-hammer chisel

Thin Wall Sections

The outline of ribs, spokes, and other thin walls are drawn on sectional views. However, section lining is not added to these parts. Figure 11-22 shows an example of a typical **thin wall section** for a corner bracket.

FIGURE 11-21 An assembly section drawing depicting multiple sectioned parts of a model airplane motor (Reprinted from TECHNICAL DRAWING AND DESIGN by Goetsch and Nelson, © 1986 Delmar Publishers Inc.)

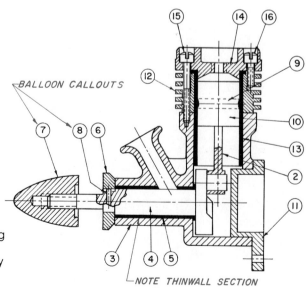

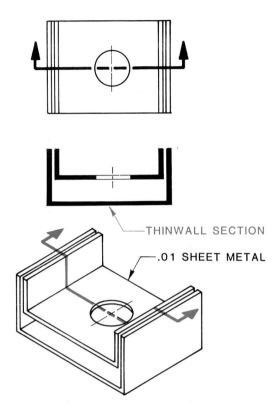

FIGURE 11-22 Thin wall section for a corner bracket

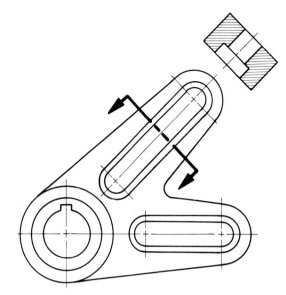

FIGURE 11-23 Auxiliary section for a locking lever arm

Materials too thin to section with crosshatching, such as paper and sheetmetal, are sectioned solid.

Auxiliary Sections

Most sectional views are aligned with the normal planes of projection. When a section must be drawn of an auxiliary view, follow the guidelines covered in Chapter 12. **Auxiliary sections** should be aligned with the projection lines of the auxiliary views **(Fig. 11–23)**.

Enlarged Section

Sectional details are often too small to be interpreted and dimensioned. In such a case, **enlarged sections** are prepared to clarify details and for ease of dimensioning **(Fig. 11–24)**.

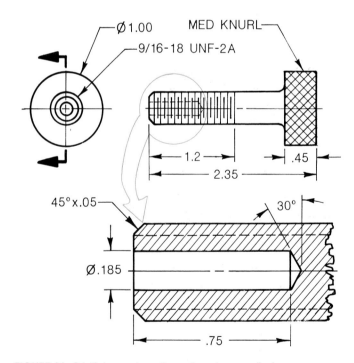

FIGURE 11-24 Enlarged section of a screw adjuster

COMPUTER-AIDED DRAFTING & DESIGN

COMPUTER-AIDED SECTIONAL VIEWS

CAD simplifies the drawing of sectional views. An advantage of all CAD systems is the ability to store drawing elements in the system and rapidly recall them when needed. Drawing elements may be standard symbols, line types, styles of text, or in this case, sectioning patterns.

Figure 11–25 shows a few of the available sectioning patterns. These are called hatches and the CAD command for inserting them is HATCH. These hatch patterns may be put into any CLOSED geometric shape drawn by the CAD operator. The shape may be a circle, or any shape made up of lines and arcs. **It is necessary to be sure the shape is closed and has no gaps.** If a gap appears, the hatch pattern will "escape" through the gap and go where it is not wanted. A layer is often created on the drawing for the construction of the shapes to be hatched, and to fill them with the desired hatch pattern **(Fig. 11–28)**.

Figures 11–26 through 11–29 show the steps for drawing a simple full section of a hub.

Step 1. Draw the desired views in **Fig. 11–26**. In the view to be sectioned, use solid lines in the place of hidden lines. The views have been created with the LINE and CIRCLE commands.

Step 2. Create a new layer **(Fig. 11–27)**. In this case it has been called "sect." Draw the shapes to be sectioned. Using the LAYER command, turn the other layers off so only the areas to be sectioned appear on the monitor.

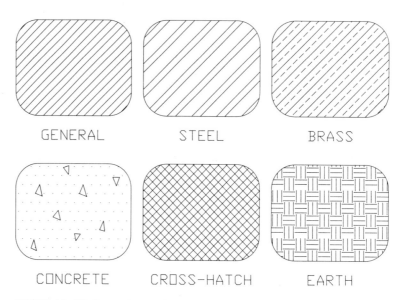

FIGURE 11–25 Examples of CAD-generated sectioning patterns

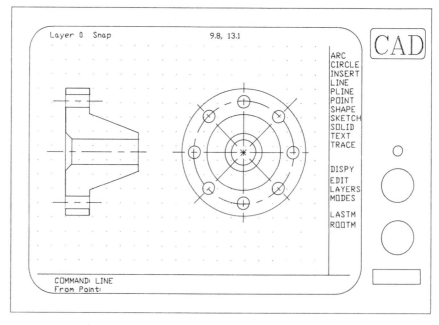

FIGURE 11–26 Lay out a new drawing or call up an existing drawing from the CAD data bank to be sectioned.

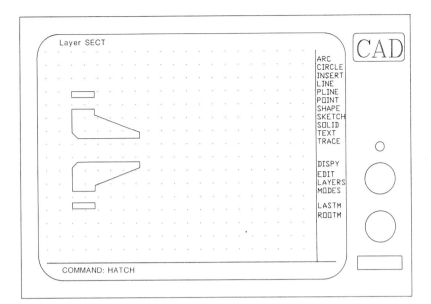

FIGURE 11-27 Create a new layer of the areas to be crosshatched.

Step 3. Using the HATCH command **(Fig. 11-28)**, insert the hatch patterns into the described areas. The GENERAL hatch pattern has been used in this drawing.

Step 4. Using the LAYER command **(Fig. 11-29)**, turn on all of the layers and the geometry for the sectional drawing.

To finish the drawing, dimensions should be added.

FIGURE 11-28 Insert the required crosshatching pattern.

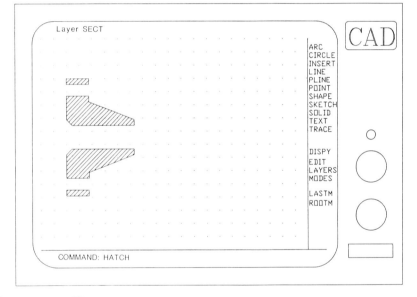

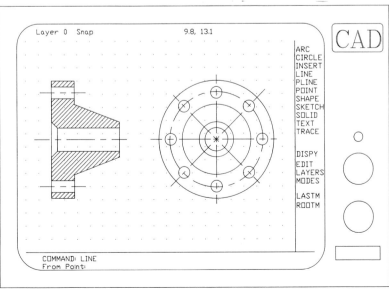

FIGURE 11-29 Turn on all layers for the completed full section drawing.

EXERCISES

1. Sketch the sections in **Fig. 11–30**.
2. With instruments, draw the sections noted in **Fig. 11–31**. Have your instructor add new cutting planes for additional sections to be drawn in the working drawings.
3. Draw the working drawings and the isometric sections for **Figs. 11–32 and 11–33**.

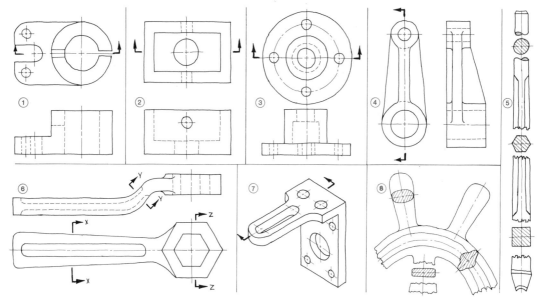

FIGURE 11–30 Sketch the sectional views for each exercise.

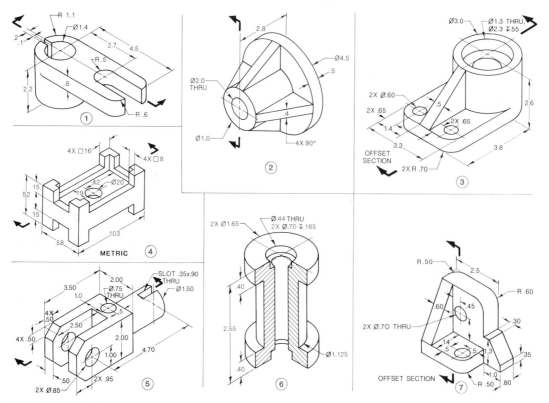

FIGURE 11–31 With drawing instruments, draw the sectional views as required.

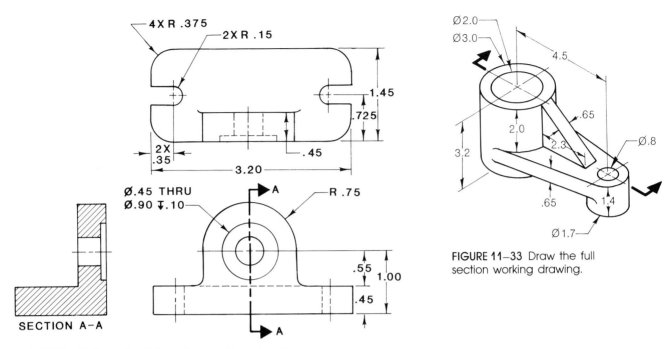

FIGURE 11–32 Draw the full section working drawing.

FIGURE 11–33 Draw the full section working drawing.

4. Make the following ECOs (Refer to Engineering Change Orders, Chapter 17) to **Fig. 11–34**:
   1. Change the base dimension 1.60" to 1.50".
   2. Change the base dimension 2.10" to 2.00".
   3. Change the counterbore depth to .200".
   4. Change the vertical section to a horizontal section through the bearing.
5. Design a tape dispenser for a standard size roll of scotch tape. Draw a set of working drawings, including a full sectional drawing.
6. Design a coffee cup. Draw the working drawings, including a sectional drawing.
7. Design a stand to support a video monitor over a microcomputer. Research the dimensions for any computer and monitor. Draw a set of working drawings, including sectional drawings.

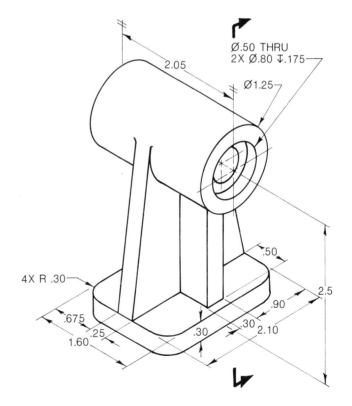

FIGURE 11–34 Make the ECOs (Engineering Change Orders, refer to Chapter 17) listed in Exercise 4.

190 Drafting in a Computer Age

DATA TABLE					
LINE	X	Y	go to	X	Y
1-2	1	1	—	1	12
2-3	1	12	—	7	12
3-4	7	12	—	7	1
4-1	7	1	—	1	1
NEW START					
5-6	6	3	—	2	3
6-7	2	3	—	4	10
7-5	4	10	—	6	3
NEW START (CUTTING PLANE)					
8-9	4	12	—	4	1
NEW START					
10-11	10	1	—	10	12
11-12	10	12	—	15	12
12-13	15	12	—	15	1
13-10	15	1	—	10	1
NEW START					
14-15	15	3	—	10	3
NEW START					
16-17	10	10	—	15	10
finish					

FIGURE 11-36 Using the coordinate grid system, layout the sectional drawing on ¼" grid paper. Add crosshatching and the cutting plane.

FIGURE 11-35 Lay out the sectional working drawings on a CAD system.

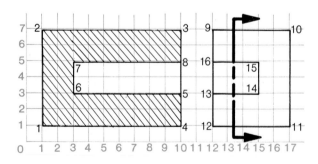

DATA TABLE					
LINE	X	Y	go to	X	Y
1-2	1	1	—	1	7
2-3	1	7	—	10	7
3-4	10	7	—		
4-5					
5					

FIGURE 11-37 Complete the data table. Do not digitize the coordinates of the crosshatching into the data table.

8. With a CAD system, draw the sections in **Fig. 11-35**.
9. Using the cartesian coordinate system, complete the drawing on ¼" grid paper from the data table in **Fig. 11-36**. Add the crosshatching.
10. Using the cartesian coordinate system, complete the data table from **Fig. 11-37**. Do not digitize the crosshatching into the data table.
11. Draw the twin engine Cessna airplane and the removed sections in **Fig. 11-38**.

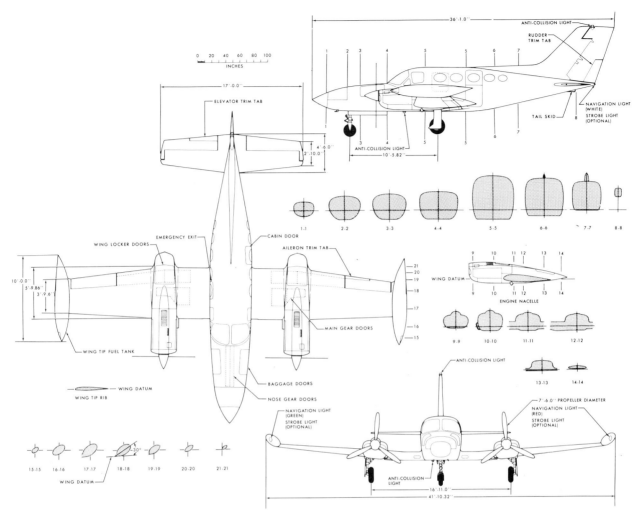

FIGURE 11–38 Draw the multiview drawings and the removed sections for the twin engine Cessna airplane. (Courtesy of Cessna Aircraft Company)

 **KEY TERMS**

Assembly section
Auxiliary section
Broken-out section
Crosshatching
Cutting plane line

Enlarged section
Full section
Half section
Offset section
Removed section

Revolved section
Section lining
Sectional view
Thin wall section

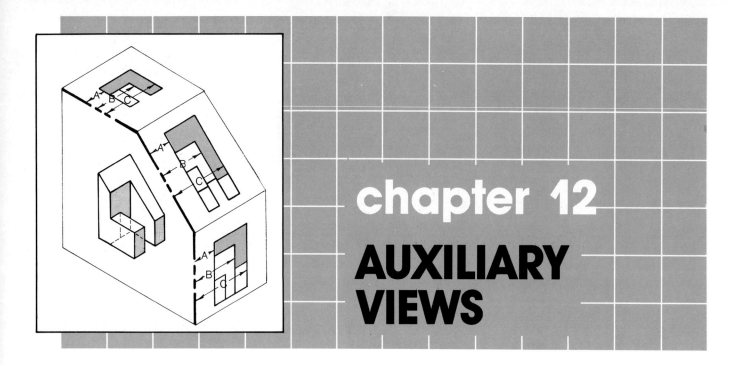

chapter 12
AUXILIARY VIEWS

OBJECTIVES

The student will be able to:

- project foreshortened lines
- project an auxiliary view
- project an auxiliary view from the front view
- project an auxiliary view from the top view
- project an auxiliary view from the side view
- draw primary and secondary auxiliary views
- draw a simple auxiliary view on a CAD system

INTRODUCTION

In Chapter 8 (Multiview Drawings), you learned that when each surface of an object is either parallel or perpendicular (90°) to other surfaces, one or more of the six normal orthographic views (front, top, right side, left side, back, or bottom) can be used to accurately describe the object. **Figure 12–1** shows an object of this type in which all angles are perpendicular. The three views are drawn as viewed through the normal orthographic planes of projection. Each projection plane is parallel to the corresponding face of the object. In this drawing, all lines and surfaces are shown as true size.

When an inclined surface is viewed from a point perpendicular to the surface plane, the true shape of the inclined surface will appear **(Fig. 12–2)**. These views are known as **auxiliary views.**

FORESHORTENING

Many objects contain surfaces that are not parallel or perpendicular to other surfaces. When objects of this type are drawn using the normal planes of projection, any inclined or sloped surface will be **foreshortened.** A foreshortened line or surface is one which appears shorter than actual size, because the viewing angle is not perpendicular to the line of sight. For example, hold a 12″ ruler perpendicular to your line of sight **(Fig. 12–3)**. Note the true vertical distance covered by the ruler. Now, tilt the ruler from the bottom to move

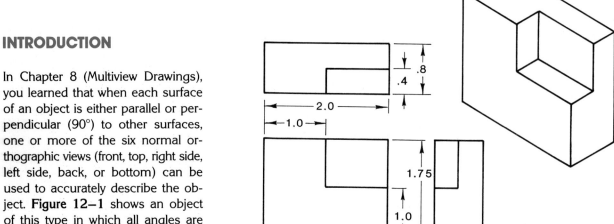

FIGURE 12–1 All surfaces and edges (lines) that are perpendicular and parallel to the planes of projection will appear true size.

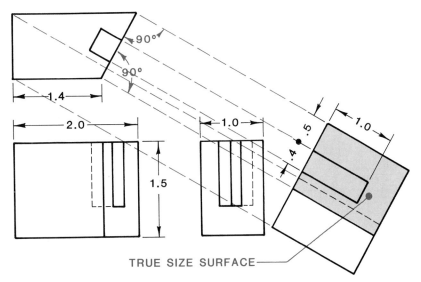

FIGURE 12-2 An auxiliary view is projected perpendicular to its inclined surface.

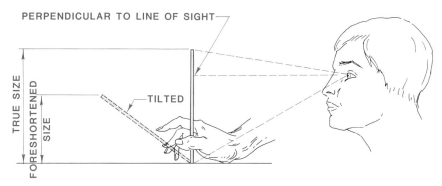

FIGURE 12-3 An object will only appear true size when the line of sight is perpendicular to that object.

the top of the ruler away from you. The ruler no longer appears 12″ long, nor does it cover the same amount of vertical distance. It appears foreshortened.

Foreshortened lines are also shown in **Fig. 12-4**. When the music box lid is closed (A), the front view shows all true dimensions (width and height). Likewise, when the lid is fully opened to 90° (B), the 2″ x 4″ lid measures exactly 2″ x 4″ on the front orthographic view. When the front view of the music box is drawn with the lid partially opened, the 4″ width doesn't change, but the 2″ depth is foreshortened (C). You can also partially open your book cover and observe the same effect as illustrated by the music box views.

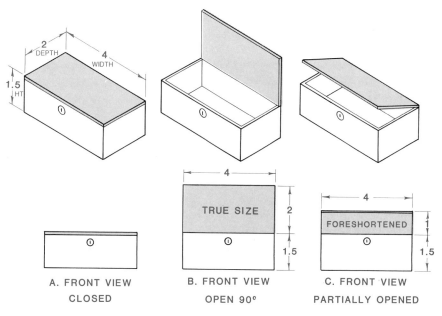

FIGURE 12-4 The apparent size of a surface will change when viewed at different angles.

194 Drafting in a Computer Age

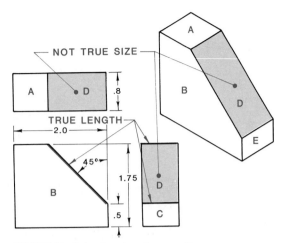

FIGURE 12–5 Inclined surfaces will appear foreshortened (not true size).

To understand this concept better, look at **Fig. 12–5**. Surface D is foreshortened in both the top and right-side views because this surface is not perpendicular to either the top or right-side planes of projection. Surfaces A, B, and C are shown in true size because these surfaces are perpendicular to the normal orthographic planes of projection. Only the true depth of surface D is shown on the top and left side views. Only the true length of surface D is shown, as an inclined line representing the edge, on the front view. Even when all six orthographic planes are used **(Fig. 12–6)**, the true shape of surface D in Fig. 12–5 will not appear. This surface appears only, in foreshortened form, on the top and right-side views. It appears hidden, in foreshortened form, in the bottom and left-side views.

AUXILIARY PLANES

The plane through which the inclined surface is viewed is the **auxiliary plane.** The auxiliary plane must always be parallel to the inclined surface to be drawn. **Figure 12–7** shows the use of an auxiliary plane compared to the normal orthographic planes of projection. You can see that viewing a surface through a parallel auxiliary plane eliminates foreshortened lines, and reveals the

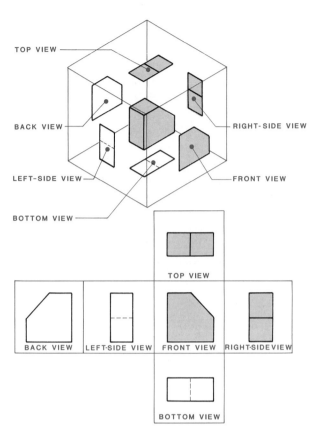

FIGURE 12–6 The true size of the inclined surface is not shown in any of the six orthographic projections.

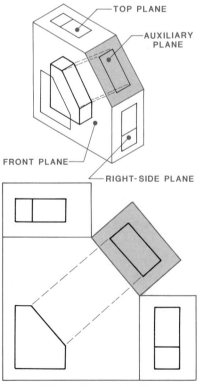

FIGURE 12–7
Planes of projection

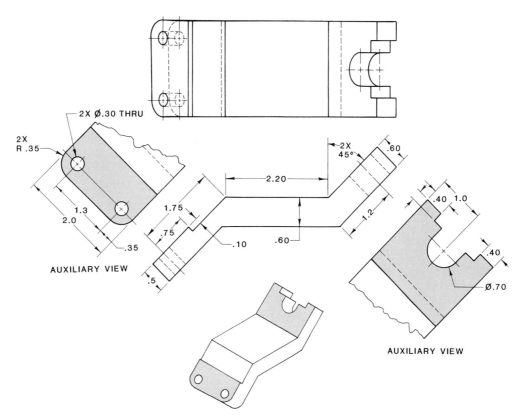

FIGURE 12-8 An auxiliary view shows the offset surface in true size and form.

true geometric shape and size of the surface. Viewing an inclined surface through an auxiliary plane is similar to viewing the sides of a right-angled object through one of the six planes of an orthographic projection box.

The auxiliary planes can also be envisioned as **hinged planes** that are folded from the normal orthographic planes. Figure 12-7 shows the auxiliary plane on a projection box, and an auxiliary plane folded from the normal orthographic planes. Both methods reveal the true shape of the inclined surface which cannot be seen through any of the normal orthographic planes.

Figure 12-8 shows an orthographic drawing containing an auxiliary view of a manufactured jig guide. In this drawing, the auxiliary views are needed because the true shape of the angled surfaces does not show in any of the orthographic views. The front view does show the exact shape and size of this part. The top view shows round holes as ellipses, rather than true circles. It also shows foreshortened overall lengths. An auxiliary view can show the true shape of this part with the proper relationship between holes and radii.

PROJECTION

When drawing auxiliary views, always project perpendicularly from a **true size edge** on a normal view. In **Fig. 12-9**, the true length of the in-

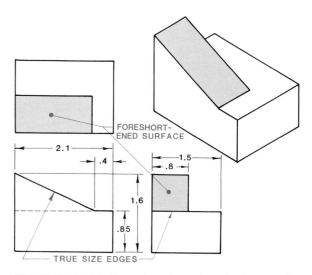

FIGURE 12-9 Only the edge view of an inclined surface will appear true length.

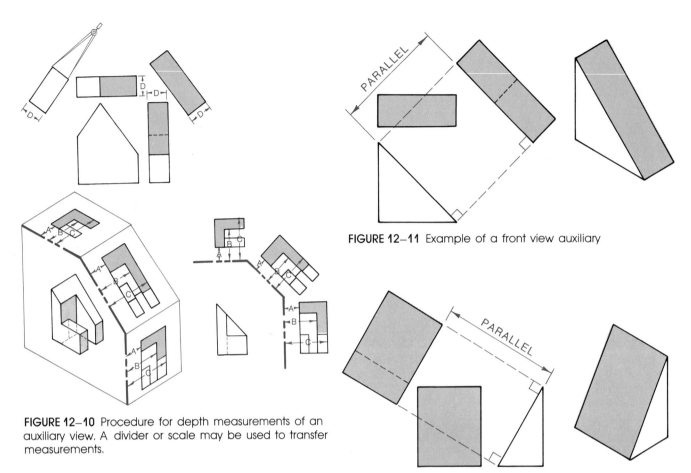

FIGURE 12-10 Procedure for depth measurements of an auxiliary view. A divider or scale may be used to transfer measurements.

FIGURE 12-11 Example of a front view auxiliary

FIGURE 12-12 An example of a side view auxiliary

clined surface is shown as an inclined line representing the surface edge on the front view. The length of this line is true (not foreshortened), therefore, the auxiliary view length can be projected from this line. The true depth of the inclined surface is shown on the top and side views. This depth can be transferred at 90° angles from the length projection line to form the outline of the auxiliary view. Two procedures for transferring width dimensions for an auxiliary view are shown in **Fig. 12-10**.

PRIMARY AUXILIARY VIEWS

Primary auxiliary views are drawings projected directly from any of the six normal orthographic views. The three most common auxiliary views are projected from the front, side, or top normal views. **Primary auxiliary views** are used to define either depth, height, or width details not found in true form in one of the principal views. The true height and width of the object, plus the true length of the inclined plane edge, are shown on the orthographic front view in **Fig. 12-11**. The true size surface of the sloped surface is not shown, therefore, a primary front auxiliary view is projected from the front view to show the sloped surface in true size. This view is projected perpendicularly from the inclined plane edge on the front view.

The normal side view shows the true height and depth of the object, plus the true length of the inclined plane edge **(Fig. 12-12)**. A true size surface of the sloped surface is not shown. A primary side auxiliary view is projected from the side orthographic view to show the sloped surface in true size.

True width and depth dimensions are on the orthographic top view **(Fig. 12-13)**. A primary top auxiliary view is needed to show the sloped surface in true size.

SECONDARY AUXILIARY VIEWS

When surfaces are sloped in several primary auxiliary views, a secondary auxiliary view may be necessary to show the true shape of the inclined surfaces. A **secondary auxiliary view** is a view projected from a primary auxiliary view. To draw a secondary auxiliary view, a primary auxiliary view must first be drawn. The secondary auxiliary view is then projected perpendicularly from a selected true size edge on the primary auxiliary view **(Fig. 12-14)**.

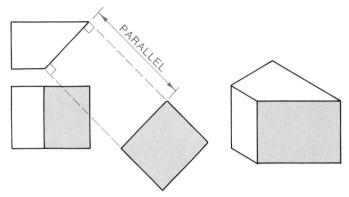

FIGURE 12-13 An example of a top view auxiliary

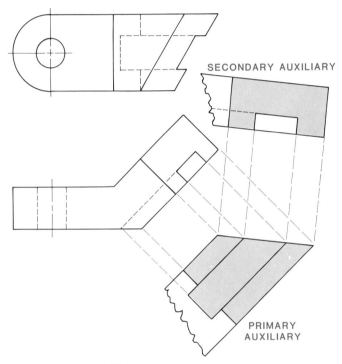

FIGURE 12-14 Multiple auxiliary views may be drawn.

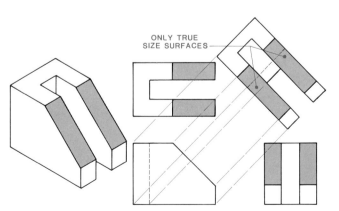

FIGURE 12-15 Only the inclined surface will appear true size in the auxiliary view, because the auxiliary is viewed perpendicular to its inclined surface.

COMPLETENESS OF AUXILIARY VIEWS

In principal orthographic views, all parallel or perpendicular surfaces are shown in true scale, but inclined surfaces are distorted or foreshortened. In an auxiliary view, just the opposite is true. Because the auxiliary plane is parallel to the inclined surface, it will not be parallel to the normal right angle surfaces of the object. Therefore, all surfaces, other than the inclined plane, will appear foreshortened when viewed from the auxiliary plane **(Figs. 12-15 and 12-16)**. For this reason, only the actual inclined surface is usually shown as a partial auxiliary view **(Fig. 12-17A)**. **Figure 12-17B** shows a partial auxiliary view of a digital clock.

Other Angles

In addition to auxiliary views which are projected from normal orthographic views or other auxiliary views, auxiliary views can be projected from any surface. In all cases, the view must be projected perpendicular to the surface to be shown to insure a true shape description. **Figure 12-18** shows a nut with auxiliary views projected from surfaces that are not normal orthographic views.

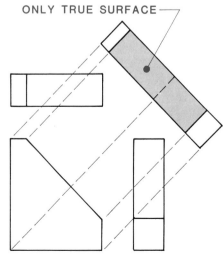

FIGURE 12-16 A full auxiliary drawing

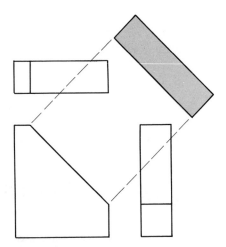

FIGURE 12–17A Partial auxiliary drawing

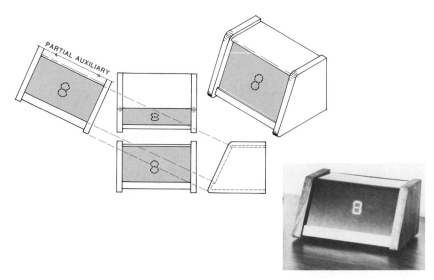

FIGURE 12–17B An auxiliary drawing of a single number digital clock (Courtesy of Michael Robbins)

Notice that the auxiliary planes of projection are all parallel to the surfaces to be drawn. The auxiliary projections are perpendicular to the inclined surface.

Hidden Lines

Normally, hidden (dashed) lines are omitted from auxiliary views as their use may clutter the drawing and make interpretation difficult. However, if needed for clarity, hidden lines can be added to auxiliary views.

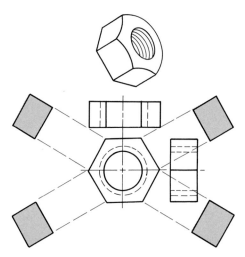

FIGURE 12–18 The number of auxiliary projections are unlimited.

Holes and Ellipses

When a hole is projected from an inclined plane on an orthographic drawing, it will appear as an ellipse (**Fig. 12–19**). To show the hole in its true form, an auxiliary view must be drawn.

Dimensioning

A basic rule for dimensioning is to dimension only true sizes. Never dimension a foreshortened line or surface. Draw the auxiliary view, and place the required dimensions on the true size auxiliary view (**Fig. 12–20**).

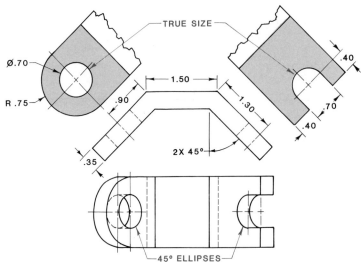

FIGURE 12–19 Projecting circles and ellipses

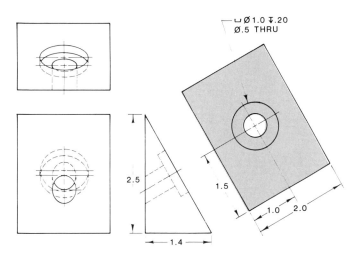

FIGURE 12-20 Dimension only true sizes.

PROCEDURES

The following illustrations will show the procedure to develop several types of auxiliary views.

Figure 12-21 shows the steps for an auxiliary view of a wedge stop.

Figure 12-22 develops the auxiliary view of one surface of a rectangular pyramid.

Figure 12-23 develops the auxiliary view of a truncated hexagonal prism.

Figure 12-24 is the last example, showing the steps to draw the auxiliary view of a truncated cylinder.

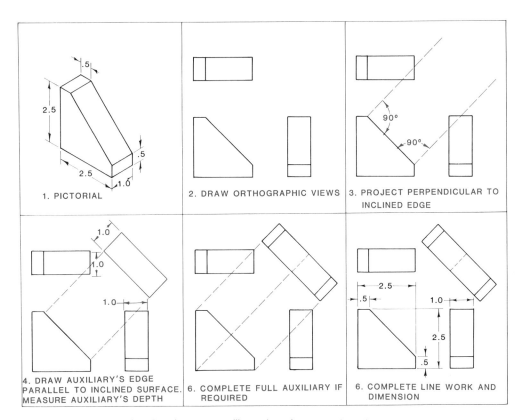

FIGURE 12-21 Steps for drawing an auxiliary view for a wedge stop

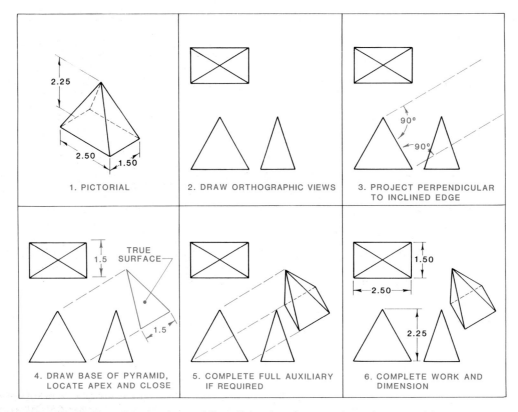

FIGURE 12-22 Steps for drawing a full auxiliary view for a rectangular pyramid

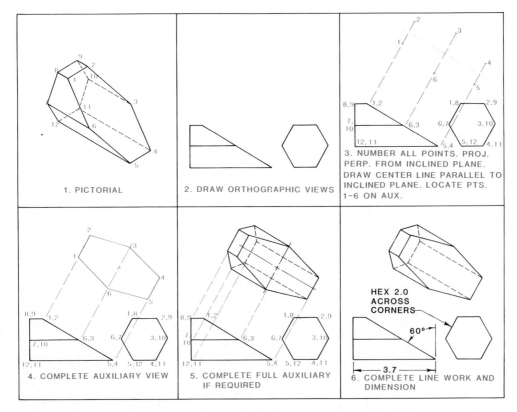

FIGURE 12-23 Steps for drawing an auxiliary view for a truncated hexagonal prism

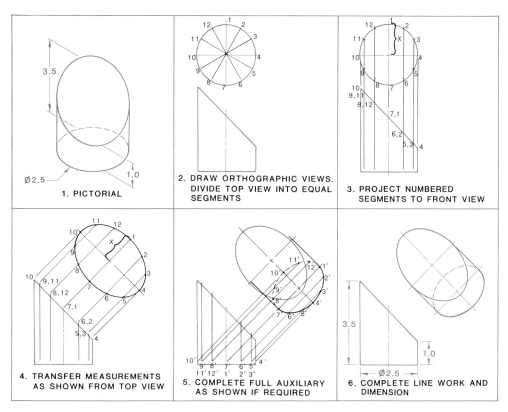

FIGURE 12-24 Steps for drawing an auxiliary view for a truncated cylinder

REVOLUTIONS

Up to now, the positioning of an object's multiview drawing for clarity may be done with three procedures. Orthographic projection is the basis for all multiview drawings (Chapter 8). All projections are perpendicular between adjacent views. Descriptive geometry is the theory of doing graphical solutions of points, lines, and planes in space using the principles of orthographic projection (Chapter 13). Auxiliary projections can be drawn at any angle off a primary orthographic view or auxiliary view, as explained earlier in this chapter.

A **revolution drawing** is a fourth procedure used to reposition a view from the primary orthographic drawing, and project its repositioned view **(Fig. 12-25)**. The drafter must decide the most advantageous viewing positions.

Any view can be rotated about a theoretical axis placed on the view **(Fig. 12-26)** or on a corner of the

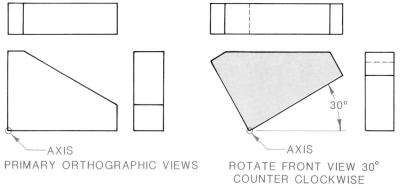

FIGURE 12-25 Rotation of a view using a corner as an axis

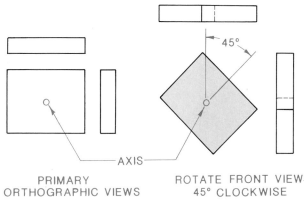

FIGURE 12-26 Rotation of a view with the axis on the view

view (Fig. 12–25). The views may be rotated any number of degrees clockwise or counter-clockwise.

The first view projected from a rotated primary orthographic view is called a **primary revolution (Fig. 12–27)**. Rotating a primary revolution and projecting a new view is called a **successive revolution.** Successive revolutions may be used to rotate an object to any desired position **(Fig. 12–28)**. There is no limitation on the amount of successive revolutions that can be drawn.

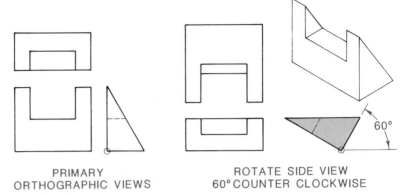

FIGURE 12–27 The first rotation of a primary orthographic view is called a primary revolution.

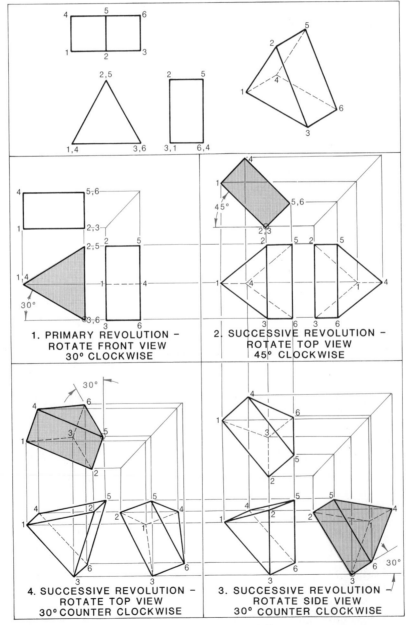

FIGURE 12–28 When drawing successive revolutions, any view may be rotated any number of degrees to achieve specific placement of revolved views.

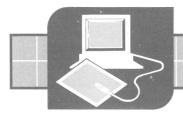

COMPUTER-AIDED DRAFTING & DESIGN

COMPUTER-AIDED AUXILIARY VIEWS

Developing auxiliary views with CAD is similar to the layout of orthographic views. Auxiliary views are composed of lines, circles, arcs, and dimensions, as are other normal views of projection. The major difference is that the auxiliary views are projected from an inclined plane. Another difference is that it is common for auxiliary view drawings to include ellipses. The construction of ellipses on CAD systems varies greatly from one system to another.

The CAD operator must be able to accurately draw parallel lines at various angles (perpendicular to the edge view of the inclined plane). One way to accurately draw inclined lines on a CAD system is to establish a grid at the angle of the desired lines. This grid method is used in the following examples:

Figures 12–29 through 12–32 illustrate the construction of a simple auxiliary drawing. The starting point is the orthographic projection of the object **(Fig. 12–29)**. Note that

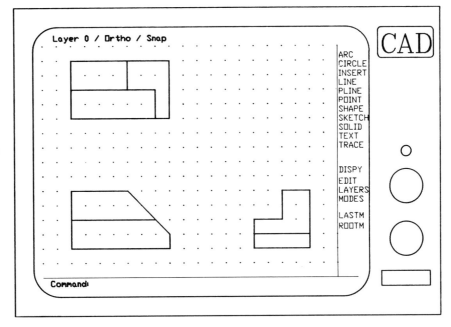

FIGURE 12–29 Draw or call up an existing multiview working drawing stored in the data base.

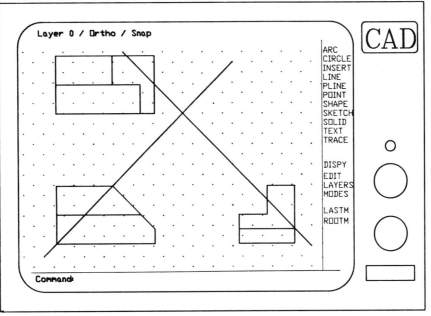

FIGURE 12–30 Reset the grids to be parallel to the inclined surface. Note that the cursor lines will match the grids.

204 Drafting in a Computer Age

in this illustration, the grid pattern is horizontal and vertical. The next step is to establish the inclined grid to match the inclined plane of the object **(Fig. 12–30)**. The auxiliary view is completed by projecting from the inclined plane using the new grid **(Fig. 12–31)**. This makes the drawing of the auxiliary view as easy as a normal view. To finish the drawing, replace the inclined grid with a normal grid and add the dimensions **(Fig. 12–32)**.

A more complex auxiliary drawing is illustrated in Figs. 12–33

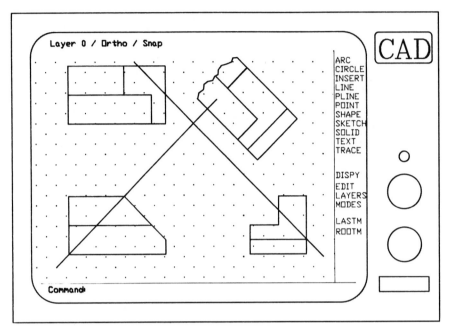

FIGURE 12–31 Transfer the depth dimensions, and with the LINE command, complete the auxiliary.

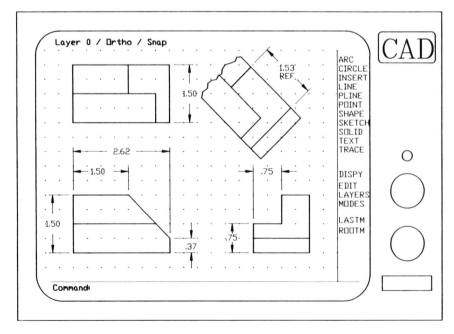

FIGURE 12–32 Remove the projection lines, reset grid to NORMAL, and add dimensions with the AUTOMATIC DIMENSIONING.

through 12–37. The auxiliary view is projected from the top view and the object contains a hole. The hole will project as ellipses on two of the views. The procedure is as follows:

Step 1. Construct the normal orthographic views. Use all the available CAD aids. These include GRID, SNAP, and ORTHO, as well as the basic entities **(Fig. 12–33)**.

Step 2. Change the grid to match the angle of the inclined plane. Note that the screen cursor also changed and matches the new grid angles **(Fig. 12–34)**.

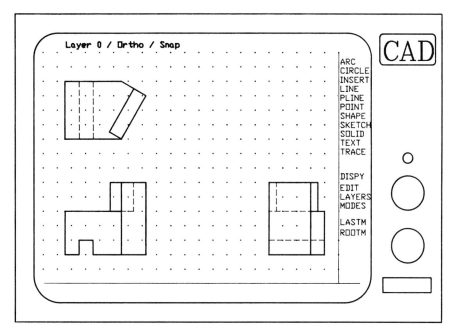

FIGURE 12–33 Draw or call up the stored multiview drawing.

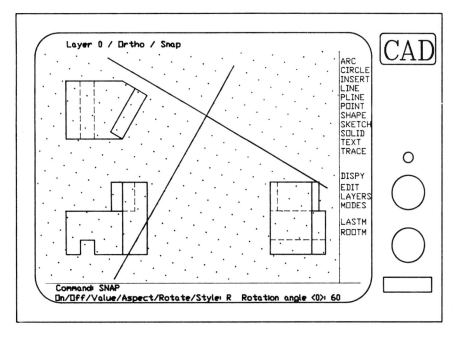

FIGURE 12–34 Reset the grids to be parallel to the inclined surface.

Step 3. Construct the auxiliary view and add details to the view, including hidden lines and the hole **(Fig. 12–35)**.

Step 4. Project the hole (as ellipses) back to the normal views. The method of constructing the ellipses is usually a subfunction of the CIRCLE entity in the CAD software **(Fig. 12–36)**.

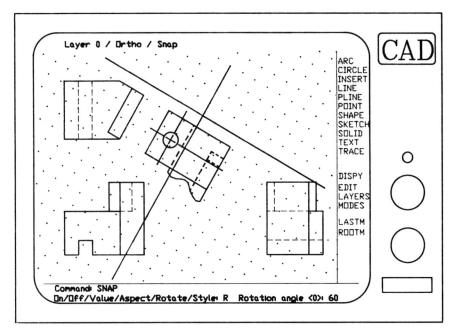

FIGURE 12–35 Construct the auxiliary view with the CIRCLE, LINE, HIDDEN LINE, and CENTER LINE commands.

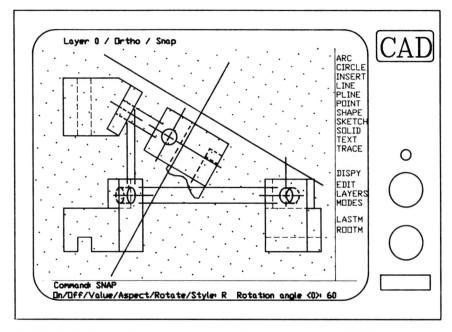

FIGURE 12–36 Project the hole back to multiviews. Go to the ELLIPSE command and place on the top and side views.

Step 5. Turn off the construction lines used to place the ellipses. Return to the normal grid and add the necessary dimensions to the drawing **(Fig. 12–37)**.

In Fig. 12–37, the grid has also been turned off to make the drawing easier to read. The final drawing will appear as in Fig. 12–37 when plotted.

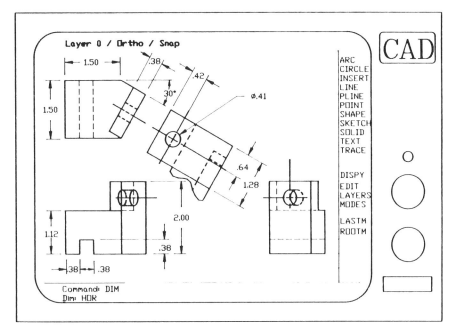

FIGURE 12–37 Remove the projection lines. Go to the AUTOMATIC DIMENSIONING command and dimension. Turn off the grids. Check the drawing and make a hard copy.

EXERCISES

1. Sketch the multiview drawings in **Fig. 12–38**. Sketch the auxiliary views for each problem.
2. Using your drawing instruments, draw the multiview drawings of the octagon, hexagon, and cylinder in **Fig. 12–39**. Draw their auxiliary views. Numbering all the corners will help with the development of the auxiliary views.
3. With drafting instruments, draw the multiview drawings and their auxiliary views as needed for **Figs. 12–40 through 12–51**.
4. Using the concepts of the cartesian coordinate system, make a freehand sketch on $1/4''$ grid paper from the information in the data table in **Fig. 12–52**.
5. Make up a data table for the cartesian coordinate system of the drawing in **Fig. 12–53**.
6. With a CAD system, duplicate as closely as possible the procedures in **Figs. 12–40 through 12–45**.

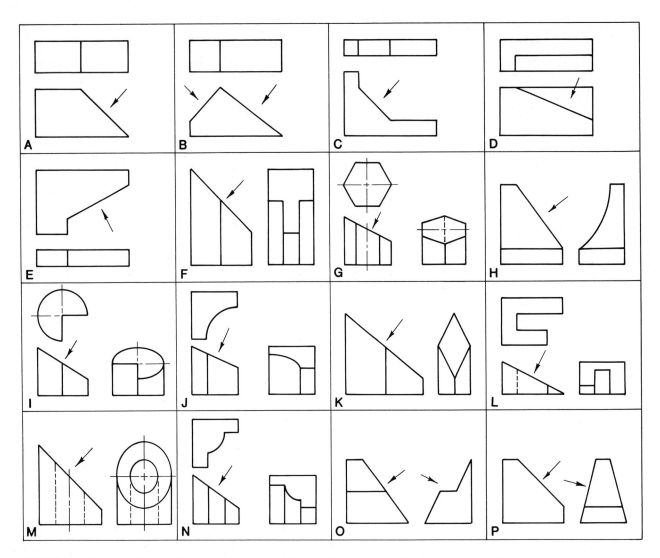

FIGURE 12–38 Sketch the multiview drawings and the auxiliary solution for each exercise.

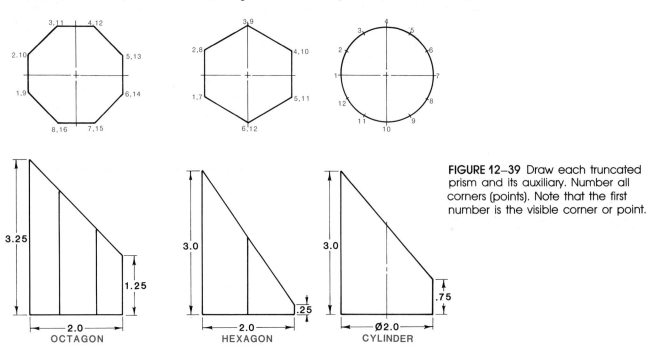

FIGURE 12–39 Draw each truncated prism and its auxiliary. Number all corners (points). Note that the first number is the visible corner or point.

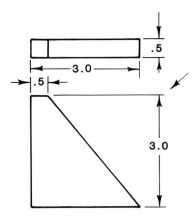

FIGURE 12-40 Draw the multiview drawings and auxiliary view.

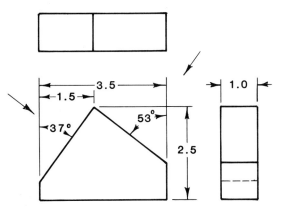

FIGURE 12-41 Draw the multiview drawings and two auxiliary views.

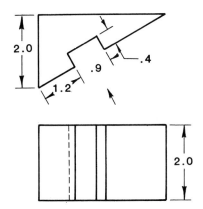

FIGURE 12-42 Draw three-view multiview drawings, isometric, and auxiliary views.

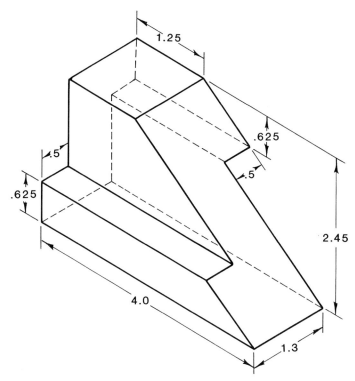

FIGURE 12-43 Draw the multiview drawing and auxiliary view of the inclined surface.

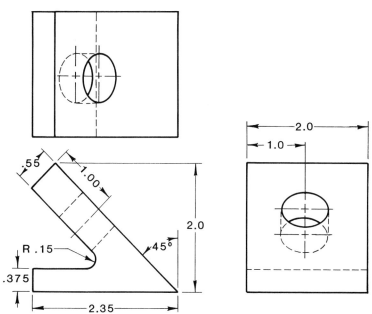

FIGURE 12-44 Draw the multiview drawing and an auxiliary view of the inclined surface.

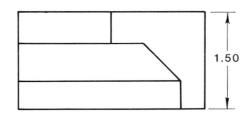

FIGURE 12–45 Draw the multiview drawing and auxiliary views of the front and side views.

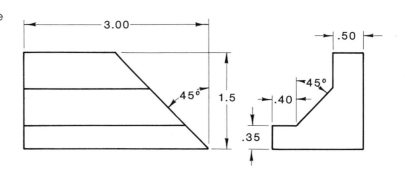

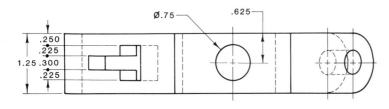

FIGURE 12–46 Draw the multiview drawing and two partial auxiliary views for the angled clamping jig.

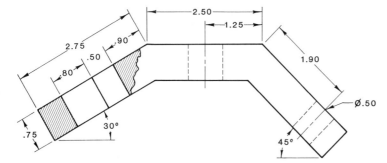

FIGURE 12–47 Draw the multiview drawing and a partial auxiliary view for the angle slot's face.

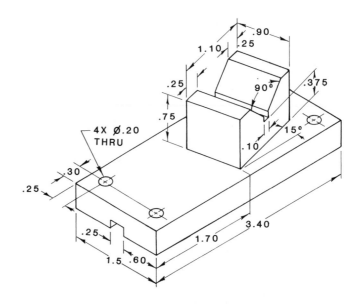

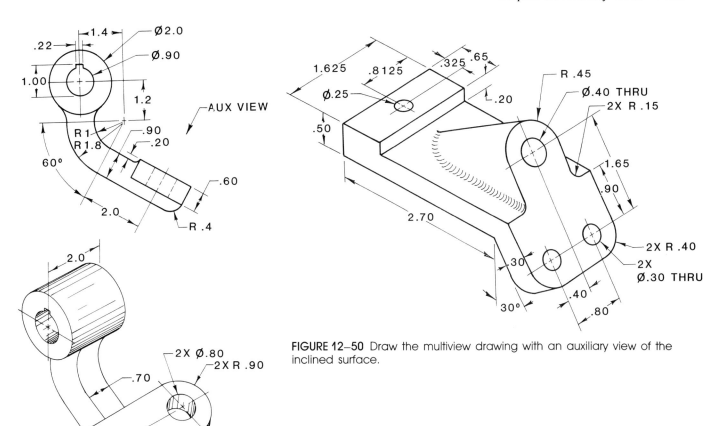

FIGURE 12-50 Draw the multiview drawing with an auxiliary view of the inclined surface.

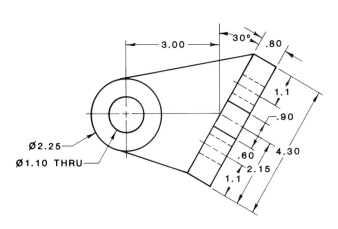

FIGURE 12-48 Draw the multiview drawing with an auxiliary view of the base pad.

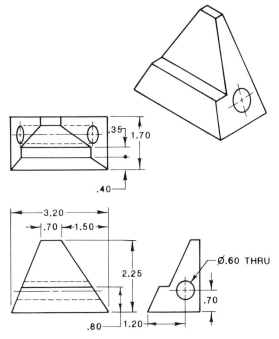

FIGURE 12-51 Draw the multiview drawing and right-side auxiliary view of the angle insert guide.

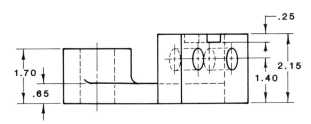

FIGURE 12-49 Draw the multiview drawing with an auxiliary view of the angled base. Select the best direction for auxiliary view.

212   Drafting in a Computer Age

DATA TABLE					
LN	X	Y	GO TO	X	Y
1	0	0	—	0	7
2	0	7	—	4	3
3	4	3	—	6	3
4	6	3	—	6	0
5	6	0	—	0	0
			NEW START		
1	1	8	—	2.5	9.5
2	2.5	9.5	—	6.5	5.5
3	6.5	5.5	—	5	4
4	5	4	—	1	8
			NEW START		
1	8	7	—	10	7
2	10	7	—	10	0
3	10	0	—	8	0
4	8	0	—	8	7
			NEW START		
5	8	3	—	10	3
			FINISH		

FIGURE 12–52 Lay out the multiview drawing and the partial auxiliary view from the data table on 1/4" grid paper.

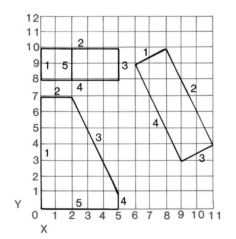

FIGURE 12–53 Complete the data table.

DATA TABLE					
LN	X	Y	GO TO	X	Y
1	0	0	—	7	0
2	7	0	—		
3			—		
4			—		
5			—		
			NEW START		
1			—		
2			—		
3			—		
4			—		
5			NEW START		
1			—		
2			—		
3			—		
4			—		
			FINISH		

 **KEY TERMS**

Auxiliary plane
Auxiliary view
Foreshortened
Hinged planes

Primary auxiliary view
Primary revolution
Revolution drawing

Secondary auxiliary view
Successive revolutions
True size edge

the world of CAD

What do you plan to do when you graduate from high school? Are you planning to go to work, to college, or both? What field will you pursue? I know these are tough questions to answer—I faced the same questions ten years ago.

Today, I am the Training Manager for Versacad Corporation, an independent company of Prime Computer, Inc., and I love my job. Versacad Corporation manufactures and sells VersaCAD software. Perhaps you have heard of VersaCAD or used it in a class? VersaCAD is a very popular computer-assisted design software package used in many schools and companies throughout the world. Our training department teaches people like you how to use VersaCAD.

Ten years ago, I would not have predicted that today I would be managing the training department for a CAD company. Let me tell you how I got here.

After I received a B.A. degree in Writing Arts, I realized the rest of the world was entering the computer age. In order to improve my computer literacy, I went back to college. I took math, computer programming, and for some reason that I can't remember, I took a drafting course. Did I like drafting? Sort of. Was I good at it? Sort of. Am I glad now that I took it? YOU BET I AM!

By developing technical skills and combining them with my writing skills, I became a Technical Writer. Technical writers are people who write technical manuals such as computer manuals, appliance-maintenance manuals, car-repair manuals, etc. The company I worked for specialized in technical manuals for the computer industry. One of my first assignments was to work with a team of people who were developing a CAD system. I had never been exposed to CAD. When I sat down to learn it, I was fascinated by how easy it was to do everything that I had struggled to do manually in my drafting class.

As a Technical Writer, I spent over four years writing manuals for many different types of software—not just CAD. With each new project, I learned more about the computer industry.

I wanted to teach people how to use computers to help them in their jobs. In order to write manuals that were more like teaching guides than dry reference manuals, I took graduate courses in education and training.

Equipped with my newly acquired training skills, I embarked on a new career. At that time, Versacad Corporation was looking for some-

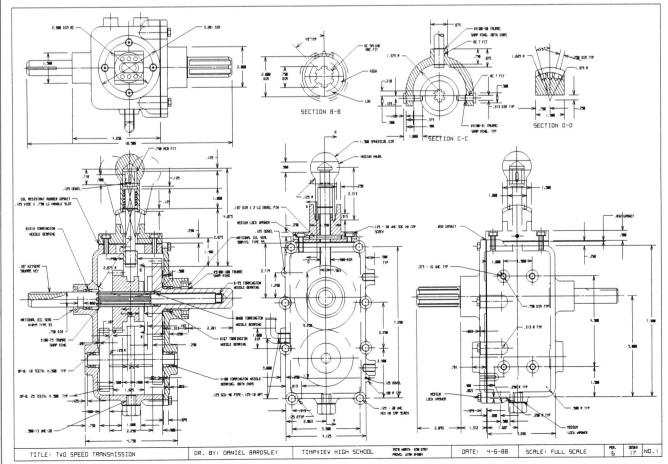

FIGURE 1 The winner of Versacad Corporation's student drawing contest in 1988 was this drawing of a two-speed transmission by Daniel Bardsley of Provo, Utah.

one to develop and teach courses on Versacad products. I applied for the job, and here I am, a manager of a successful training department for a great company and loving every minute of it.

But how can CAD make a difference in your life? The major industries that use CAD are mechanical, manufacturing, civil, structural, heating/ventilating/air conditioning, architectural, and aerospace. What do all of these industries do? Design and build everything from jet engines and high-rises to freeways and space ships. CAD is also used for designing furniture, plumbing fixtures, playground equipment, amusement park rides, sailboats, circuit boards, musical instruments—the list goes on and on.

Why is CAD so popular? Because very few people create every drawing perfect the first time. Do you ever have to make changes to your drawings? Do you ever make mistakes? Do you ever wish you could modify an existing drawing to create a new drawing? It's easy with CAD because CAD automates the drawing process. You can rotate, copy, erase, etc. And there are CAD systems for every kind of computer from personal computers through to mainframes.

What types of careers are available in CAD? You can work for a company that manufactures and sells a CAD product, like I do, or you can work for one of the many companies, both large and small, that use CAD. At Versacad Corporation, we employ programmers, customer assistance representatives, teachers, writers, marketing and publicity representatives and salespersons, to name a few.

Companies that use CAD employ CAD operators who create accurate drawings from engineers' drafts, programmers who tailor the CAD program specifically for the company, artists who use CAD to develop publicity and presentation materials, system administrators who organize and track the drawings, and computer technicians who maintain the computers CAD runs on. These are just a few of the CAD-related jobs available.

A rewarding career in CAD is waiting for you. Should you go for it? YOU BET!

Each year, Versacad Corporation holds a student drawing contest. The winner of 1988's contest (and a $1000 Grand Prize) was Daniel Bardsley of Timpview High School, Provo, Utah, who drew the two-speed transmission (Figure 1). Among the 24 runners-up was Brad Hamilton of Burney High School, California, with a map (Figure 2).

(Courtesy of Maria Morrissey, Versacad Corporation, Huntington Beach, CA)

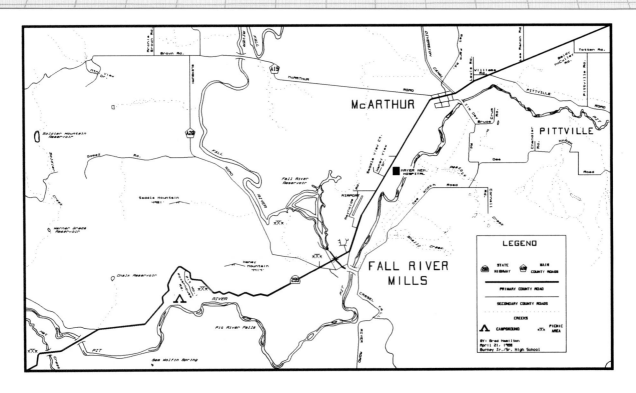

FIGURE 2 Among the 24 runners-up in Versacad Corporation's student drawing contest in 1988 was this map by Brad Hamilton of Burney, California.

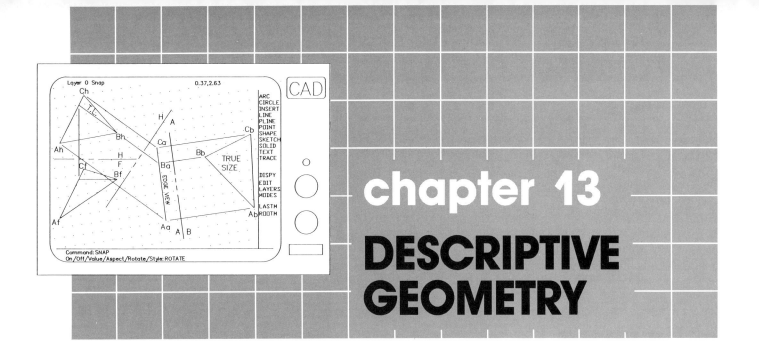

chapter 13
DESCRIPTIVE GEOMETRY

OBJECTIVES

The student will be able to:

- draw a point in space onto various planes of projection
- draw a line in space onto various planes of projection
- draw a point view of a line
- draw the true length of a line
- draw an edge view of a surface
- draw a true size surface
- lay out and solve a descriptive geometry exercise on a CAD system

INTRODUCTION

Descriptive geometry is a study of the three-dimensional relationship of points, lines, angles, and surfaces as drawn on two-dimensional surfaces. The principles of descriptive geometry are used to graphically solve problems, and gain information about geometric forms and their position in space. Since descriptive geometry is directly related to orthographic and auxiliary projection, a basic understanding of these concepts is a vital prerequisite to understanding the material presented in this chapter. **Figure 13–1** graphically defines the geometric terms commonly used in the development of graphic solutions using descriptive geometry. Most graphic problems can be solved by finding the true length and point projection of lines, and the true size and edge view of planes.

POINTS IN SPACE

Geometric forms are all constructed from a series of points. Understanding the precise location of a single point is the first step in understanding the principles of descriptive geometry.

When a **point in space** is theoretically surrounded by the **frontal plane, horizontal plane,** and **profile plane,** it may be projected perpendicularly to the face of the planes **(Fig. 13–2)**. It is extremely important to understand why X = X and

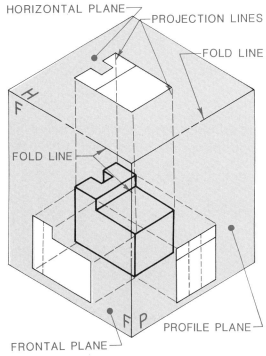

FIGURE 13–1 Graphic terms for descriptive geometry

213

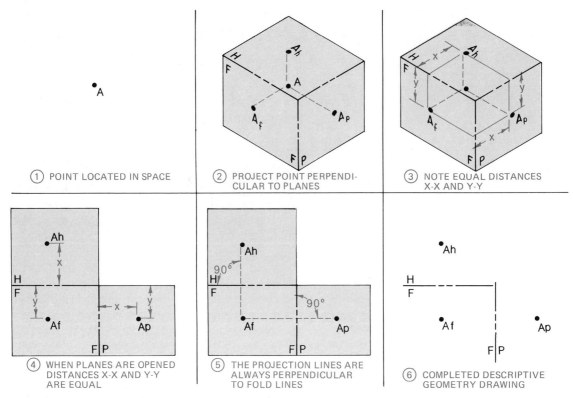

FIGURE 13-2 Descriptive geometry graphics layout for a single point in space

Y = Y. Study Fig. 13-2 until it is clear. When the planes are opened flat on to a two-dimensional surface, the distances of X = X and Y = Y remain the same. It is possible to solve the location of a point to a third plane when the point's location is known on two planes. This principle in Fig. 13-2 is the basis of descriptive geometry. When you can solve for one point, you can solve for many points, and points make up drawings.

FIGURE 13-3 An auxiliary plane

GEOMETRIC PLANES

Planes used in descriptive geometry include the normal orthographic planes, plus auxiliary planes. There are only six normal orthographic planes; however, the number of auxiliary planes is unlimited. An **auxiliary plane (Fig. 13-3)** is a plane that lies on any angle other than the normal orthographic planes. **Figure 13-4** shows several examples of auxiliary planes projected from orthographic

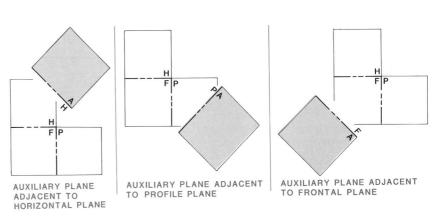

FIGURE 13-4 An auxiliary plane may be taken off any plane at any position.

Chapter 13 Descriptive Geometry 215

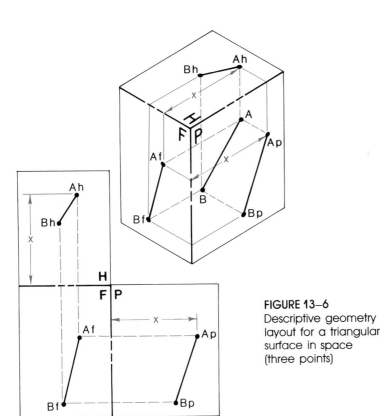

FIGURE 13–5 Descriptive geometry layout for a line in space (two points)

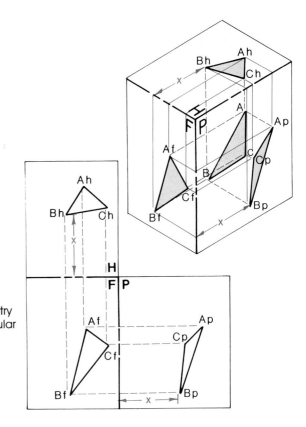

FIGURE 13–6
Descriptive geometry layout for a triangular surface in space (three points)

planes. Any angle may be used with a **fold line** of an auxiliary plane.

GEOMETRIC LINES

Think of a line as a series of connected points. A straight line is the shortest distance between any two points. Only two points are needed to create a straight line, since all points located on a straight line are aligned. **Figure 13–5** shows the projection of two points in space to create a straight line.

The connection of three points in space creates a **triangle (Fig. 13–6)**. It logically follows that any two-dimensional plane surface or three-dimensional solid object can be created by the projection and connection of multiple points in space **(Fig. 13–7)**.

Curved lines are comprised of a series of unaligned points. A curved line is actually a series of very short straight lines connecting closely

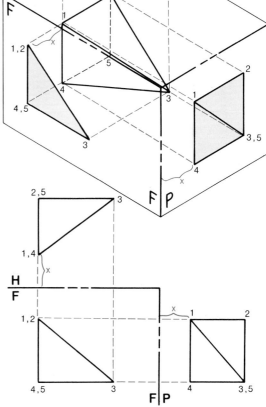

FIGURE 13–7 Descriptive geometry layout for a five-point solid in space

spaced points. Closely spaced points create smooth curves, and widely spaced points create ragged curves.

True Line Length

Lines perpendicular and parallel to normal orthographic planes are all **true length.** Lines on auxiliary planes that are oblique or inclined, are not true length when viewed from a normal plane of projection. However, lines on auxiliary planes are true length when viewed perpendicular to the adjacent plane.

When a fold line is parallel to a line in space, the line on that plane will be true size. Therefore, to find the true length of a line on any plane, locate the fold line parallel to the line. Project the line to the adjacent plane where the lines becomes true length **(Fig. 13–8).**

Point View of a Line

Lines on a plane are represented by points on all other perpendicular planes. Think of a point on a line as a cross section of the line. The **point view** of a line can be drawn by placing the fold line perpendicular to the true length line. Project the two points of the line which will appear as a single point **(Fig. 13–9).** In Fig. 13–9, the fold line is perpendicular to the true length line. All X and Y distances are equal.

Edge View

To find the **edge view** of a plane, follow the steps in **Fig. 13–10.** Establish a true length line within the plane's surface. Locate a point view of the true length line using the procedures in Fig. 13–9. Locate the third point and connect the points with a line. This creates an edge view of the triangular plane.

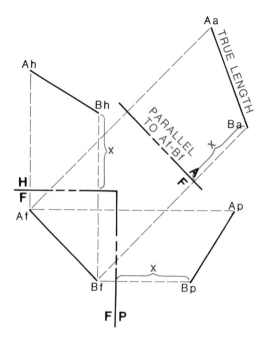

FIGURE 13–8 When the fold line is parallel to a line in space, the line on that projected plane will be true size.

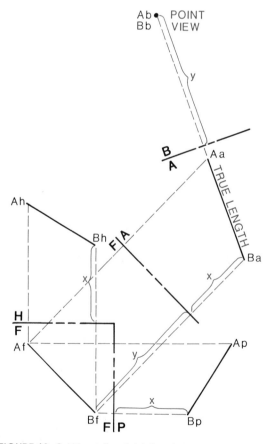

FIGURE 13–9 When the fold line is perpendicular to the true length line, the projection will be a point view. Note that all X distances are equal and all Y distances are equal.

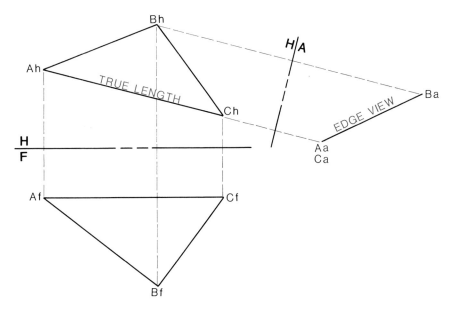

FIGURE 13-10 The fold line F–H is parallel to edge Af–Cf; therefore, edge Ah–Ch is true length on the horizontal plane. Fold line H–A is perpendicular to the true length line Ah–Ch; therefore, true length edge Aa–Ca will appear as a point view. The surface Aa–Ba–Ca will appear as an edge view on auxiliary plane A.

True Size of a Surface

Drawing the true size of a plane is an important graphic solution procedure used in descriptive geometry. To draw the **true size** of a surface, find an edge view of the surface. Draw a fold line parallel to the edge view. Project the surface to the new plane and it will measure true size and shape **(Fig. 13–11)**.

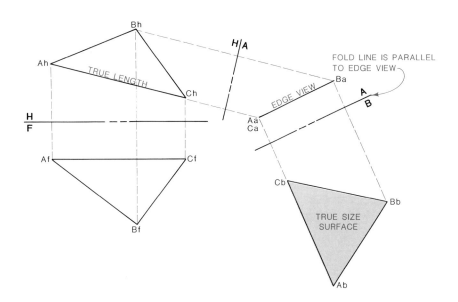

FIGURE 13-11 Once the edge view is drawn, the placement of a fold line parallel to it will show the true size of the surface on auxiliary plane B.

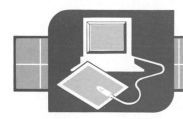

COMPUTER-AIDED DRAFTING & DESIGN

COMPUTER-AIDED DESCRIPTIVE GEOMETRY

Since solving descriptive geometry problems consists primarily of drawing series of auxiliary views, CAD procedures are similar to those used in manually drawing auxiliary views. CAD is an effective tool for solving descriptive geometry problems because of the accuracy that can be achieved. CAD operators must be extremely careful in drawing projections to insure that all geometric data is accurate. If the CAD operator's accuracy is correct, the result will be accurate. To aid in the construction of the various geometric projections, the grid should be rotated to align with the reference (fold) line (Figs. 13–14 and 13–15, page 219). Note that the grid is aligned with the reference line (fold line).

To find the true size of a plane on a CAD system, follow these steps:

Step 1. Draw the problem. Extreme care must be taken to insure the problem is set up accurately. Label the views **(Fig. 13–12)**.

Step 2. Draw a horizontal line in the frontal view, and find its resulting true length line in the horizontal view **(Fig. 13–13)**.

Step 3. Draw a reference line perpendicular to the true length line. Rotate the grid so it is aligned with the new reference line. Using projection lines perpendicular to the reference line and transferring points to the new view, construct the edge view. Label the edge view **(Fig. 13–14)**.

Step 4. Draw a reference line parallel to the edge view. Rotate the grid so it is aligned with the new reference line. Draw projection lines and transfer points to construct the true size of the plane. Label the true size view **(Fig. 13–15)**.

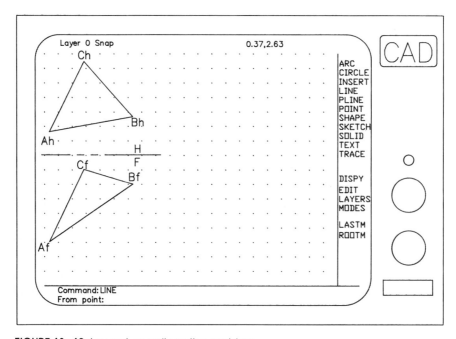

FIGURE 13–12 Lay out or call up the problem.

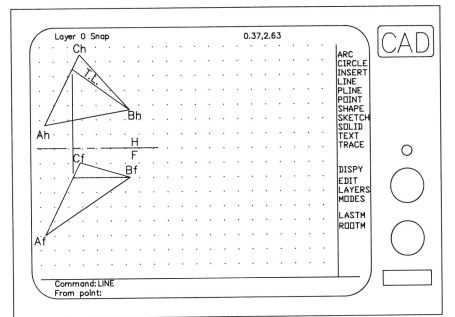

FIGURE 13–13 Find the true length of the horizontal line in the front plane as shown.

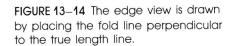

FIGURE 13–14 The edge view is drawn by placing the fold line perpendicular to the true length line.

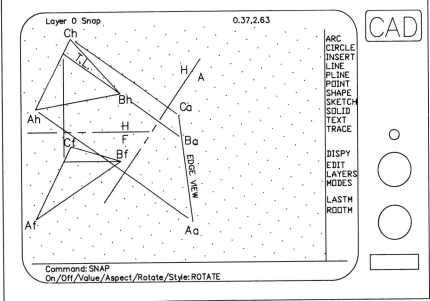

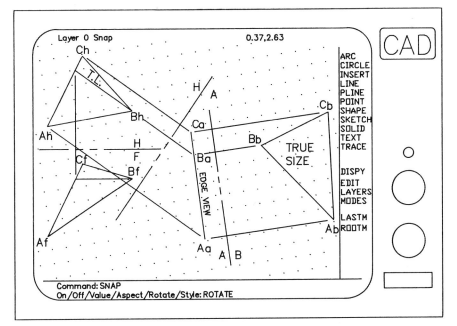

FIGURE 13–15 The true size of the surface is drawn by placing the fold line parallel to the edge view.

EXERCISES

1. Sketch the graphic solutions for the exercises shown in **Fig. 13–16**. Sketch the exercises two or three times larger than shown.
2. With instruments, find the graphic solutions for the problems in **Fig. 13–17**. Triple the size of the exercises with a scale or dividers.
3. Solve the problems in Fig. 13–17 using a CAD system.
4. Find the point in space on the profile plane if the point on the frontal plane is in the exact center of the plane, and the point on the horizontal plane is .75″ above the fold line. Draw all planes 2″ square.
5. Make the following ECOs (Engineering Change Orders—refer to Chapter 17) to **Fig. 13–18**: Change the length of tunnel **A—B** to 35′ (shorten from bottom end). Change the angle of tunnel **A—B** to 15° (rotate the tunnel clockwise from the lower end).
6. Complete the horizontal plane for **Fig. 13–19**. Record the horizontal plane's data into the data table using .1″ grids.
7. Complete the profile plane for **Fig. 13–20**. Record the profile's data into the data table using .1″ grids.

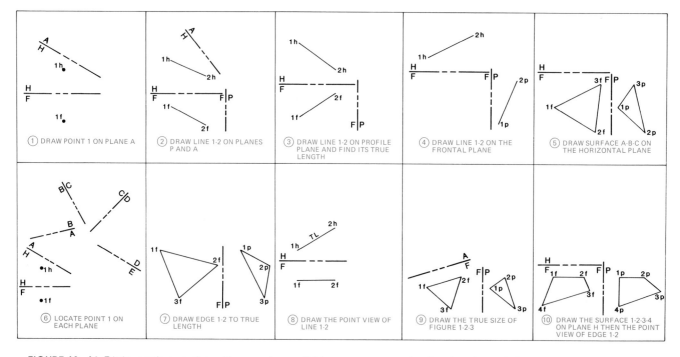

FIGURE 13–16 Triple each exercise with a scale or dividers on a separate drawing format. Sketch or draw the solutions with instruments.

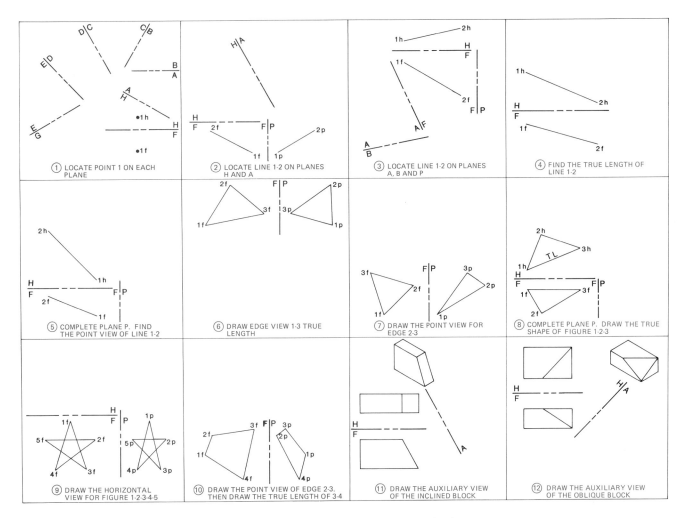

FIGURE 13–17 With drawing instruments, triple the size of each exercise and solve.

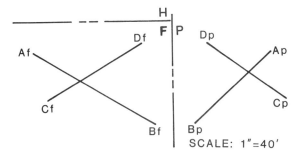

FIGURE 13–18 Lay out the drawing and make the ECOs as specified in Exercise 5.

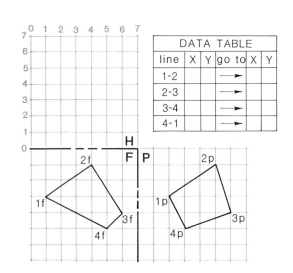

FIGURE 13–19 Complete the drawing on the horizontal plane, and record the coordinate points into the data table.

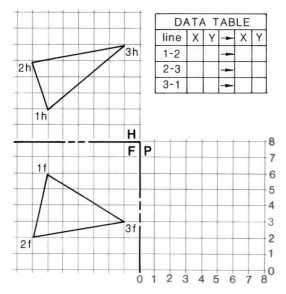

FIGURE 13-20 Complete the drawing on the profile plane, and record the coordinate points into the data table.

KEY TERMS

Auxiliary plane
Curved line
Descriptive geometry
Edge view
Fold line

Frontal plane
Horizontal plane
Point in space
Point view

Profile plane
Triangle
True length
True size

the world of CAD

From network television to motion pictures to theme park design to the Broadway stage, computer-aided design is one of the newest and most important design tools for the entertainment industry. The UCLA School of Theater, Film and Television is adapting an AutoCAD program to fit specialized applications in this diversified field. Scenic designers, lighting designers, and costume designers are being trained to design scenery, make complex lighting plots, and draft costume patterns with the aid of the computer and CAD technology. This recent introduction of CAD into the entertainment media is yielding design techniques never before possible.

CAD AND SCENIC DESIGN

Some of these techniques are being applied to the preproduction planning phase. In director/designer conferences, the script, its themes, and how the design of the production should support these elements are discussed. A floor plan (or staging plan as it is called in film and television) is one of the drawings that the theatrical scenic designer and television or motion picture art director uses to work through these issues. In the past, each time the designer and director met, the designer needed to manually draft a new plan to incorporate any desired changes. With the computer, the designer and director move the scenic elements around as they are discussing the possibilities. Repositioning walls, stairs, and furniture is accomplished quickly. There is no waiting for another stage plan to be revised. The electronic process greatly accelerates the early design phases allowing the designer and director to get a head start on the later stages of preproduction, Figure 1.

The finalized digitized stage plan is used to determine which camera positions and angles are best. Symbols for cameras and different lens angles are layered over an existing stage plan in order to determine the best shots and positions of scenic elements. The hundreds or thousands of "set-ups" (the positioning of cameras and lighting for each new shot) are planned out quickly and accurately. No longer is time and money wasted as cast and crew wait for shots to be determined on location.

CAD technology is speeding up the drafting process and assuring more precise drawings in another area of scenery—stock scenery. These are walls, doors, windows, columns, and other scenic elements that have been built for past productions, then saved and stored for future use in different films or plays. The manual drafter has to go to the "scene dock" where these units are kept, locate each piece, and measure it before redrafting it. At UCLA, a catalog of stock scenery and other scenic elements is being developed on a series of disks called AutoCAT. Each designer receives a book containing a plotted drawing of each scenic element complete with dimensions and other important information such as construction materials, weight, etc. Inserted in this catalog of stock drawings is a disk with the same information. The designer looks through the catalog to choose perhaps a door for a set wall in a new set design for an upcoming production. Rather than redraft the door for the new construction drawings, the designer copies the door drawing from the disk and easily inserts the exact same door drawing into the new design. The designer can just as easily try out other door styles that might work in the new design. These different combinations can be printed out and shown to the other members of the production team. This is a new technique unique to CAD that saves hours of searching, measuring, and drafting time, and insures an accurate drawing. And all of this is accomplished without leaving the workstation.

CAD AND LIGHTING DESIGN

CAD is also having an impact in the field of lighting design. One facet of this is the light plot. This drawing is a floor plan of the set in the theater or studio that shows the positions of the hundreds of lighting instruments essential to light the actors and the scenery, Figure 2. In addition to the plot, a hook-up sheet, or instrument schedule, is developed. This instrument schedule lists each instrument, its I.D. number, lamp wattage, circuit number, the color of the filter to be placed in the instrument, the lens configuration, focal length, lamp reflector configuration, what dimmer the instrument is to be plugged into,

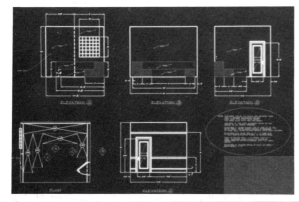

FIGURE 1
Set elevations

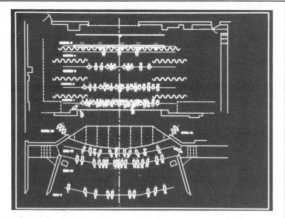
FIGURE 2 Lighting plot

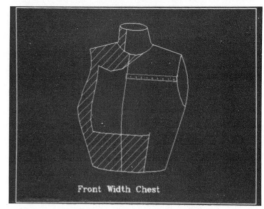
FIGURE 3 Help screen from PATTERN MAKER

and where on stage the instrument is to be pointed.

The task of precisely drawing these hundreds of tiny lighting instrument symbols and developing this complex chart without error is overwhelming and time-consuming. By adapting AutoCAD's symbol attribute capabilities to this task, UCLA has eliminated much of the tedium and the possibility for error. Their designers use a library of symbols that represent the dozens of different lighting instruments available to them. Each symbol has attributed to it all information required for the instrument schedule. After an instrument is placed in its desired location on the plot, the information for that instrument is typed in. This information is automatically written onto the plot inside and adjacent to the instrument symbol. After repeating this process for each instrument, a precision-drawn, highly accurate light plot is printed. With the completed plot and the aid of a data base, information about each instrument is printed out in the correct instrument schedule form. The designer can have this form instantly. Since the source of the information on both the plot and the instrument schedule is the same initial input there are no discrepencies.

The application of CAD technology to lighting design has become so popular that many software developers have created special lighting design programs that further streamline the process. These AutoCAD-based programs introduce new commands, contain pre-packaged symbol libraries, and have built-in dedicated data bases with instrument schedule templates ready to go. Many tasks that are confusing on a lighting plot are taken care of by these programs, simplifying the designer's task.

CAD AND COSTUME DESIGN

Computer-aided design is only beginning to affect costume designing. Costume designers work with inexact sketches and bolts of flowing fabric. If CAD is to find a way of helping the costume designer, it will not find an answer in the more traditional theatrical CAD applications involving orthographic projections and mathematical computations.

The UCLA Laboratory for Technology in the Arts looked into an application. Focused on was the basic costume fabric pattern, or "flat", which is at the core of all costume construction. The making of a useful costume fabric pattern is a combination of skills that include taking the actor's measurements, and factoring in the actor's movements during the production. It involves a thorough knowledge of period costume detail and how it translates into a pattern. It also involves years of specialized training and experience in drafting the hundreds of pieces of fabric that make up theatrical costumes.

PATTERN MAKER is the program that was developed at UCLA by the Laboratory for Technology in the Arts for the costume designer. PATTERN MAKER can store hundreds of flats representing the pieces of fabric that make up a costume. The costume designer calls up the flats needed for a particular costume. It might be a pair of men's 18th century trousers or maybe a woman's skirt from the 1950s. Each flat has a "slide" showing a generic pattern and the measurements needed for AutoCAD to construct the pattern. The designer obtains these measurements from the actor and inputs them to PATTERN MAKER which quickly draws the actual pattern on the screen, Figure 3. With the streamlined commands in the PATTERN MAKER program, the designer alters the pattern in any manner. Cuts are made in the fabric with electronic scissors, button holes are placed on a shirt, or any number of other refinements are plotted out automatically in full scale on a large format pen plotter. This application promises to open up the world of complex pattern detail and the rich history of clothes to any costume designer.

Computer-aided design is a standard tool in other fields such as architecture and engineering. Only now in the late 1980s is it beginning to be a familiar instrument for the scenic, lighting, or costume designer. Increasingly, more designers in all areas of the entertainment arts are discovering the tremendous benefits that CAD offers.

(Courtesy of Rich Rose, Associate Professor of Design and Production, UCLA School of Theater, Film and Television)

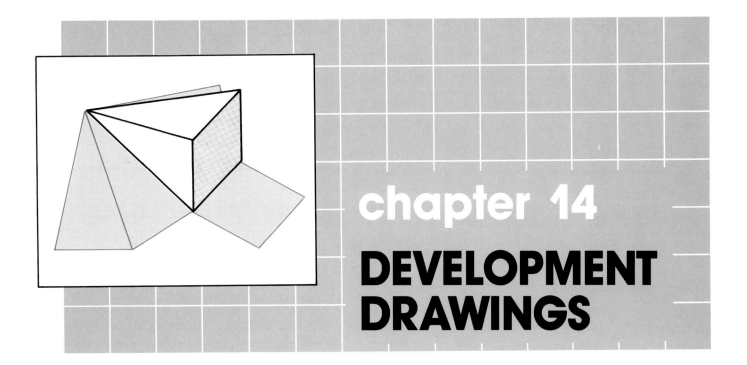

chapter 14
DEVELOPMENT DRAWINGS

OBJECTIVES

The student will be able to:

- recognize the basic geometric shapes
- find true length lines
- draw patterns using the parallel line method
- draw patterns using the radial line method
- draw patterns using the triangulation method
- draw patterns using a CAD system

INTRODUCTION

Products with thin, continuous surfaces, such as sheetmetal, paper, and plastic sheets, are manufactured with the aid of flat **patterns.** The drawings used to develop these patterns are called **development drawings,** flat surface patterns, surface developments, stretchout drawings, and sheetmetal patterns. This field of drafting is called **pattern drafting.** Pattern drafting is used extensively in the automotive, aircraft, packaging, and HVAC industries (**Fig. 14–1**).

Although finished products are three-dimensional, development drawings (patterns) are drawn on a two-dimensional surface. All dimensions on a pattern drawing must be true size, since the pattern is used as a template to cut material to an exact size and shape. The material, once cut in the shape of the pattern, is formed into a three-dimensional shape and fastened into its final form. **Figure 14–2** shows a container box pattern outlined as a flat surface. **Figure 14–3** shows the container material cut to the outline of the pat-

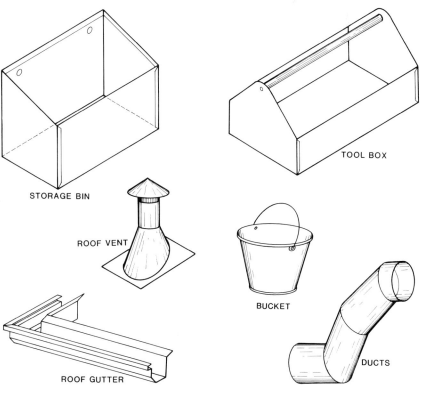

FIGURE 14–1 Many products are manufactured from pattern developments.

223

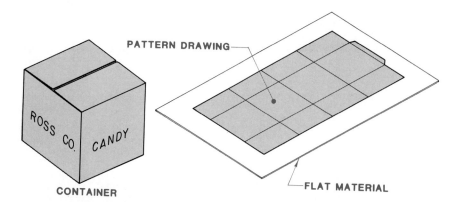

FIGURE 14-2 A container box and its pattern on a flat material. All dimensions are true size on a pattern drawing.

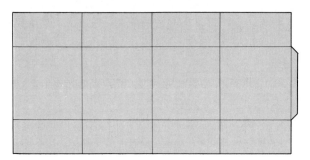

FIGURE 14-3 Cut out pattern

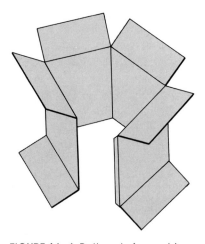

FIGURE 14-4 Pattern is formed by folding

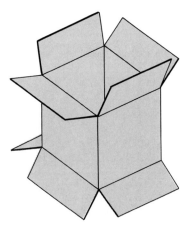

FIGURE 14-5 Fasten together by taping.

tern. **Figure 14-4** shows how the material is folded into a three-dimensional form and fastened into position **(Fig. 14-5)**.

SURFACE FORMS

Surface planes, for which development drawings are prepared, include flat, curved, warped, regular solid, and basic geometric forms. Plane (flat) surfaces are simple to lay out because there are no inclined or oblique surfaces to project **(Fig. 14-6)**.

Surfaces with a single curve include cylinders and cones **(Fig. 14-7)**. The development of pattern drawings for **single-curve surfaces** is also simple, since all lines are either parallel or radiate from a common apex. **Warped surfaces** are curved in two directions which makes pattern layout more complex. Warped surfaces include spheres, paraboloid, and hyperboloid forms **(Fig. 14-8)**. The surfaces of regular solids, such as the tetrahedron, hexahedron, octahedron, dodecahedron, and icosahedron, form patterns as shown in **Fig. 14-9**. The surfaces of basic forms, such as prisms, pyramids, cylinders, and cones, are created through the development of pattern shapes as shown in **Fig. 14-10**.

FIGURE 14-6 The flat or plane surface

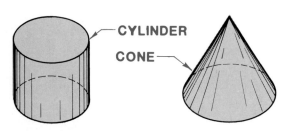

FIGURE 14-7 Examples of single-curve surfaces

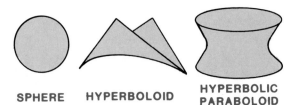

FIGURE 14-8 Examples of warped (double-curve) surfaces

Chapter 14 Development Drawings   225

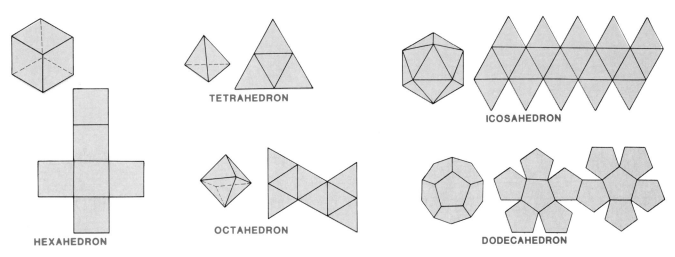

FIGURE 14-9 The five regular solids

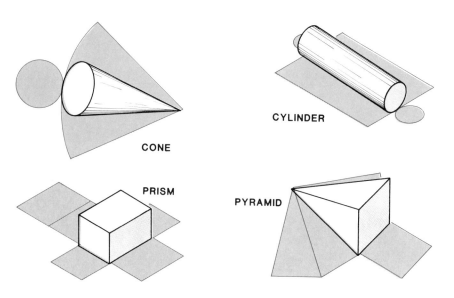

FIGURE 14-10 Basic forms and their patterns

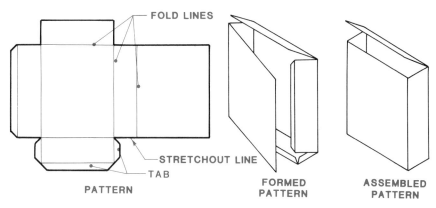

FIGURE 14-11 Terminology for a pattern drawing

PATTERN DRAWING TERMINOLOGY

Pattern drawings are used as the guide for cutting the surface outline of an object from a sheet. Pattern drawings are also used to show the location of lines for bending, folding, or rolling flat forms into three dimensional shapes. In addition, they are used for outlining the position of seams and hems for fastening. This is all accomplished through **stretch-out lines, fold (bend) lines,** and **tab lines** (Fig. 14–11). Tabs are used to form hems and seams. **Hems** are formed along the edges of sheet metal pieces to seal in the sharp edges. **Seams** are used to connect two or more pieces of material together **(Fig. 14–12).**

In addition to the basic outline of the pattern, additional material must be added to provide a base for permanent connections. The type of connection depends on the size and type of material, plus the fastening method. Common methods for fastening flat material include soldering, welding, riveting, gluing, stapling, and sewing.

All lines on a pattern drawing must be true size, otherwise the size and shape of the object cut from the pattern will be incorrect. All lines that fall on normal orthographic planes,

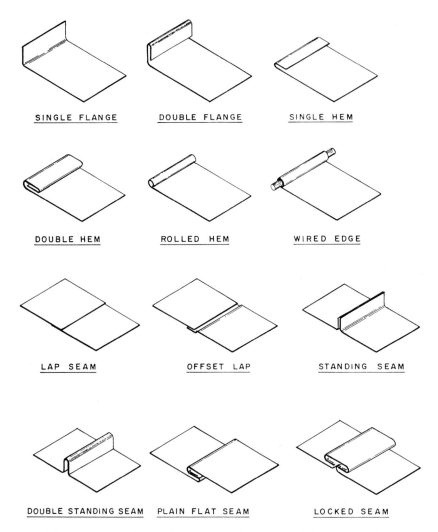

FIGURE 14–12 Various types of seams and hems (Reprinted from TECHNICAL DRAWING AND DESIGN by Goetsch and Nelson, © 1986 Delmar Publishers Inc.)

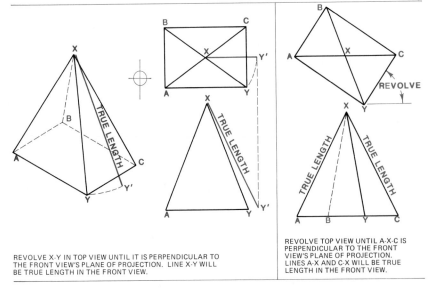

FIGURE 14–13 Two procedures used to find the true length of a line

perpendicular to the plane of projection are **true length**. All lines on inclined or oblique surfaces are not true length (**foreshortened**) on normal orthographic views. The true length of these foreshortened lines must be determined before they can be used in a pattern drawing. The true length of foreshortened lines can be obtained by revolvement, using one of two methods (**Fig. 14–13**).

There are three basic methods of developing pattern drawings depending on the geometric shape of the finished form. These include parallel line development, radial line development, and triangulation.

Chapter 14 Development Drawings   227

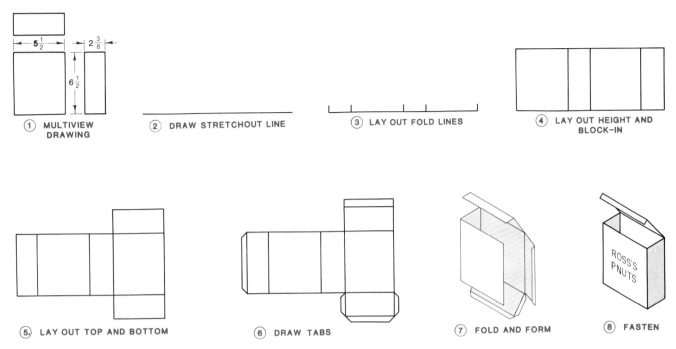

FIGURE 14–14 Steps for parallel line development of a rectangular prism

PARALLEL LINE DEVELOPMENT

Parallel line development is a method used to develop patterns for objects that contain parallel lines, such as prisms and cylinders.

Prisms

All lines in a multiview drawing of a prism are true length. Therefore, parallel lines are extended from the top and bottom of the front view which represents the true height of the pattern **(Fig. 14–14)**. These lines are called **stretchout lines** because they represent the pattern's stretched out form. The true length of the pattern is determined by measuring, projecting, or transferring the width of each side to the pattern. A fold line marks the location of each corner. The fold lines are extended upward using construction lines, and the height of each corner is projected from the front view to the stretchout lines. Lastly, the tabs necessary for making hems and seams are added.

Cylinders

In developing a pattern for a cylinder, parallel stretchout lines are projected from the top and bottom of the front view **(Fig. 14–15)**. The length of the stretchout line is equal to the circumference of the cylinder. The circumference is determined by multiplying the diameter by 3.1416 (π), or by stepping off the **chordal distances** around the circumference of the top view.

Curved Surfaces

When the object lines of a geometric form are parallel, the pattern outline of irregular cuts in its surface can also be determined using parallel line

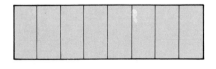

FIGURE 14–15 Steps for parallel line development of a cylinder

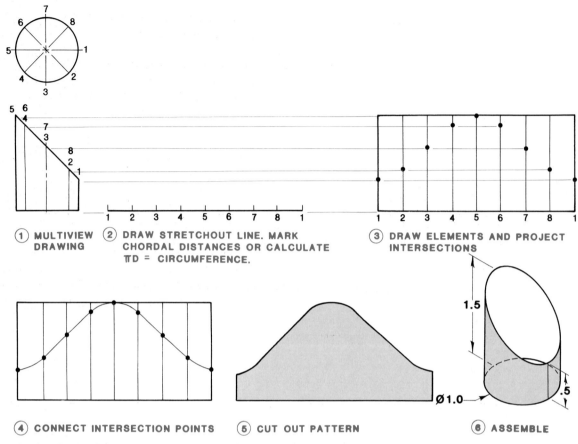

FIGURE 14-16 Steps for parallel line development of a truncated cylinder

development methods. Curved surfaces may consist of a series of small flat surfaces called **elements**. Closely spaced elements create smooth curves and widely spaced elements create flat-sided curves.

In projecting an irregular outline on a parallel surface, such as a cylinder, first draw a stretchout line and calculate the circumference **(Fig. 14-16)**. Next, divide the circumference into an equal number of **chords**, and transfer these chordal distances to the stretchout line. Finally, measure or project the height of each element on a multiview front or side view, and transfer the heights to similar elements on the stretchout line. Connecting the intersecting points creates the outline of the completed pattern.

RADIAL LINE DEVELOPMENT

Geometric forms, such as pyramids and cones, do not contain parallel lines. These forms contain lines that radiate to a common apex and are developed through the use of **radial lines**. Since the inclined vertical edges of cones and pyramids are not parallel, stretchout lines are not parallel. In radial line developments, stretchout lines are arcs, and element lines become triangles that radiate from an apex to the **stretchout arc** (Fig. 14-17).

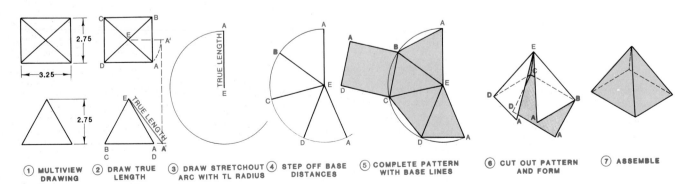

FIGURE 14-17 Steps for radial line development of a rectangular pyramid

Pyramids

The orthographic top view of a pyramid reveals true width and length measurements. The orthographic front or side view contains receding object lines that are not true length, depending on the orientation of the view. To find the true length of a pyramid's side, the pyramid's corners must be revolved to convert inclined lines into normal orthographic true height lines. This is shown in Fig. 14–17 by rotating corner A to position A′, aligning corner A–E on a normal orthographic plane. This true length line (A′–E) is the true length of the radius between the apex of the stretchout line. Once the stretchout arc is drawn, the base distances (which are true length in the top view) are located on the arc and connected to the apex. The pattern is then completed with the addition of the base.

Truncated and Triangular Pyramids. To develop a pattern for a truncated (cut) pyramid, follow the first five steps demonstrated in Fig. 14–17. Then, determine the true length of the lines connecting the base corners with the corners of the cutoff by revolving a corner **(Fig. 14–18)**. This involves finding the true length of lines H–C, G–D, F–E, and I–B. These lengths are projected to the true length line A–H′, then transferred to finish the outline of the truncated portion of the pyramid.

The development of patterns for triangular pyramids is similar to square pyramids. The exception is that only three base sides are projected on the stretchout line **(Fig. 14–19)** after the true length for the stretchout arc is found.

Cones

Unlike pyramids, the elements of a cone are all equal in length from base

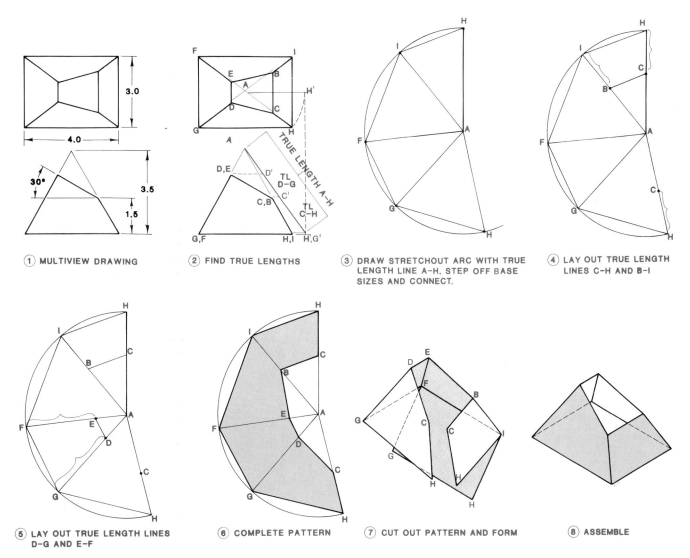

FIGURE 14–18 Steps for radial line development of a truncated pyramid

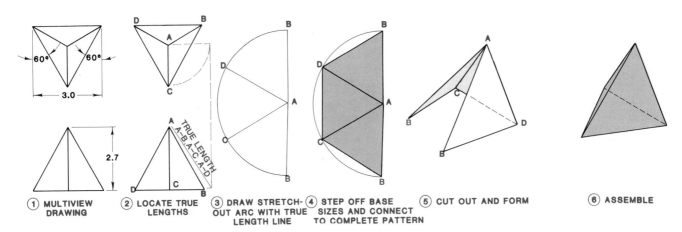

FIGURE 14-19 Steps for radial line development of a triangular pyramid

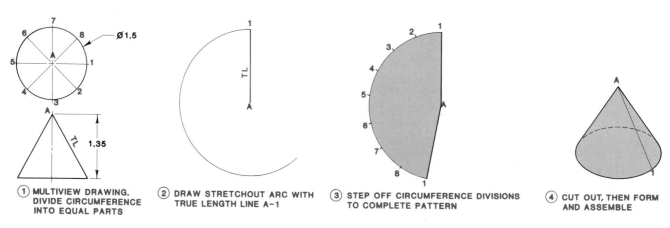

FIGURE 14-20A Steps for radial line development of a cone

to apex. The side and front view of a cone are identical, and contain true length object lines **(Fig. 14-20A)**. The radial stretchout arc for the development of a cone pattern is equal to the distance from the base to the apex, as found on the side view. The top view of a cone reveals the true circumference, but the element lines are foreshortened.

To complete a pattern drawing of a cone, draw an arc with a radius equal to the distance from the base to the true length apex **(Fig. 14-20A)**. Divide the circumference of the base into evenly spaced distances, and transfer these distances to the radial stretchout line. These points are then connected to the apex forming triangular elements. There are eight elements in Fig. 14-20A. More elements will create a smoother and more accurate cone.

A simple, symmetrically shaped cone, without cuts or intersections, can also be plotted mathematically by determining the angle of the stretchout and the circumference of the base. The angle of the stretchout is determined by multiplying the base diameter by 180, divided by the true length of the side. The length of the stretchout line is determined by multiplying the base diameter by 3.1416 (π) **(Fig. 14-20B)**.

FIGURE 14-20B Computing the stretchout angle for the development of a cone's pattern

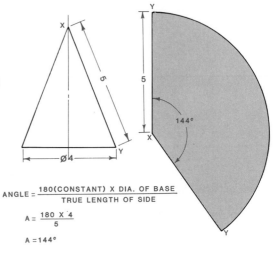

$$\text{ANGLE} = \frac{180(\text{CONSTANT}) \times \text{DIA. OF BASE}}{\text{TRUE LENGTH OF SIDE}}$$

$$A = \frac{180 \times 4}{5}$$

$$A = 144°$$

Truncated Cones. Truncated cones are developed similar to truncated pyramids. The top opening is divided into equal parts since the truncated portion intersects a curved surface **(Fig. 14–21)**. These points are then projected to a side element to show the true lengths. After a true length radial stretchout is drawn, each true distance from base to the intersection point is transferred to equivalent points on the stretchout arc. Connecting these points completes the pattern.

Triangulation Development

Parallel line and radial line developments are used to develop patterns for objects that contain flat or single-curved surfaces. These methods cannot be used to develop patterns for warped surfaces, such as spheres, paraboloids, or hyperboloids, including oblique pyramids, oblique cones, and square to round intersections. Since warped surfaces are curved in two directions, the even distribution of elements used in parallel and radial line development is not possible. The process of triangulation is used to develop patterns for warped surfaces.

Triangulation development is a method of dividing a warped surface into a series of triangles, and transferring the true size of each triangle to a flat pattern. When flat triangulated patterns are bent or folded into a three-dimensional form, the desired warped form is created. Triangulation is an approximate method of pattern development. The use of

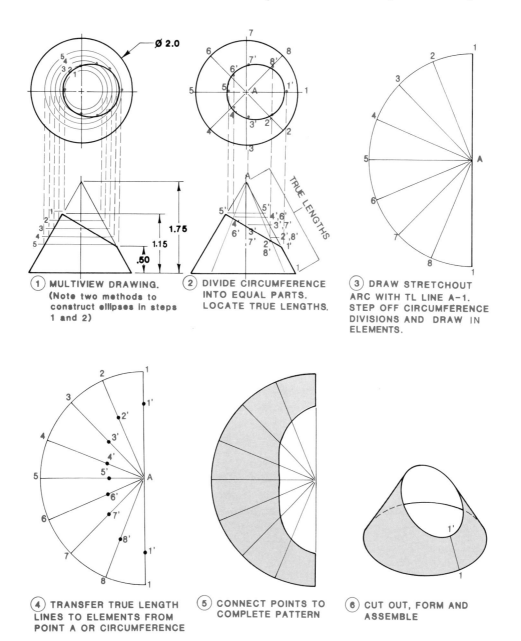

FIGURE 14–21 Steps for radial line development of a truncated cone

more and smaller triangles creates a closer approximation of the desired form than a few large triangles.

The process of triangulation involves dividing the top and front views of an object into a series of triangles. This is done by dividing the top and bottom outlines into a series of parts and connecting them with element lines on both views. The intersections of element lines are evenly spaced on the top view. They converge at common points on the bottom view, creating a succession of triangles (**Fig. 14–22**). Next, true

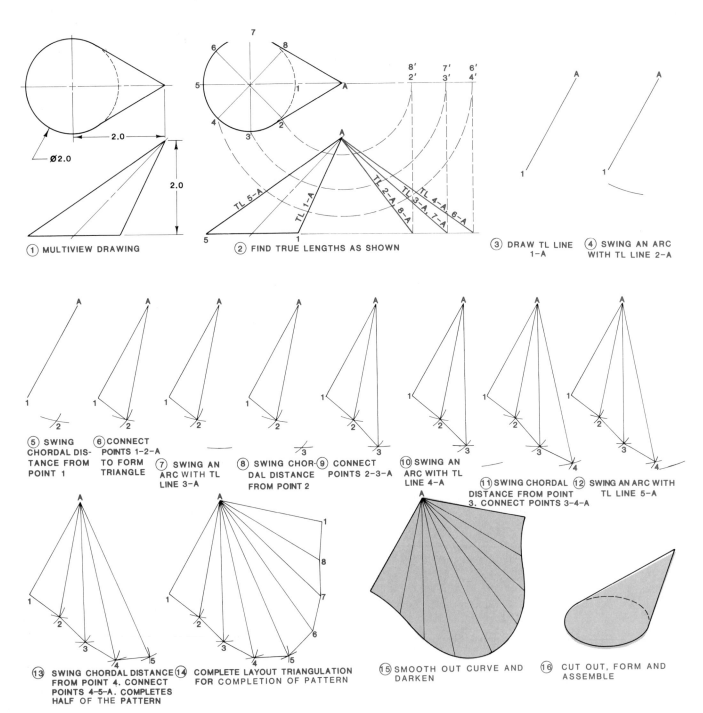

FIGURE 14–22 Steps for triangulation development of an oblique cone

lengths are determined for each element line. These distances are transferred progressively to the stretchout. In the development of triangulated patterns, one triangle at a time is added to the stretchout. Each element line is connected with the chordal divisions layed out on the circumference of the cone's base in the top view.

Triangulation can also be used to develop flat or single-curved surfaces. Since the process is very time-consuming, triangulation is normally used only where necessary. Figure 14–22 shows the development of an oblique cone pattern through triangulation methods. **Figure 14–23** shows the triangulation of an oblique pyramid pattern.

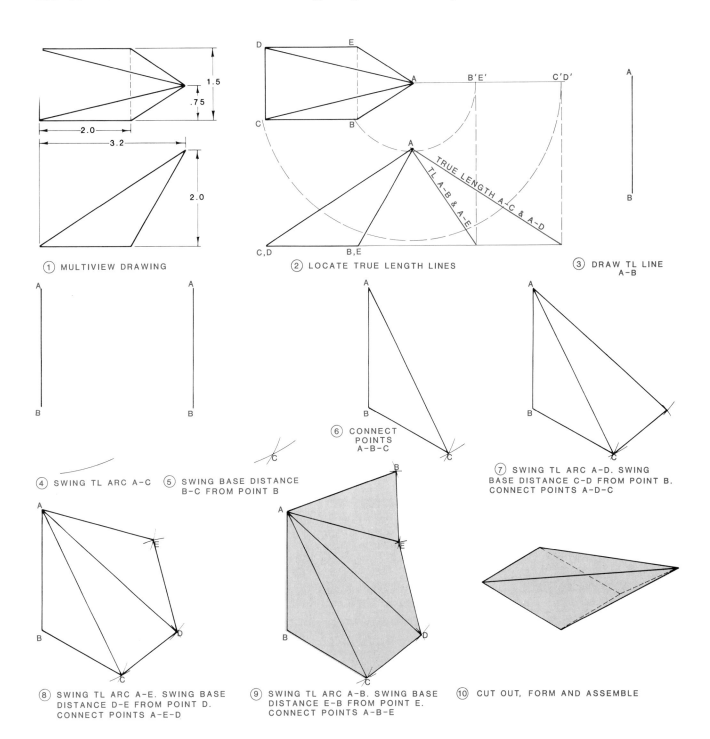

FIGURE 14–23 Steps for triangulation development of an oblique pyramid

When flat surfaces meet and the intersection lines are straight, the related pattern can be developed using parallel line development. When single-curved surfaces meet, the intersection lines can be developed using radial line development methods. However, when warped surfaces meet, or when different geometric forms intersect at varying angles, triangulation must be used to develop the pattern for both parts.

Figure 14–24 shows the development of a pattern drawing for a square form intersecting a round form. This same procedure can be used to develop patterns for any intersecting geometric forms.

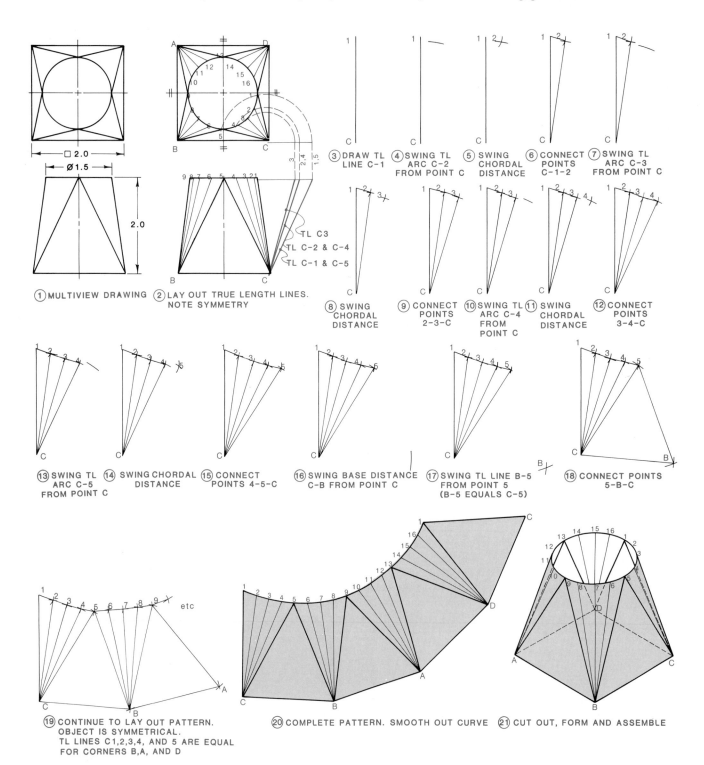

FIGURE 14–24 Development of a pattern drawing for a square form that intersects a curved form

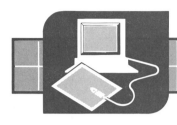

COMPUTER-AIDED DRAFTING & DESIGN

COMPUTER-AIDED DEVELOPMENT DRAWINGS

The application of CAD to draw developments is widespread. The automotive industry uses special CAD software to develop automotive body parts. These programs are CAD/CAM programs that actually tell the sheet metal working machines where to make cuts and folds. The garment industry uses CAD/CAM programs to develop clothing patterns. These CAD/CAM programs run complex laser cutting machines that accurately cut through many layers of material at one time.

The CAD examples in this chapter are much simpler applications of CAD for drawing developments. When drawing developments on a CAD system, the operator uses many of the CAD commands to speed up and increase the accuracy of the developments.

Parallel Line Development

Step 1. Draw and label the necessary orthographic views. Note the circle has been divided into 16 parts. This may be done with the ARRAY command **(Fig. 14–25)**.

Step 2. Draw the stretchout line along the base of the cylinder. Its length should be the circumference of the circle (C = πD). Divide the line into 16 equal parts **(Fig. 14–26)**.

Step 3. Switch layers to a construction line layer and project each segment to the development view. Switch to a new layer and use the PLINE command to connect the proper intersections. Only do half of the development. Use the CURVE FITTING command to have the curve automatically smoothed out **(Fig. 14–27)**.

Step 4. Use the MIRROR command to finish the curve **(Fig. 14–28)**.

Step 5. Freeze (remove) the construction line layer and the development is clearly shown **(Fig. 14–29)**.

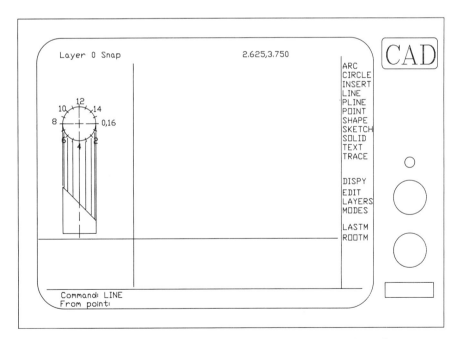

FIGURE 14–25 Lay out the multiview drawings and divide the circumference into equal parts.

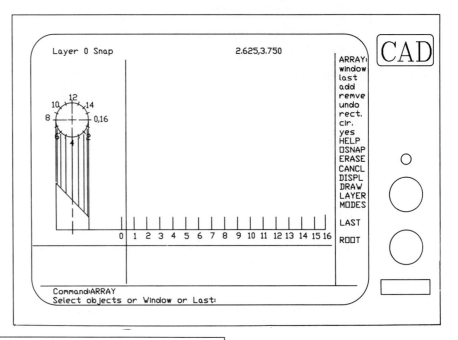

FIGURE 14–26 Lay out the stretchout line for the circumference's length and divide into 16 equal parts.

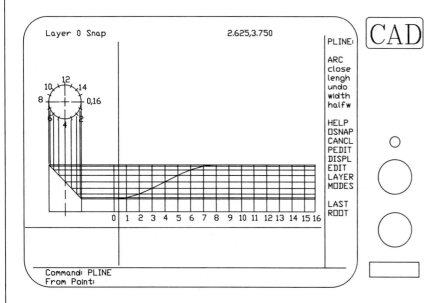

FIGURE 14–27 Complete half a pattern.

FIGURE 14–28 Complete the symmetrical half with the MIRROR command.

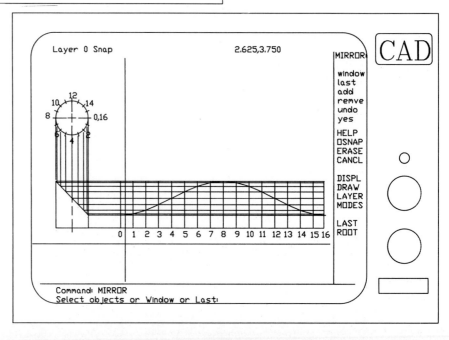

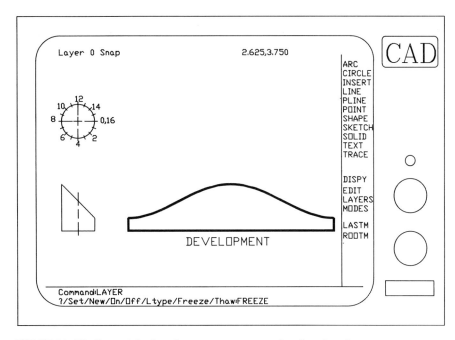

FIGURE 14–29 Completed pattern—remove construction line layers.

Radial Line Development

Step 1. Draw and label the necessary orthographic views. Note the horizontal (top) view has been rotated so the true length of the edge of the pyramid is shown in the frontal view as true size **(Fig. 14–30)**.

Step 2. Pick any convenient point on the screen and label it "V" for the vertex of the pyramid. The location of the point is not critical as it can easily be moved if needed. Using a construction layer, draw the stretchout arc V–1 and arc V–X **(Fig. 14–31)**.

Step 3. Lay out the segments, 1–2, 2–3, 3–4, and 4–1 along the arc. The true lengths of these segments are in the top view. Draw lines from V to 1, 2, 3 and 4. For accuracy, use OSNAP (intersection) **(Fig. 14–32)**.

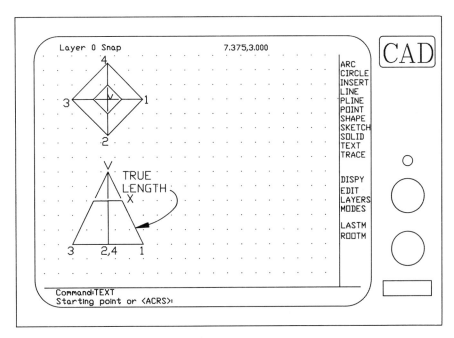

FIGURE 14–30 Draw and label multiview drawings.

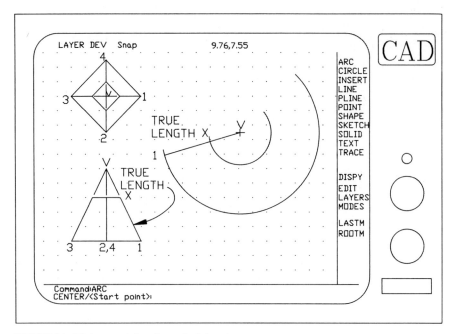

FIGURE 14–31 Lay out the two stretchout arcs.

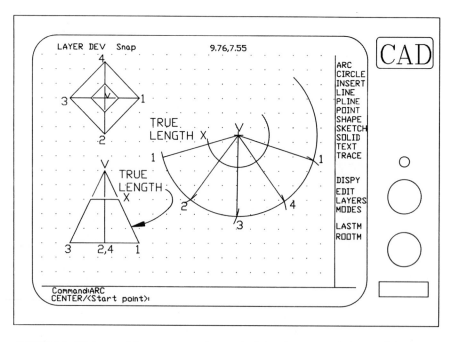

FIGURE 14–32 Lay out base segments on stretchout arcs and connect to apex V.

Step 4. Using OSNAP (intersection), draw straight lines connecting 1–2, 2–3, 3-4, and 4–1 **(Fig. 14–33)**.

Step 5. Freeze the construction line layers and the development is complete **(Fig. 14–34)**.

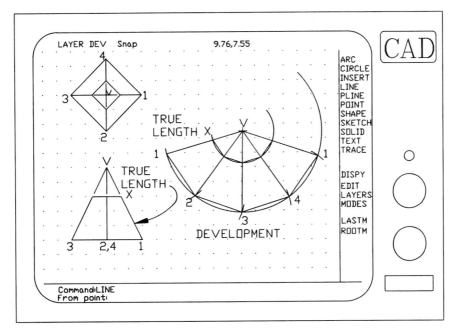

FIGURE 14–33 Connect the base points.

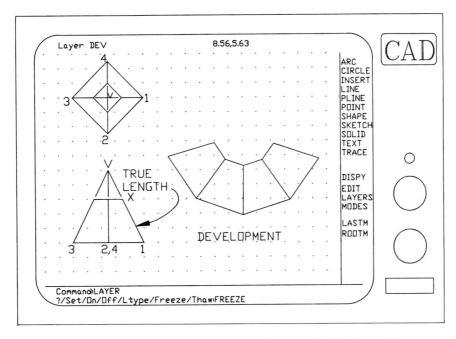

FIGURE 14–34 Complete pattern by removing the construction layout layers.

EXERCISES

1. Sketch the approximate pattern of each item in **Fig. 14–35** using parallel line development.
2. Sketch the approximate pattern of each item in **Fig. 14–36** using radial line development.
3. Sketch the approximate pattern of each item in **Fig. 14–37** using triangulation development.
4. With instruments or a CAD system, draw the patterns for **Figs. 14–38** through **14–51**.
5. Make the following ECO (Engineering Change Order—refer to Chapter 17) changes for **Fig. 14–52**:
   a. Change the vertex offset from 1.1" to 1.25".
   b. Increase the base diameter to 4.00".
   c. Change the overall height to 2.5".
6. Design a heavy cardboard storage case to hold 100 5.25" square floppy disks in its holder; the thickness is .075". Draw and cut out a paper test pattern. The disks must be loosely packed.
7. Design a cardboard carrying case for 8 standard-size cans of soda.
8. Design a cardboard carrying case or envelop to carry an 18" x 24" drafting board, a 24" T square, and assorted drafting supplies, such as pencils, erasers, triangles, irregular curves, and compasses.
9. Using the cartesian coordinate system, complete the pattern on $1/4$" grid paper from the data table in **Fig. 14–53**.
10. Using the cartesian coordinate system, complete the data table of the pattern in **Fig. 14–54**.
11. With a CAD system, draw the patterns for Figs. 14–38 through 14–46.

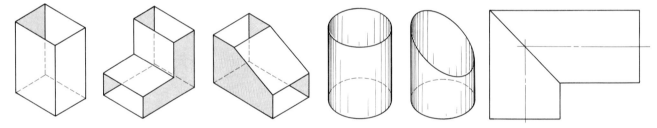

FIGURE 14–35 Sketch an approximate parallel line pattern for each item.

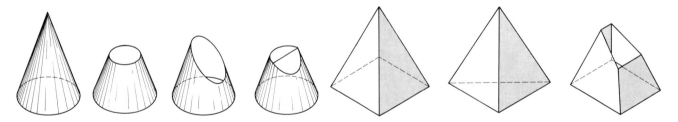

FIGURE 14–36 Sketch an approximate radial line pattern for each item.

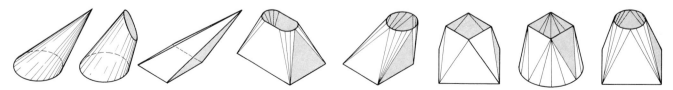

FIGURE 14–37 Sketch an approximate triangulation pattern for each item.

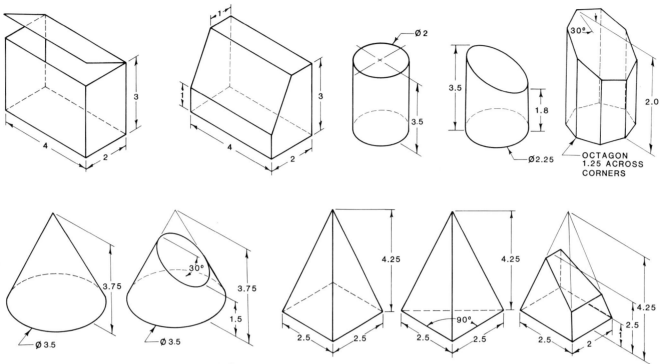

FIGURE 14-38 Draw the pattern for each item.

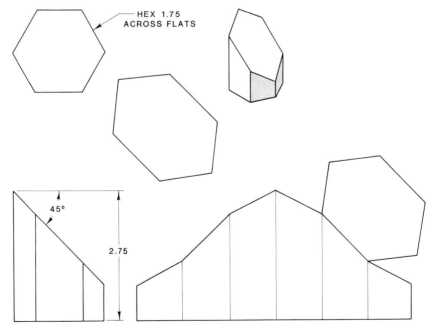

FIGURE 14-39 Draw the multiview, auxiliary view, and pattern drawings.

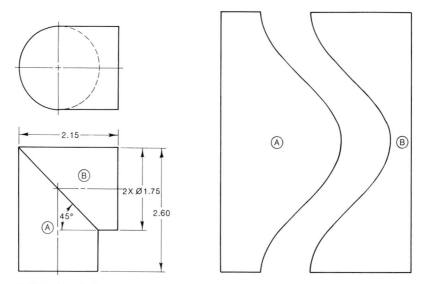

FIGURE 14-40 Redraw the multiview drawing and patterns for the 90° sheet metal elbow.

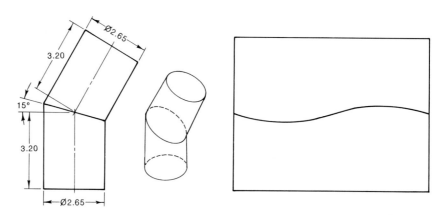

FIGURE 14-41 Draw the multiview drawing and patterns for the elbow.

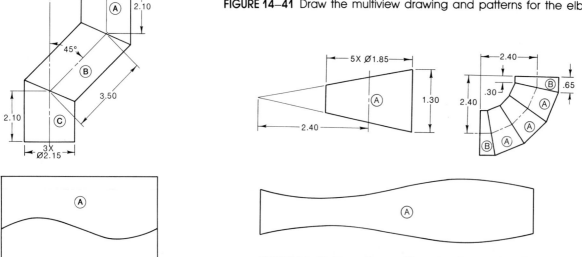

FIGURE 14-43 Draw the working drawing and patterns A and B to make the five-piece elbow. Pattern A is already drawn.

FIGURE 14-42 Draw the working drawing and patterns for the three-piece elbow.

Chapter 14 Development Drawings   **243**

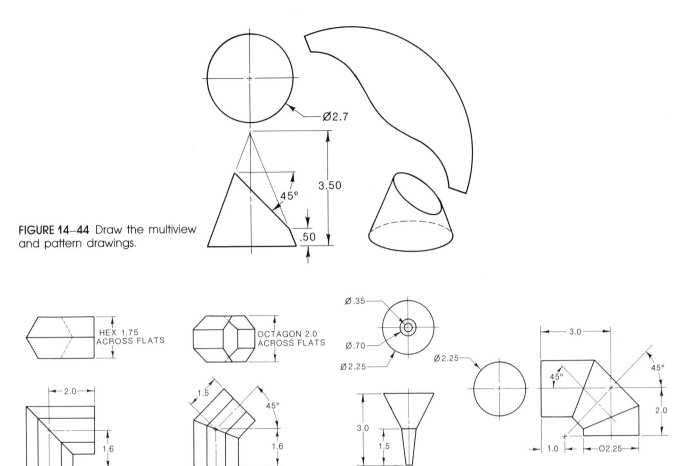

FIGURE 14-44 Draw the multiview and pattern drawings.

FIGURE 14-45 Develop the multiview drawings and the multiple patterns required for each item.

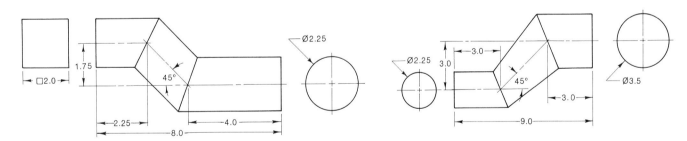

FIGURE 14-46 Develop the multiview drawings and the multiple patterns required for each transitional item.

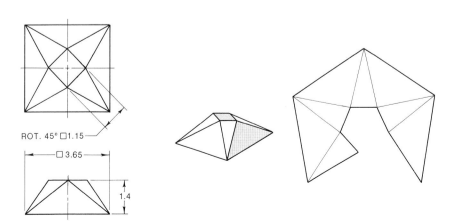

FIGURE 14-47 Draw the multiview and pattern drawings.

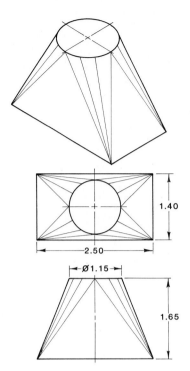

FIGURE 14-48 Redraw the multiview and pattern drawings.

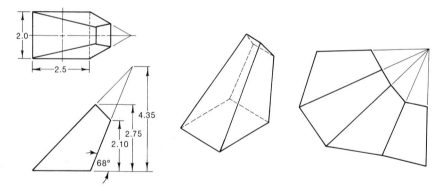

FIGURE 14-49 Redraw the multiview and pattern drawings.

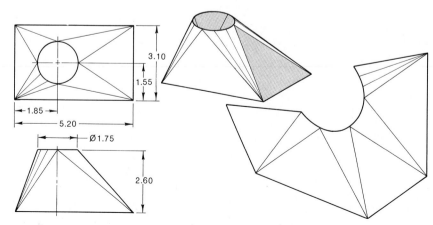

FIGURE 14-50 Redraw the multiview and pattern drawings.

Chapter 14 Development Drawings   **245**

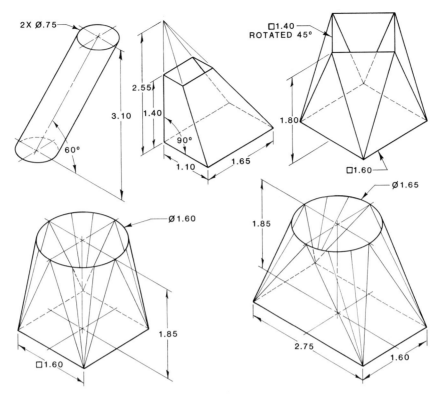

FIGURE 14–51 Develop the multiview and pattern for each item

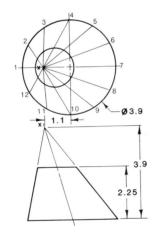

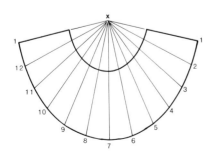

FIGURE 14–52 Make the ECOs (Engineering Change Orders—refer to Chapter 17), as specified in Exercise 5 and complete the multiview and pattern.

DATA TABLE					
LINE	X	Y	go to	X	Y
1	1	3	–	4	5
2	4	5	–	8	9
3	8	9	–	12	11
4	12	11	–	16	9
5	16	9	–	20	5
6	20	5	–	24	3
7	24	3	–	24	0
8	24	0	–	1	0
9	1	0	–	1	3

FIGURE 14–53 Lay out the pattern from the coordinate data table on ¼" grid paper. Cut out the pattern and form it to shape to check the layout.

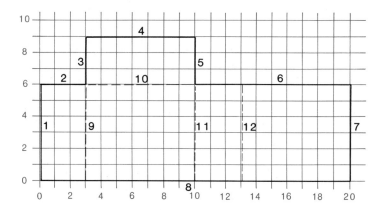

FIGURE 14-54 Complete the data table from the box's pattern.

 **KEY TERMS:**

Chordal distances
Chords
Development drawing
Elements
Fold (bend) lines
Foreshortened
Hems

Parallel line development
Pattern
Pattern drafting
Radial line development
Single-curve surface
Seams

Stretchout arc
Stretchout line
Tab lines
Triangulation development
True length
Warped surface

the world of CAD

Sverdrup Corporation, St. Louis, Missouri, is a leading provider of professional services for development, design, construction, and operation of capital facilities. Projects range from office buildings to stadiums and convention centers; from airports to the space shuttle launch complex; from state-of-the-art breweries and testing facilities to award-winning bridges and tunnels throughout the world.

Sverdrup applies CADD to every discipline they address: architecture, transportation, public works, environmental sciences, industrial design, and advanced technology. Sverdrup has discovered that CADD is a valuable tool not only for design but for coordinating, controlling, and communicating information for the duration of a project.

From its initial use of an Intergraph PDP-11/70 system in 1981 to its current projects using Intergraph VAX 751, 785, and 200 systems with InterAct and CLIPPER workstations, Sverdrup continues to integrate CADD into its design routine.

In 1982, after several days of heavy rainfall, a street collapsed at a major intersection in the city of St. Louis, creating a hole 40 feet wide, 40 feet deep, and 100 feet long, Figure 1. This collapse was caused by the cave-in of an eight-foot high, horseshoe-shaped brick sewer running below the street.

Due to the critical time frame involved in finding both temporary and permanent solutions to the situation, a composite map of the utilities in the area was generated using Sverdrup's Intergraph CADD system, Figure 2. The map was used to locate utilities to be cut off, plan the work for a temporary cofferdam, and then get the information out to the site as quickly as possible. Plan and profile sheets, as well as associated details, were then generated by the system to be used in the repair work.

FIGURE 1 The view looking up Fillmore Avenue in south St. Louis shows the collapsed storm sewer and subsequent street cave-in.

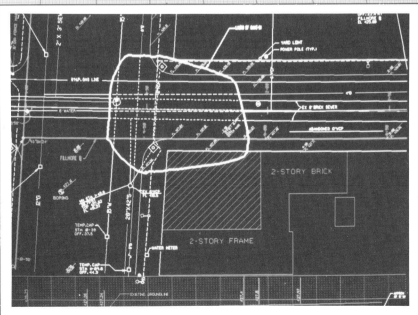

FIGURE 2 Sverdrup generated this map to speed repair of the street and sewer system.

FIGURE 3 Sverdrup solved several problems for St. Louis' Barnes Hospital by developing an enclosed pedestrian bridge with Intergraph CADD.

CADD's ability to work three-dimensionally contributed to an ingenious design solution for one of the nation's largest, most respected teaching and research hospitals.

In 1986, Barnes Hospital in St. Louis had three related problems. First, a busy boulevard separated the 1.7 million-square-foot health care complex from its major parking facility. Second, visitors and patients crossing the roadway and entering the hospital's ground floor did not directly reach the main lobby and reception area. Third, any sort of construction linking the hospital and underground parking garage had to meet stringent design standards, since it would be built partially in St. Louis' famous Forest Park.

"CADD helped us develop an enclosed pedestrian bridge that answers all three concerns," says Sverdrup architect Mark Gustus. "A new bank of elevators an escalators within the garage takes newcomers up to the bridge level, which then leads them across the street directly to Barnes' central reception area. The bridge itself is an intricate truss of steel tubes clad in transparent glass, which provides a pleasant view for both hospital visitors and Forest Park visitors." See Figures 3 and 4.

FIGURE 4 Interior view of Barnes Hospital pedestrian bridge.

"CADD enabled us to try more imaginative, complex designs because it works in three dimensions all at once," says Gustus. "Ordinary paper drawings aren't much help, and manmade models are time-consuming. CADD's gaming capabilities, along with its ability to rotate and animate an image, made it an ideal medium for a project like this."

(Courtesy of Louise Wiedermann, CADD System Manager for Sverdrup Corporation. Sverdrup is engaged in the Development Design, Construction, and Operation of Capital Facilities.)

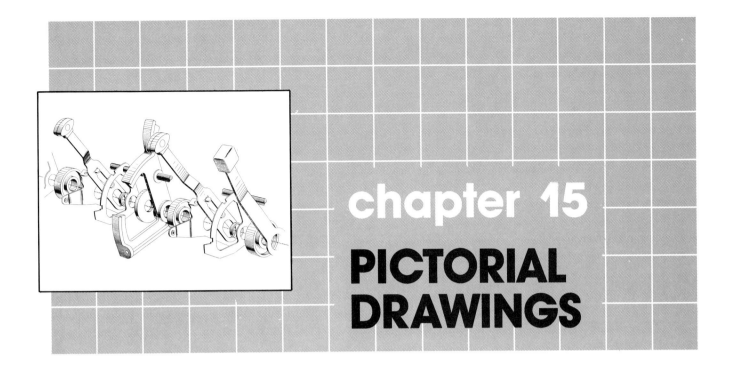

chapter 15
PICTORIAL DRAWINGS

OBJECTIVES

The student will be able to:

- block in an isometric square
- draw nonisometric lines
- construct isometric circles
- draw isometric circles with a template
- draw isometric drawings with axes in various positions
- center an isometric drawing
- plot an isometric curve
- draw isometric radii and rounds
- draw various types of isometric sections
- draw an isometric assembly drawing
- draw an isometric exploded view drawing
- dimension an isometric drawing with the aligned system
- dimension an isometric drawing with the unidirectional system
- sketch an isometric on isometric grid paper
- draw a dimetric drawing
- draw a trimetric drawing
- draw a cabinet drawing
- draw a cavalier drawing
- draw circles and curves on an oblique drawing
- draw simple pictorial drawings on a CAD system
- draw a one-, two-, and three-point perspective drawing

INTRODUCTION

Pictorial drawings have been used throughout history to communicate ideas that cannot be described easily in words. Early attempts to show depth and detail were crude and often distorted the appearance of objects. **Figure 15–1** shows primitive pictorial drawings which lack correct proportion and realism. Methods and techniques used to prepare realistic pictorial drawings have been continually refined and improved through the years. **Figure 15–2** shows a contemporary pictorial drawing which is difficult to distinguish from a photograph.

FIGURE 15–1 Primitive drawings usually appeared two-dimensional and distorted.

247

248 Drafting in a Computer Age

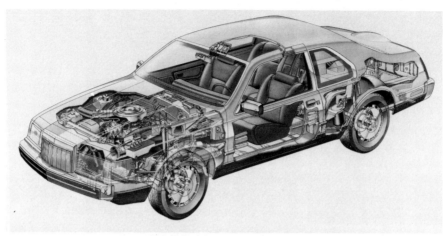

FIGURE 15-2 An artistic version gives a three-dimensional effect and realistic appearance. (Courtesy of Ford Motor Company)

In addition to their function as an art form, **pictorial drawings** are used today by engineers, architects, and all types of designers and drafters to illustrate the size and shape of an object. Pictorial drawings are multidimensional; that is, they show more than one side of an object **(Fig. 15-3)**. Pictorial drawings can show width, height, and length on one drawing. For this reason, they are easier to understand and interpret than orthographic multiview drawings which show only one face of an object on each view. Since pictorial drawings are not always dimensionally accurate, they are rarely used as detailed working drawings for manufacturing or construction purposes.

In engineering and architecture, pictorial drawings are used to improve the visualization skills of nontechnical personnel who may have difficulty reading orthographic drawings. Pictorial drawings are used extensively

- as idea recorders in the design process,
- to clarify assembly and repair instructions,
- for new product sales training, and
- for sales promotion.

Once a product is manufactured, a photograph may be used for these purposes. Only pictorial drawings or models can be used to show

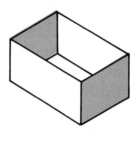
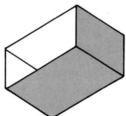
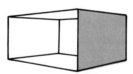

FIGURE 15-3 Pictorial drawings will show two or more sides on each drawing.

the complete appearance of a product before manufacture. Unlike photographs, pictorial drawings can be used to create sectional or **exploded views** which show details or the interior of an object. **Figure 15-4A** shows a photograph of a technical pen set. **Figure 15-4B** shows a drawing of an inking tip with sections removed to promote clarity.

FIGURE 15-4A Photograph of a technical pen set (Courtesy of J. S. Staedtler, Inc.)

FIGURE 15-4B Sectional drawing of a technical pen (Courtesy of J. S. Staedtler, Inc.)

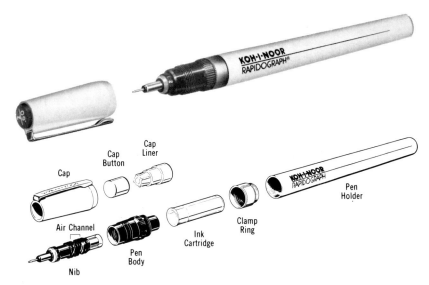

FIGURE 15-4C A photograph and an exploded pictorial line drawing (Courtesy of Koh-I-Noor)

Figure 15-4C shows a photograph of a technical pen and an exploded view line drawing of the same pen.

In the fields of engineering and architecture, there are three basic types of pictorial drawings—axonometric, oblique, and perspective. **Figure 15-5A** compares these three types applied to the same object. **Figure 15-5B** shows four pictorial drawings for a single-number digital clock. **Axonometric drawings** are most frequently used for engineering drawings. **Oblique drawings** are used mostly for simple thin objects or for progressive design sketches. **Perspective drawings** are used extensively to create realistic renderings of objects such as buildings, cars, and boats.

AXONOMETRIC DRAWINGS

In axonometric drawings, the major characteristic is the receding lines that are always parallel. Axonometric drawings show all three faces (front, side, and top) of an object in one view. There are three types of axonometric drawings—**isometric**, **dimetric**, and **trimetric** (Fig. 15-5A and B).

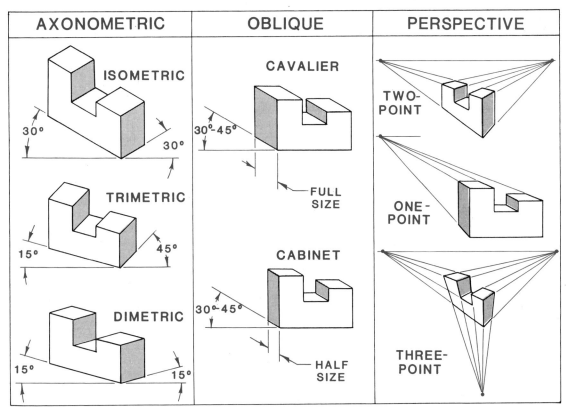

FIGURE 15-5A Various types of pictorial drawings

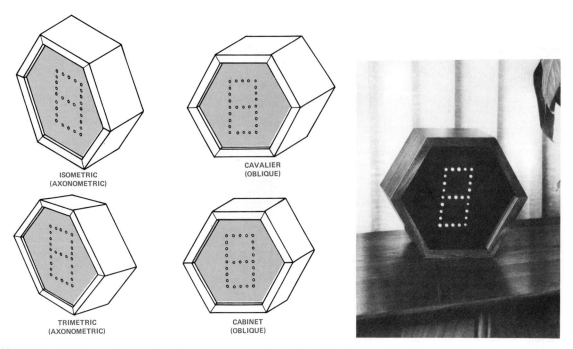

FIGURE 15–5B Various types of pictorial representation of a digital clock (Courtesy of Michael Robbins)

Isometric Drawings

The most popular form of axonometric drawing used in engineering is the isometric. Isometric drawings are easier to draw, detail, and dimension than any other type of pictorial drawing. This is because the three 30° planes are shown true size **(Fig. 15–6)**. To visualize an object in true isometric form, the following steps are necessary **(Fig. 15–7)**.

Step 1. Draw the three orthographic views.

Step 2. Revolve the top view 45° and draw the remaining two views.

Step 3. Revolve the side view in Step 2 35°. Project and draw the remaining two views.

The front view will be a true isometric projection. Note that the vertical lines are not true length. The dimensions will appear foreshortened because of the 35° revolution. Regular isometric drawings will use true size dimensions for speed and convenience.

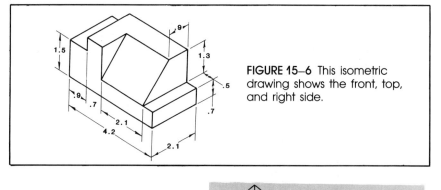

FIGURE 15–6 This isometric drawing shows the front, top, and right side.

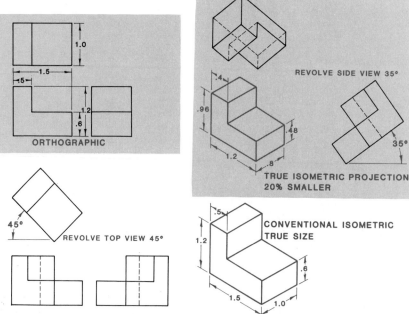

FIGURE 15–7 The true isometric drawing and the conventional isometric drawing

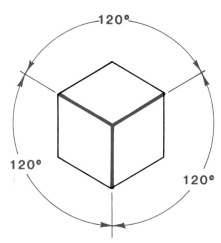

FIGURE 15-8 The major axes for an isometric drawing are spaced at 120°.

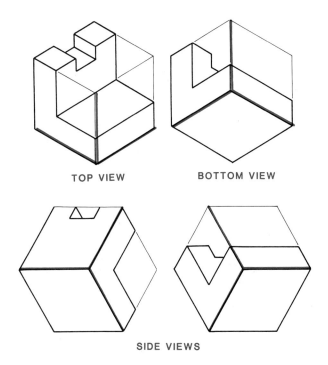

FIGURE 15-9 Alternate positions for isometric axes

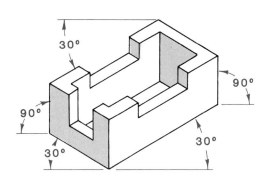

FIGURE 15-10 Isometric lines are 30° from the horizontal.

Isometric Axis. The **isometric axis** is the basis for all isometric drawings. This axis consists of three lines, 120° apart. These three lines represent the front corner and the two receding edges of an object **(Fig. 15-8)**. Although the three-axis lines are always 120° apart, the axis can be positioned to show either the top, bottom, or side views as the dominant view **(Fig. 15-9)**.

Isometric Lines. All lines drawn parallel to any of the isometric axis lines are **isometric lines.** These lines are always true size. To prepare an isometric drawing, all isometric lines are drawn 30° or vertical from the horizontal **(Fig. 15-10)**.

Nonisometric Lines. As you learned in developing auxiliary views, objects may contain lines that are not parallel to the major axes of the object. These lines, representing sloping surfaces, are called **nonisometric lines.** Nonisometric lines are lines that are not parallel to isometric lines. The angle and length of nonisometric lines can only be established by locating points of intersection with isometric lines. This is because actual measurements can only be made on true size isometric lines.

Before drawing nonisometric lines, first lightly draw all isometric lines, completing the outline of the object as it would appear if no surfaces were slanted **(Fig. 15-11)**. Then, transfer the distances of each corner from an orthographic view to one of the faces of the isometric drawing. Connect these points on the isometric lines to form the nonisometric lines. Project 30° isometric lines from these points until the lines intersect the lines representing the back surface of the object. Erase the isometric construction lines and complete the isometric drawing as shown.

Centering Isometric Drawings. To center an isometric drawing, follow the steps in **Fig. 15-12A**:

Step 1. Check the overall dimensions of the object to be drawn.

Step 2. Outline the area needed for the drawing.

Step 3. Locate the center of the area.

Step 4. Draw a vertical line upward from the centerpoint of the area

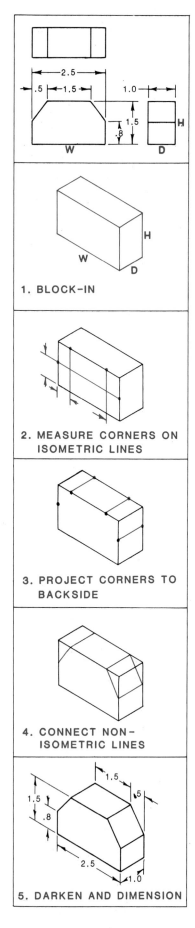

1. BLOCK-IN
2. MEASURE CORNERS ON ISOMETRIC LINES
3. PROJECT CORNERS TO BACKSIDE
4. CONNECT NON-ISOMETRIC LINES
5. DARKEN AND DIMENSION

FIGURE 15-11 Steps in drawing an isometric drawing

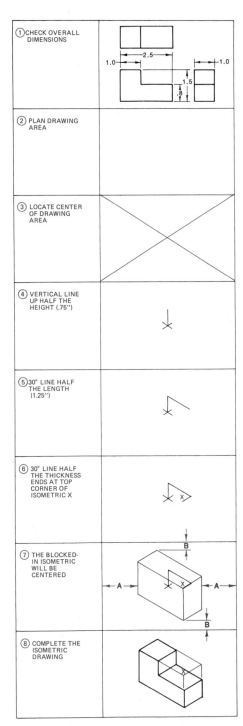

① CHECK OVERALL DIMENSIONS
② PLAN DRAWING AREA
③ LOCATE CENTER OF DRAWING AREA
④ VERTICAL LINE UP HALF THE HEIGHT (.75″)
⑤ 30° LINE HALF THE LENGTH (1.25″)
⑥ 30° LINE HALF THE THICKNESS ENDS AT TOP CORNER OF ISOMETRIC X
⑦ THE BLOCKED-IN ISOMETRIC WILL BE CENTERED
⑧ COMPLETE THE ISOMETRIC DRAWING

FIGURE 15-12A Steps to center an isometric drawing

equal to one half the height of the object (.75″).

Step 5. From the top of this line, draw a line 30° downward to the right, equal to one half the length (1.25″) of the object.

Step 6. From the right end of this line, draw a line 30° downward to the left, equal to one half the thickness of the object (.5″). The left end of this line represents the front, top corner (X) of the object.

Step 7. Extend a light vertical line down from corner X to represent the front edge of the object. The length of this line is the height of the object at its highest point (1.5″). Next, extend 30° lines upward to the left and right to represent the top right and top left edges of the object. This creates the isometric axis. These lines should be true size overall dimensions of the object. Now, **block in** all remaining isometric lines. The blocked-in isometric will be centered horizontally and vertically.

Step 8. Complete the isometric drawing. **Figure 15-12B** reviews the steps to draw the isometric drawings.

Isometric Circles. Circles will appear as **ellipses** on isometric drawings because they are located on **receding planes** that are not perpendicular to the line of sight. Ellipses are oriented differently on each of the three isometric planes **(Fig. 15-13)**. To draw **isometric circles** (which are actually 35° ellipses), follow the steps outlined in Fig. 15-13:

Step 1. Locate the center of the circle. Then, draw centerlines that are parallel to the isometric lines of the object.

Step 2. Establish the size of the circle by drawing an isometric square with sides equal to the diameter of the circle.

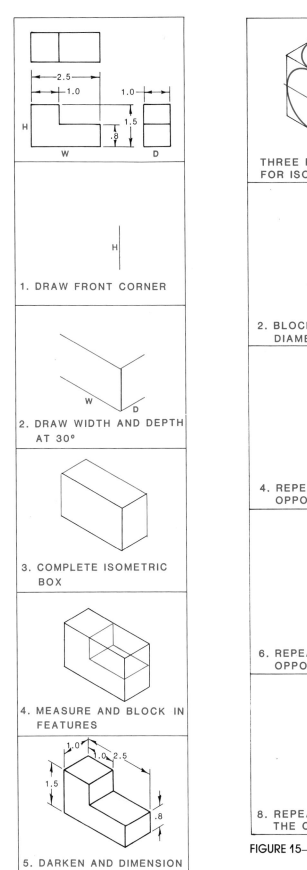

FIGURE 15–12B Steps in drawing an isometric drawing

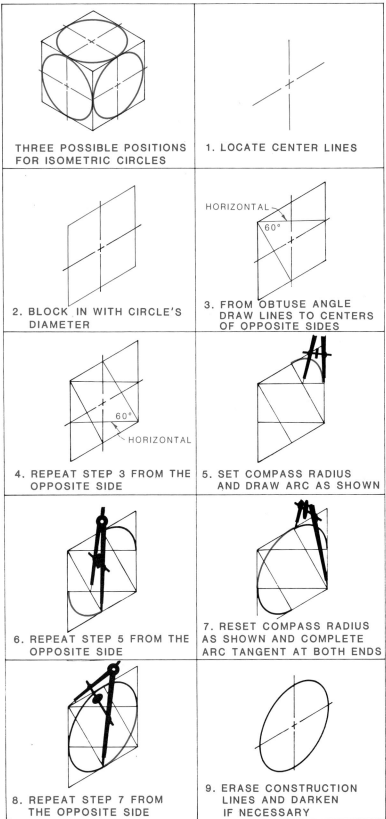

FIGURE 15–13 Steps in constructing an isometric circle

Step 3. Draw two lines from the vertex of the obtuse angle of the square to the point where the center lines intersect the opposite edges of the square. One line will be horizontal and the other will be 60°.

Step 4. Repeat the above procedure in reverse from the obtuse angle on the opposite side of the square.

Step 5. Place the point of a compass on the intersection of these two lines. Set the compass opening to align with the intersection of the centerline and the outer edge of the square. Draw an arc from this point to the corresponding point on the adjacent side of the square.

Step 6. Repeat this procedure on the opposite end of the square.

Step 7. Place the compass point on the obtuse angle corner of the square. Draw a large arc on the opposite side to connect with the edges of the small axis drawn in steps 5 and 6 above. Be certain to draw smooth tangent points.

Step 8. Repeat this procedure on the opposite side of the square to complete the isometric circle.

Step 9. Erase all construction lines.

Fillets and Rounds. **Fillets** are continuous internal (concave) arcs. **Rounds** are continuous external (convex) arcs on the surface of an object. Since arcs are partial circles, to draw an isometric arc, draw part of an isometric circle. To visualize this, observe the circles drawn in the corners of the object in **Fig. 15–14**. To draw **isometric arcs,** use the same procedure as drawing isometric circles on the three different planes. Draw only the part of the circle that describes the fillets or rounds.

Ellipse Templates. Ellipse templates eliminate much of the tedious

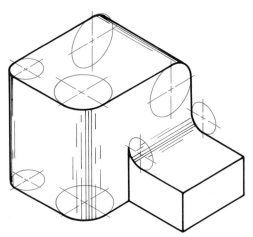

FIGURE 15–14 Positioning of isometric circles

work of drawing isometric circles and arcs. An **ellipse template (Fig. 15–15)** is the most efficient method to draw isometric circles. The isometric ellipse template uses 35° ellipses. To use an isometric template, first locate and draw the intersecting centerlines of the circle. Choose the correct isometric circle size. Align the marks on the edge of the isometric template opening with the centerlines on the drawing. Trace around the inside of the hole with a pencil.

Ellipse on Nonisometric Planes. The template and construction method of drawing ellipses can only be used on isometric planes that are at 30°. These methods cannot be used on nonisometric planes that do not align with the isometric axis. Ellipses, or any projection or depression on a nonisometric plane, must be plotted directly from a normal orthographic view.

To plot ellipses on nonisometric surfaces, first prepare an ortho-

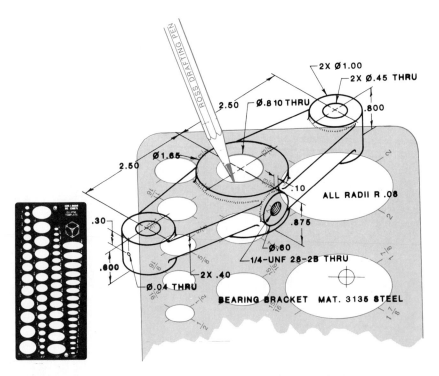

FIGURE 15–15 Line up the diagonals on the isometric template's circle on the drawing's isometric center lines, and trace around to complete the isometric circle.

graphic two-view drawing. Show the sloped surface profiled on one view, and the true shape of the circle on the other. Follow the steps outlined in **Fig. 15–16**:

Step 1. Divide the circle into eight or more equal parts as shown on the front view. Number each equally spaced mark on the circumference of the true circle. The accuracy and smoothness of the final ellipse will increase as the number of circle divisions is increased. Project these points horizontally to intersect the line representing the edge of the inclined plane on the front view. Label the points of intersection on the profile plane. Draw the isometric form. Draw a centerline representing the vertical center of the ellipse on the inclined surface.

Step 2. Measure the X and Y distances to find the position of the horizontal centerline on the inclined surface. Label the intersection of the vertical and horizontal centerline with a 0.

Step 3. Measure the distance from 0 to 3, and from 0 to 7 on the front view. Transfer these distances to the horizontal centerline on the inclined surface as shown.

Step 4. Determine the distance from 0 to 1, and from 0 to 5 on the side view sloping edge. Transfer or project these distances to the vertical centerline on the inclined surface as shown.

Step 5. Measure the horizontal and vertical X–Y distances from the inclined surface for points 8 and 2, then 6 and 4. If more points are used, lines representing all of the points would be drawn at this time. Draw a dot where these lines intersect with corresponding numbered lines measured from the vertical centerline. Lightly sketch the ellipse.

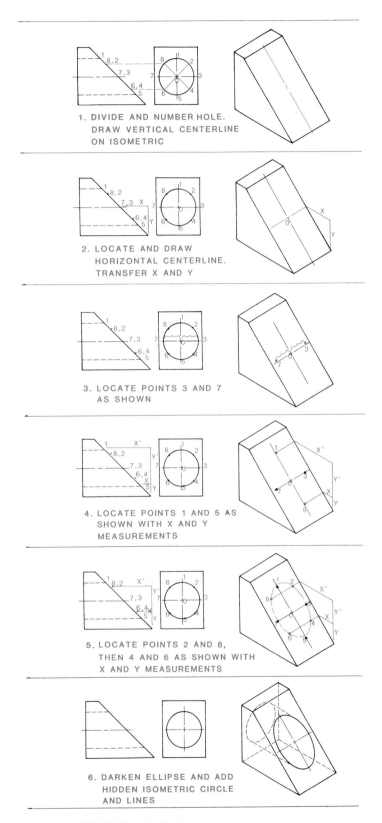

FIGURE 15–16 Plotting a nonisometric circle

Step 6. Darken with an ellipse template or irregular curve to form the ellipse as shown.

Plotting Curves. Curved nonisometric lines are drawn by transferring points from an orthographic view to an isometric plane **(Fig. 15–17)**:

Step 1. Draw two orthographic views of the object containing the curved surface.

Step 2. Draw random horizontal and vertical lines covering the curved areas to form a square grid pattern. The closer the grid line spacing, the greater the accuracy of the final drawing. Mark each point of intersection between the grid lines and the curved line with a dot.

Step 3. Draw the isometric plane containing the curved areas. Draw the same grid spacing on the isometric plane. Transfer the position of the dots from the orthographic view to similar points on the isometric grid.

Step 4. Connect the points with a smooth, freehand-sketched line.

Step 5. From each point on the curve, project a 30° line upward to the right.

Step 6. Measure and transfer the thickness of the object at each point, to points representing the back side of the curved surfaces. If the thickness of the object is the same throughout, these distances will all be equal.

Step 7. Connect the points on the back surface of the curved areas with a freehand-sketched line to complete the layout.

Step 8. Complete the drawing with instruments and an irregular curve.

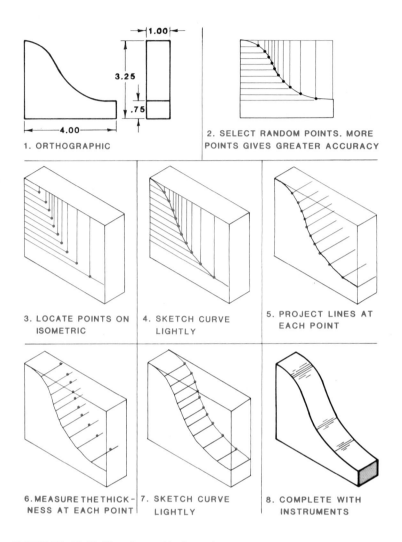

FIGURE 15–17 Plotting isometric irregular curves

Isometric Sections. Isometric drawings are normally used to show only the exterior appearance of an object. When the inside of an object must be shown in pictorial form, an isometric section is also prepared. The two basic types of isometric sections are the full isometric section and the half isometric section.

Isometric sections, like orthographic sections, follow a cutting plane line which defines the edge of the area to be sectioned. In a **full isometric section,** the cutting plane extends across the entire width or height of an object **(Fig. 15–18)**. A full isometric section cuts the object in half to show the inside half of the object. In a **half isometric section,** the cutting plane extends only halfway through the object. Half sections show only a one-fourth cutaway view of the object **(Fig. 15–19)**.

Symmetrical objects are usually half-sectioned to their centerline.

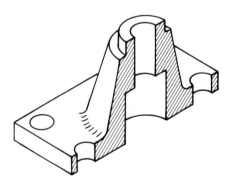

FIGURE 15–18 Isometric full section

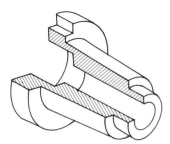

FIGURE 15-19
Isometric half section

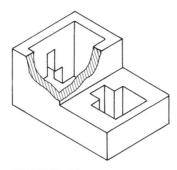

FIGURE 15-20
Isometric broken-out section

When objects are not symmetrical, or do not contain a centerline, a separate cutaway isometric section may be prepared. In **Fig. 15-20**, part of the wall is removed to reveal details on the back surface. **Cutaway views** are also used to show the shape of an object which may be difficult to interpret from the isometric drawing. The drawing of the nuclear reactor vessel includes a cutaway section to expose the shape of the interior of the vessel **(Fig. 15-21)**.

Sometimes cutaway sections are removed, to show the shape of an object without changing the form of the isometric drawing **(Fig. 15-22)**. There are two methods of drawing isometric sections. The first method involves drawing the complete object lightly **(Fig. 15-23)**. The cutting plane line or outline of the cutaway area is drawn. Lines representing the cutaway area are erased to show the section. In the second method, only the sectioned area part is drawn **(Fig. 15-24)**.

Isometric Exploded Drawings. Isometric drawings are ideal for showing various parts of a multipart

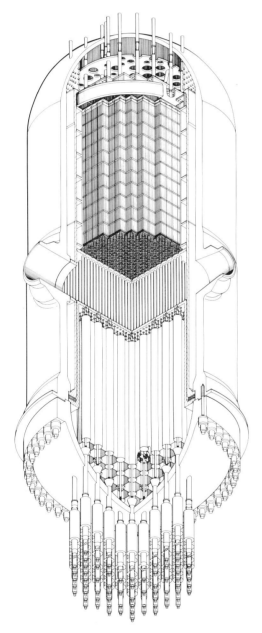

FIGURE 15-21 Isometric cutaway sectional drawing of a nuclear reactor vessel (Courtesy of Combustion Engineering, Inc.)

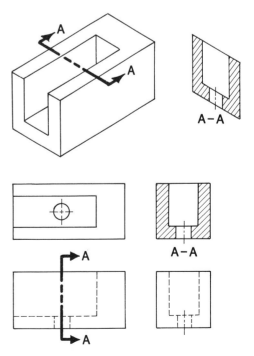

FIGURE 15-22 Isometric removed section

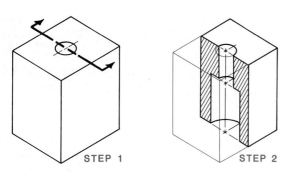

FIGURE 15-23 Isometric section

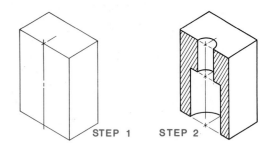

FIGURE 15-24 Lay out only the isometric section.

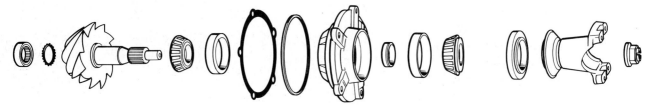

FIGURE 15-25 Example of an exploded-view isometric drawing used for assembly instruction

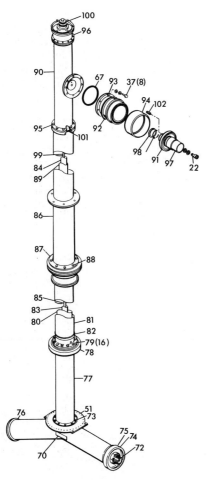

FIGURE 15-26 Exploded-view drawings are used to label parts for identification.

component in related position **(Fig. 15-25)**. Drawings of this type are exploded-view drawings because the parts are separated (disassembled). The parts are aligned and related to their assembled position. Exploded views allow each part to be drawn showing their full shape. This eliminates hidden areas that do not show when the parts are assembled.

Exploded view assembly drawings also allow for the labeling and identification parts **(Fig. 15-26)**. In very complex assembly drawings, connecting lines show the relationship of parts when assembled **(Fig. 15-27)**. This technique provides a guide for the assembly of parts to produce the finished component. Exploded views may be accompanied by an assembly drawing of the component with the parts all assembled **(Fig. 15-28)**.

Isometric Dimensioning. Since isometric drawings are normally not used as working drawings, detailed dimensions are not found on most isometric drawings. However, when isometric drawings of very basic objects are used for manufacturing or construction purposes, dimensions, notes, and specifications are included.

Chapter 15 Pictorial Drawings 259

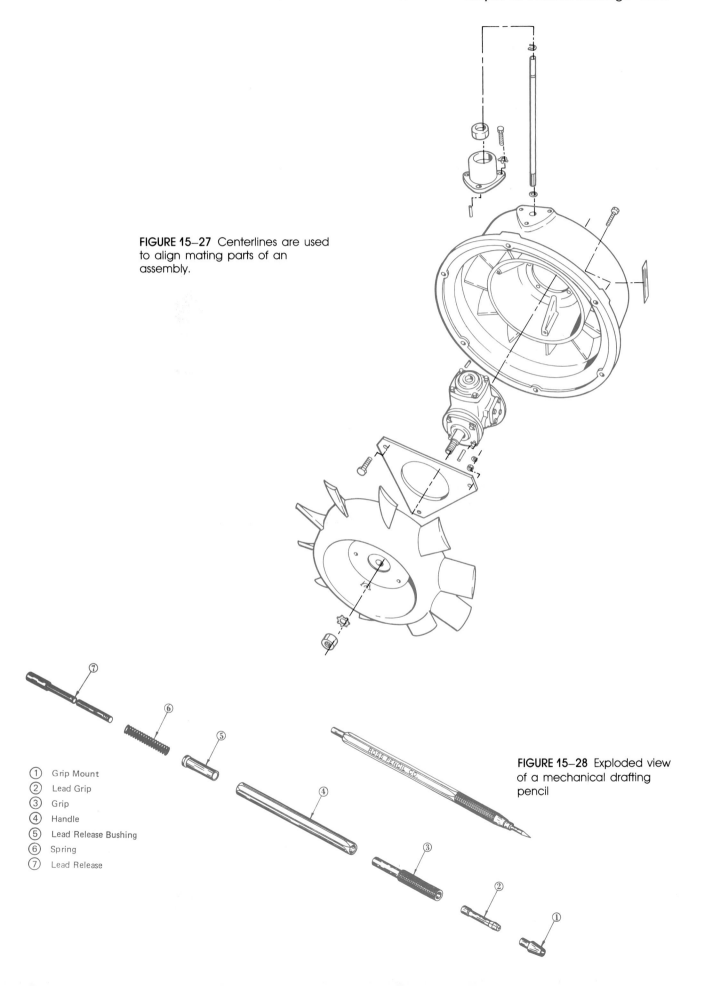

FIGURE 15–27 Centerlines are used to align mating parts of an assembly.

① Grip Mount
② Lead Grip
③ Grip
④ Handle
⑤ Lead Release Bushing
⑥ Spring
⑦ Lead Release

FIGURE 15–28 Exploded view of a mechanical drafting pencil

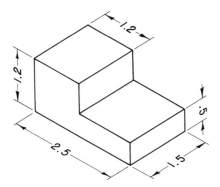

FIGURE 15–29 Aligned isometric dimensions

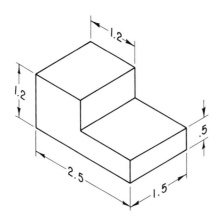

FIGURE 15–30 Unidirectional isometric dimensions

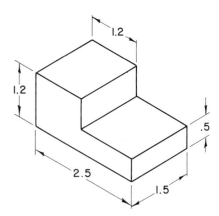

FIGURE 15–31 Vertical plane dimensioning on an isometric drawing

Dimension lines, extension lines, arrowheads, letters, and numerals on isometric drawings are similar to those on orthographic drawings. Isometric dimensions and extension lines all follow the isometric axis. Arrowheads are also drawn in the same plane as dimension and extension lines **(Fig. 15–29)**.

There are two methods of dimensioning isometric drawings—the aligned method and the unidirectional method. In **aligned dimensioning,** all isometric letters, numerals, and arrowheads are aligned with isometric planes (Fig. 15–29).

Unidirectional dimensioning (Fig. 15–30) is generally preferred for engineering working drawings. In the unidirectional method, all isometric numerals and letters are positioned vertically. A variation of unidirectional dimensioning is **vertical plane dimensioning (Fig. 15–31)**. In this method, all dimension and extension lines are drawn in the isometric plane, with the numerals placed horizontally. Straight horizontal lettering is simpler and faster to use on isometric drawings. It is the method preferred in this book because the numbers and text generated for the multiview drawings may also be used for the pictorial drawings.

Isometric Grids. To save time in preparing isometric drawings, isometric grid paper is often used **(Fig. 15–32)**. The grid lines are printed

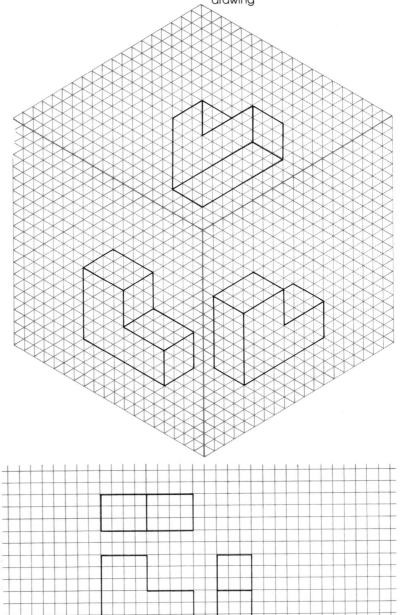

FIGURE 15–32 Isometric grid paper

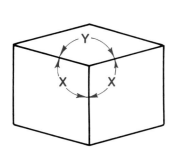

FIGURE 15-33 In a dimetric drawing the angles may vary, but two of the angles must be equal.

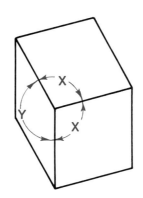

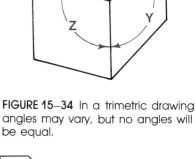

FIGURE 15-34 In a trimetric drawing angles may vary, but no angles will be equal.

on 30° angles representing the isometric axis. Grid lines are often printed in nonreproducible blue. This enables an isometric drawing to be prepared directly on the grid paper without the grid lines appearing on the final print. The grid paper may be used as an underlay guide in preparing isometric drawings. The use of grids is a great timesaver for sketching and isometric layout work.

Dimetric Drawings

In dimetric drawings, the receding angles may vary, but two of the three faces and axes are always equal **(Fig. 15-33)**. The method of drawing dimetric pictorials is similar to preparing isometric drawings. Once the axis angles are established, all lines extending in the same direction must be kept parallel.

Trimetric Drawings

In isometric drawings, all three angles around the isometric axis are the same. In dimetric drawings, two of the angles are the same. In trimetric drawings, all three of the angles around the axis are different **(Fig. 15-34)**. Trimetric drawings are prepared similar to isometric and dimetric drawings, except that each pictorial plane is projected on a different angle. All lines in each of the three planes must be parallel. This three-angle difference makes the preparation of trimetric drawings very time-consuming. Trimetric drawings are also difficult to dimension since each dimension and extension line must be drawn parallel to each of the three different planes of projection.

The angles of trimetric drawings and, to a lesser extent dimetric drawings, can be adjusted to any convenient angle to provide more realism and emphasize one side while making another side recede. This is closer to the way objects are normally viewed in isometric drawings.

OBLIQUE DRAWINGS

In axonometric drawings (isometric, dimetric, and trimetric), the object is drawn as viewed from its nearest edge. In oblique drawings, the object is drawn as viewed from one of its orthographic faces. The front face of an oblique drawing is parallel to the front plane of projection. This means that the front face is drawn to exact size and shape. The other two surfaces are projected from this front

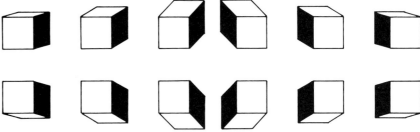

FIGURE 15-35 Alternate positions for oblique drawings. Receding lines are usually at 15°, 30°, or 45°.

face at any convenient angle (usually 30° or 45°) and in any direction **(Fig. 15-35)**. The angle and direction depend on the nature of the object and the planes to be emphasized. There are two types of oblique drawings—cavalier and cabinet.

Cavalier Drawings

When the lengths of the receding angles of an oblique drawing are drawn to full scale, the drawing is a **cavalier oblique drawing (Fig. 15-36)**. If the length of the receding lines is excessive, a cavalier drawing will appear distorted.

Because the front face is both geometrically and dimensionally accurate, the front face can be dimensioned the same as in an orthographic drawing. The receding planes can be dimensioned as in isometric drawings, with the dimension and extension lines drawn parallel to the angle of the plane.

Cavalier drawings are especially appropriate for objects with a shal-

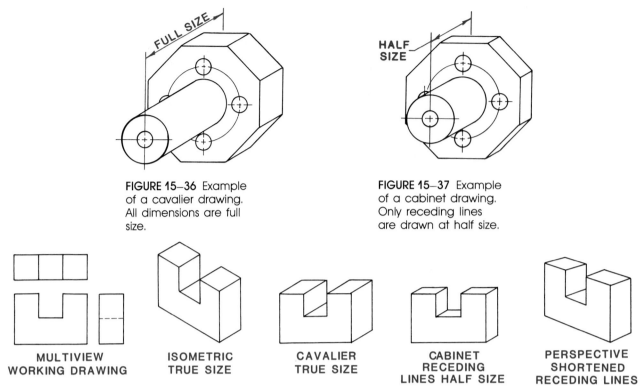

FIGURE 15-36 Example of a cavalier drawing. All dimensions are full size.

FIGURE 15-37 Example of a cabinet drawing. Only receding lines are drawn at half size.

FIGURE 15-38 Comparison of pictorial drawings

low depth dimension, since depth distortion is minimized. The largest area is usually selected as the front plane to avoid as much distortion as possible on the receding planes. Since the front plane of a cavalier drawing is geometrically accurate, the front face should contain the greatest amount of detail or irregularity.

Cabinet Drawings

To reduce the amount of visual distortion, long receding lines are sometimes reduced by one half, but dimensioned true size. When this is done, the drawing is known as a **cabinet oblique drawing (Fig. 15-37)**. This creates a more realistic drawing, but eliminates the use of true scale dimensions on the receding planes.

Figure 15-38 shows a comparison of the types of pictorial drawings. The drafter must be familiar with all pictorial drawings methods as to select the best possible type for a set of working drawings.

Oblique Circles

Because the front face of an oblique drawing is not distorted, angles, circles, and arcs drawn on this surface are true and accurate. For this reason, the front face is chosen to show the circular end view of a cylinder hole or other curved surface **(Fig. 15-39A)**. Circles and arcs shown on the receding planes of an oblique drawing will appear as ellipses. This is similar to the way they appear on axonometric receding planes **(Fig. 15-39B)**.

Both isometric circles and oblique circles on receding planes are drawn as ellipses. When plotting circles on oblique receding planes, follow the steps in **Fig. 15-40**:

Step 1. Locate the center of the circle on the inclined plane, and draw vertical and horizontal centerlines paralleling the angle of the plane.

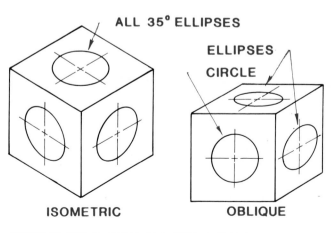

FIGURE 15-39A Isometric and oblique circles

Chapter 15 Pictorial Drawings **263**

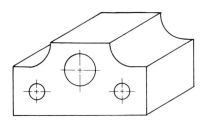
EFFICIENT LAYOUT

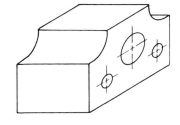
POOR LAYOUT

FIGURE 15–39B Layout for isometric and oblique circles

Step 2. Draw a square about the centerlines representing the diameter of the circle. Draw the diagonals.

Step 3. Transfer points of the true circle's intersections with the diagonals to the oblique surface.

Step 4. Darken with an irregular curve or ellipse template.

The accuracy of the ellipse on an oblique drawing is not critical. A faster method of drawing an ellipse is shown in **Fig. 15–41**.

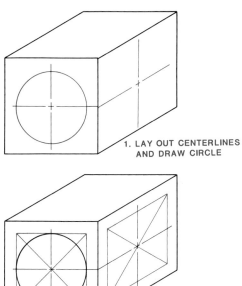
1. LAY OUT CENTERLINES AND DRAW CIRCLE

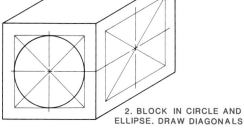
2. BLOCK IN CIRCLE AND ELLIPSE. DRAW DIAGONALS

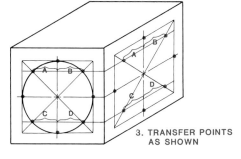
3. TRANSFER POINTS AS SHOWN

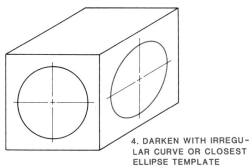
4. DARKEN WITH IRREGULAR CURVE OR CLOSEST ELLIPSE TEMPLATE

FIGURE 15–40 Constructing oblique circles

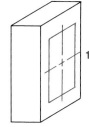
1. DRAW CENTERLINES AND BLOCK IN CIRCLE (DEPTH FOR CABINET DRAWINGS ARE HALF SIZE)

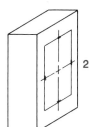
2. CIRCLE IS TANGENT AT CENTER LINES

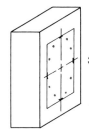
3. ESTIMATE POINTS

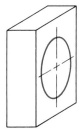
4. DARKEN WITH CLOSEST ELLIPSE IN AN ELLIPSE TEMPLATE

FIGURE 15–41 Estimation procedure for drawing an ellipse

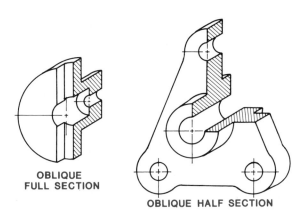

FIGURE 15-42 Oblique sections

Oblique Sections

Oblique full sections and oblique half sections are drawn in a similar fashion as isometric sections. When drawing oblique sections, they are usually cut through the receding plane **(Fig. 15-42)**. To draw an oblique section, draw the front orthographic view. Project the receding lines, beginning at the points of intersection between the cutting plane line and front plane lines. Extend these lines on the established angle to the depth of each part of the object.

PERSPECTIVE DRAWINGS

The receding sides of axonometric and oblique drawings are parallel. This creates visual distortion because the human eye does not perceive objects in this manner. The human eye sees close-up objects in true size, detail, and color. Distant objects are seen as smaller, less detailed, and paler in color. The combination of these factors gives the eye the ability to perceive the dimension of depth. Look at the perspective drawing in **Fig. 15-43**. Notice how the lines of the road are not parallel in the picture. Yet, in reality the two sides of the road are actually parallel. The camera, like the eye, sees objects in perspective. That is, parts of the object further from the eye appear to get progressively smaller as the viewing distance is increased. Perspective drawings, unlike axonometric drawings, are prepared with the more distant areas proportionally smaller to create the illusion of depth **(Fig. 15-44)**.

Perspective Uses

As previously mentioned, axonometric and oblique drawings of large objects appear greatly distorted because the large receding sides are parallel. For this reason, perspective drawings are often prepared for large objects, such as buildings, cars, boats, and machinery. **Figure 15-45** is a perspective drawing of a building housing a steam generator. Notice the increased realism found in the perspective drawing.

FIGURE 15-43 A perspective view gives the illusion of depth.

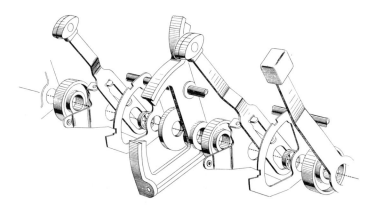

FIGURE 15-44 Machined parts drawn in perspective will appear realistic to the eye. (Courtesy of General Motors Corporation, Engineering Standards)

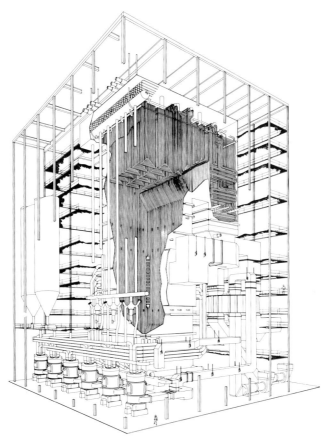

FIGURE 15-45 Perspective drawings show depth with the diminishing lines. (Courtesy of Combustion Engineering, Inc.)

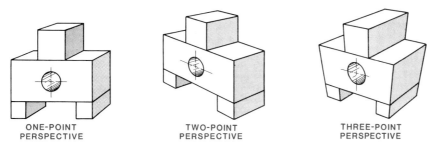

FIGURE 15-46 Examples of one-, two-, and three-point perspective drawings

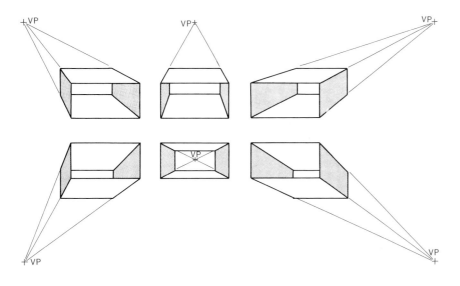

Perspective Types

There are three basic types of perspective drawings: one-point perspective, two-point perspective, and three-point perspective **(Fig. 15–46)**. The type of drawing used depends on the size and shape of the object, and the preferred viewing angle.

One-Point Perspective

A **one-point perspective** (parallel) drawing is similar to an oblique drawing except the receding lines are not parallel. The receding lines of a one-point perspective drawing meet at a distant location called a **vanishing point (Fig. 15–47)**. Like oblique drawings, the front face is drawn at true scale. The receding planes can be projected in any direction from this true front plane.

The vanishing point can be located any distance from the horizon. When the distance from the horizon to the vanishing point is short, the angles of the receding sides will be more acute **(Fig. 15–48)**.

Before preparing a one-point perspective drawing, the concepts of station point, picture plane, and lines of vision must be understood **(Fig. 15–49)**. The **station point** represents the location of the observer. The final drawing will therefore appear as viewed from the station point. The **picture plane** represents the vertical plane upon which the object is viewed. The distance from the station point to the picture plane determines the size of the drawing. An object on a picture plane closer to the station point will appear smaller than the same object on a picture plane further from the station point. **Lines of vision** are imaginary lines connecting the station point with the picture plane and object.

FIGURE 15-47 Receding lines can be projected to vanishing points at any position.

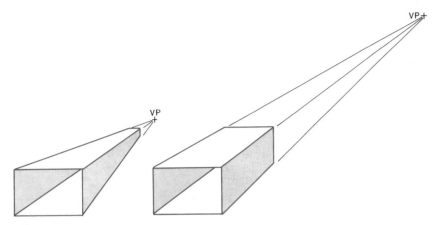

FIGURE 15–48 The distance or position that the vanishing point is placed from the drawing will affect its final form.

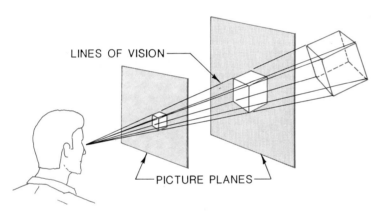

FIGURE 15–49 The location of the picture plane controls the size of the perspective drawing.

To draw a one-point perspective of the exterior of an object, follow the steps outlined in **Fig. 15–50**:

Step 1. Draw the front and top orthographic views of the object. Draw horizontal lines representing the picture plane, horizon, and ground line. The **horizon** is a horizontal line upon which the vanishing points are placed. The **ground line** is the base line upon which the front plane is placed.

Step 2. Locate the station point; that is, the exact position from which the object (top view) will be viewed. The final drawing will appear as viewed from this point. Connect each back corner of the top view with the station point.

Step 3. Project a vertical line from the station point up to intersect the horizon line. This point of intersection is the vanishing point.

Step 4. Connect each corner of the front view with lines to the vanishing point.

Step 5. Establish points of intersection between each of the station point connecting lines and the lines to the vanishing point. Draw vertical lines down from these points

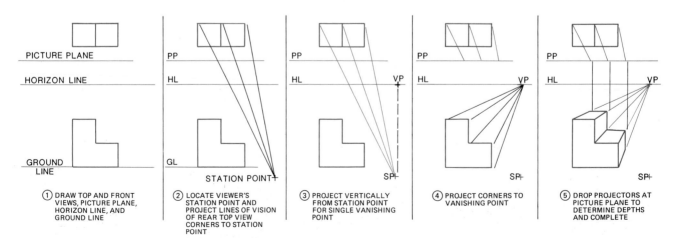

FIGURE 15–50 Steps for drawing with one-point perspective

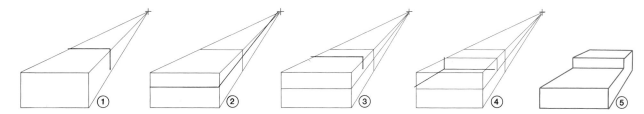

FIGURE 15–51 Laying out and blocking in a one-point perspective

to intersect lines that connect similar points on the front plane with the vanishing point. Draw horizontal and vertical lines parallel with the lines on the front plane to represent the back of the object. To finish the drawing, connect all visable lines with an object line and erase all construction lines. Shade if required.

If part of the front plane's detail is missing, lightly block in the entire front plane, including the empty areas **(Fig. 15–51)**. This provides a base area from which to project lines to the vanishing point. Project light lines from the corners of the front face to the vanishing point. Draw the outline of the object on the receding sides using heavy lines. Erase the construction lines to complete the drawing.

One of the most common uses for a one-point perspective drawing is to show the interior view of a building **(Fig. 15–52)**.

Two-Point Perspective

Two-point perspective (angular) drawings **(Fig. 15–53)** are most popular because they provide the greatest amount of realism with a minimum amount of complex projection. Two-point perspective drawings are prepared when viewed from a corner, not from a front orthographic face. Each of the two sides are projected to separate vanishing points on the horizon. **Figure 15–54** shows how the two vanishing points relate to the object, picture plane, horizon and station point.

Vertical Position Alternatives

The vertical position of the horizon line in relation to the object affects the amount of each surface on the final drawing **(Fig. 15–55)**. When the horizon is placed below the object, the bottom and two sides show (producing a worms-eye view). When the horizon is placed above the ob-

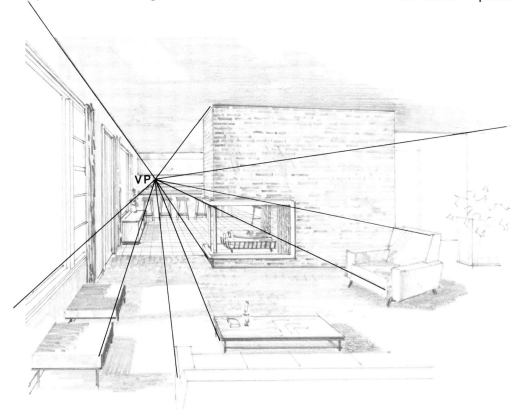

FIGURE 15–52 Example of a one-point interior perspective (Courtesy of Home Planners, Inc.)

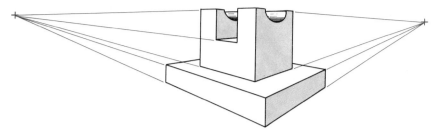

FIGURE 15-53 Example of a two-point perspective

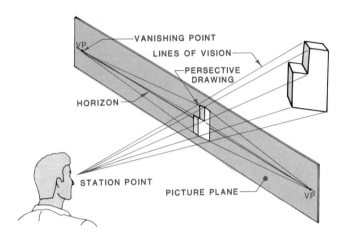

FIGURE 15-54 Deciding factors that make up a two-point perspective drawing

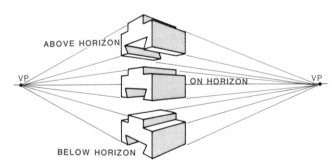

FIGURE 15-55 The vertical position of the viewer affects the perspective drawing.

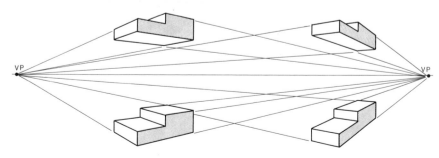

FIGURE 15-56 As the viewer's position changes, the perspective drawing will change.

ject the top and two sides show (producing a birds-eye view). When the horizon extends through the object, only the two sides show. This view more closely relates to an eye-level view, especially for large objects that are higher than eye level. The vertical position of the horizon is selected to provide the most appropriate viewing angle.

Station Point Position

Just as the placement of the horizon greatly affects the vertical view of the object, the position of the station point changes the horizontal view of the object. Moving the station point closer to one vanishing point, and farther away from the other, results in the differences shown in **Fig. 15-56**.

Vanishing Point Location

In addition to the vertical location of the horizon and the horizontal location of the station point, the final appearance of the drawing is greatly affected by the placement of the vanishing points on the horizon. **Figure 15-57** shows a perspective drawing with widely spaced vanishing points compared with a drawing with closely spaced vanishing points. To insure the base angle is more than 90°, spread the vanishing points apart on the horizon line and/or draw the perspective layout closer to the horizon line **(Fig. 15-58)**. The station point should be located below the picture plane at a distance of approximately two times the width of the object (Fig. 15-58).

Estimating Procedure

Since perspective drawings are not used as working drawings, drawing to precise dimensional limits is usually not required. When only rough proportions are sufficient, a two-point

Chapter 15 Pictorial Drawings **269**

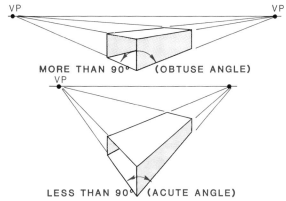

FIGURE 15–57 Placing the vanishing points too close together will cause the drawing to appear distorted.

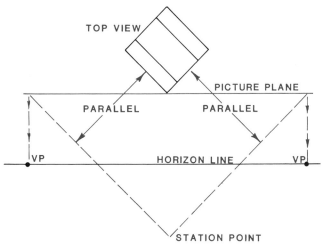

FIGURE 15–58 Locating the vanishing point's position on the horizon

perspective drawing can be prepared by using the following estimating procedures **(Fig. 15–59)**:

Step 1. Draw a horizontal line and locate the position of two vanishing points. Draw a vertical line representing the leading edge of the object above or below the horizon, depending on the view required.

Step 2. Draw lines connecting the front corner ends with each vanishing point.

Step 3. Estimate the length of the object compared to the height of the front corner line. Mark the estimated length and depth with vertical lines drawn between the receding lines.

Step 4. Draw lines connecting the back receding corners to the opposite vanishing point.

Step 5. Draw details by estimating the position of features on the receding planes.

Step 6. Connect intersecting points with lines extending to the opposite vanishing point.

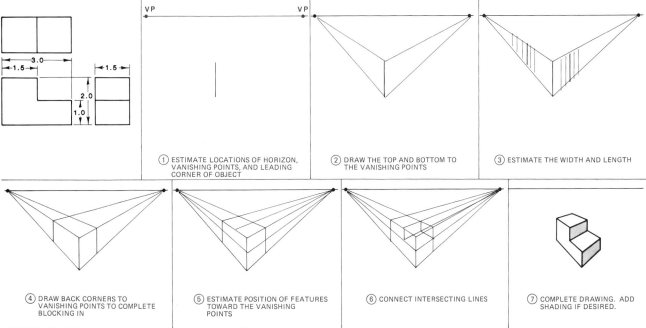

FIGURE 15–59 Drawing a two-point perspective drawing using an estimating procedure

Step 7. Complete by erasing the construction lines, darkening the object lines, and shading.

Construction Procedure

When a more dimensionally accurate two-point perspective drawing is required, the relationship of the station point, vanishing point, horizon, and picture plane must be precisely controlled. The following is the procedure for constructing a two-point perspective drawing which will closely approximate the true appearance of the object: **(Fig. 15–60)**

Step 1. Draw horizontal lines representing the position of the picture plane, horizon, and ground line.

Step 2. Draw the orthographic top view of the object, rotated to the desired viewing angle above the picture plane line. Draw the orthographic front view of the object on the ground line.

Step 3. Locate the station point, and project lines of sight to the picture plane parallel to the sides of the top view. Draw vertical lines downward from the picture plane to intersect the horizon to locate the vanishing points.

Step 4. Draw lines connecting all corners of the object with the station point.

Step 5. From the points of intersection between the projected lines and the picture plane line, project vertical lines downward to intersect the ground line. Mark the line representing the front corner line.

Step 6. From the side view, project the object's height with horizontal lines to intersect the front corner line.

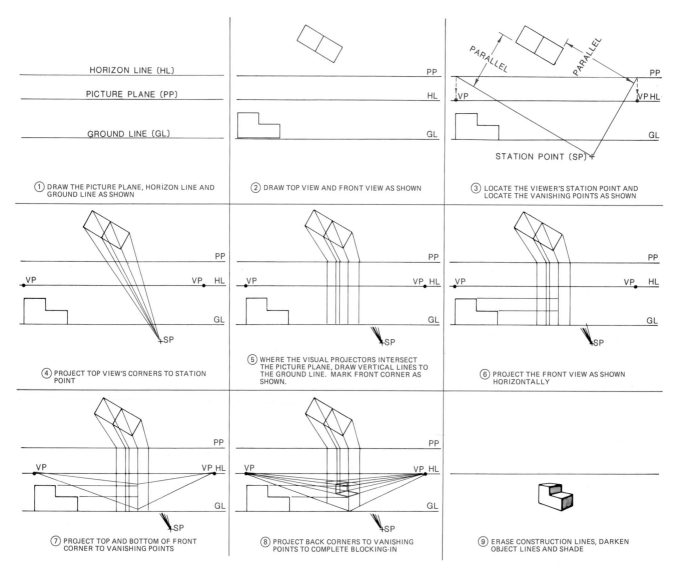

FIGURE 15–60 Projection procedure for a two-point perspective drawing

Step 7. Establish the height of the front corner line by marking the intersection between the vertical and horizontal projection lines. Project the lines from the bottom and top of the front corner line to each vanishing point. Draw vertical lines representing the width and depth of the object between the receding lines.

Step 8. Project lines from the back corners to the opposite vanishing points. Locate the position of depressions, cutaway areas, extensions, and holes on the planes. Project lines to the vanishing points to establish the final shape. Adjust the edges for any fillets and round.

Step 9. Darken the major object lines and erase all construction lines. Shade and render if desired.

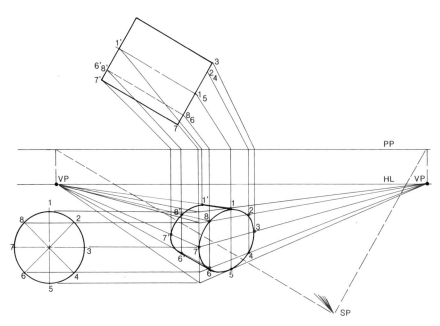

FIGURE 15-61 Projection procedure to draw circles in two-point perspective drawing

Perspective Ellipses

Circles on perspective drawings, as in axonometric drawings, are shown as ellipses. Since, the receding lines of a perspective drawing are not parallel, ellipses are not symmetrical.

To be totally accurate, ellipses must be projected as shown in **Fig. 15-61**. Small perspective ellipses can usually be drawn using an ellipse template. To use an ellipse template, follow the same procedure for using a template on axonometric drawings. The exception is that all centerlines and circle construction lines are projected to vanishing points. Once the outline of circle is established, choose the appropriate ellipse template opening size. Position the opening as close as possible to align with the blocked-in area **(Fig. 15-62)**.

Perspective Grids

Perspective grid paper, like isometric grid sheets, are often used as a guide in preparing perspective drawings. Grid sheets are available for one-, two-, and three-point perspective drawings, although two-point grids are the most popular. Grids are also available for either interior or exterior perspective drawings. Grids **(Fig. 15-63)** are plotted on specific pre-set positions of the vanishing point, station point, and picture plane. This limits the options in selecting the exact vertical and horizontal position of the view and angles of the receding sides. Although excellent timesavers, the theory of perspective projection must be thoroughly understood to effectively use perspective grids.

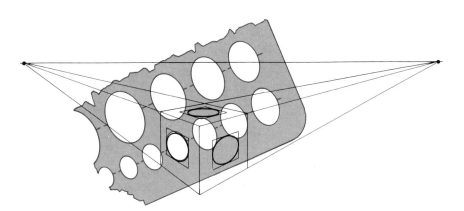

FIGURE 15-62 To draw an estimated perspective circle, block in a perspective square and find the closest ellipse on an ellipse template.

Three-Point Perspective

When a two-point perspective drawing contains long, narrow, vertical planes that extend far from the horizon, the object will appear distorted. This optical illusion is created because the vertical lines are parallel. This is not consistant with the normal way the eye sees the object. **Figure 15-64** shows a distorted two-point perspective drawing of this type. **Three-point perspective** drawings are used to correct this illusion by providing a third vanishing point **(Fig.**

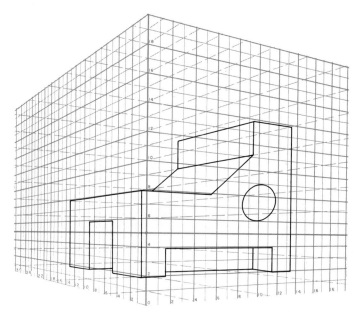

FIGURE 15–63 Drawing on a perspective grid is fast and will help to produce well-proportioned perspective drawings.

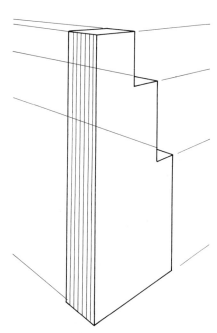

FIGURE 15–64 Tall structures will appear distorted with a two-point perspective drawing.

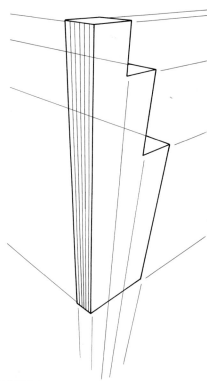

FIGURE 15–65 Three-point perspective drawings for tall structures adds realism.

15–65). There are no parallel lines on a three-point perspective drawing since all lines connect to one of the three vanishing points. Lines representing horizontal surfaces are connected to vanishing point one and two, just as in two-point perspective drawings. All lines representing vertical planes are connected to vanishing point number three.

In preparing three-point perspective drawings, follow the same procedures as outlined in drawing two-point perspectives. Instead of drawing vertical parallel lines representing the width and depth of the object, draw vertical lines to a third vanishing point. The third vanishing point is located below the object on a line projected vertically down from the front corner line. If the viewer is at the base of the building, the third vanishing point is placed above the object. This gives the added illusion of depth to the drawing. **Figure 15–66** is an industrial perspective drawing of an automobile engine.

Chapter 15 Pictorial Drawings **273**

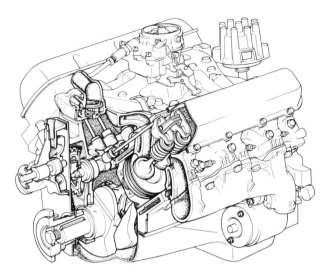

FIGURE 15-66 A perspective drawing of a V-8 automobile engine (Courtesy of General Motors Corporation, Engineering Standards)

COMPUTER-AIDED DRAFTING & DESIGN

COMPUTER-AIDED PICTORIAL DRAWINGS

Drawing pictorial drawings with a CAD system varies greatly due to the drawing software available. Many of the less expensive software CAD systems that are run on personal computers are primarily two-dimensional systems, and have limited pictorial capabilities.

There are systems that are called two and one-half dimensional systems. These systems draw **wire frame** three-dimensional drawings. **Figure 15-67** is an example of a wire frame drawing of a 12-button puck.

More expensive systems are truly three-dimensional. These systems do wire frame drawings as well as **solid modeling (Fig. 15-68)**. Some of these systems automatically construct the pictorial drawing as the orthographic drawing is being drawn.

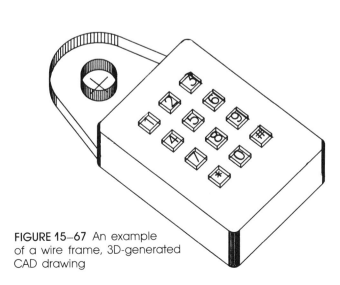

FIGURE 15-67 An example of a wire frame, 3D-generated CAD drawing

FIGURE 15-68 An example of solid modeling 3D-generated CAD drawing

Simple Object

Step 1. Set an isometric grid **(Fig. 15–69)**. Notice that this grid differs from a normal grid pattern in that it is at 30° angles. The cursor also changes to match the isometric grid.

Step 2. Set the SNAP so the screen cursor snaps to the grid points **(Fig. 15–70)**. Start drawing the basic shape using the LINE command. Since the cursor is on the isometric axis rather than the normal horizontal and vertical axes, the lines that are drawn are on the proper isometric axis. This greatly simplifies the drawing of isometrics.

Step 3. Move the cursor and draw lines on the isometric axes, to complete the drawing **(Fig. 15–71)**.

Object with Circles

Step 1. Set the isometric grid, and turn on the SNAP **(Fig. 15–72)**.

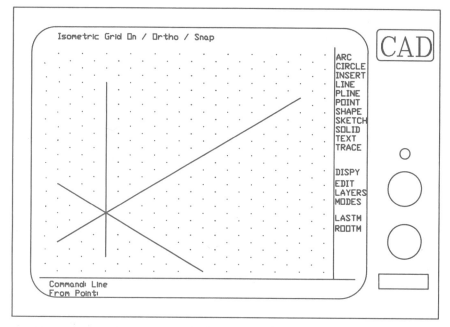

FIGURE 15–69 Reset the grids to an isometric grid. Note that the cursor will match the isometric grids.

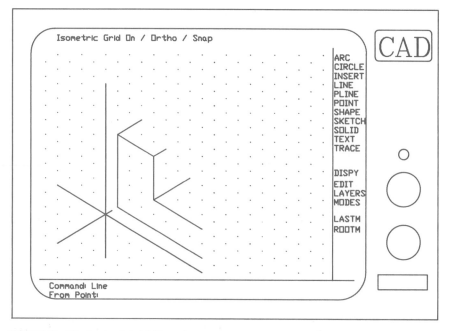

FIGURE 15–70 Go to the LINE command to lay out isometric lines.

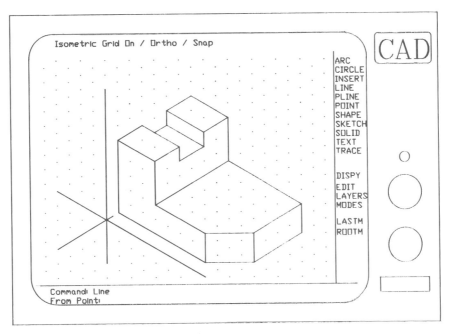

FIGURE 15–71 Complete the isometric drawing with the LINE command and grid SNAP. (Each line end is on a grid point.)

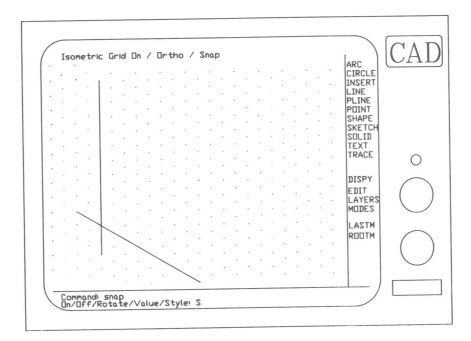

FIGURE 15–72 Set the isometric grids and grid SNAP.

276 Drafting in a Computer Age

Step 2. Draw the basic shape using the LINE command **(Fig. 15–73)**.

Step 3. Locate the centerlines for the ellipses (use a construction layer if available) **(Fig. 15–74)**.

Step 4. Insert the ellipses on the proper centerlines **(Fig. 15–75)**. Erase and edit as is necessary to make clean tangencies.

Step 5. Turn off the construction layer or erase the construction lines **(Fig. 15–76)**.

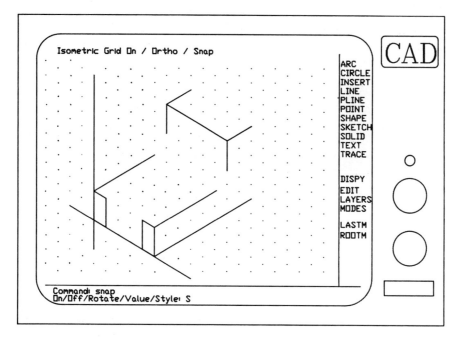

FIGURE 15–73 Go to LINE command and draw the isometric form.

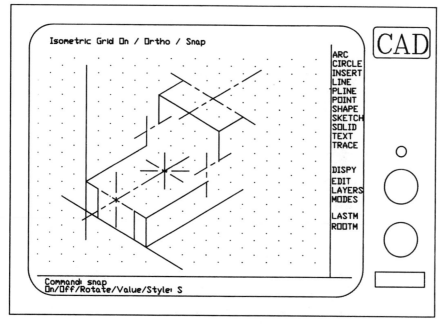

FIGURE 15–74 Go to center lines with the LINE command, and locate the center lines for the isometric circles.

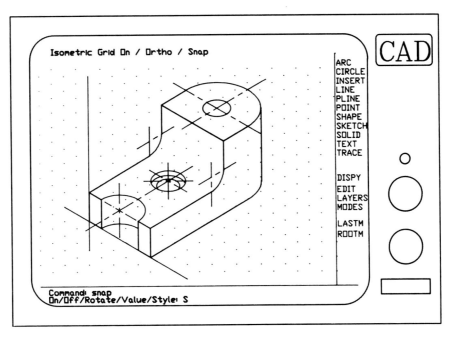

FIGURE 15-75 Go to the ELLIPSE command and set an isometric ellipse. Insert ellipses and edit a clean tangency.

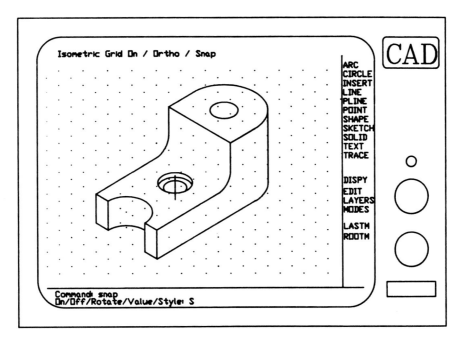

FIGURE 15-76 Remove the construction lines.

EXERCISES

1. Sketch the isometric drawings in **Fig. 15–77**. Make your sketches approximately two times larger. If available, make your sketches on isometric grid paper.

2. Sketch the isometric drawing of each multiview drawing in **Fig. 15–78**. If available, make your sketches on isometric grid paper.

3. With instruments, redraw **Figs. 15–79** through **15–84**. Complete the isometric dimensioning.

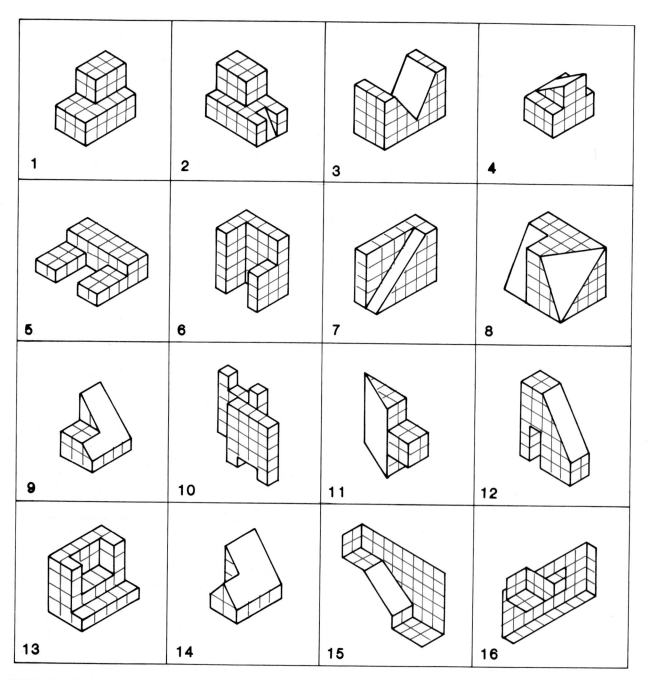

FIGURE 15–77 Sketch each isometric exercise. Each square is ½″.

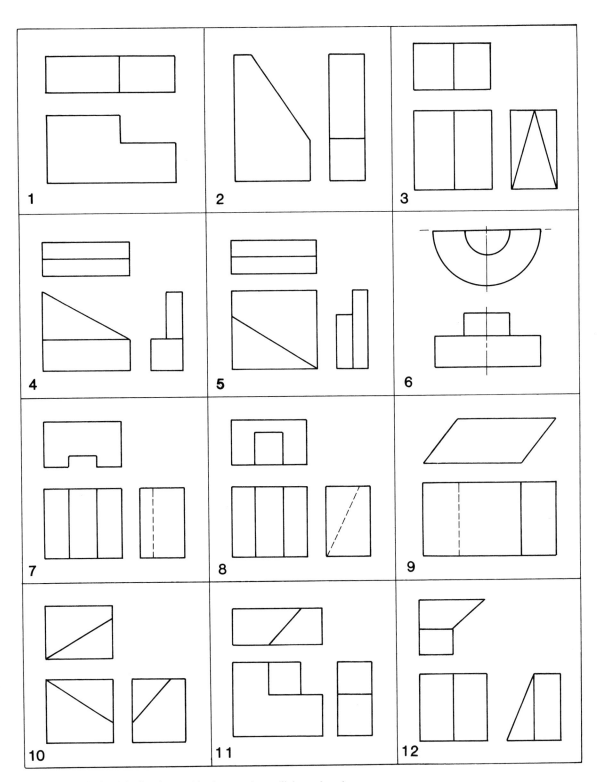

FIGURE 15-78 Sketch the isometric for each multiview drawing.

280 Drafting in a Computer Age

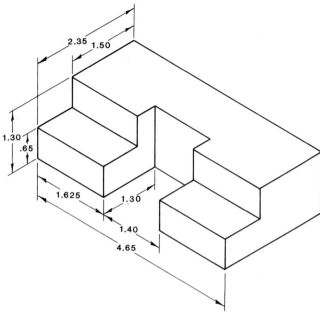

FIGURE 15–79 With drawing instruments, redraw the isometric drawing with dimensions.

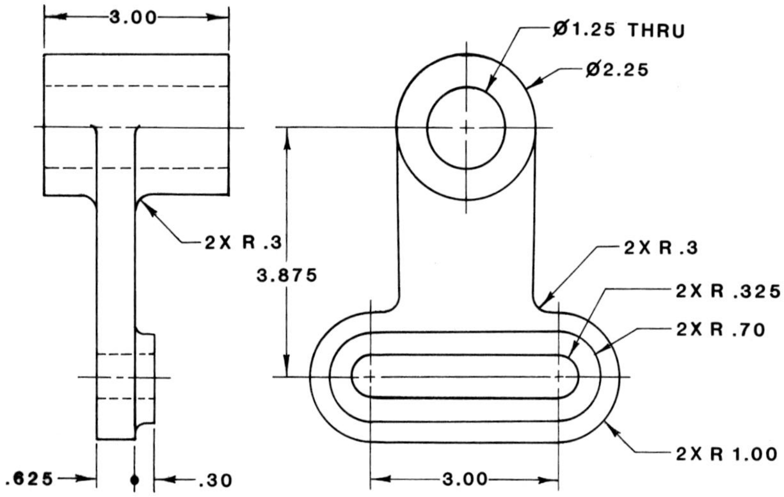

FIGURE 15–80 Redraw the isometric drawing with dimensions.

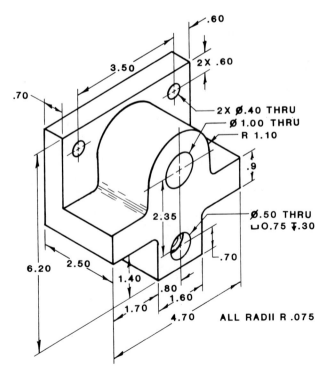

FIGURE 15–81 Redraw the isometric drawing with dimensions.

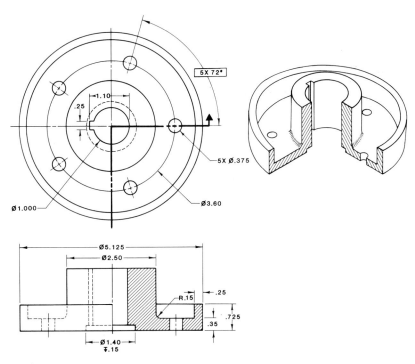

FIGURE 15–82 Redraw the isometric half section with dimensions.

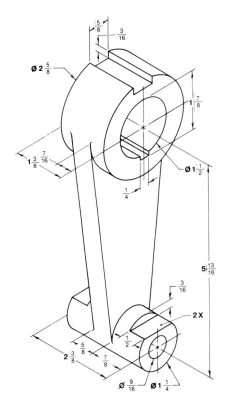

FIGURE 15–83 Redraw the isometric drawing with dimensions.

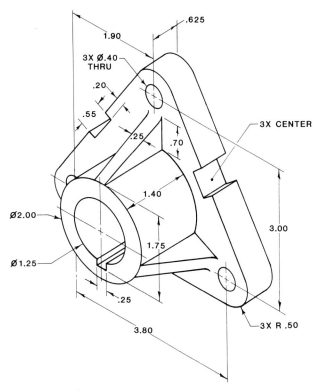

FIGURE 15–84 Redraw the isometric drawing with dimensions.

4. Using the concepts of the cartesian coordinate system, make a freehand sketch on ¼" grid paper from the information in the data table in **Fig. 15–85**.

5. Complete the data table which has been started in **Fig. 15–86**.

6. With a CAD system, follow the steps in Figs. 15–69 through 15–71 as closely as possible.

7. With a CAD system, follow the steps in Figs. 15–72 through 15–76 as closely as possible.

8. With a CAD system, draw Fig. 15–79 through Fig. 15–84.

DATA TABLE					
LN	X	Y	go to	X	Y
1	4	0	–	0	4
2	0	4	–	0	9
3	0	9	–	4	11
4	4	11	–	8	7
5	8	7	–	8	2
6	8	2	–	4	0
7	4	0	–	4	5
8	4	5	–	8	7
			new start		
9	4	5	–	0	9
			FINISH		

FIGURE 15–85 Lay out the drawing from the data table on ¼" grid paper.

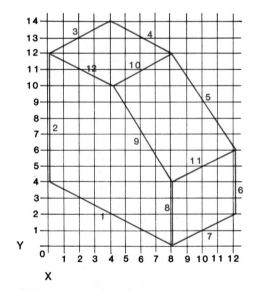

DATA TABLE					
LN	X	Y	go to	X	Y
1	8	0	–	0	4
2	0	4	–		
3			–		
4			–		
5			–		
6			–		
7			–		
8			–		
9			–		
10			–		
			new start		
11			–		
			FINISH		

FIGURE 15–86 Complete the data table of the drawing.

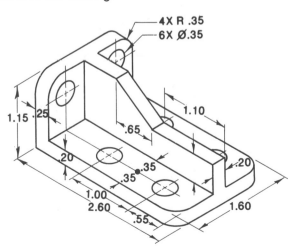

FIGURE 15–87 Draw the isometric drawing of the jig guide support. Make the following ECOs (Engineering Change Orders—refer to Chapter 17):
1. Reduce .20" web to .15".
2. Reduce 6 holes to .25".
3. Reduce four-corner radii to .25".
4. Increase height from 1.15" to 1.25".
5. Increase base thickness to .25".

 **KEY TERMS**

Aligned dimensionally
Axonometric drawing
Block in
Cabinet drawing
Cavalier drawing
Cutaway view
Dimetric
Ellipse template
Ellipses
Exploded view
Fillets
Full isometric sections
Ground line

Half isometric section
Horizon
Isometric
Isometric circles
Isometric arcs
Isometric axis
Isometric lines
Lines of vision
Nonisometric lines
Oblique drawing
One-point perspective
Perspective drawing

Pictorial drawing
Picture point
Receding planes
Rounds
Solid modeling
Station point
Three-point perspective
Trimetric
Two-point perspective
Unidirectional dimensioning
Vanishing point
Wire frame drawing

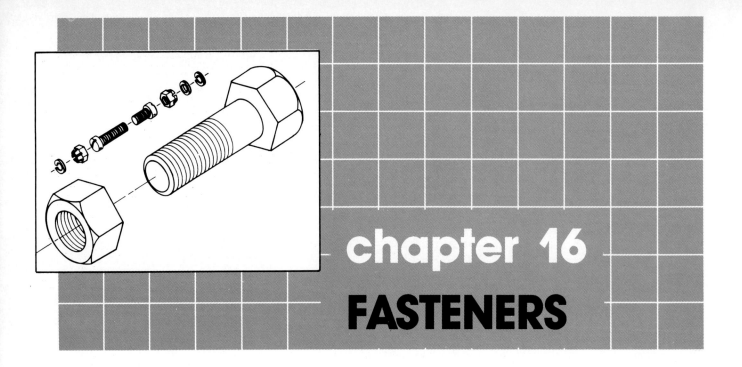

chapter 16
FASTENERS

OBJECTIVES

The student will be able to:

- draw American standard internal and external thread symbols (simplified, schematic, and detailed)
- draw ISO metric internal and external thread symbols (simplified, schematic, and detailed)
- read and understand fastener tables
- recognize the various types of fasteners used in industry

INTRODUCTION

A fastener is any device used to join separate parts or materials. There are two general types of fasteners—removable and permanent. The use of removable fasteners, such as screws, pins, keys, nuts, and bolts, allow attached parts or materials to be separated and reattached. Permanent fasteners such as rivets, nails, and staples, and fastening methods, such as soldering, welding, and gluing, affix parts or materials that never need separating. The **American National Standards Institute (ANSI)**, the **International Organization for Standardization (ISO),** and the **Industrial Fasteners Institute (IFI)** establish the standards for fastening methods and devices.

THREADED FASTENERS

The functioning of all threaded fasteners is based on the principle of the inclined plane. An **inclined plane** is a simple machine that accomplishes work by reducing force, by increasing distance. A screw thread is a spiral inclined plane. An evenly spaced spiral curve around a cylinder is called a **helix. Figure 16–1** shows how an inclined plane is applied to a cylinder to create a helex. The **lead** is the lateral distance a screw moves along the axis with one 360° revolution. The **lead per revo-

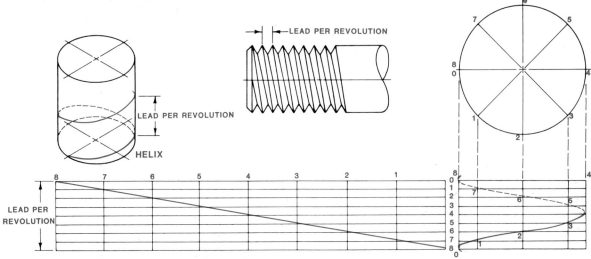

FIGURE 16–1 The thread's form is a helex. One revolution equals the lead.

lution of the inclined plane is equal to the circumference of the cylinder.

Threaded Fastener Nomenclature

To work effectively with threaded fasteners using conventional drafting instruments or a CAD system, the following standard nomenclature must be understood **(Figs. 16–2 and 16–3)**:

- **Angle of thread**—The included angle formed between the major diameter of a thread and the depth of thread.
- **Axis of Fastener**—The cylindrical center of a fastener shaft.
- **Screw Thread Series**—Screws with similar combinations of pitches and diameters are divided into several related series. These are specified in the American National Standard Unified screw threads for screws, bolts, nuts, and other fasteners in ANSI Bl.1 (UN) and in ISO for metric threads. The following is the symbol designation for thread series:
  Course thread series—UNC, NC
  Fine thread series—UNF, NF
  Extra fine thread series—UNEF, NEF
- **Class of Thread Fits**—Threads are divided into the following three tolerance fit classes, 1A, 2A, and 3A (for external threads) and 1B, 2B, and 3B (for external threads):
  Class 1A and 1B—loose fits
  Class 2A and 2B—general purpose fits
  Class 3A and 3B—tight fits
- **Crest**—The point of widest diameter of a screw thread which lies on the major diameter plane.
- **Thread Depth**—The distance from crest (major diameter) to the root (minor diameter) measured perpendicular to the axis.
- **External Thread**—Threads on the outside surface of a cylinder as on a bolt or screw.
- **Internal Thread**—Threads on the inside surface of a hole as in a nut.

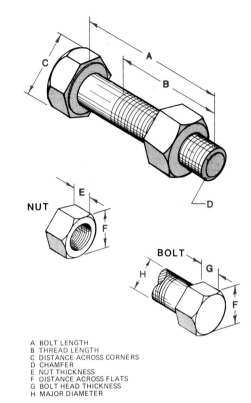

A BOLT LENGTH
B THREAD LENGTH
C DISTANCE ACROSS CORNERS
D CHAMFER
E NUT THICKNESS
F DISTANCE ACROSS FLATS
G BOLT HEAD THICKNESS
H MAJOR DIAMETER

FIGURE 16–2 Nomenclature for nuts and bolts

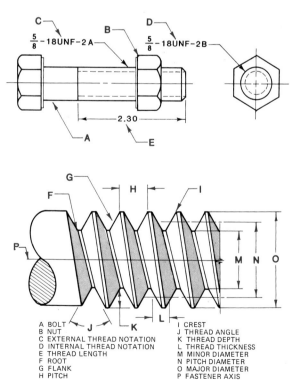

A BOLT
B NUT
C EXTERNAL THREAD NOTATION
D INTERNAL THREAD NOTATION
E THREAD LENGTH
F ROOT
G FLANK
H PITCH
I CREST
J THREAD ANGLE
K THREAD DEPTH
L THREAD THICKNESS
M MINOR DIAMETER
N PITCH DIAMETER
O MAJOR DIAMETER
P FASTENER AXIS

FIGURE 16–3 Nomenclature for threaded nuts and bolts

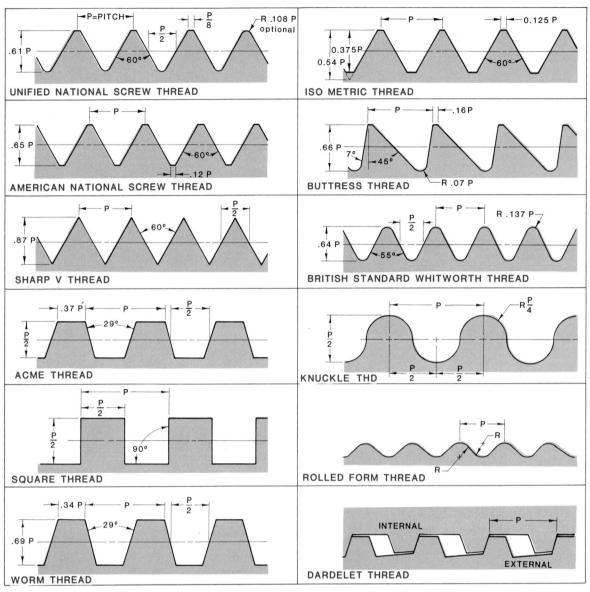

FIGURE 16–4 Thread forms

- **Lead**—The lateral distance the axis of a fastener moves during one 360° revolution of the fastener.
- **Major Diameter**—The outside diameter of the threaded portion of a fastener measured from crest to crest.
- **Minor Diameter**—The diameter of the thread measured from root to root perpendicular to the axis.
- **Root**—The deepest point or valley of a thread.
- **Thread Forms**—The shape of the thread in profile section **(Fig. 16–4)**.
- **Right-Hand Thread**—As viewed from the head, a right hand thread turns clockwise into an object or threaded opening.
- **Left-Hand Thread**—As viewed from the head, a left hand thread turns counterclockwise into an object or threaded opening.
- **Single Thread**—Threads with one continuous helical pattern in which the lead (one full turn) is equal to the single thread pitch **(Fig. 16–5)**.
- **Multiple Threads**—Threads that contain more than a single helical pattern side by side. For example, a double thread (Fig. 16–5) includes two side-by-side helical threads; the lead is two times the pitch. In a triple threaded fastener, the lead is three times the pitch. Multiple threaded fasteners are used where few turns are desired to move the axis a maximum distance.
- **Pitch**—Pitch is the lateral distance between two related points on a thread (crest to crest or root to root). Pitch is expressed in the number of **threads per inch (TPI)**.

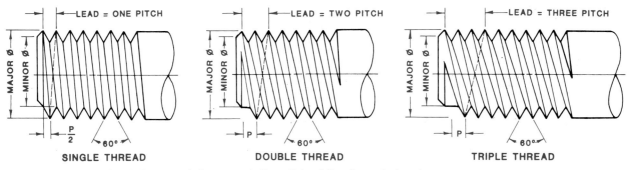

FIGURE 16-5 One revolution equals the pitch of the threaded rod.

- **Pitch Diameter**—The pitch diameter is the diameter of an imaginary line draw halfway between the major diameter and minor diameter **(Fig. 16-6)**.

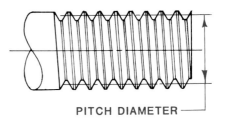

FIGURE 16-6 The pitch diameter lies between the major and minor diameters.

Thread Symbolism

A true orthographic view or section of a threaded fastener is very time-consuming and complex to draw. Since most fasteners shown on working drawings are standard, a true orthographic or pictorial representation is not necessary. Simplified and schematic symbols are used on most working drawings. Simplified symbols show only the alignment of major and minor diameters, and the length of the threaded portion of the fastener. Schematic symbols represent crest and root lines with solid and hidden lines perpendicular to the axis.

Figure 16-7 shows a comparison of a simplified schematic, and pictorial symbol for external threads. **Figure 16-8** shows simplified symbols for internal threads. When threads are shown in section, the three types of symbols are drawn **(Fig. 16-9)**; however, the simplified symbol is most widely used and recommended.

Thread Notations

Regardless of the thread symbol used, all essential information cannot be shown graphically or with

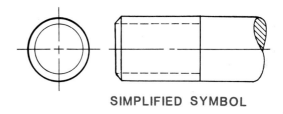
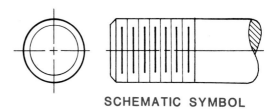
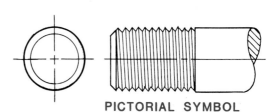

FIGURE 16-7 Symbols for drawing external threads. The simplified symbol is recommended for working drawings.

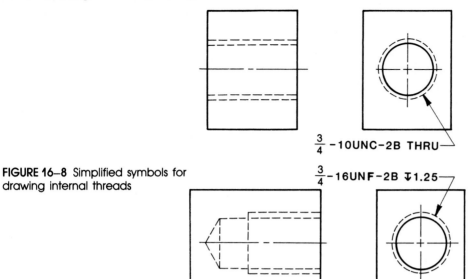

FIGURE 16-8 Simplified symbols for drawing internal threads

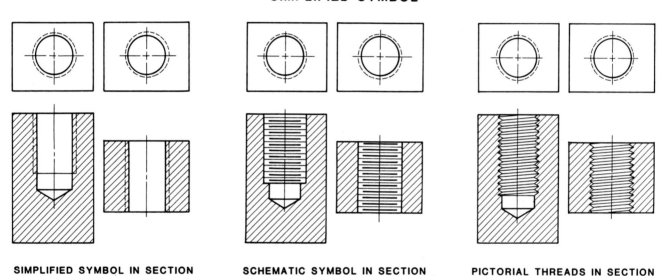

SIMPLIFIED SYMBOL IN SECTION   SCHEMATIC SYMBOL IN SECTION   PICTORIAL THREADS IN SECTION

FIGURE 16-9 Drawing symbols for sectioned internal threads

normal dimensioning methods effectively. Notations are used to show the type, class, series, form, TPI, and major diameter of fasteners on working drawings. **Figure 16-10** shows the notation method used to specify these features for internal threads. **Figure 16-11** shows the notation method used for external threads.

DRAFTING PROCEDURES

A pictorial drawing of an external thread is drawn according to the following steps **(Fig. 16-12)**. (The

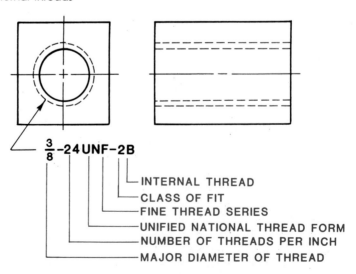

FIGURE 16-10 Placement of internal thread notation's data

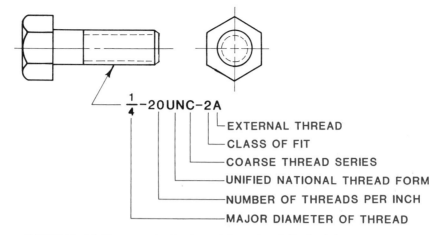

FIGURE 16-11 Placement of external thread notation's data

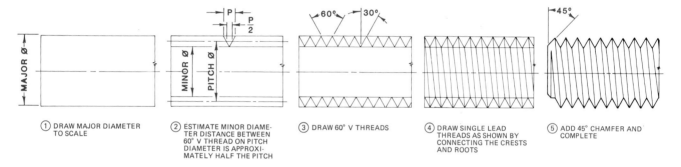

FIGURE 16-12 Steps to draw external pictorial (detailed) V thread symbol

major diameter should be drawn to scale, but all other measurements can be estimated.)

Step 1. Draw the major diameter and axis.

Step 2. Draw the minor diameter and the pitch diameter lines. Draw two 60° lines upward from the minor diameter line (root). As a check, the distance between the 60° lines on the pitch circle line should be about one-half the space of the 60° lines at the major diameter (crest).

Step 3. Continue to draw the 60° V threads for the full length of the threaded rod at the top and bottom.

Step 4. If the threaded is a single lead, connect a line from a peak angle (vertex) on the major diameter to a peak angle on the opposite side, one thread over. Repeat for all the major (crest) and minor (root) diameters to complete the pictorial thread layout.

Step 5. Add a 45° chamfer and erase all construction lines. Darken to complete.

Simplified internal thread symbols are drawn according to the following steps (Fig. 16-13):

Step 1. Draw the drill depth.

Step 2. Draw the major diameter and depth of the thread.

Simplified external thread symbols are drawn according to the following steps (Fig. 16-14):

Step 1. Draw the outline and centerline of the fastener. The object lines, which parallel the axis centerline, represent the major diameter and are drawn to scale.

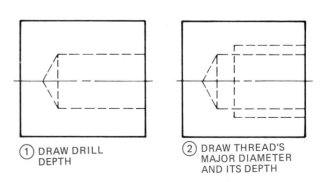

FIGURE 16-13 Steps to draw simplified internal thread symbol

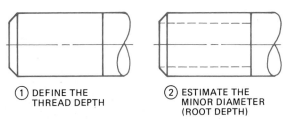

① DEFINE THE THREAD DEPTH

② ESTIMATE THE MINOR DIAMETER (ROOT DEPTH)

FIGURE 16-14 Steps to draw simplified external thread symbol

Step 2. Draw a line representing the threaded length.

Step 3. Draw a 45° chamfer on the opposite end.

Step 4. With construction lines, project a 60° included angle from the major diameter (crest) in order to determine the thread depth (root).

Step 5. At this point (thread depth), draw dashed lines parallel to the axis to represent the minor diameter.

Schematic internal thread symbols are drawn according to the following steps (Fig. 16-15):

Step 1. The tap drill outline and major thread diameter lines are drawn to the established depth.

Step 2. Lines are drawn representing the thread pitch from crest to crest.

Step 3. Parallel lines are drawn connecting the two minor diameter lines which coincide with the tap drill diameter lines.

Step 4. External schematic thread symbols are drawn. This involves drawing the chamfer and perpendicular lines representing pitch spacing to major and minor diameter lines (Fig. 16-16).

TYPES OF THREADED FASTENING DEVICES

There is a vast number of threaded fasteners ranging from the most common (Fig. 16-17) to very specialized. Threaded fasteners consist of a body, head, and point. **Figure 16-18** shows a variety of fastener heads, and **Fig. 16-19** shows drive configurations used on fastener heads. Various types of fastener points are shown in **Fig. 16-20**.

In designing the type and size of fasteners for a project, information concerning the class and dimensions is necessary to insure the fastener will fit and perform the desired function. To find this detailed data, ANSI, ISO, IFI, and manufacturer's charts are used. **Tables 16-1, 16-2,** and **16-3** show samples of this type of data for threaded fasteners.

Metric Thread Series

ANSI conventions cover inch standards, so they are not compatible with ISO metric series. Today, the United

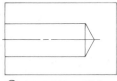 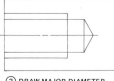 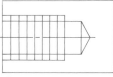 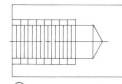 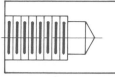

① DRAW TAP DRILL HOLE DIAMETER TO REQUIRED DEPTH
② DRAW MAJOR DIAMETER FOR THREADS TO REQUIRED DEPTH
③ ESTIMATE PITCH SPACING FOR MAJOR DIAMETER OF THREAD (CREST TO CREST)
④ DRAW MINOR DIAMETER OF THREAD (ROOT TO ROOT)
⑤ COMPLETE INTERNAL SCHEMATIC THREAD SYMBOL AS SHOWN

FIGURE 16-15 Steps to draw schematic internal thread symbols

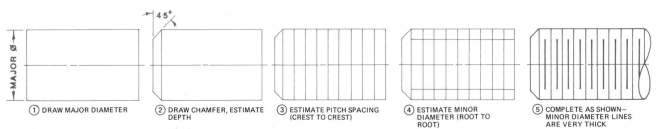

① DRAW MAJOR DIAMETER
② DRAW CHAMFER, ESTIMATE DEPTH
③ ESTIMATE PITCH SPACING (CREST TO CREST)
④ ESTIMATE MINOR DIAMETER (ROOT TO ROOT)
⑤ COMPLETE AS SHOWN— MINOR DIAMETER LINES ARE VERY THICK

FIGURE 16-16 Steps to draw schematic external thread symbol

FIGURE 16-17 Common threaded fasteners

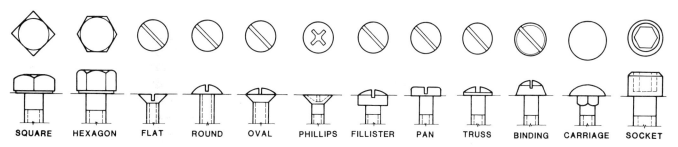

FIGURE 16-18 Fastener heads

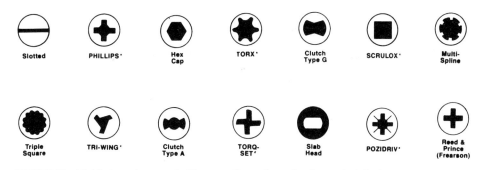

FIGURE 16-19 Various types of drive configurations for threaded fasteners (Reprinted from TECHNICAL DRAWING AND DESIGN by Goetsch and Nelson, © 1986 Delmar Publishers Inc.)

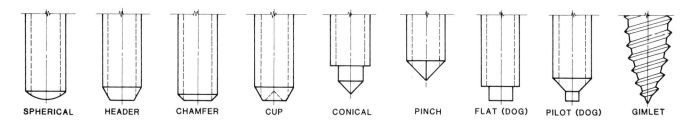

FIGURE 16-20 Fastener points

Major Diameter	Number of Threads per Inch	
	Course UNC	Fine UNF
0 (.060)		80
1 (.073)	64	72
2 (.086)	56	64
3 (.099)	48	56
4 (.112)	40	48
5 (.125)	40	44
6 (.138)	32	40
8 (.164)	32	36
10 (.190)	24	32
12 (.216)	24	28
1/4	20	28
5/16	18	24
3/8	16	24
7/16	14	20
1/2	13	20
9/16	12	18
5/8	11	18
3/4	10	16
7/8	9	14
1	8	12
1 1/8	7	12
1 1/4	7	12
1 3/8	6	12
1 1/2	6	12
1 3/4	5	—
2	4 1/2	—
2 1/4	4 1/2	—
2 1/2	4	—
2 3/4	4	—
3	4	—
3 1/4	4	—
3 1/2	4	—

TABLE 1 Thread table for commonly used ANSI fasteners

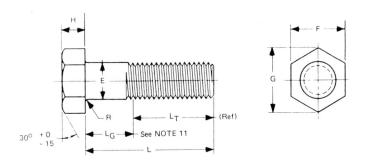

Nominal Size or Basic Product Dia (17)		E Body Dia (7)	F Width Across Flats (4)			G Width Across Corners		H Height			R Radius of Fillet		L_T Thread Length For Bolt Lengths (11)	
													6 in. and shorter	over 6 in.
		Max	Basic	Max	Min	Max	Min	Basic	Max	Min	Max	Min	Basic	Basic
1/4	0.2500	0.260	7/16	0.438	0.425	0.505	0.484	11/64	0.188	0.150	0.03	0.01	0.750	1.000
5/16	0.3125	0.324	1/2	0.500	0.484	0.577	0.552	7/32	0.235	0.195	0.03	0.01	0.875	1.125
3/8	0.3750	0.388	9/16	0.562	0.544	0.650	0.620	1/4	0.268	0.226	0.03	0.01	1.000	1.250
7/16	0.4375	0.452	5/8	0.625	0.603	0.722	0.687	19/64	0.316	0.272	0.03	0.01	1.125	1.375
1/2	0.5000	0.515	3/4	0.750	0.725	0.866	0.826	11/32	0.364	0.302	0.03	0.01	1.250	1.500
5/8	0.6250	0.642	15/16	0.928	0.906	1.083	1.033	27/64	0.444	0.378	0.06	0.02	1.500	1.750
3/4	0.7500	0.768	1 1/8	1.125	1.088	1.299	1.240	1/2	0.524	0.455	0.06	0.02	1.750	2.000
7/8	0.8750	0.895	1 5/16	1.312	1.269	1.516	1.447	37/64	0.604	0.531	0.06	0.02	2.000	2.250
1	1.0000	1.022	1 1/2	1.500	1.450	1.732	1.653	43/64	0.700	0.591	0.09	0.03	2.250	2.500
1 1/8	1.1250	1.149	1 11/16	1.688	1.631	1.949	1.859	3/4	0.780	0.658	0.09	0.03	2.500	2.750
1 1/4	1.2500	1.277	1 7/8	1.875	1.812	2.165	2.066	27/32	0.876	0.749	0.09	0.03	2.750	3.000
1 3/8	1.3750	1.404	2 1/16	2.062	1.994	2.382	2.273	29/32	0.940	0.810	0.09	0.03	3.000	3.250
1 1/2	1.5000	1.531	2 1/4	2.250	2.175	2.598	2.480	1	1.036	0.902	0.09	0.03	3.250	3.500
1 3/4	1.7500	1.785	2 5/8	2.625	2.538	3.031	2.893	1 5/32	1.196	1.054	0.12	0.04	3.750	4.000
2	2.0000	2.039	3	3.000	2.900	3.464	3.306	1 11/32	1.388	1.175	0.12	0.04	4.250	4.500
2 1/4	2.2500	2.305	3 3/8	3.375	3.262	3.897	3.719	1 1/2	1.548	1.327	0.19	0.06	4.750	5.000
2 1/2	2.5000	2.559	3 3/4	3.750	3.625	4.330	4.133	1 21/32	1.708	1.479	0.19	0.06	5.250	5.500
2 3/4	2.7500	2.827	4 1/8	4.125	3.988	4.763	4.546	1 13/16	1.869	1.632	0.19	0.06	5.750	6.000
3	3.0000	3.081	4 1/2	4.500	4.350	5.196	4.959	2	2.060	1.815	0.19	0.06	6.250	6.500
3 1/4	3.2500	3.335	4 7/8	4.875	4.712	5.629	5.372	2 3/16	2.251	1.936	0.19	0.06	6.750	7.000
3 1/2	3.5000	3.589	5 1/4	5.250	5.075	6.062	5.786	2 5/16	2.380	2.057	0.19	0.06	7.250	7.500
3 3/4	3.7500	3.858	5 5/8	5.625	5.437	6.495	6.198	2 1/2	2.572	2.241	0.19	0.06	7.750	8.000
4	4.0000	4.111	6	6.000	5.800	6.928	6.612	2 11/16	2.764	2.424	0.19	0.06	8.250	8.500

TABLE 2 Hex bolt table for commonly used ANSI standard hex bolts

(Reprinted from The American Society of Mechanical Engineers—ANSI B18.2.1–1981)

Basic Major DIA & Pitch	Tap Drill DIA	INTERNAL THREADS Minor DIA MAX	Minor DIA MIN	EXTERNAL THREADS Major DIA MAX	Major DIA MIN	Clearance Hole
M1.6 × 0.35	1.25	1.321	1.221	1.576	1.491	1.9
M2 × 0.4	1.60	1.679	1.567	1.976	1.881	2.4
M2.5 × 0.45	2.05	2.138	2.013	2.476	2.013	2.9
M3 × 0.5	2.50	2.599	2.459	2.976	2.870	3.4
M3.5 × 0.6	2.90	3.010	2.850	3.476	3.351	4.0
M4 × 0.7	3.30	3.422	3.242	3.976	3.836	4.5
M5 × 0.8	4.20	4.334	4.134	4.976	4.826	5.5
M6 × 1	5.00	5.153	4.917	5.974	5.794	6.6
M8 × 1.25	6.80	6.912	6.647	7.972	7.760	9.0
M10 × 1.5	8.50	8.676	8.376	9.968	9.732	11.0
M12 × 1.75	10.20	10.441	10.106	11.966	11.701	13.5
M14 × 2	12.00	12.210	11.835	13.962	13.682	15.5
M16 × 2	14.00	14.210	13.835	15.962	15.682	17.5
M20 × 2.5	17.50	17.744	17.294	19.958	19.623	22.0
M24 × 3	21.00	21.252	20.752	23.952	23.577	26.0
M30 × 3.5	26.50	26.771	26.211	29.947	29.522	33.0
M36 × 4	32.00	32.270	31.670	35.940	35.465	39.0
M42 × 4.5	37.50	37.799	37.129	41.937	41.437	45.0
M48 × 5	43.00	43.297	42.587	47.929	47.399	52.0
M56 × 5.5	50.50	50.796	50.046	55.925	55.365	62.0
M64 × 6	58.00	58.305	57.505	63.920	63.320	70.0
M72 × 6	66.00	66.305	65.505	71.920	71.320	78.0
M80 × 6	74.00	74.305	73.505	79.920	79.320	86.0
M90 × 6	84.00	84.305	83.505	89.920	89.320	96.0
M100 × 6	94.00	94.305	93.505	99.920	99.320	107.0

TABLE 3 Thread table for commonly used ISO metric fasteners

States is the only country which uses an inch-based thread system, although the IFI has established U.S. metric standards for some fasteners. A comparison of U.S. standard (UN) threads and metric threads is shown in **Fig. 16–21**. ISO metric notations differ from ANSI notations **(Fig. 16–22)**. Metric notations include pitch in millimeters, major diameter in millimeters, and the ISO thread symbol.

Pipe Threads

Pipes have either straight or tapered threads **(Fig. 16–23)**. They are drawn with either detailed or simplified symbols. In either case, the pipe thread notation includes the nominal diameter of the pipe in inches, threads per inch, standard used, pipe designation, and an entry indicating whether the pipe is straight or tapered.

FASTENER TEMPLATES

The normal laying out and drawing of fasteners with instruments is a long and costly process. Consequently,

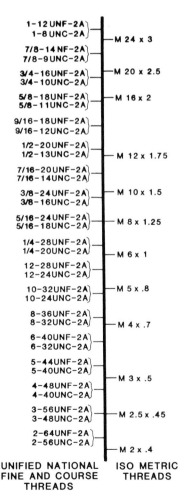

FIGURE 16–21 Comparison of typical ANSI's UN thread series to ISO metric threads

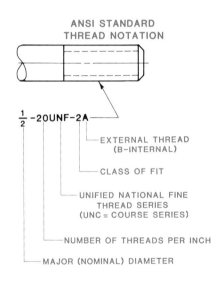

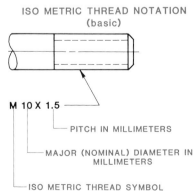

FIGURE 16–22 Comparison of ANSI's standard inch thread note to ISO metric thread note

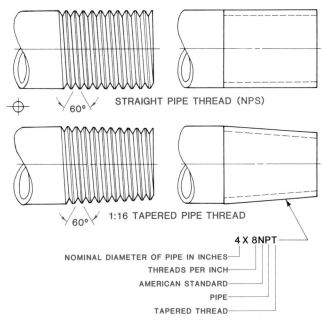

FIGURE 16–23 Pictorial and simplified representation of pipe threads and notations

most fasteners are drawn with templates using simplified fastener symbols. Templates provide outlines for two- and three-dimensional fastener drawings such as the bolt and nut template **(Figs. 16–24 and 16–25)**.

Templates provide the outline of fastener components; however, combinations of template openings must be used to draw all the features of a fastener symbol. **Figure 16–26** shows the steps used to draw a bolt using a template. To use a template, first establish centerline **(Fig. 16–27A)**. In Fig. 16–27A a centerline positions the hexagon nut opening, providing a guide for the addition of concentric circles with a hole template. The use of some template openings may require reversing a template outline to draw

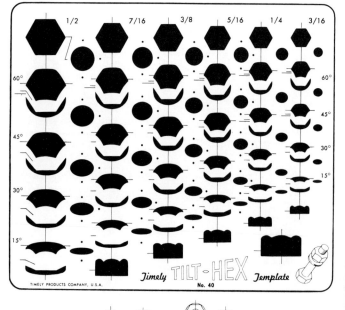

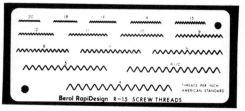

FIGURE 16–24 Two- and three-dimensional bolt and nut drawing template (Courtesy of Timely Products)

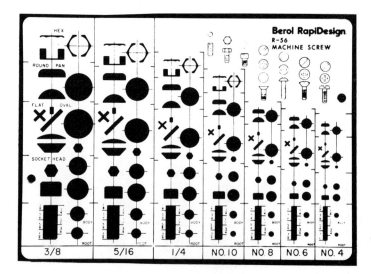

FIGURE 16–25 Various types of working drawing templates for screw threads and machine screws (Courtesy of Berol USA/RapiDesign)

Chapter 16 Fasteners **295**

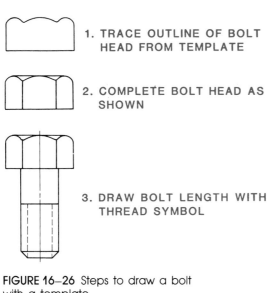

FIGURE 16–26 Steps to draw a bolt with a template

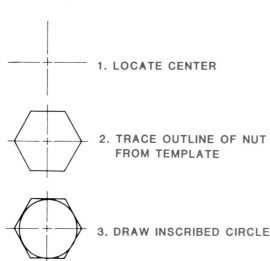

FIGURE 16–27A Steps to draw the face view of a nut with a template

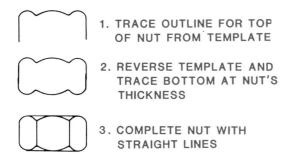

FIGURE 16–27B Steps to draw a side view of a jamb nut with a template

identical sides of a symbol. This procedure is illustrated in the layout of the jamb nut shown in **Fig. 16–27B**.

Templates do not always provide the exact outline of every fastener size and configuration. Some contain the most common sizes with the most common heads, bodies, and points. Most templates contain separate heads which can be combined with different lengths and points to produce the symbol needed. A machine screw template is used to draw the flat head and fillister machine screws in **Fig. 16–28**.

Pictorial representations of fasteners are also accomplished with the use of templates **(Fig. 16–29)**. In using pictorial templates, the drafter must make sure that the template angle of projection is the same as the angle in the drawing. Most technical drawing pictorial templates include isometric angles, although ellipse templates cover a wide range of angles. Isometric fastener templates contain symbol outlines designed for the receding 30° planes and the top plane. These templates

1. TRACE A FLATHEAD AND FILLISTER SCREW HEADS FROM A TEMPLATE

2. DRAW BODY WITH CHAMFERED END

3. DRAW THREAD SYMBOLS

4. TRACE TOP VIEWS FROM TEMPLATE IF NECESSARY

FIGURE 16–28 Steps to draw a flat head and fillister head machine screw with a template

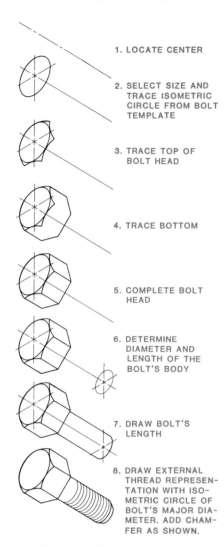

FIGURE 16-29 Steps to draw an isometric hex bolt with a template

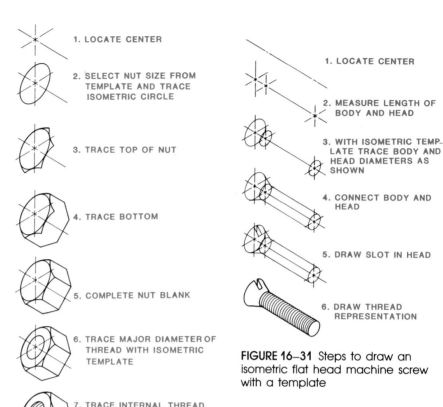

FIGURE 16-30 Steps to draw an isometric hex nut with a template

FIGURE 16-31 Steps to draw an isometric flat head machine screw with a template

provide isometric views of unique features which, used with centerlines and isometric circles (35°), produce the final drawing.

Figures 16-29 and **16-30** show the procedure to draw an isometric hex nut and bolt. **Figure 16-31** shows the application of this method to drawing an isometric flat head machine screw. There are four isometric planes available on most templates: top, bottom, right side, and left side **(Fig. 16-32)**. Usually, the right-plane outlines are provided and the template is reversed for the left receding plane. A wide variety of fasteners can be drawn pictorially with fastener and hole templates **(Fig. 16-33)**.

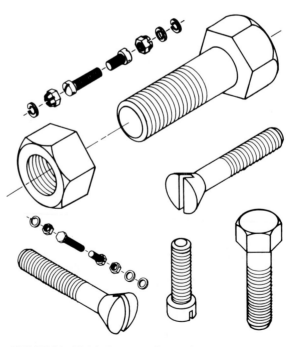

FIGURE 16-32 Various positions of isometric planes

Chapter 16 Fasteners **297**

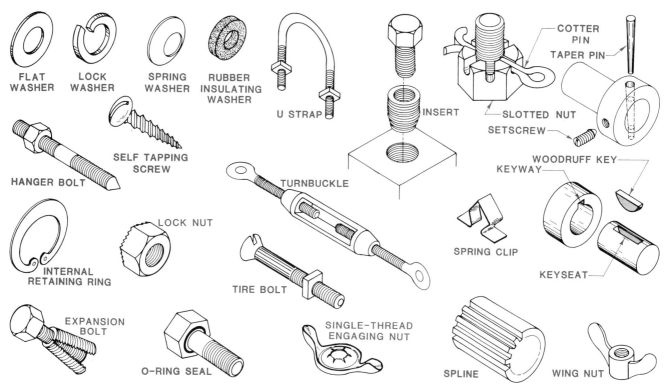

FIGURE 16-33 Miscellaneous types of fasteners

THREADED PERMANENT FASTENERS

Most permanent fastening systems do not use fastening devices, but bond materials through heat, pressure, or chemical action. These include the use of adhesives, bonding agents, brazing, soldering, and welding (see Chapter 19). The major fastening devices used for permanent fastening are rivets.

Rivets are available in many different sizes and configurations. Large rivets are used primarily in heavy construction. Small rivets are used to join relatively thin materials, such as sheets and plates. Rivets have no threads, but hold materials together when one head is flattened. Rivet specifications include the type of material, head shape, shaft diameter, and length **(Fig. 16-34)**. The symbols in **Fig. 16-35** describe common rivet types.

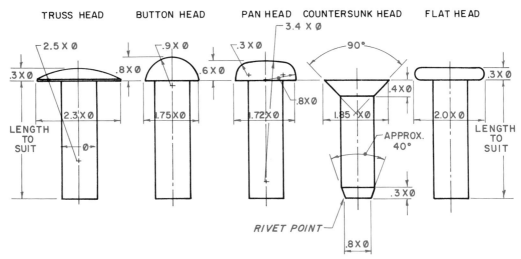

FIGURE 16-34 American Standard small rivets (Reprinted from TECHNICAL DRAWING AND DESIGN by Goetsch and Nelson, © 1986 Delmar Publishers Inc.)

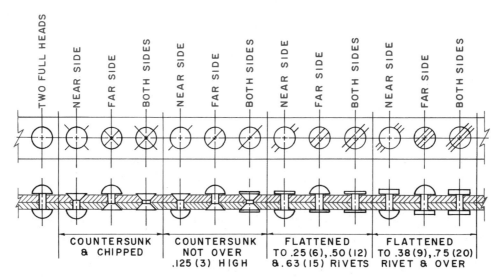

FIGURE 16-35 Rivet symbols for working drawings (Reprinted from TECHNICAL DRAWING AND DESIGN by Goetsch and Nelson, © 1986 Delmar Publishers Inc.)

COMPUTER-AIDED DRAFTING & DESIGN

FASTENER DRAWINGS

Since fasteners are standard items, they are manufactured according to standard specifications. The drafter or designer selects the appropriate fasteners for the design need. These standard fasteners include the full range of nuts, bolts, washers, cotter pins, rivets, and other fastening devices.

Because fasteners are standardized, drawing them on a CAD system is easy. Each type of fastener is drawn only once and saved in the library as a block for insertion or copying as needed. Standard overlay template systems for graphics tablets that include the most common fasteners are available. These template overlays may be purchased and added to existing CAD software. They make drawing detailed fasteners as easy as drawing a single line or circle. The template system in **Fig. 16-36** includes a master

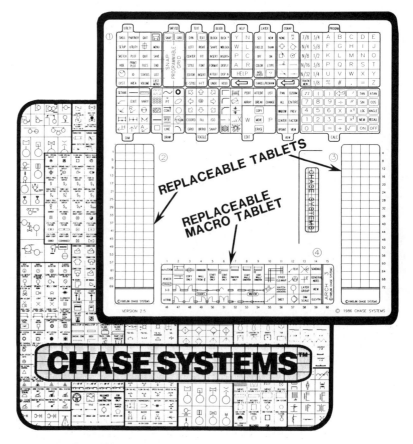

FIGURE 16-36 Template overlays for graphics tablet (Courtesy of Chase Systems)

Chapter 16 Fasteners 299

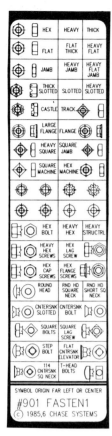

FIGURE 16–37 Replaceable fastener overlay for the graphics tablet (Courtesy of Chase Systems)

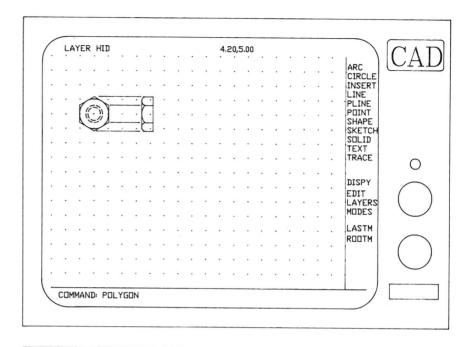

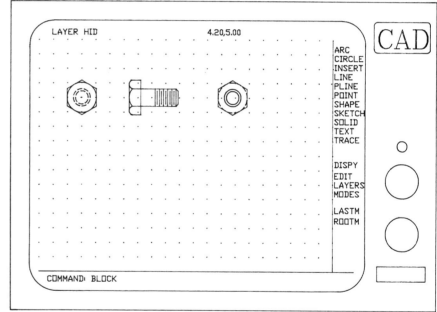

FIGURE 16–38 Steps to draw a bolt with a CAD system

template for drawing and editing. It also includes areas for replaceable tablets, such as fasteners. **(Fig. 16–37)**. Note that Fig. 16–37 includes standard hex nuts, castle nuts, various washers, and bolts with different head configurations.

To use a system such as this, the CAD operator simply places the puck crosshairs over the desired symbol, presses the button, and the symbol appears on the monitor. By using the puck the operator may move the symbol to the desired location, press the puck button again, and the symbol is drawn. To change the size of the symbol, the scale may be keyed in after selecting the symbol, or the SCALE command may be used after inserting the symbol onto the drawing.

If a standard template system is not available, the CAD operator may easily create and save the standard symbols for future use. The following steps show how to create and insert a standard symbol into a drawing of a $1/4$–20 hex bolt, 1.25″ long:

Creating a Symbol

Step 1. Obtain accurate information about the fastener. Three sources may be used: a text, such as this one, the Machinists' Handbook, or a vendor's catalog. Once the fastener is found, locate the sizes for a 1″ hex bolt.

Step 2. Draw the 1″ hex bolt **(Fig. 16–38)**. This is critical because it allows the operator to easily insert it into a drawing at any desired size. If the bolt is created at 1″, and the operator inserts it at a $1/4$ scale, the result is a $1/4″$ bolt. If it is inserted at a scale of 3, the result is a 3″ bolt. The head is drawn to specifications and the shaft is added. The head is copied to the other side and the hidden lines corrected.

Step 3. Make the bolt a data block (stored in the library) so it can be used in other drawings. When making a data block, the CAD operator gives it a meaningful name (such as HEXBOLT) so it is easily remembered.

Inserting a Symbol

The task is to insert the side view of a $1/4$–20 hex bolt, 1.25″ long into a drawing. The HEXBOLT symbol has been created and made into a block—it was created at a size of 1″.

Step 1. The command is INSERT. The name of the block to be inserted is HEXBOLT and the scale to be inserted is .25 **(Fig. 16–39)**. This results in a $1/4$″ bolt.

Step 2. Erase the end views; they are not desired for this drawing **(Fig. 16–40)**.

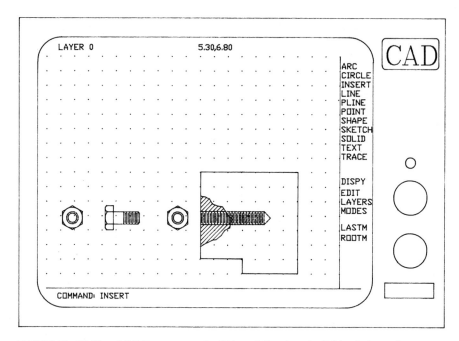

FIGURE 16–39 The INSERT command will insert the hex bolt block from the library into the drawing.

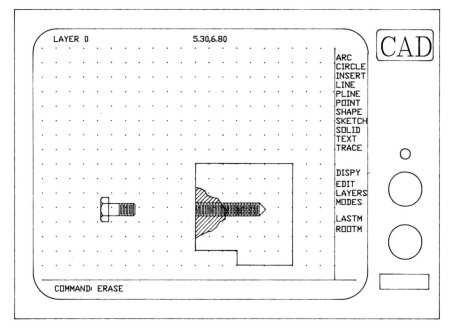

FIGURE 16–40 Erase the end views.

Step 3. Use the STRETCH command to lengthen the bolt to 1.25" **(Fig. 16–41)**.

Step 4. The desired bolt is accurately drawn to the proper scale and may now be dimensioned, moved, or copied to any location or locations on the drawing **(Fig. 16–42)**.

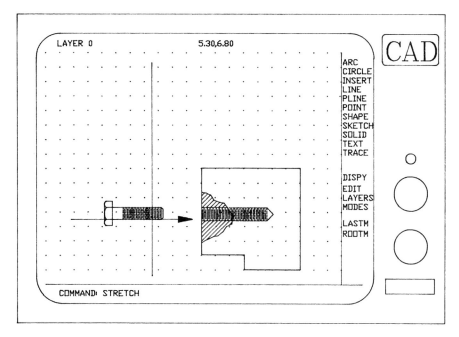

FIGURE 16–41 The STRETCH command will stretch the bolt to a 1.25" length.

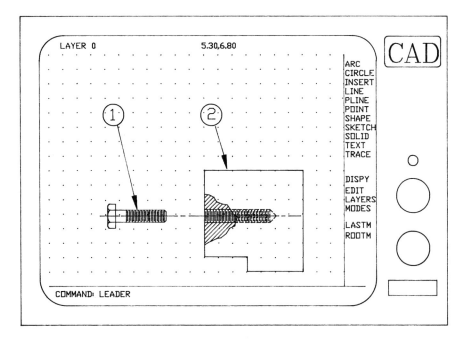

FIGURE 16–42 Complete the drawing with notations and dimensions as needed.

EXERCISES

1. Sketch the thread symbols and add the thread notations for each example in **Figs. 16-43** and **16-44**.
2. Draw the fasteners with instruments, and add the thread symbols and thread notations as specified in **Figs. 16-45** through **16-47**.
3. Draw the multiview working drawings with a CAD system, and add the thread symbols and thread notations for **Figs. 16-48** through **16-50**.
4. Perform the following ECOs for **Fig. 16-51**:
   - Change all UNF threads to UNC.
   - Add a 5th threaded fastener to the flange.
   - Add an external UNC thread 1.5" from the top of the shaft.
   - Add an internal UNF thread 2.0" from the top of the shaft.
5. Redesign and complete the plate with a hook or holder unit to be fastened with threaded fasteners

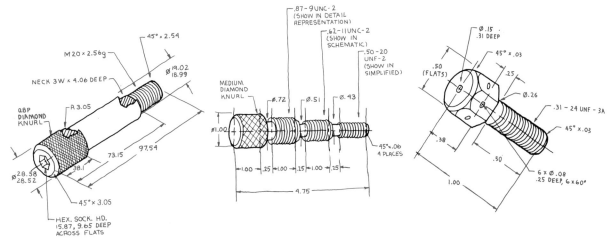

FIGURE 16-43 Sketch a pictorial of each item freehand or on a CAD system. Draw a multiview working drawing with metric or inch/decimal dimensions. (Reprinted from MECHANICAL DRAFTING by Madsen, Shumaker, and Stewart, © 1986 Delmar Publishers Inc.)

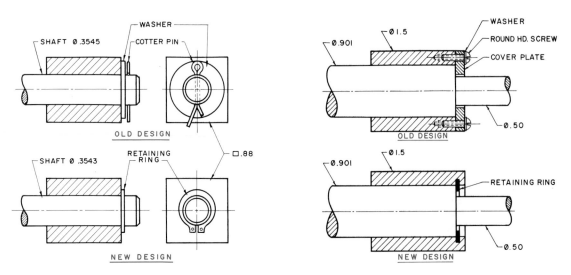

FIGURE 16-44 Sketch the multiview drawings and isometric views of the fastening systems. (Reprinted from TECHNICAL DRAWING AND DESIGN by Goetsch and Nelson, © 1986 Delmar Publishers Inc.)

Chapter 16 Fasteners **303**

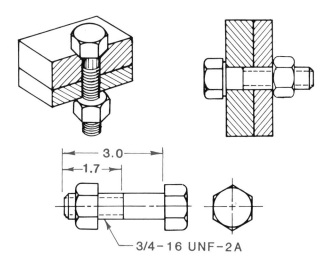

FIGURE 16–45 Draw a working drawing and isometric view of the hex bolt and nut with instruments or a CAD system.

FIGURE 16–46 On an A-size format lay out the fasteners as shown. Fastener data is:
Hex bolt $1^3/_8$–UNC x 3.0 long
Lock washer fits a $^7/_8$" diameter cap screw
Square nut 1–12 UNF 2B
Square head cap screw $^3/_4$–10UNC x 3.0 long.
(Reprinted from TECHNICAL DRAWING AND DESIGN by Goetsch and Nelson, © 1986 Delmar Publishers Inc.)

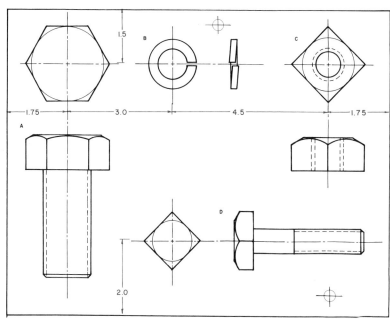

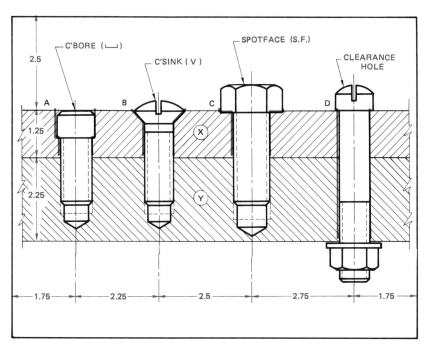

FIGURE 16–47 Draw or sketch each fastener and add a thread notation. (Reprinted from TECHNICAL DRAWING AND DESIGN by Goetsch and Nelson, © 1986 Delmar Publishers Inc.)

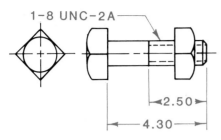

FIGURE 16-48 Draw with instruments or with a CAD system a working drawing and isometric view of the square head bolt and nut.

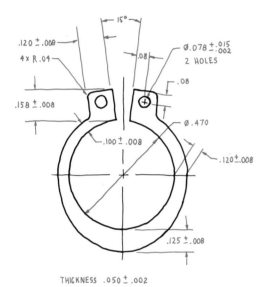

FIGURE 16-49 Draw the working drawings of the steel retaining ring. (Reprinted from TECHNICAL DRAWING AND DESIGN by Goetsch and Nelson, © 1986 Delmar Publishers Inc.)

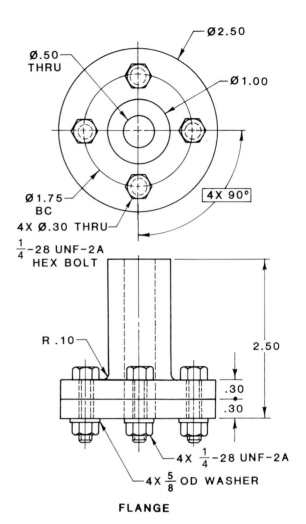

FIGURE 16-51 Redraw and the make the ECOs (Engineering Change Orders—refer Chapter 17), specified in Exercise 4.

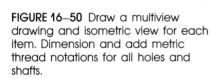

FIGURE 16-50 Draw a multiview drawing and isometric view for each item. Dimension and add metric thread notations for all holes and shafts.

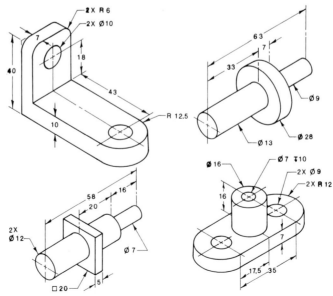

to the guide in **Fig. 16–52**. A .75" diameter rope supporting several hundred pounds, will pass through the hook or holder unit.
6. Research and record several different types and/or sizes of each fastener shown in Fig. 16–33 (page 297).
7. With 1/4" grid paper, lay out the drawing from the coordinate data table in **Fig. 16–53**. Add a UNC thread notation and dimensions.
8. Write a coordinate data table for the end view of the hex nut in **Fig. 16–54**. Plot a minimum of 8 coordinates for each circle.

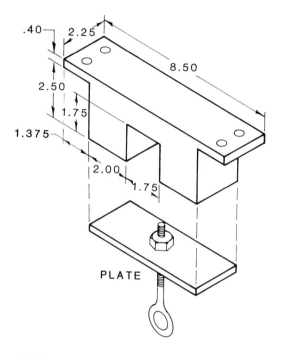

FIGURE 16–52 Redesign this holder unit following the instructions in Exercise 5.

COORDINATE DATA TABLE					
LN	X	Y	go to	X	Y
1	5	2	→	2	2
2	2	2	→	1	3
3	1	3	→	1	11
4	1	11	→	2	12
5	2	12	→	5	12
6	5	12	→	5	2
			new start		
7	5	7	→	2	7
8	2	7	→	1	6
			new start		
9	2	7	→	1	8
			new start		
10	5	10	→	20	10
11	20	10	→	21	9
12	21	9	→	21	5
13	21	5	→	20	4
14	20	4	→	5	4
			new start		
15	8	4	→	8	10
			new start		
16	21	5	→	8	5
			new start		
17	21	9	→	8	9
			FINISH		

FIGURE 16–53 Lay out the drawing on 1/4" grid paper with the coordinates in the data table. Add a UNC thread notation and dimension.

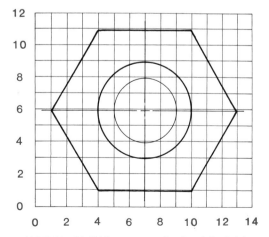

FIGURE 16–54 Write a coordinate data table for the end view of the hex bolt.

KEY TERMS

American National Standards Institute (ANSI)
Angle of thread
Axis of fastener
Crest
External thread
Helix
Inclined plane
Industrial Fasteners Institute (IFI)
Internal thread
International Organization of Standards (ISO)
Lead
Lead per revolution
Left-hand thread
Major diameter
Minor diameter
Multiple threads
Pitch
Pitch diameter
Right-hand thread
Rivets
Root
Screw thread series
Single thread
Thread depth
Thread form
Thread fit classes
Threads per inch (TPI)

the world of CAD

One of America's premiere manufacturers of race car chassis and driveline components, Mark Williams Enterprises, Inc., (M-W) has been an enthusiastic user of AutoCad for computer-generated design and drafting applications since the early 1980s. As the program continues to be enhanced over the years, so do the number of ways it is used by M-W.

An excellent example of how computerized drafting has influenced automobile racing is in the form of M-W's popular "kit car" chassis. Williams has created a number of chassis designs (for different classes of competition) and done all the work on a personal computer running AutoDesk's popular AutoCad program, Figures 1 and 2. Detailed blueprints are output on a large plotter, Figure 3, and supplied to do-it-yourself enthusiasts who wish to build their own chassis, Figure 4.

By using AutoCad, instead of traditional hand-drawing methods, Mark Williams is able to easily update the chassis blueprints to accommodate rule changes or new component technology. The program also allows Williams to provide detailed "blow-ups" of the important assembly areas with relative ease.

In addition to using AutoCad to show automotive enthusiasts across the nation how to build their own cars, M-W uses computer drafting technology extensively in-house to expedite the manufacture of the firm's various driveline and chassis components.

For each and every part manufactured by Mark Williams, there is a detailed AutoCad-created drawing that shows all critical dimensions. In most instances, the parts are made using computer-controlled machinery. A program routine must be developed based on the drawings, so their accuracy is of the utmost importance!

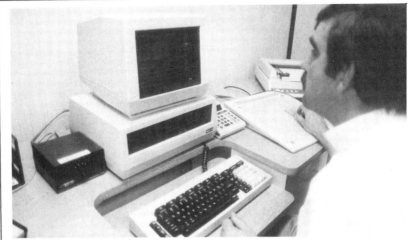

FIGURE 1 Race car chassis designer Mark Williams is shown working on his PC setup with AutoCad.

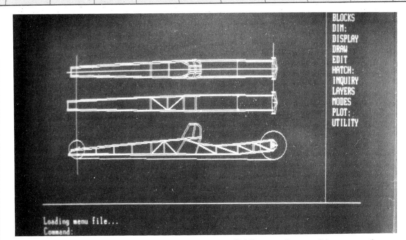

FIGURE 2 Close-up of the screen that Williams is working on shows three views of a dragster chassis.

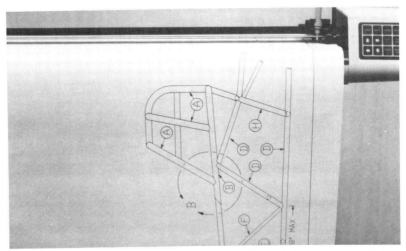

FIGURE 3 Plotter output shows close-up view of dragster chassis in driver's area.

FIGURE 4 Example of completely assembled dragster chassis (less body). With addition of motor, etc., it is ready for competition.

A recent addition to the AutoCad program (Release 10) enables a drawing to be generated in various degrees of rotation. This feature makes it easy for a machinist to envision how a component will look from virtually any angle.

The advent of desktop publishing technology has greatly expanded the use of computerized drafting programs through the use of drawings in a wide variety of documents.

For example, Mark Williams uses Xerox Ventura Publisher and a personal computer to produce detailed instruction sheets. Most all popular desktop publishing programs, including Ventura, support AutoCad (and other similar computer-drafting programs) and can import drawings. This enables M-W to write the instruction sheets, and bring in the appropriate drawings (scaled to size) for the document without the hassles of conventional cut-and-paste methods. The final product is output on a laser printer and duplicated as required.

Another excellent "marriage" of computer-aided drafting and desktop publishing technology is in the production of newsletters. Here, Mark Williams is able to design a new race car component and promptly send the information to thousands of racers across the nation. Again, the very same drawings used to aid in the manufacture of the parts are imported from AutoCad into the desktop publishing environment and merged with additional text and graphics, Figure 5.

At Mark Williams Enterprises, Inc., the technical advantages of computerized drafting are put to a very practical—and visible—application on the nation's race tracks. At 200-plus miles-per-hour, there's no room for inaccuracy!

(Courtesy of Bill Holland/Holland Communications, Inc., North Hollywood, CA and Mark Williams Enterprises, Inc., Louisville, CO)

Information on Mark Williams Products, Applications and Improvements

VOLUME 1, NUMBER 1   765 South Pierce Avenue   Louisville, Colorado

MW NEWS, THE FIRST EFFORT

The purpose of this news letter is three fold. The first is to increase the communication between MW people and our appreciated customers. You, the customer, will benefit by receiving timely inside information on MW products and their usage.

Second, to provide a window that will enable you to take a look at the Mark Williams organization. The 25 years of steady growth of this company is a great source of pride for all of the 24 employees involved with the effort to produce a quality product. We hope to allow you to get to know the people and the spirit that makes the organization what it is today.

The third is to help you become familiar with...

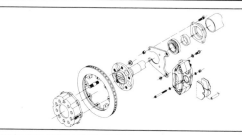

NEW BRAKE KITS EXPAND OUR LINE

MW has added several new applications for the outboard disc brake kits. The goal is to produce kits for all the popular housing ends used in today's street/strip applications. In addition to the Olds and Symmetrical type housing ends, new kits are now available for:

- Ford housing end with 1/2" diameter backing plate bolt size, P/N 71500.
- Ford housing ends with 3/8" backing plate bolts (stock Mustang ends),

FIGURE 5 Combining computer-aided drafting and desktop publishing allows Mark Williams to easily send newsletters to thousands of racers across the nation.

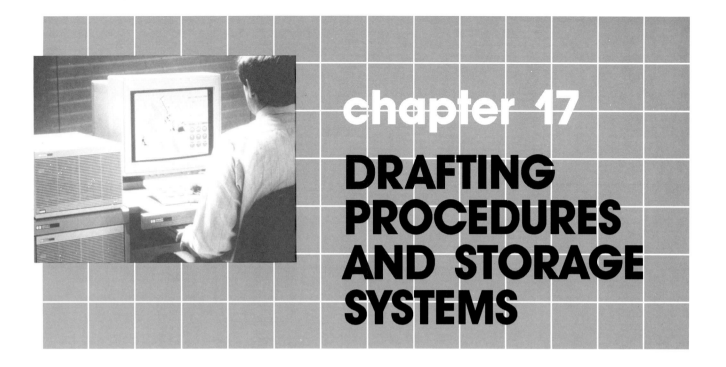

chapter 17

DRAFTING PROCEDURES AND STORAGE SYSTEMS

OBJECTIVES

The student will be able to:

- make revisions on existing working drawings
- use a microfilm reader
- check other students' working drawings
- prepare a working drawing using several overlays
- write a progress report covering the preparation of working drawings on a specific item
- lay out a drawing format with the following information: borders, title block, parts list, revision list, drawing identification number, parts identification numbers, and drawing zones
- use a photograph as the basis for a working drawing

INTRODUCTION

The drafting and design function is the core of the entire product development process. The process of product development includes the following phases:

1. Ideas and concept generation
2. Preliminary design
3. Design revisions and refinements
4. Working drawing development
5. Checking, revision, and modifications of drawings
6. Preparation of related documents, contractors specifications and estimates
7. Reproduction of drawings and documents
8. Production—manufacturing or construction
9. Product marketing and distribution

To complete the drafting functions relating to these phases, preliminary design drawings and later working drawings must be drawn, recorded, revised, checked, reproduced, stored, and retrieved through a variety of systems and procedures.

FUNCTIONAL DRAFTING

The preparation of the most accurate and complete set of technical working drawings in the least amount of time is a primary goal in the development of every product. To accomplish this, traditional instrument and board systems, or CAD systems may be used, depending on the availability of equipment and complexity of the product. Regardless of the standard system, many procedures are used to accelerate the production of drawings by organizing data and eliminating repetitive tasks. In manual drafting, these include the use of pin graphics, scissor drafting, reformatting, photodrafting, and standardized formats.

Pin Graphics

Pin Graphics is a system of binding transparent drawing film or paper together with a **pin bar** so the various layers of the drawings will **register** (align) **(Fig. 17–1)**. This function is known as **layering** when performed on a CAD system.

Scissor Drafting

Scissor Drafting is a system of cutting out unwanted areas of a drawing, making a copy, and then redrawing or pasting the new drawing segments in the blanked-out area **(Fig. 17–2)**. The revised copy is then used as the master copy.

Reformatting

Reformatting is an intermediate copy made where changes may be dry

307

308 Drafting in a Computer Age

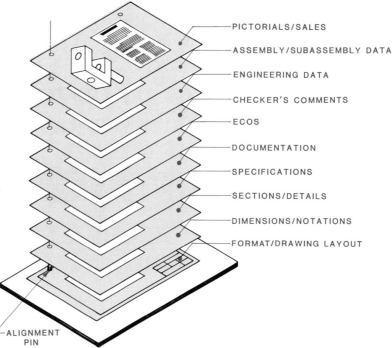

FIGURE 17–1 This is an example of how pin graphics (layering) may split up a working drawing into its separate elements of engineering, design, drawing, and documentation.

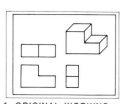
1. ORIGINAL WORKING DRAWING

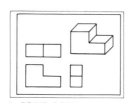
2. PRINT COPY

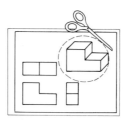
3. CUT OUT UNWANTED DRAWING AREA FROM COPY

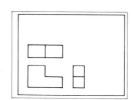
4. USE CUT OUT COPY TO PRINT A SECOND COPY

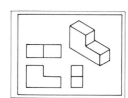
5. ON SECOND COPY ADD NEW DRAWING – THIS COPY IS NEW ORIGINAL

FIGURE 17–2 Procedures for scissor drafting

erased, wet erased, or cut out for revision. The revised copy is then used as the master copy.

Photodrafting

Photodrafting is a system of copying a photograph with a high-resolution copy machine, and adding dimensions, linework, and notes for use as a working drawing **(Fig. 17–3)**.

Standardized Formats

Preprinted formats eliminate the repetition of lines and headings on each drawing **(Fig. 17–4)**. This format can be preprinted on drawing surfaces or applied as a preprinted

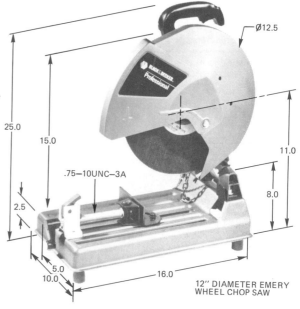

FIGURE 17–3 Photodrafting uses a photograph as the base for a working drawing. (Courtesy of Black and Decker U.S. Power Tools Group)

Chapter 17 Drafting Procedures and Storage Systems **309**

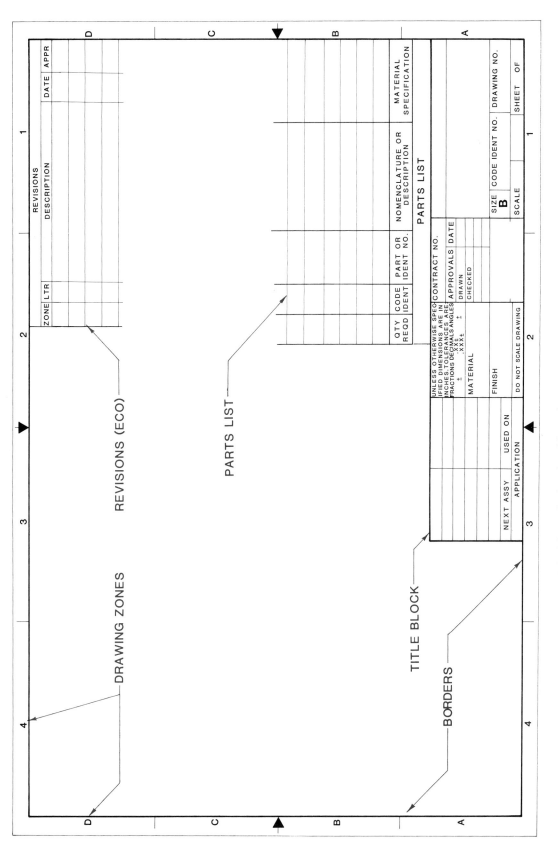

FIGURE 17-4 Example of a standardized preprinted format of a sheet layout

310 Drafting in a Computer Age

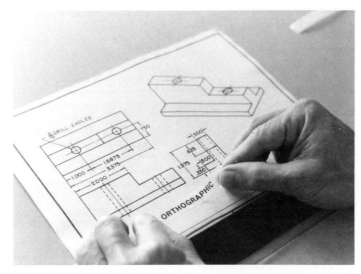

FIGURE 17-5 Transparent film stick-ons are available in many formats, or may be designed to fit the drafter's needs

transparent overlay **(Fig. 17-5)**. The following items are normally included in standard formats:

- title block and borders
- revision list
- parts list
- company's name
- drawing identification number
- part identification number
- drawing zones
- tolerances
- assembly application
- material
- finish
- approvals
- scale
- microfilm targets
- format size

REVISIONS

Revisions to working drawings are made for a variety of reasons, including design changes (**modifications**) and the correction of errors. Revisions may involve the change of a simple line, dimension, or note, or it may require a redraw of an entire part or component. In making changes and corrections, only those areas of a drawing which are affected are revised.

Each revision must be recorded in the title block. **Figure 17-6** shows the entry of five changes, A through E. Changes begin at the bottom of the **revision list**, and successive changes are added on top of the previous change. Changes are also accompanied with a document requesting and authorizing each change. These are called **Engineering Change Orders** (ECOs), **Engineering Change Requests** (ECRs), or **Engineering Change Notices** (ECNs).

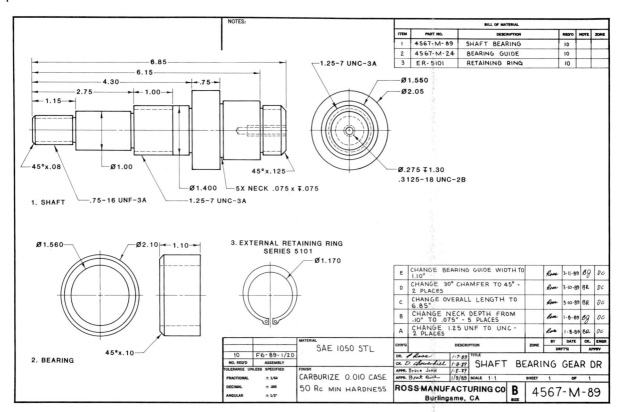

FIGURE 17-6 Engineering change orders (ECOs) listed on a working drawing

Chapter 17 Drafting Procedures and Storage Systems   311

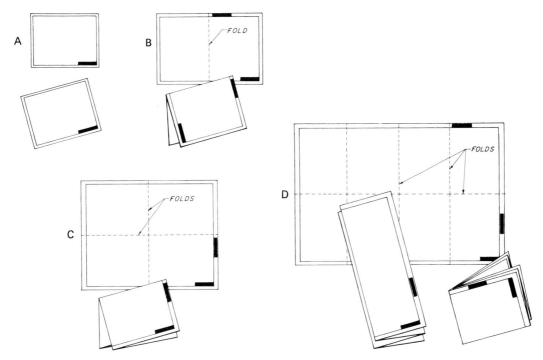

FIGURE 17–7 Procedure for folding standardized drawing formats of prints to A size for filing (Reprinted from TECHNICAL DRAWING AND DESIGN by Goetsch and Nelson, © 1986 Delmar Publishers Inc.)

STORAGE SYSTEMS

The ability to organize, store, and quickly retrieve original drawings, prints, and related documents is critical to the efficient functioning of any drafting operation. Original drawings should be stored flat in a cool, dry, and fireproof area. Copies can be folded to A size with the drawing identification number shown on the top **(Fig. 17–7)**. All standard drawing sheets can be folded to A size using the procedure in Fig. 17–7.

To conserve space, drawings may be microfilmed for storage. **Microfilmed** drawings can be placed in individual aperture filing cards, rolls, or placed in flat sheets, called **microfiche (Fig. 17–8)**. Microfilmed drawings may be stored manually in file cabinets or entered into a computerized automatic storage and retrieval system.

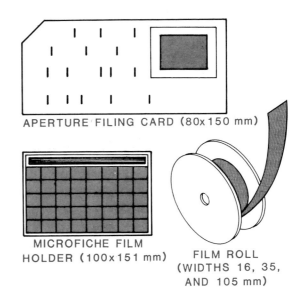

FIGURE 17–8 Microfilmed drawings and documents may be stored in various ways.

COMPUTER-AIDED DRAFTING & DESIGN

CAD STORAGE SYSTEMS

Drawings completed on a CAD system may be stored on a floppy disk, hard disk, drum, or magnetic tape. For hard copy drawings and computer-stored CAD drawings, a backup copy should be stored in a separate location to insure against loss. The CAD operator should save work every hour onto a backup disk in case of a power loss or surge that might erase all the work done. This way, only one hour's effort would need to be repeated. Because of the CAD software's special capabilities for block storage, many types of working drawings and pictorial drawings can be saved for instantaneous retrieval for reference or utilization.

CAD DATA BASES

In addition to performing CAD functions, the same computer is capable of doing word processing, organizing and manipulating data into **spread sheets,** calculating, accounting and management, engineering calculations, as well as a host of other tasks that help with the documentation of engineering drawings. Many of those functions require the existence of a data base.

As a CAD drawing is being generated, the computer is developing a data base. A **data base** is a body of information that is gathered and sorted to be used for future tasks. In drafting, the data base created by a CAD drawing may be used to accurately identify and describe segments of a drawing, calculate areas, volumes, angles and clearances, create parts lists, calculate product costs, and write programs for automated machining.

The following material covers some of the data accumulated as a CAD operator produces a drawing and how this data may be used. As technology advances, more uses will be discovered for CAD and its data base.

Recording Elements

Whenever a point, line, circle, arc, dimension, or any other element is drawn on a CAD drawing, the element is accurately recorded in a data base for that drawing **(Fig. 17–9)**. Note that Fig. 17-9 contains three elements—two circles and a line. The circles were drawn using the CIRCLE command, and specifying their centers and radii. The line was drawn tangent to the circles, and its length and angle are not known. By using the LIST command, selecting the three elements (the small boxes on the drawing are selectors) and pressing the RETURN key the properties of each of the elements are displayed on the monitor **(Fig. 17–10)**. This is the data for these three elements. The data for the circles include the location of the center point, radius, circumferences, and areas. The line data includes location of the start and end points of the line, its length, angle, and the change (**delta**) in the X and Y directions. This is more information than the CAD operator knew when drawing the elements. It may be valuable design data needed in the future.

Calculating

It was noted above that a computer may be used for calculating. The fact is that it is a very powerful calculator. Built into most CAD software are formulas for doing standard calculations. The CAD operator gives the computer the information as shown

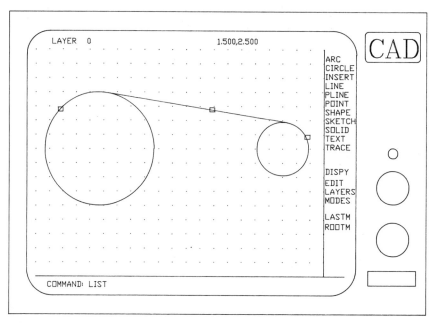

FIGURE 17–9 Using the LIST command to display the data base of each drawing entity

in Figs. 17–9 and 17–10. Using this information and the built-in formulas, the computer can very rapidly perform valuable calculations.

An example of this is the AREA command **(Figs. 17–11 through 17–13)**. The drawing includes three shapes—a rectangle and a triangle, plus a more complex polygon. Dimensions were put on the drawings so the reader of this text may manually do the calculations to verify the data. The computer does not use the dimensions when calculating, it uses the data base for each element of geometry. In Fig. 17–11, the rectangle has been selected (note the square selector boxes). The area and perimeter of the feature is instantly displayed in the prompt area below the drawing. The same procedure was used to get the area and perimeter of the triangle and polygon in Figs. 17–12 and 17–13. The calculation for the area and perimeter for Fig. 17–13 involves geometry and trigonometry, and the computer does the calculations instantly.

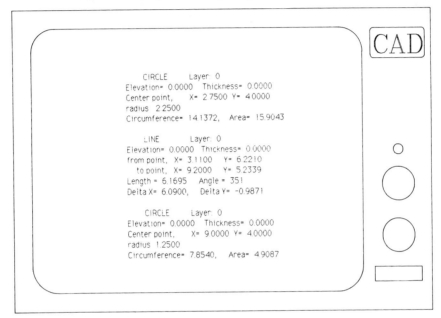

FIGURE 17–10 The data base for each entity in Figure 17–9

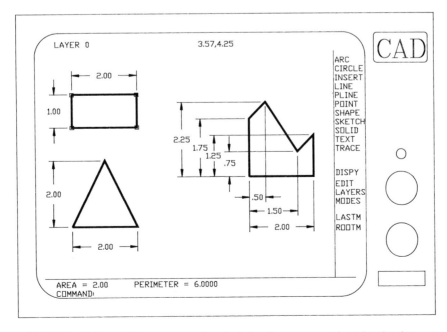

FIGURE 17–11 The AREA command calculates the area and perimeter for the rectangle.

314 Drafting in a Computer Age

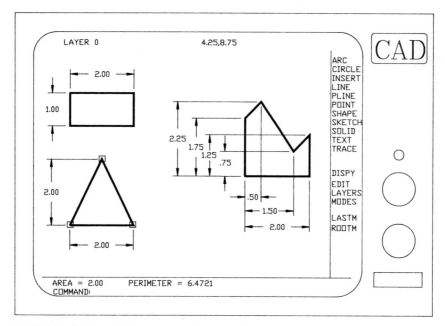

FIGURE 17-12 The AREA command calculates the area and perimeter for the triangle.

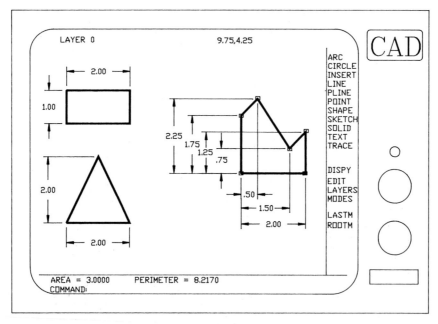

FIGURE 17-13 The AREA command calculates the area and perimeter for the irregular polygon.

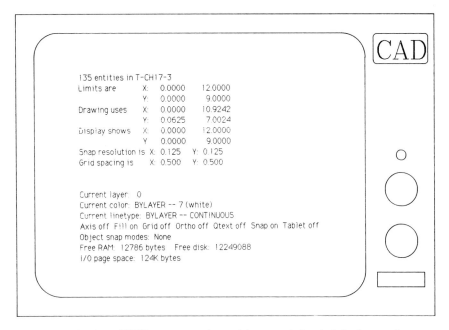

FIGURE 17-14 The STATUS command provides specialized data for each entity.

Drawing Status

A different type of data base established for each drawing is the status of the drawing. The STATUS command gives a readout of the essential parameters of a drawing. **Figure 17-14** is an example of a STATUS readout of the drawing in Fig. 17-13. This information includes the number of entities (lines, dimensions, etc.) in the drawing, drawing sizes, snap and grid spacings, color and linetype used, as well as the amount of storage space remaining in the computer for future drawings. This information is displayed by typing the STATUS command and pressing the RETURN key.

Time Records

Managers, designers, and engineers need to know how much time it takes to produce a CAD drawing initially and how much time is spent making revisions to the drawings. They are responsible for the costs and the flow of drawings through a company's system and this information aids them with cost estimating, billing, and employee evaluation. The TIME command gives this data **(Fig. 17-15)**. This information includes the dates when the drawing was created and edited, and the total time it

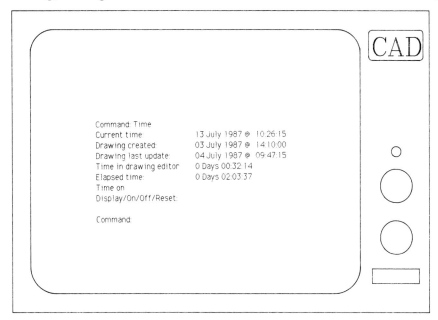

FIGURE 17-15 The TIME command keeps records on the drawings in progress.

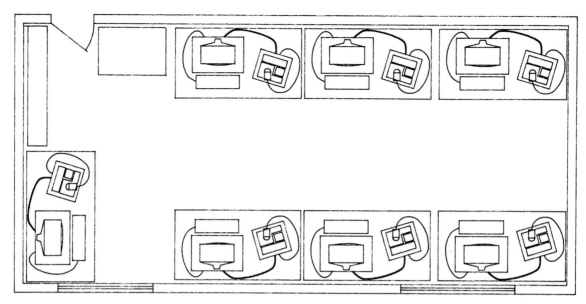

FIGURE 17-16 Seven-station, computer-aided drafting lab. Note that one lab was drawn and duplicated, revolved, and placed into the other six positions.

took to create and edit the drawing. Added to this data may be the name or code number of the CAD operator. Figure 17-15 shows the time for the gear drawing at the end of Chapter 20. The entire drawing, including editing, took 2 hours and 3 minutes—that should please any drafting manager.

Parts List

A real aid to the CAD operator is the automatic generation of a **parts list.** This may be accomplished for drawings that include several of the same items, such as fasteners or standard company parts. The example given here is a floor plan of a CAD lab containing seven CAD stations **(Fig. 17-16)**. Each CAD station is the same. By the use of hidden blocks, data may be added to the components. **Figure 17-17** shows one of the CAD stations with the names and part numbers superimposed over the parts. This information was placed on the drawing in a block that may be turned on or off. It will be used as a data base for generating a parts list. The CAD station has been inserted into Fig. 17-16 and the block has been turned off (it was inserted seven times). **Figure 17-18** shows how the drawing would look with the data blocks turned on. Note that the bookcase and the cabinet also have data assigned to them. By using the data in the blocks, the parts list in **Fig. 17-19** is automatically generated. It is important to notice that the parts list contains more information than the blocks. The parts list includes the part number, quantity (derived by the number of times the blocks were inserted), description, vendor number, and stock information. This computerized parts list may be part of the company's inventory data.

CNC Machining

Another use for the drawing data base is to program for **computer-numer-ically-controlled** (**CNC**) machining. The geometry recorded in the computer data base by the drawing, as shown with the LIST command in Fig. 17-9, may be sent directly to a CNC center for manufacturing of the part. The CNC center uses this CAD-generated geometry to select the material from inventory, and to cut, mill or turn the part as required.

Before sending the data to a CNC center, the CAD operator may use a CNC program to check the machining process **(Fig. 17-20)**. The CNC program is not going to use the dimensions on the drawing, but will use the geometry; therefore, the dimension layer is turned off **(Fig. 17-21)**. This brings up an extremely important point. **When doing a CAD drawing that is going to be CNC machined, it is imperative that**

FIGURE 17-17 Single CAD workstation with its parts data displayed

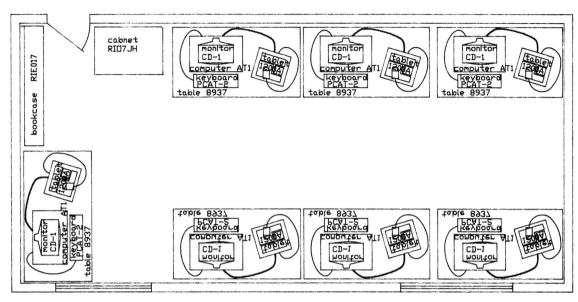

FIGURE 17–18 Full CAD lab with its parts data displayed

PART NO.	QTY	DESCRIPTION	VENDOR NO.	IN STOCK
RIE017	1	BOOKCASE 5'X8'	107	YES
R107JH	1	CABNET, LOCKING	107	YES
8937	7	TABLE 3'X5'	107	5
AT1	7	COMPUTER, 386	84	YES
CD-1	7	MONITOR, COLOR	84	YES
PCAT-2	7	KEYBOARD, W/FKEYS	84	YES
1200A	7	GRAPHIC TAB W/PUCK	85	NO

FIGURE 17–19 Computer-generated parts lists for the CAD lab

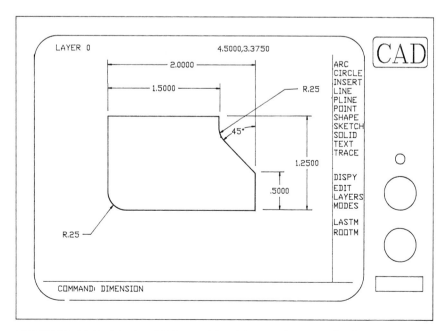

FIGURE 17–20 Part to be CNC machined (computer-numerically controlled)

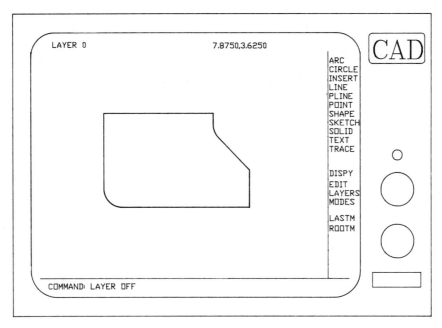

FIGURE 17-21 CNC drawing part with the dimension layer turned off

the part is drawn correctly. The geometry of the part is more important than the dimensions that are on the part. If a change to a part size is made, the part must be changed as well as the dimension.

Once the dimensions have been turned off, another program is loaded into the CAD program (note the change in the screen menu) and the cutter path (dashed circles) is displayed on the monitor **(Fig. 17-22)**. The start and finish of the cut (S/F) and the direction of the cut is also shown. This visual view of the part and the cutter path indicates to the operator if the path and cutter diameter are going to work in order to cut this part.

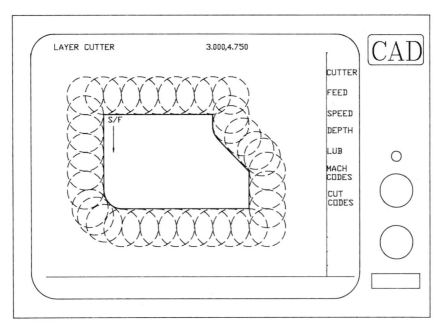

FIGURE 17-22 CNC tool part around the part. Note the CNC menu screen commands.

EXERCISES

1. Draw the working drawings in **Figs. 17–23** through **17–25** with drafting instruments or with a CAD system. Make the revisions as noted in each caption.
2. Design a slip-proof handle for the threaded rod (note ECOs) in **Fig. 17–26**.

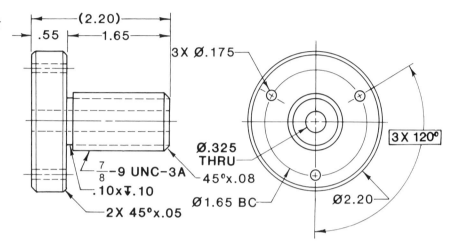

FIGURE 17–23 Draw the threaded set screw with drafting instruments or a CAD system. Make the following ECOs:
1. Change the length of the threaded shaft 1.75″.
2. Change the UNC thread to UNF.
3. Change the diameter of the bolt circle to 1.50″.
4. Change the diameter of the three holes to .20″.
5. Reduce the neck diameter below the minor diameter of the thread (↧ .15).

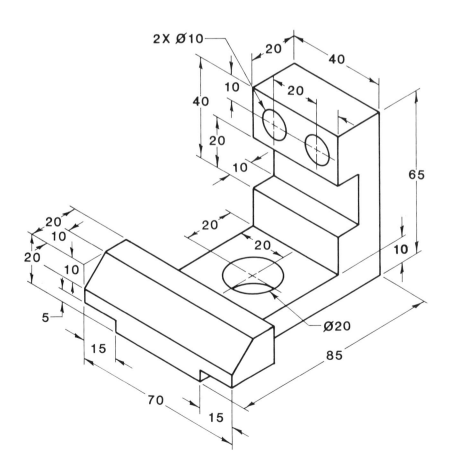

FIGURE 17–24 Redesign the metric jig guide with the following ECOs:
1. Change the 20 mm base hole to .75″ with UNC threads through.
2. Change the two 10-mm holes to .375″ with UNF threads
3. Change the base height of 10 mm to .50″.
4. Change the width of the base from 70 mm to 60 mm. Reduce the 15 mm overhang equally.
5. Redraw and convert to inch/decimal dimensions.

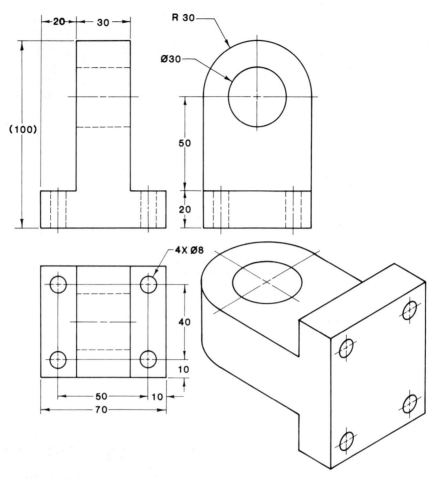

FIGURE 17-25 Redraw the (first-angle projection) T support working drawing and make the following ECOs:
1. With the scissor drafting technique, change to third-angle projection.
2. Convert metric dimensions to inch/decimal.
3. Change 30 mm to 25 mm. Add UNC threads through.
4. Add UNF threads through the four 8-mm holes.
5. Change stem of T support to 25 mm.

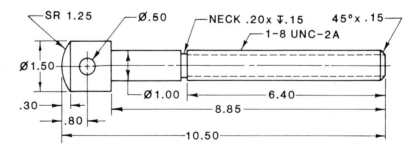

FIGURE 17-26 Redraw the threaded rod, make the following ECOs and record them on the format's revision list:
1. Change the spherical radius to 1.00".
2. Change the thread length to 6.50".
3. Change the neck depth to .175".
4. Change the thread fit to class 3.
5. Change the .50" hole in the handle to .30" diameter through.
6. Convert dimensions to metric.

3. Draw **Fig. 17–27**, making the ECO changes. Write a progress report covering the drafting time and any specific design recommendations.
4. Write the ECOs for the U-bolt in **Fig. 17–28** to improve its design functions. Redraw the drawing, listing the ECOs in a revision list.
5. Take black and white photographs with your camera of a few simple machine parts (use a very light background). Dimension the prints with technical ink pens on a vellum overlay.

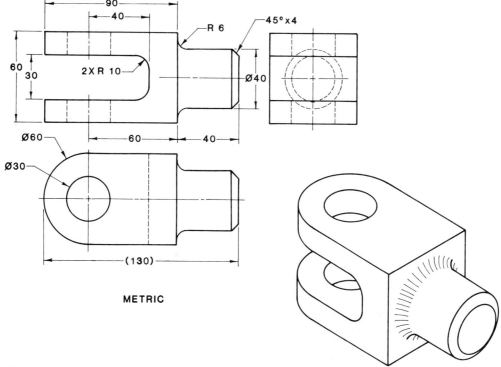

FIGURE 17–27 Redraw the (first-angle projection) bearing bracket, the isometric drawing, and make the following ECOs:
1. With the scissor drafting technique, change to a third-angle projection.
2. Convert the metric dimensions to inch/decimal.
3. Add 1" depth of UNC threads to round extrusion.
4. Add UNC threads through to all internal holes.
5. Reduce all fillets by 50%.
6. Reduce 45° chamfer by 50%.

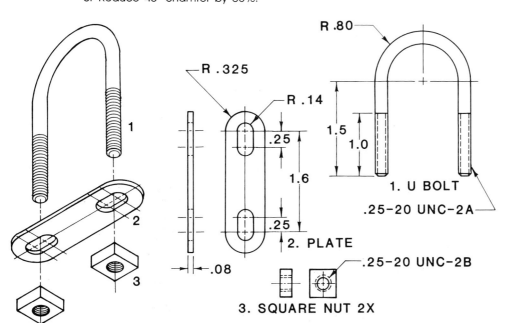

FIGURE 17–28 Redesign the U-bolt so it can strap a 1.5" diameter pipe. Convert it to metric dimensions.

KEY TERMS

Computer-numerically controlled (CNC)
Data base
Delta
Engineering Change Order (ECO)
Engineering Change Notice (ECN)
Engineering Change Request (ECR)
Layering

Microfiche
Microfilm
Modifications
Parts list
Photodrafting
Pin bar

Pin graphics
Reformatting
Register
Revision list
Scissor drafting
Spread sheets

the world of CAD

The New York State Geological Survey is presently working on a project titled "Prospecting for Sand, Gravel, and Heavy Minerals in the New York State Offshore." The main goal of this project is to locate sand, gravel, and mineral resources that exist in the State's waters so their extent and worth may be determined. The need for sand and gravel for use in concrete has become critical in the southeastern New York metropolitan area; a company based in New Jersey is presently mining sand and gravel offshore. Economically important heavy minerals, which now are imported for the most part, include titanium minerals, zircon, monazite, aluminosilicates, staurolite, garnet, corundum, and several others usually found in only trace concentrations. AutoCAD is being used to create maps that display both the concentrations and geographic distribution of gravel and the economically important heavy minerals in New York State's waters.

To display mineral resources data that is comprehensive and easy to interpret, a base map of the study area is required. The United States Geological Survey (USGS) has been digitizing maps of the United States at varying scales for the last several years. This digital map data is available now for use in CAD systems. Earlier phases of this project required digitizing outlines of the north shore of Long Island and the south shore of Connecticut, manually to create a base map. The availability of the USGS digital map data for CAD represents a great time savings for geologists since it eliminates the need to hand digitize outlines of the study area. This new base map data is very detailed, including political boundaries, lakes rivers, roads, railroads, and all types of geographic features similar to the information present on the USGS topographic maps. It can be used by many different projects as the outline of New York State, its counties, lakes, and other features do not change. With this source of base map data, geologists can concentrate on developing methods of analysis and presentation for the information that is being found.

Once the base map data has been entered, the data specific to the project is prepared to be added to the plot. Typically, a spread sheet program is used to capture the data. For this offshore minerals project, the data is a list of core locations, longitude, and latitude describing where the core sample was taken, and the weight of the gravel and of the twelve types of economic heavy minerals that need to be quantified. After the data is entered, the result is a large

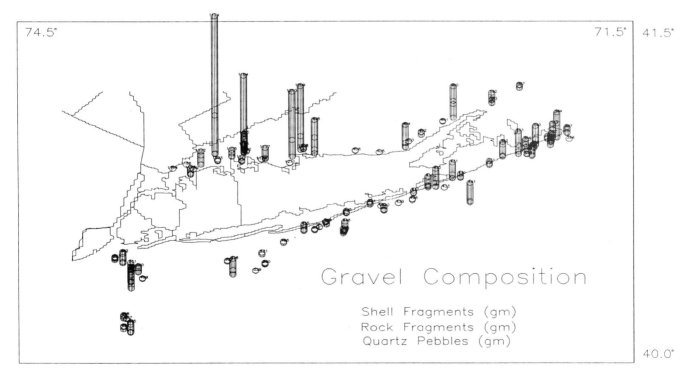

FIGURE 1 The location of samples taken from the north and south shores of Long Island. The height of each cylinder is proportional to the amount of each of the minerals found there.

and complicated table. Using this table to identify the areas where these minerals are concentrated in is not an easy task. However, AutoCAD has incorporated an implementation of the LISP programming language, AutoLISP, which can be used to write programs to read directly from data files, calculate the appropriate scaling factors, and issue drawing commands. This eliminates the need for the manual compilation of the data, scaling of each of the fractions to be displayed, and having to repeat these processes each time new data is to be plotted on the map. The program plots a cylinder at the location where the core was drilled and makes its height proportional to the amount of each of the minerals found there, Figure 1. Colors are used to represent the different species of minerals.

Using this system, a detailed diagram can be produced from this data in a matter of hours rather than the days to weeks it would take for it to be drafted by hand. After a plot has been produced in this way, it can be reproduced many times and the size of the finished drawing can be altered to suit a specific application. Once the specific programs to plot the data have been written, much less time elapses between later data collection and analytical parts of the project, and the production of the diagrams. Working with digital rather than paper maps allow for new data to be added to an existing drawing, so the picture grows as the analyses are completed.

(Courtesy of J. R. Albanese, New York State Geological Survey, Albany, NY)

the world of CAD

At Cornell University, six advanced students in the Apparel Design program in the College of Human Ecology used CAD software to design children's apparel. They were continuing the work begun at the University of Tennessee. The faculty at Tennessee adapted an AutoCAD software program, originally intended for interior design, for apparel design. Methods of fashion illustration, digitizing, grading, and marker making were developed, and with the addition of pattern making and alteration techniques, the system now offers manufacturers a full range of operations.

The students at Cornell University initially spent 6 to 9 hours of group instruction to learn basic AutoCAD commands before beginning their designs. A complete line of children's wear was designed by each student.

The first assignment was to use AutoCAD to create a sketch of a child's basic figure to serve as a model for the clothing. This basic sketch could be altered to create male and female figures of various shapes and sizes. The figures were stored in the computer for future use.

Although some professional designers fear that the computer will inhibit their creativity, the students found that the computer simplified and even enhanced the design process. "As I worked with the designs, I went through the normal design process of trying different proportions and adjusting details. This process was much easier on the computer than sketching each version by hand," says Susan Perkins Ashdown, a lecturer in Costume Technology in Cornell's Theatre Arts Department. "It was so quick. And the finished version of each drawing was clean and precise."

Continuing, Ashdown says, "The variations-on-a-theme had elements that were exact copies of one another. I was able to use one motif over and over with very slight modifications each time."

Other students found that the computer had a positive impact on the designs themselves. "The AutoCAD system helped to influence my designs," says Lisa Gedzelman, senior apparel design student. "It made me critically consider simplicity of line and detail without sacrificing style or originality. I carefully weighed every detail and chose simple, yet stylish features." See Figures 1 and 2.

After learning the basic computer pattern-making techniques developed by the instructors, the students created patterns for their designs. Each student developed patterns for two garments, one to fit a dress mannequin and the other to fit a live model.

The students arranged the markers on the monitor and plotted them to scale. They then fitted the muslins on the forms and on the models. All resulting alterations and modifications were performed on the computer.

FIGURE 1 Senior design student Lisa Gedzelman, above, uses AutoCAD to experiment with proportion and design features. At right, the patterns for the duck applique, toy and pocket were all made directly from the original sketch of this jumper designed by Susan Perkins Ashdown.

Students reported they were impressed by the extreme accuracy and precision with which the pattern pieces fit together. Rather than drawing and tracing pattern pieces such as facings and collars by hand, they used AutoCAD's copying capabilities. The ability to scale garment features and seams to any size or to match existing seams further increased accuracy. The computer also generated seam allowances instantly.

One of the greatest benefits of the system is the ability to copy part of an illustration, such as a pocket or motif, and make a pattern piece directly from it. From her sketch, Ashdown developed duck appliques and varied the proportions of the motif by computer. She then separated the motif into pattern pieces for the child's jumper. Later she scaled up the applique pattern to create a pattern for a stuffed toy in minutes.

AutoCAD proved to be an excellent tool for designing motifs and fabrics as well as patterns. The abil-

Susan P. Ashdown   Sandra Hsu   Lisa Gedzelman

Laura Russell   Andrea Kirchgessner   Kathy Riendeau

FIGURE 2 Examples of children's clothing designs that students have created with the aid of AutoCAD in Cornell's Apparel Design program.

ity to copy, scale, compress, and sketch a figure in a variety of ways provides an endless array of design possibilities.

During the course, faculty and students developed and refined techniques for using AutoCAD for fashion illustration, pattern-making alterations and marking. They also experimented with a grading system that is under development. These techniques are now being combined into a customized apparel package that will be marketed in the future. The AutoCAD apparel applications package contains custom screen and tablet menus that automate the operations most often used in designing and working with apparel.

To streamline the illustration process, an icon menu displays a variety of figures and places them on the screen with one key-stroke, giving any designer immediate access to figures ready for sketching. Another icon menu instantly displays and inserts motifs, such as flowers, animals, and nautical gear. The package also contains a variety of slopers, labels, buttons, buttonholes, and symbols as well as a grading menu. It can be used as is or customized to fit any manufacturer's needs.

Professionals in the apparel industry who have reviewed the apparel applications and the finished garments feel that AutoCAD has many possibilities. "The illustrations are so precise that they could be faxed along with the dimensions directly to the manufacturer," says Coty-Award-winning designer Jon Haggins. "The garments could be produced directly from the sketches. The illustrations would be excellent for anything done from specifications."

When combined with other software and equipment, AutoCAD can produce realistic enough images of products that manufacturers can reduce or eliminate the expense of making samples.

To keep pace with industry demands for computer-literate graduates, Cornell is redeveloping all of its apparel courses to include CAD and other computer applications. Although apparel design students are still required to know the basics of the field, they now use computers to construct their designs.

(Adapted from Phyllis Bell Miller and Anita Racine, "Less Means More in CAD," Bobbin, *Vol. 30, Number 1, September 1988)*

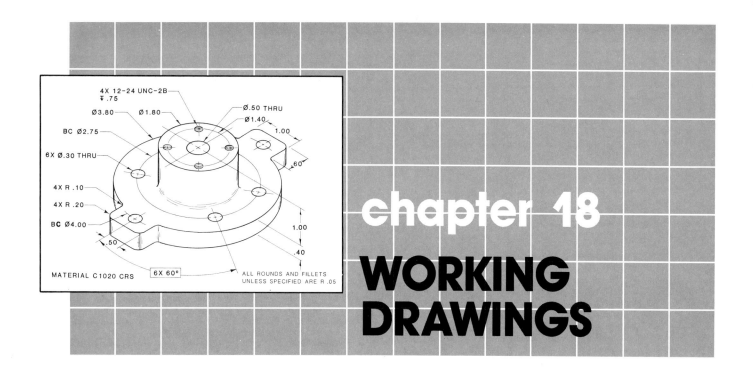

chapter 18
WORKING DRAWINGS

OBJECTIVES

Students will be able to:

- prepare detailed drawings manually
- prepare assembly drawings manually
- prepare detailed drawings on a CAD system
- prepare assembly drawings on a CAD system
- select the proper assembly drawings for a working drawing set

INTRODUCTION

Working drawings provide specific information and instructions needed to manufacture or construct products and structures. This includes detailed graphics of precise shapes, and dimensions of exact sizes. They also contain other notations for manufacturing, such as material descriptions, finishing methods, and tolerances. Working drawings are used in all phases of manufacturing, including fabrication and assembly. All information needed to manufacture a product must be included in a working drawing. The manufacturer must not be forced to guess about any detail to produce a product as designed.

WORKING DRAWINGS

There are two general categories of working drawings—detail drawings and assembly drawings. Working drawings usually include fully dimensioned multiview drawings of the assembled product, plus detail drawings for each part and component **(Fig. 18–1)**. **Detail drawings**

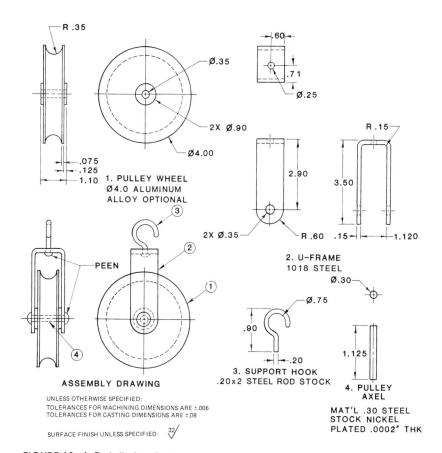

FIGURE 18–1 Detail drawings

323

324 Drafting in a Computer Age

contain information relating to the manufacture of parts. **Assembly drawings** describe the relationship of parts and the product assembly.

Pictorial drawings may be used as working drawings if the exact shape and necessary dimensions can be shown with clarity **(Fig. 18–2)**. Pictorial drawings are usually used to illustrate product assembly, and multiview drawings are used to convey manufacturing details.

Detail Drawings

A product may consist of a single part, or contain hundreds or thousands of parts **(Fig. 18–3)**. The manufacture of each part requires a working drawing containing specifications for its manufacture. These include dimensions, tolerances, scale, material, machining and casting details, fastening methods, finishing instructions, identification numbers, parts and standard purchase lists, and indexes to related part drawings. A typical working drawing of a single part is shown in **Fig. 18–4**.

Since different parts may be manufactured and/or assembled in different locations, it is imperative that uniform standards, such as ANSI, be used in the preparation of all detail drawings. Unless the entire manufacturing operation is contained in one location, detail drawings for one part should not be placed on the same sheet as drawings for another part. Even then, careful indexing and cross-indexing is necessary to avoid confusion. Drawing sheet layout standards, which outline indexing methods, are covered in chapter 6.

Specialized detail drawings are often prepared in separate sets for different manufacturing operations. Specialized sets may be prepared for foundry use, rough machining, finish machining, and surface finishing. The detail drawings show only the information needed for the particular operation. When this is done, great care must be taken to insure the individual sets are totally compat-

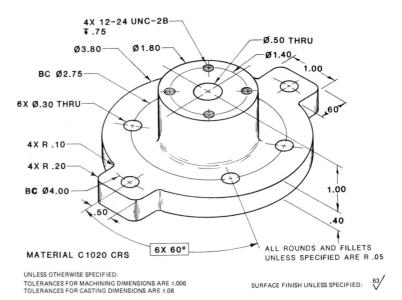

FIGURE 18–2 Pictorial working drawing of a plate cover

FIGURE 18–3 Thousands of working drawings are required to manufacture this airplane. (Courtesy of Cessna Aircraft Company)

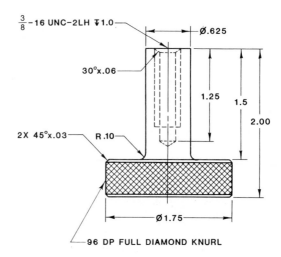

FIGURE 18–4 Detail working drawing for an adjustment knob from a lathe assembly

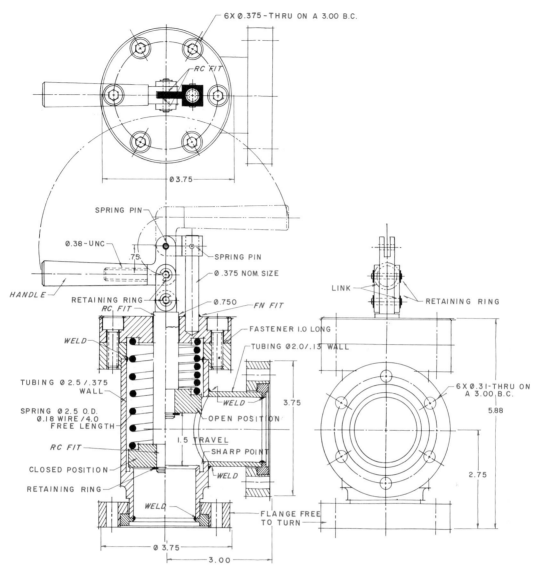

FIGURE 18-5 Assembly drawing for a pipe valve (Reprinted from TECHNICAL DRAWING AND DESIGN by Goetsch and Nelson, © 1986 Delmar Publishers Inc.)

ible, especially in the assignment of tolerances.

Assembly Drawings

A drawing which shows the relationship of two or more parts of a product is an assembly drawing **(Fig. 18–5)**. There are many different types of assembly drawings. The amount of detail and drawing type depends on its use during manufacturing. Assembly drawings do not contain as much detail as in detail drawings. Most are prepared without hidden lines to promote clarity. The following are the most common types of assembly drawings.

- **General assembly drawings** include orthographic drawings of assembled products, parts lists, and related data needed in the manufacturing process **(Fig. 18–6)**.
- **Layout assembly drawings** show rough preliminary design ideas, sometimes in sketch form. They show overall size, form, location, and relationship of only major components **(Fig. 18–7)**.
- **Design assembly drawings** are refined versions of layout assembly drawings with more accurate and refined dimensions, line work and manufacturing data **(Fig. 18–8)**.
- **Working assembly drawings** are fully dimensioned and noted drawings that are used directly in

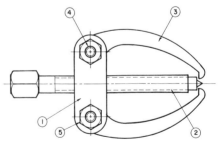

NO.	DRAWING NO.	ITEM	REQ'D.
–	–	1/4-20 UNC	–
5	PURCH	LOCK NUT-HEX	2
–	–	1/4-20 UNC X 1 1/2 LG.	–
4	PURCH	SCREW-CAP HEX HD	2
3	A661982	HOOK	2
2	A661981	SCREW-CENTER	1
1	A661983	YOKE	1

FIGURE 18–6 General assembly drawing and parts list for a gear puller (Reprinted from TECHNICAL DRAWING AND DESIGN by Goetsch and Nelson, © 1986 Delmar Publishers Inc.)

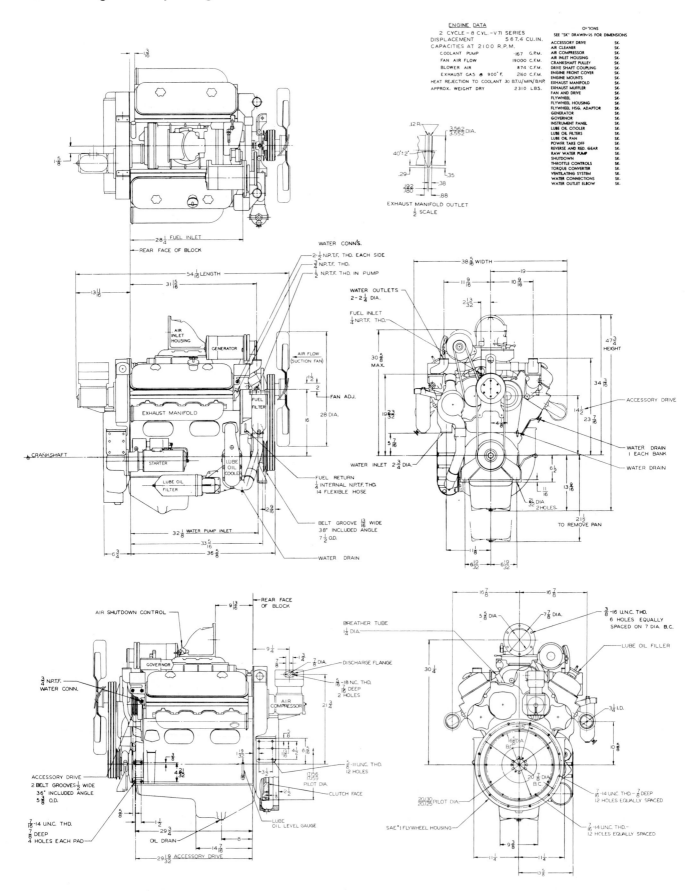

FIGURE 18-7 Layout assembly drawing for an automobile engine (Courtesy of General Motors Corporation, Engineering Standards)

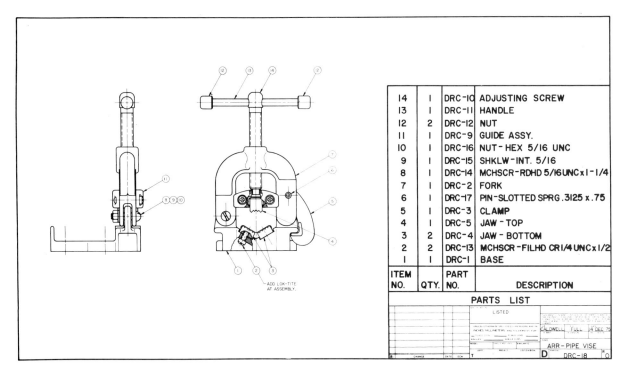

FIGURE 18-8 Design assembly drawing for a pipe vise (Reprinted from MECHANICAL DRAFTING by Madsen, Shumaker, and Stewart, © 1986 Delmar Publishers Inc.)

the manufacturing process **(Fig. 18-9)**.
- **Erection assembly drawings** are used in the construction industry and include all information necessary for the erection of structural framework. Detailed information relating to structural members is omitted **(Fig. 18-10)**.
- **Subassembly drawings** show the smallest assembled components which are part of a larger com-

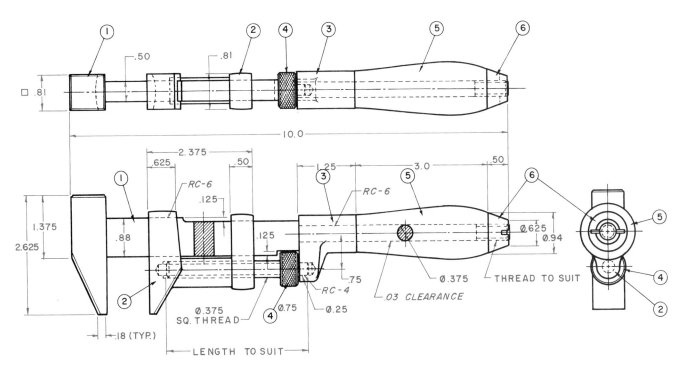

FIGURE 18-9 Working assembly drawing for a wrench (Reprinted from TECHNICAL DRAWING AND DESIGN by Goetsch and Nelson, © 1986 Delmar Publishers Inc.)

328　Drafting in a Computer Age

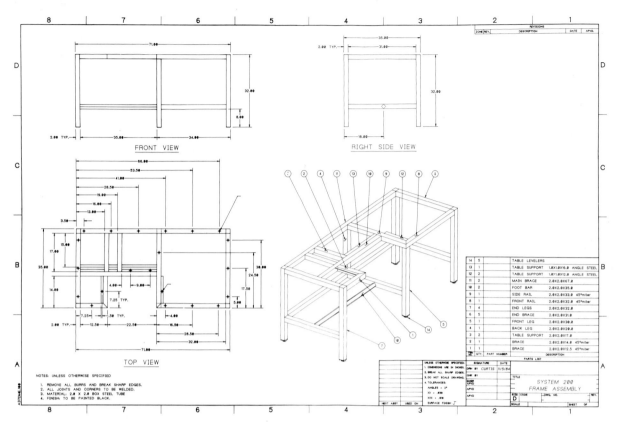

FIGURE 18–10　Erection assembly drawing of a bench frame (Courtesy of EFT Systems)

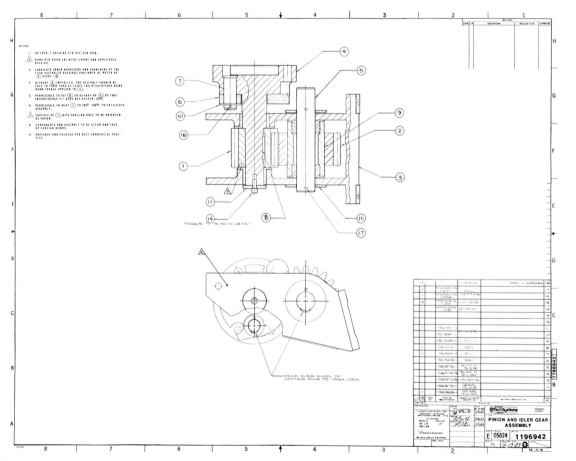

FIGURE 18–11　Subassembly drawing of a pinion and idler gear assembly (Courtesy of Aerojet Techsystems Co.)

ponent. Working drawings may include many layers of subassembly drawings, depending on the size and complexity of the finished product **(Fig. 18–11)**.
- **Pictorial assembly drawings** are substitutes for multiview assembly drawings if the product is not complex or if it is intended for nontechnical interpretation **(Fig. 18–12)**. Many pictorial assembly drawings are sectional views designed to show the interior structure and relationship of static and moving parts, and components **(Fig. 18–13)**. Pictorial assembly drawings are often drawn in exploded-view form (see Chapter 15) to show the relationship of parts before assembly.
- **Outline assembly drawings** show only the major outline of an assembly or subassembly and includes only a few basic notations. Symmetrical objects are often sparsely detailed on one half with only major outlines shown on the other half **(Fig. 18–14)**. It is also used to show only the basic positioning of subassemblies. Catalog publications often use outline assembly drawings.

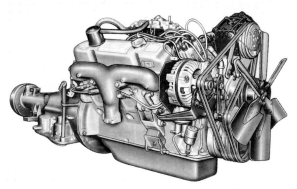

FIGURE 18–12 Pictorial assembly drawing (Courtesy of General Motors Corporation, Engineering Standards)

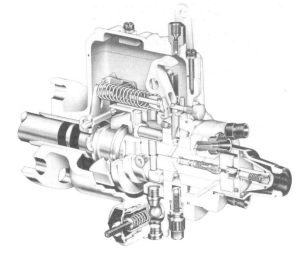

FIGURE 18–13 Cutaway pictorial assembly drawing (Courtesy of Stanadyne Diesel Systems)

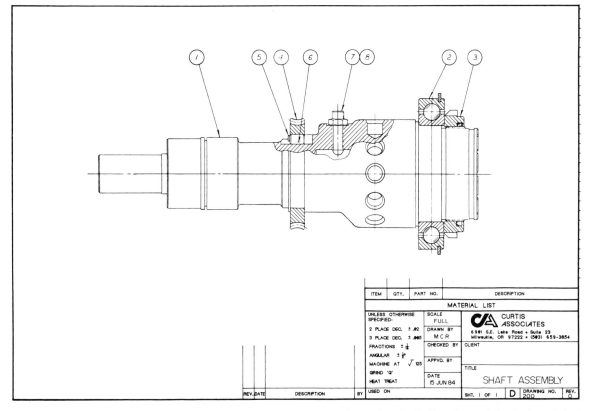

FIGURE 18–14 Outline assembly drawing with a broken-out section of a shaft (Courtesy of Curtis Associates)

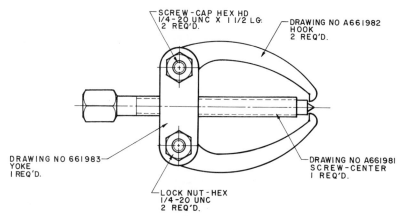

FIGURE 18-15 Operation assembly drawing for a gear puller (Reprinted from TECHNICAL DRAWING AND DESIGN by Goetsch and Nelson, © 1986 Delmar Publishers Inc.)

- **Operation assembly drawings** illustrate the mechanical operation. Multiple positions of moving parts are often shown with phantom lines, sections, and enlarged details **(Fig. 18-15)**.
- **Diagram assembly drawings** show the form and location of all subassemblies with multiview drawings. Partial cutaway sections and limited hidden lines are often used to show only the basic outline of hidden parts. Phantom lines are used to depict adjacent parts **(Fig. 18-16)**.
- **Installation assembly drawings** show how to assemble products. Only the details and dimensions needed for assembly are included **(Fig. 18-17)**. These often contain multiview drawings and a pictorial drawing as an aid to interpretation.

WORKING DRAWING DIMENSIONS

Dimensions on working drawings are used when manufacturing products to an exact specified size. Dimensions must be selected and placed to insure that no further calculations are required during any stage of the manufacturing process. All related drawings must use the same standards for tolerancing to insure proper fitting in the final assembly. **Fig. 18-18** shows a summary of the use

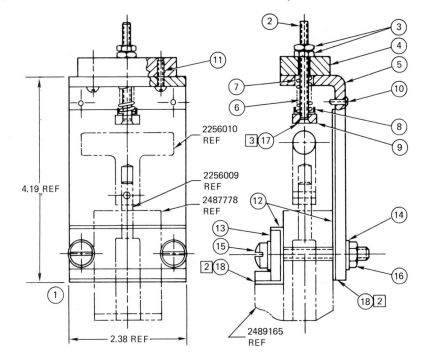

FIGURE 18-16 Diagram assembly drawing of a valve adjuster

of the U.S. customary system of measure. **Figures 18–19** through **18–24** show the metric system of measure.

Although working drawing dimensions must be extremely accurate and consistent, specific manufacturing processes should not be specified. Exactly how the product is manufactured is the role of the manufacturer, not the engineer, designer, or drafter. Only the final desired result is specified on working drawings. Outcomes may be specified by machine process notations on working drawings. In these cases, the end result is also the goal, not the process (**Fig. 18–25,** page 337). Since there are other manufacturing methods that can produce the same result, only the size and form of the final product are interpreted from the drawing—not the specific process.

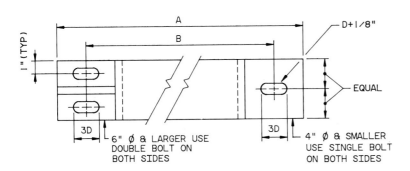
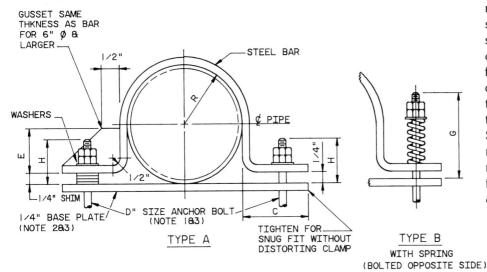

FIGURE 18–17 Installation assembly drawing of a base plate for pipes

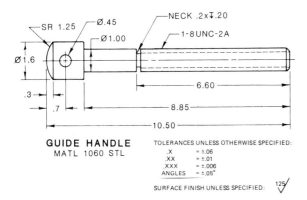

FIGURE 18–18 Working drawing of a guide handle with U.S. customary dimensions (ANSI)

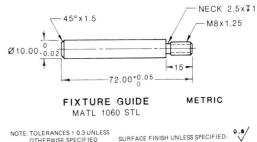

FIGURE 18–19 Working drawing of a fixture guide with metric dimensions (ANSI)

332 Drafting in a Computer Age

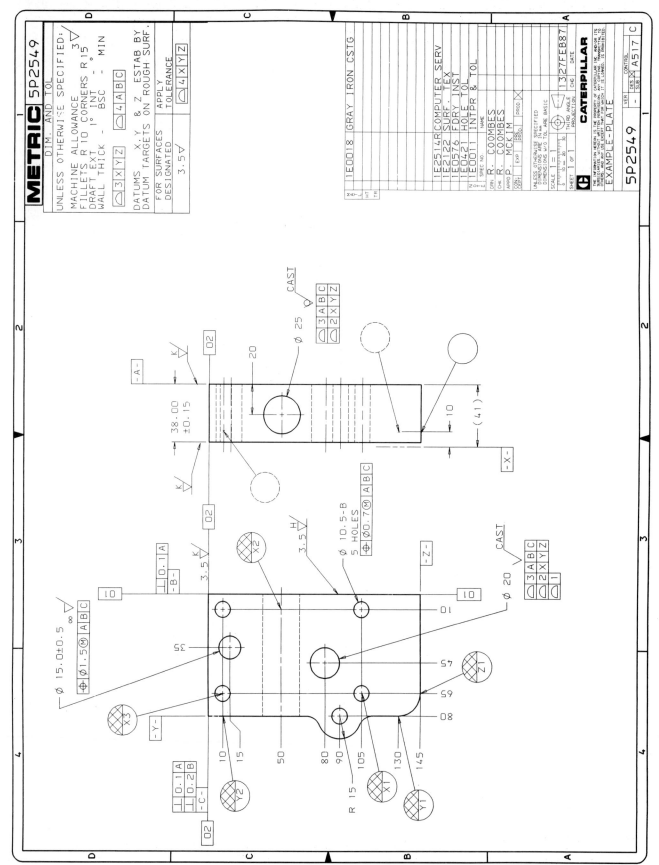

FIGURE 18–20 Working drawing with metric dimensions (Courtesy of Caterpillar Inc.)

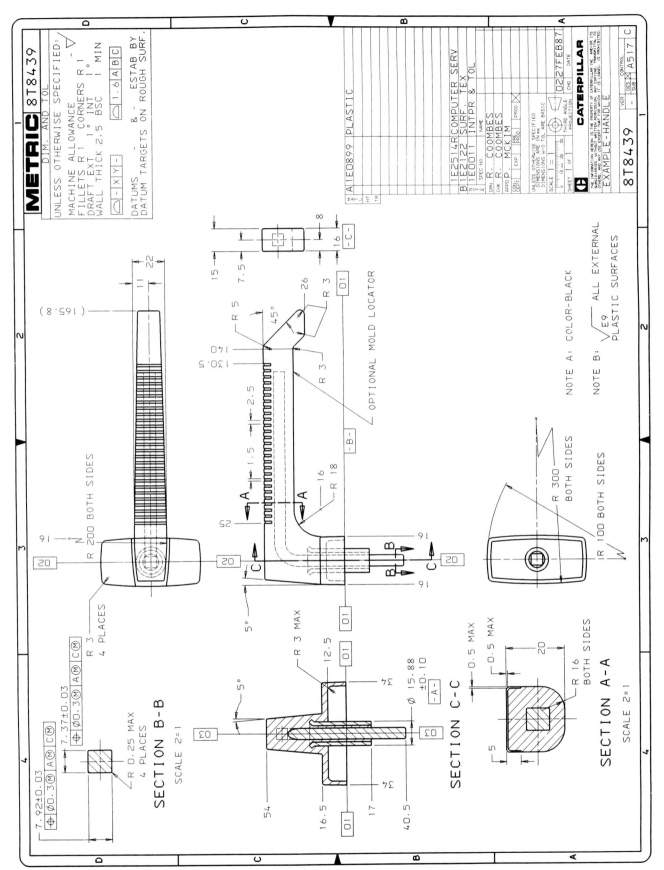

FIGURE 18-21 Working drawing with metric dimensions (Courtesy of Caterpillar Inc.)

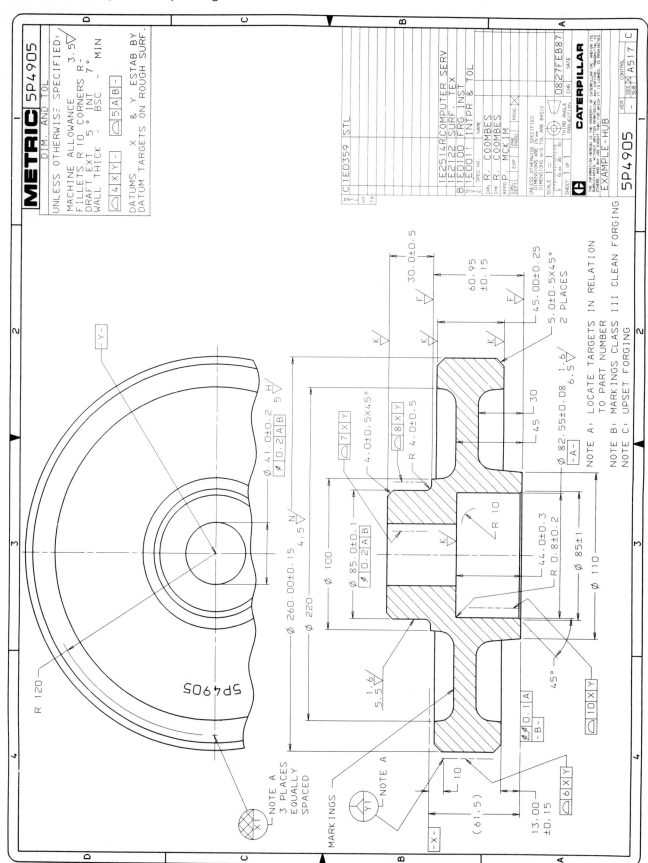

FIGURE 18-22 Working drawing with metric dimensions (Courtesy of Caterpillar Inc.)

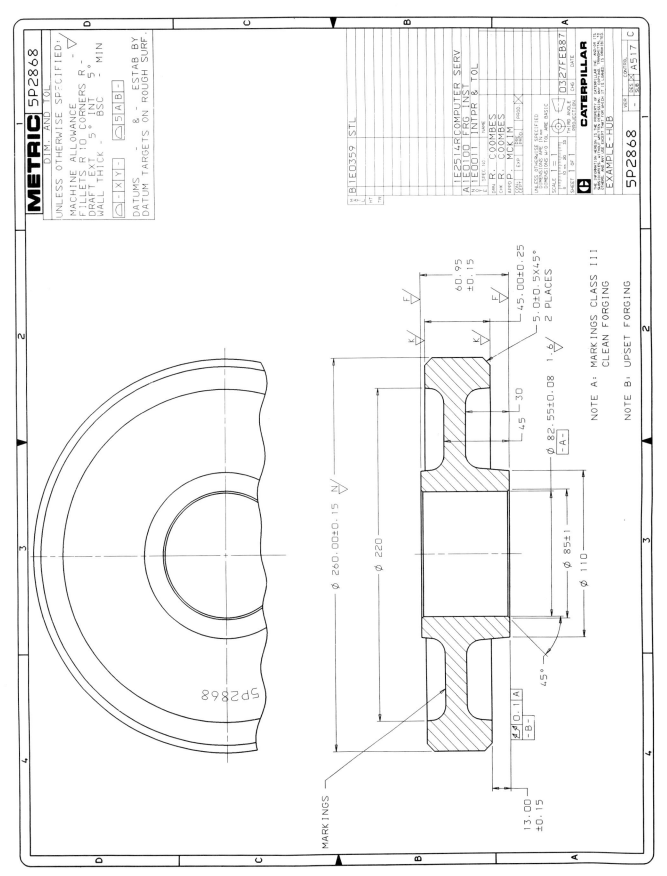

FIGURE 18–23 Working drawing with metric dimensions (Courtesy of Caterpillar Inc.)

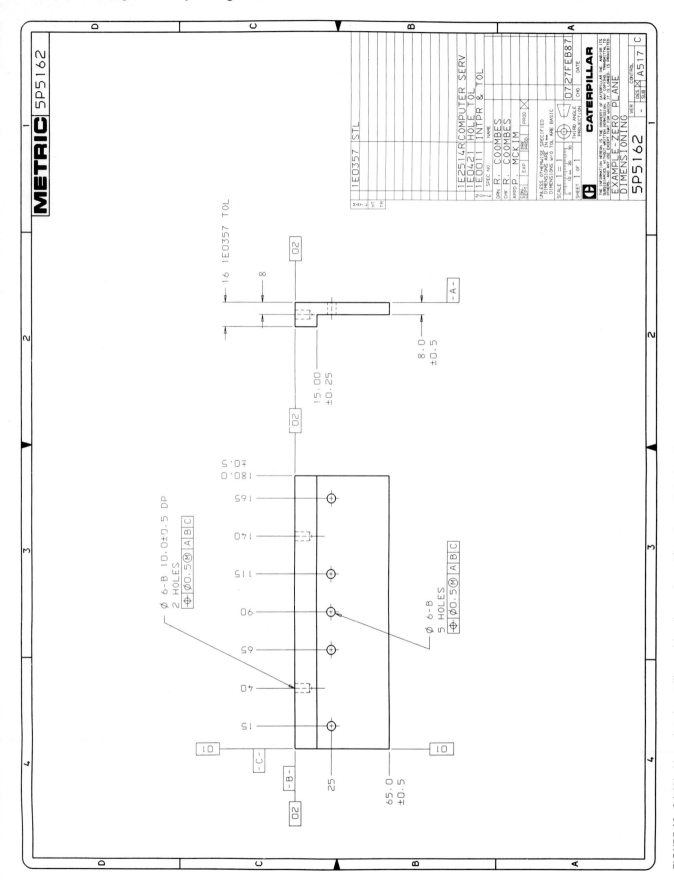

FIGURE 18–24 Working drawing with metric dimensions (Courtesy of Caterpillar Inc.)

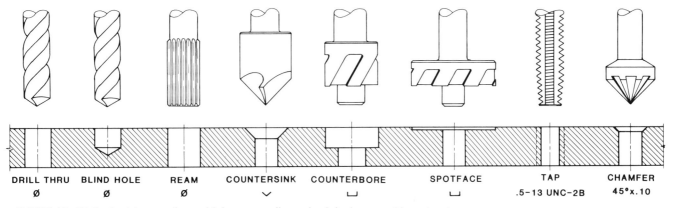

FIGURE 18-25 Typical types of machining operations depicted on working drawings

COMPUTER-AIDED DRAFTING & DESIGN

COMPUTER-AIDED WORKING DRAWINGS

The creation of working drawings incorporates most of the drafting and CAD practices covered in the previous chapters. These included multiview drawing, dimensioning, tolerancing, sectioning, auxiliary views, pictorial drawings, fasteners, and drafting conventions and formats. CAD practices also included drawing and editing techniques, rotating grids for auxiliary views, crosshatching for sectional views, and inserting fasteners, standard items, and title blocks.

A set of CAD working drawings for a collector assembly is shown in **Figs. 18–26** through **18–34**. This set consists of seven detail drawings and two assembly drawings. In studying these examples, many of the following CAD practices become apparent:

1. Multiview drawing in all of the details and one of the assembly drawings (Figs. 18–26 through 18–33).
2. Extensive dimensioning in Figs. 18–26 and 18–27.
3. Tolerancing used in Fig. 18–26. Note both bilateral and limit tolerancing have been used.
4. Sectioning, using the HATCH command, as described in Chapter 11, has been used in Figs. 18–26 and 18–27.
5. An auxiliary view (Fig. 18–27) was done by rotating the grid, as described in Chapter 12.
6. Text has been used extensively to insert notes in Figs. 18–26 and 18–27, and in the parts lists assembly drawing. Note that in Fig. 18–26, two styles of text have been used. The standard text style has been used for dimensioning and notes, and a larger style for the SECTION A–A callout.
7. A CAD shortcut was used in creating the subassembly drawing in Fig. 18–29. Note that in Part A, the shoe was detailed in Fig. 18–28. The CAD operator copied it to Fig. 18–29, mirrored it, and added Part B. This eliminated redrawing the part and saved drafting time, which is one of the advantages of CAD.
8. The spring in Fig. 18–30 is a standard fastener used by a company. The CAD operator inserted the standard spring onto the company's standard drawing format and added the necessary dimensions. The drawing was completed in minutes.
9. The parts lists are identical on the two assembly drawings. The CAD operator created it once and copied it onto the drawings. This is also true for drawing borders and title blocks for each of the drawings. These were done in the same manner as described in Chapter 6.
10. An isometric drawing, described in Chapter 15, was used to create Fig. 18–34.

The advantages of CAD become more apparent by further studying these drawings. In addition to the timesaving advantage of preparing the working drawings initially on a CAD, there is a big timesavings when making revisions. It is much easier to change CAD-generated drawings than to change drawings that have been manually drawn. Many drawing changes may simply involve correcting a dimension or a tolerance, or moving the location of a note. A change may necessitate the removal or addition of a view. Simple editing commands, such as ERASE, MOVE, and COPY make changes a snap.

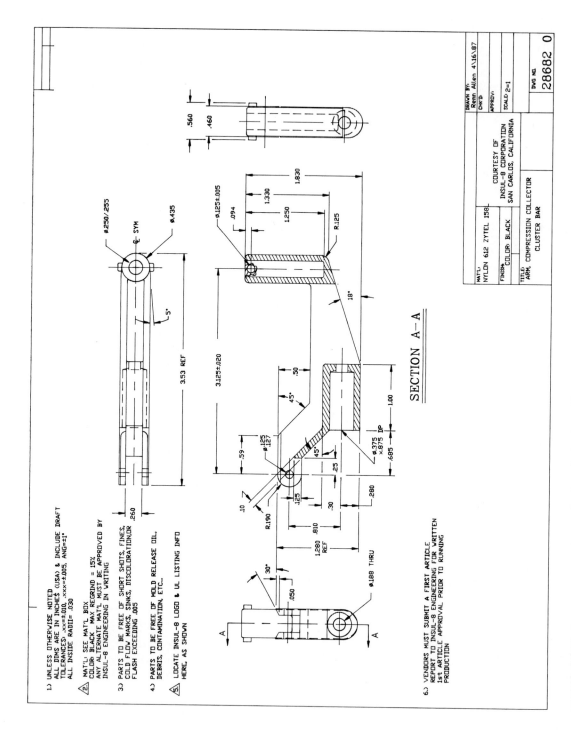

FIGURE 18–26 Arm—compression collector cluster bar (Sheet 1 of 9) (Courtesy of Ms. Renn Allen, Draftsperson, Insul-8 Corporation)

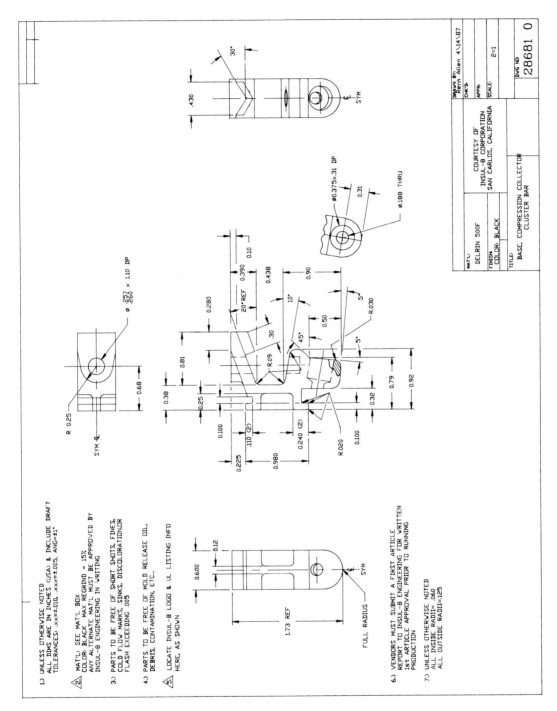

FIGURE 18–27 Base—compression collector cluster bar (Sheet 2 of 9) (Courtesy of Ms. Renn Allen, Draftsperson, Insul-8 Corporation)

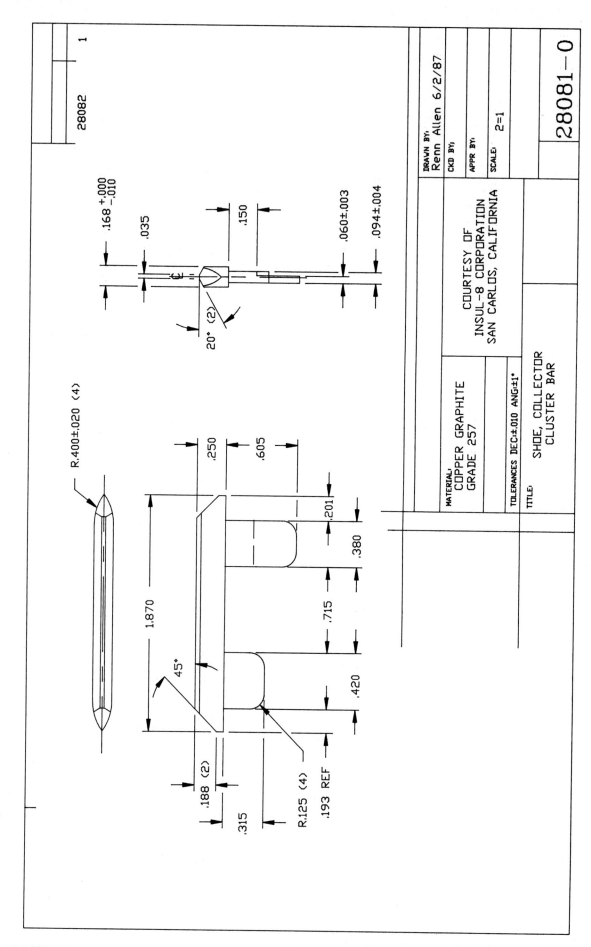

FIGURE 18-28 Shoe—collector cluster bar (Sheet 3 of 9) (Courtesy of Ms. Renn Allen, Draftsperson, Insul-8 Corporation)

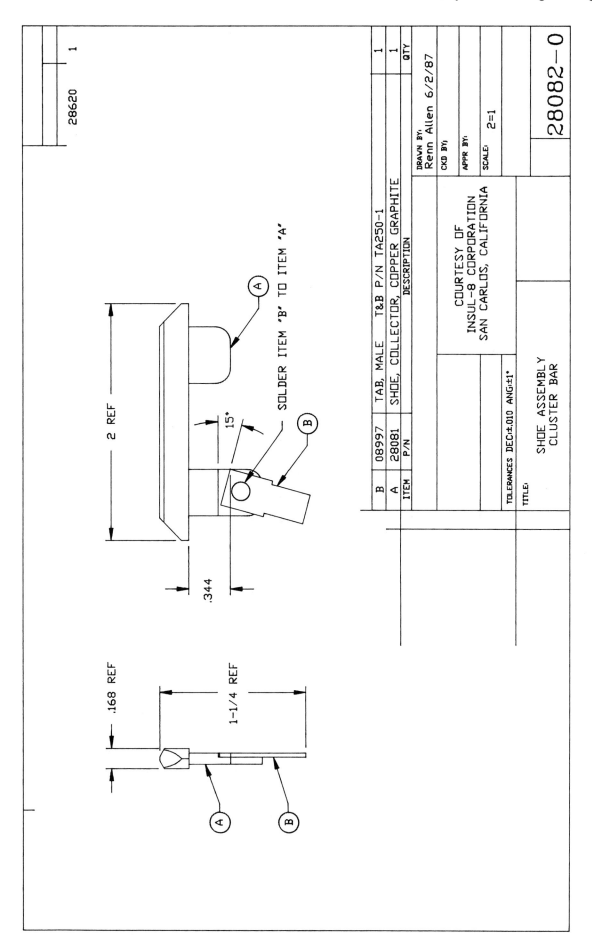

FIGURE 18-29 Shoe assembly—cluster bar (Sheet 4 of 9) (Courtesy of Ms. Renn Allen, Draftsperson, Insul-8 Corporation)

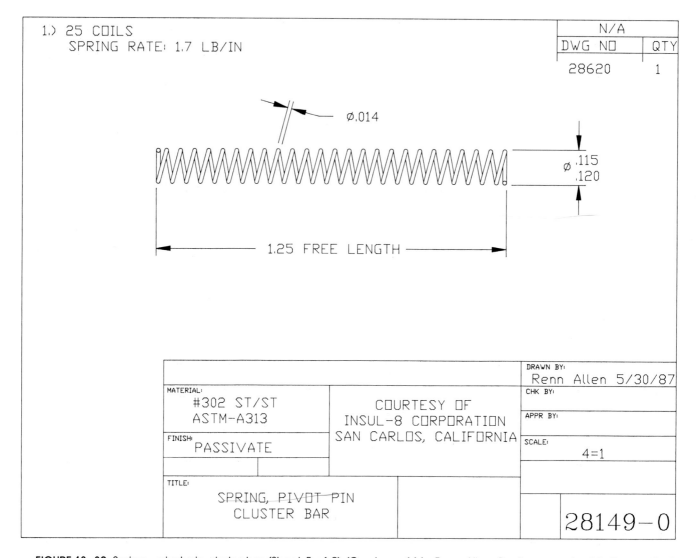

FIGURE 18-30 Spring—pivot pin cluster bar (Sheet 5 of 9) (Courtesy of Ms. Renn Allen, Draftsperson, Insul-8 Corporation)

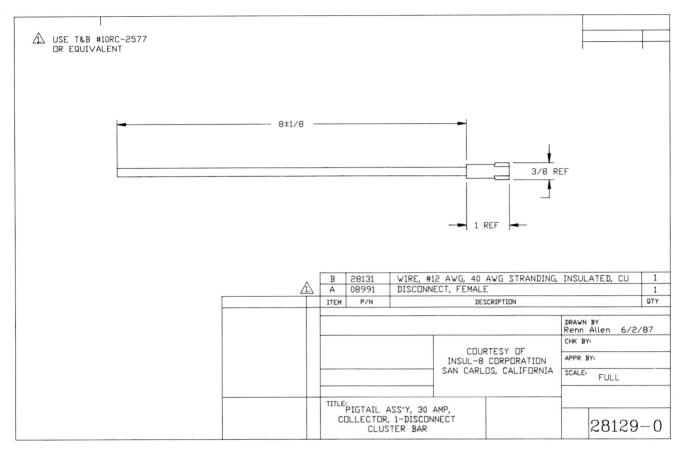

FIGURE 18–31 Pigtail assembly—30 amp, collector, 1-disconnect cluster bar (Sheet 6 of 9) (Courtesy of Ms. Renn Allen, Draftsperson, Insul-8 Corporation)

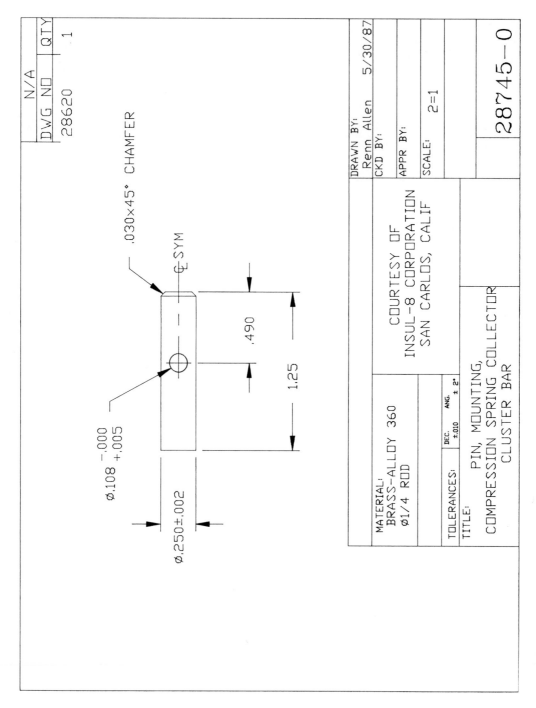

FIGURE 18-32 Mounting pin—compression spring collector cluster bar (Sheet 7 of 9) (Courtesy of Ms. Renn Allen, Draftsperson, Insul-8 Corporation)

Chapter 18 Working Drawings 345

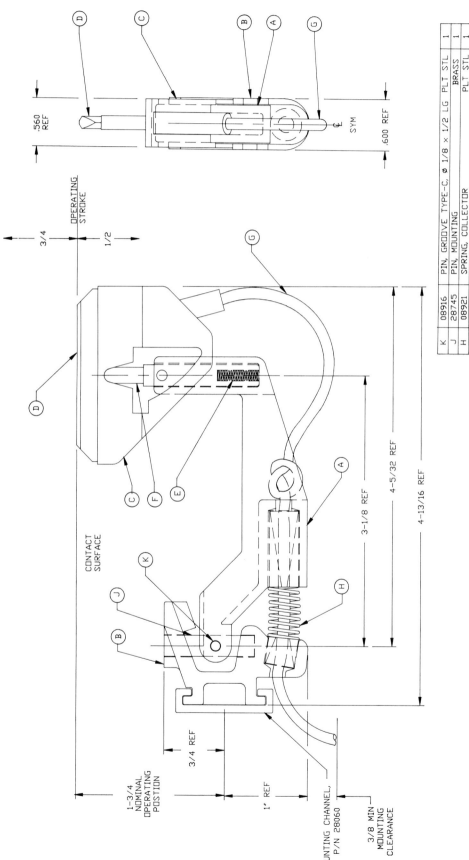

FIGURE 18-33 Collector assembly—compression spring cluster bar (Sheet 8 of 9) (Courtesy of Ms. Renn Allen, Draftsperson, Insul-8 Corporation)

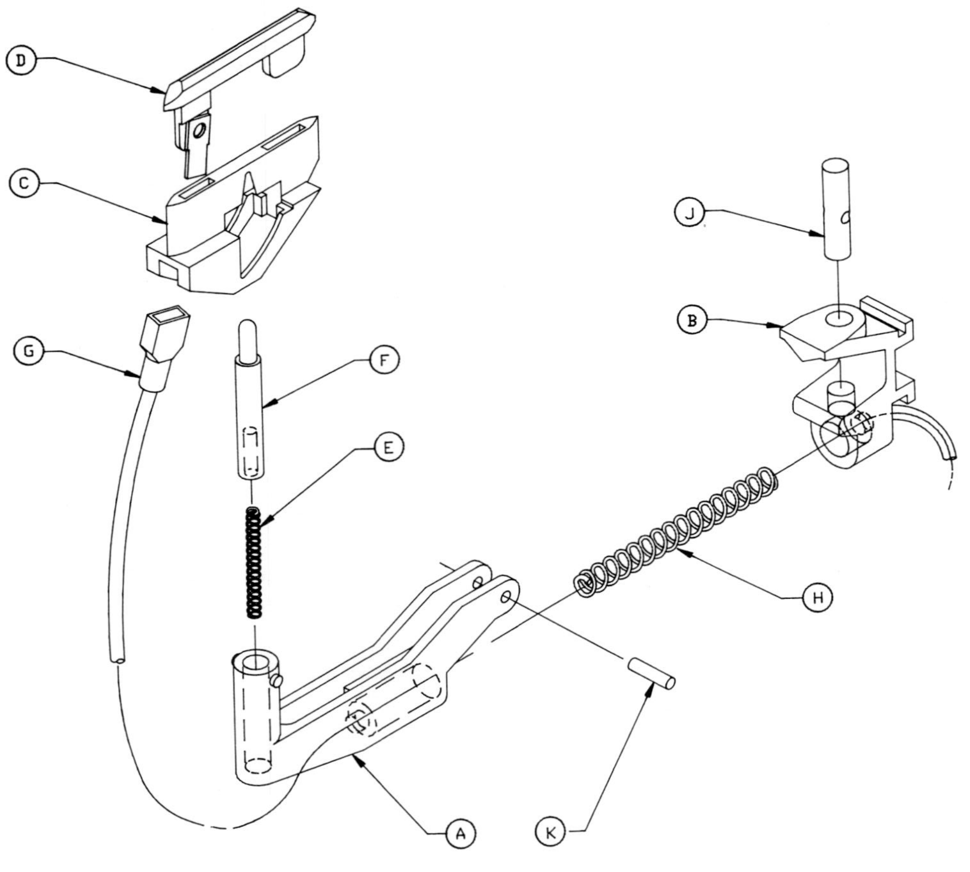

ITEM	P/N	DESCRIPTION		QTY
K	08916	PIN, GROOVE TYPE-C, Ø1/8 × 1/2 LG	PLT ST	1
J	28745	PIN, MOUNTING	BRASS	1
H	08921	SPRING, COLLECTOR	PLT ST	1
G	28129	PIGTAIL ASS'Y, 30 AMP	POLYCARBONATE	1
F	28075	PIN, PIVOT		1
E	28149	SPRING, PIVOT PIN	ST/ST	1
D	28082	SHOE ASS'Y	POLYCARBONATE	1
C	28080	HEAD, SHOE HOLDER		1
B	28681	BASE, COMPRESSION COLLECTOR	ACETAL	1
A	28682	ARM, COMPRESSION COLLECTOR	NYLON	1

COURTESY OF
INSUL-8 CORPORATION
SAN CARLOS, CALIFORNIA

DRAWN BY: Renn Allen 5/26/87
CHK'D BY:
APPR BY:
SCALE: 1=1

28620-0

TITLE: COLLECTOR ASS'Y, COMPRESSION SPRING, CLUSTER BAR

FIGURE 18-34 Pictorial collector assembly—compression spring cluster bar (Sheet 9 of 9) (Courtesy of Ms. Renn Allen, Draftsperson, Insul-8 Corporation)

EXERCISES

1. Sketch the multiview working drawings for **Figs. 18–35** through **18–37**. With instruments, draw each sketch to accurate scale.

2. With instruments, draw the multiview working drawings and the isometric drawing in **Figs. 18–38** through **18–44**.

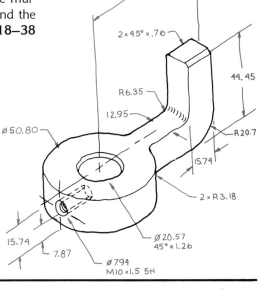

FIGURE 18–35 Sketch a multiview working drawing. With instruments, complete the scaled drawing for the lathe dog. (Reprinted from MECHANICAL DRAFTING by Madsen, Shumaker, and Stewart, © 1986 Delmar Publishers Inc.)

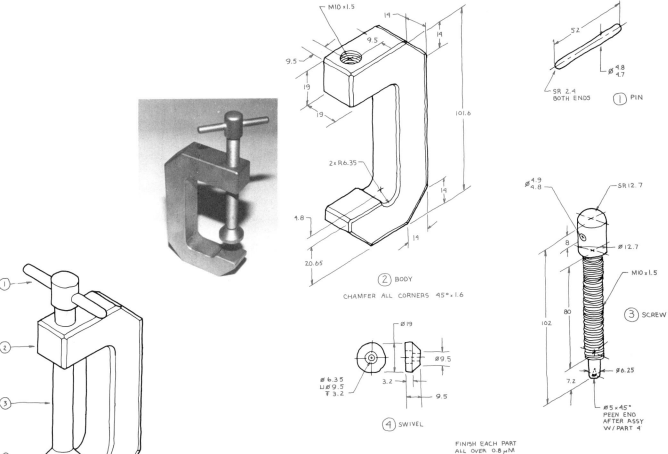

FIGURE 18–36 Sketch the detailed working drawings. With instruments, complete the scaled drawings for the C-clamp. (Reprinted from MECHANICAL DRAFTING by Madsen, Shumaker, and Stewart, © 1986 Delmar Publishers Inc.)

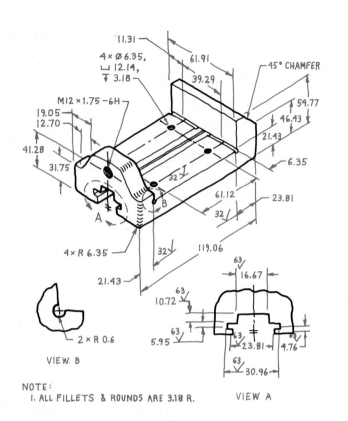

FIGURE 18–37 Sketch the multiview working drawing. With instruments, complete the scaled drawing for the vise base. (Reprinted from MECHANICAL DRAFTING by Madsen, Shumaker, and Stewart, © 1986 Delmar Publishers Inc.)

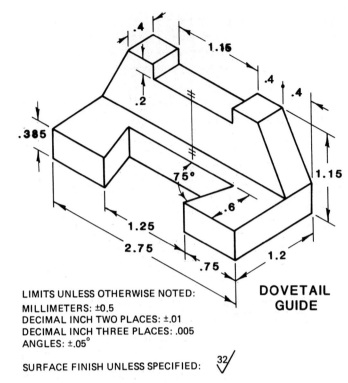

FIGURE 18–38 Draw the multiview working drawing and the isometric drawing for the dovetail guide.

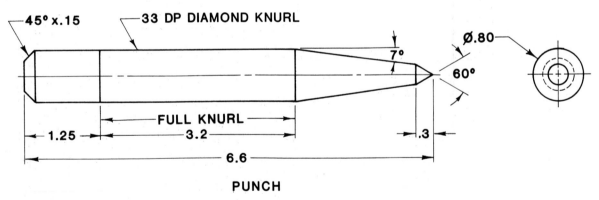

FIGURE 18–39 Draw the multiview working drawing and the isometric drawing for the punch.

Chapter 18 Working Drawings 349

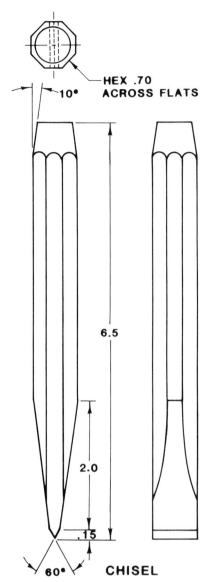

FIGURE 18–40 Draw the multiview working drawing and the isometric drawing for the chisel.

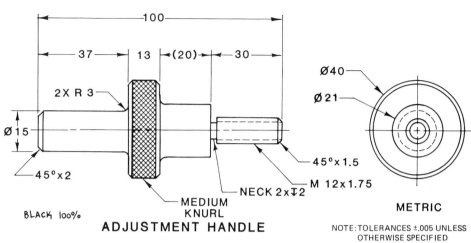

FIGURE 18–41 Draw the multiview working drawing and the isometric drawing for the adjustment handle.

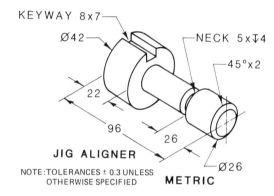

FIGURE 18–42 Draw the multiview working drawing and the isometric drawing for the jig aligner

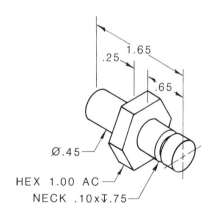

FIGURE 18–43 Draw the multiview working drawing and the isometric drawing for the fastener guide.

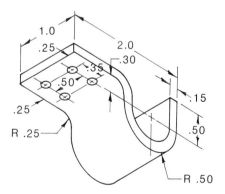

FIGURE 18–44 Draw the multiview working drawing and the isometric drawing for the dowel support.

3. With instruments, draw the multiview working drawings in **Figs. 18–45** through **18–57**.

4. With a CAD system or with drafting instruments, draw the multiview working drawings for **Figs. 18–58** through **18–62**.

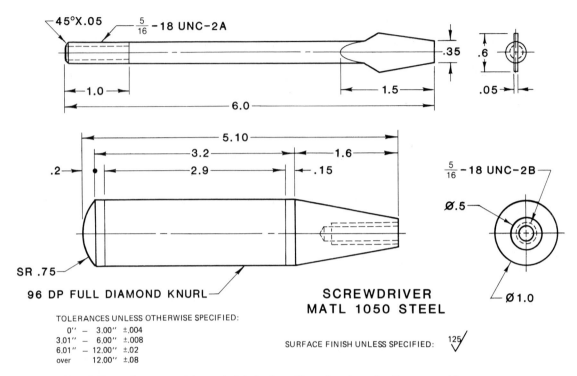

FIGURE 18–45 Draw the assembly and detailed working drawings for the screwdriver.

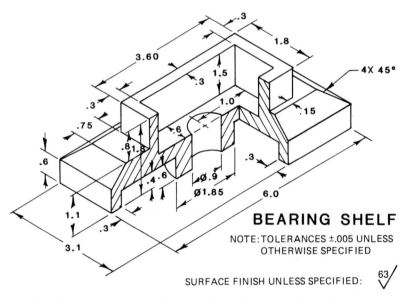

FIGURE 18–46 Draw the multiview working drawing, with sections, for the bearing shelf.

Chapter 18 Working Drawings   351

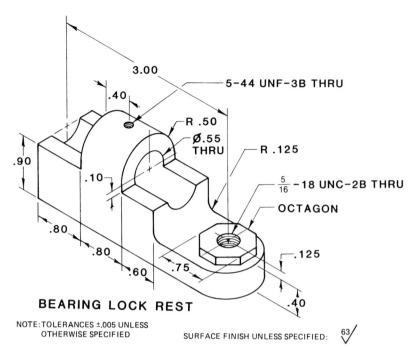

FIGURE 18–47 Draw the multiview working drawing for the bearing lock rest.

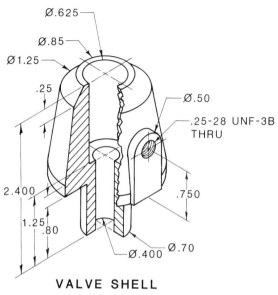

FIGURE 18–48 Draw the multiview working drawing, with sections, for the valve shell.

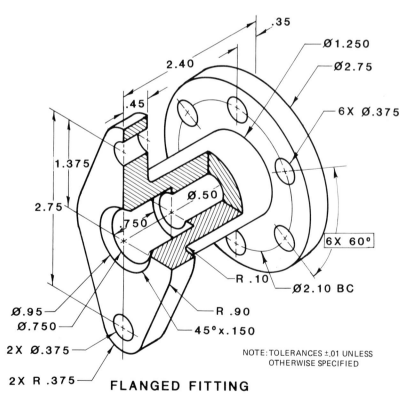

FIGURE 18–49 Draw the multiview working drawing, with sections, for the flanged fitting.

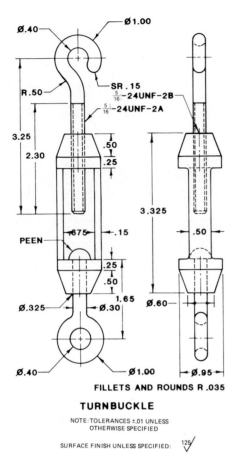

FIGURE 18–50 Draw the assembly and the detailed working drawings for the turnbuckle.

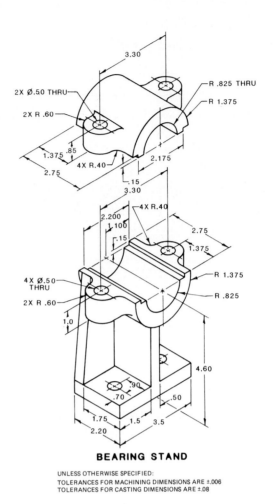

FIGURE 18–51 Draw the assembly working drawing for the bearing stand. Add bolts and nuts as needed.

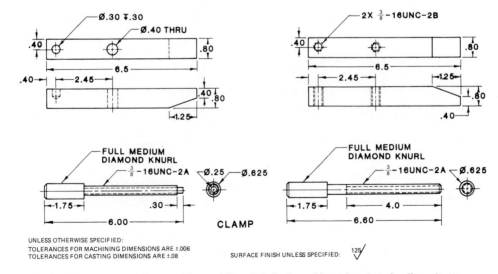

FIGURE 18–52 Draw the assembly and the detailed working drawing for the clamp.

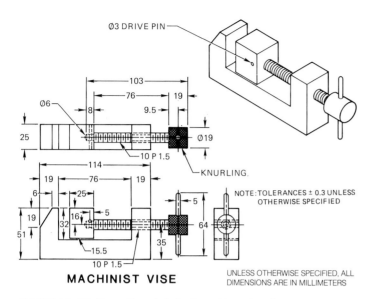

FIGURE 18–53 Draw the detailed working drawing and the isometric drawing for the machinist vise.

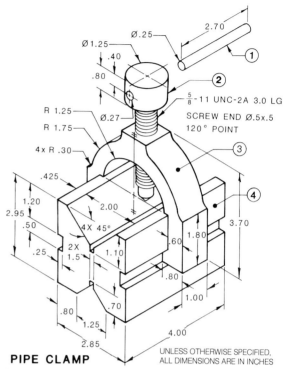

FIGURE 18–54 Draw the detailed working drawings for the pipe clamp. Include a materials list.

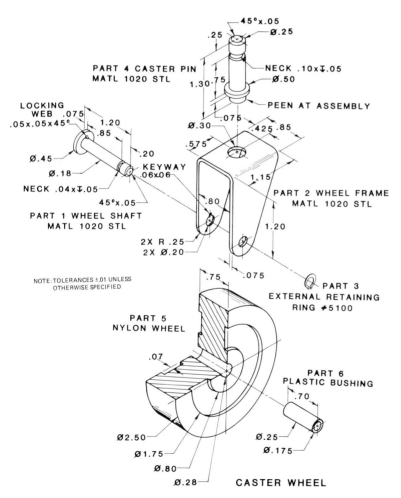

FIGURE 18–55 Draw the detailed working drawings for the caster wheel. Include a materials list.

Assembly Name: Machine Vice
PARTS LIST:

ITEM	QTY	NAME	DESCRIPTION	MATERIAL
1	2	HANDLE CAP		MS
2	1	HANDLE		MS
3	1	BODY		SAE 4320
4	1	SCREW		SAE 4320
5	1	MOVABLE JAW		SAE 1020
6	1	MOVABLE JAW PLATE		SAE 4320
7	1	FIXED JAW PLATE		SAE 4320
8	1	GUIDE		SAE 1020
9	2	MACHINE SCREW	.25-20UNC-2 × .500 SLOT FIL HD	STL
10	2	MACHINE SCREW	.190-32UNF-2 × .875 SLOT FIL HD	STL
11	1	SET SCREW	.25-20UNC-2 × .250 FULL DOG POINT	STL
12	2	MACHINE SCREW	.190-32UNF-2 × 6 SLOT FIL HD	STL

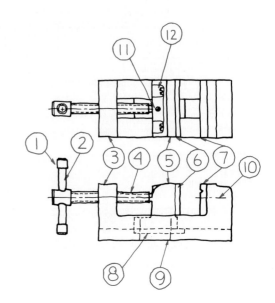

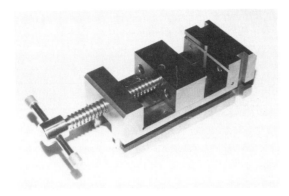

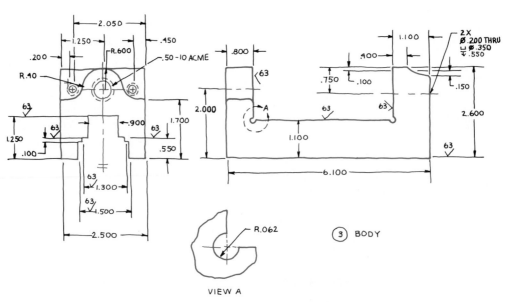

FIGURE 18–56 Draw the full set of working drawings for the machine vise. (Reprinted from MECHANICAL DRAFTING by Madsen, Shumaker, and Stewart, © 1986 Delmar Publishers Inc.)

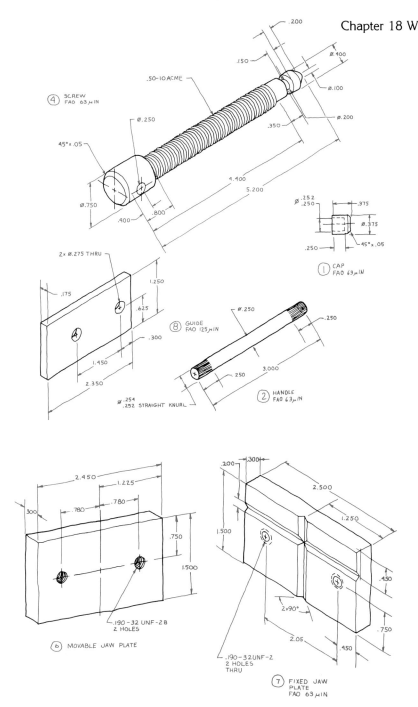

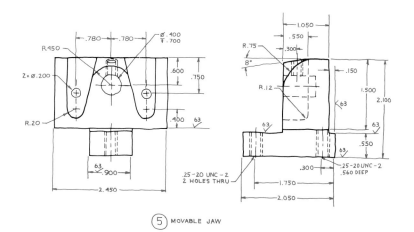

FIGURE 18–56
continued

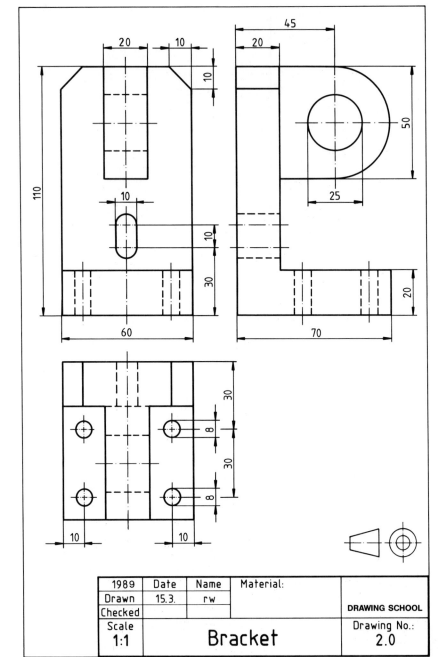

FIGURE 18–57 Redraw the first-angle projection drawing of the bracket to a third-angle projection working drawing. (Courtesy of Koh-I-Noor)

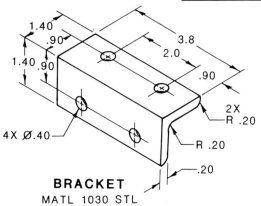

BRACKET
MATL 1030 STL

NOTE: TOLERANCES ±.005 UNLESS OTHERWISE SPECIFIED

FIGURE 18–58 Draw the multiview working drawing of a bracket with a CAD system.

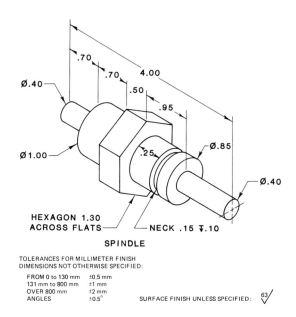

FIGURE 18-59 Draw the multiview working drawing of the spindle with a CAD system.

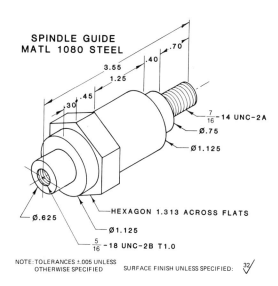

FIGURE 18-60 Draw the multiview working drawing of the spindle guide with a CAD system.

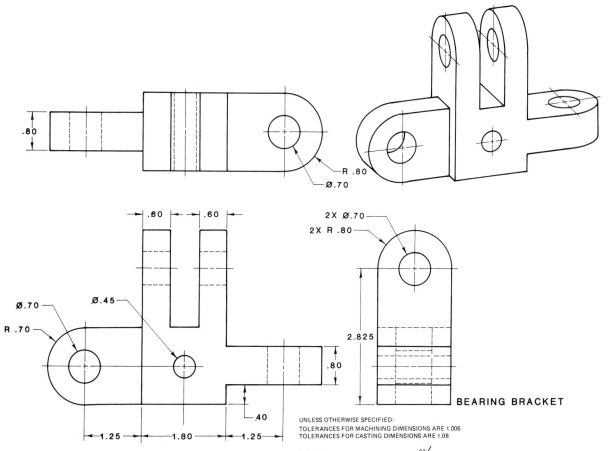

FIGURE 18-61 Redraw the multiview and isometric working drawings of the bearing bracket with a CAD system.

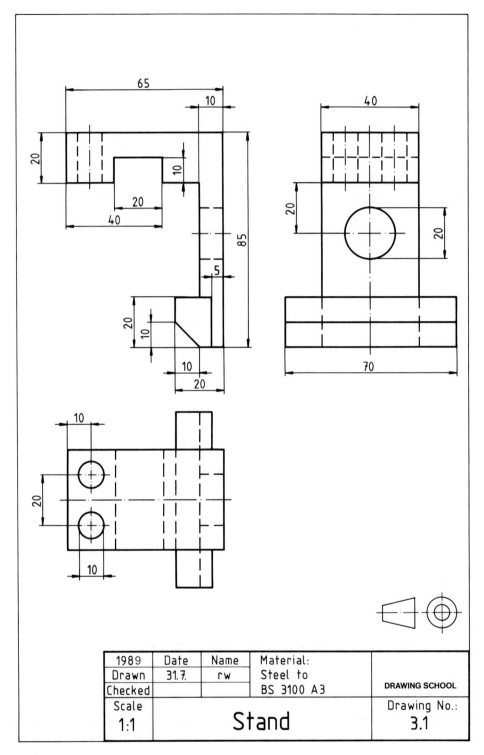

FIGURE 18–62 With a CAD system, lay out the metric first-angle projection working drawing of the stand.

5. Draw the multiview working drawings in **Figs. 18–63** through **18–67** and make all ECOs as noted. Record all ECOs onto the proper listing on your drawing format.

6. Draw the working drawing for **Fig. 18–67** using the machine operations in **Fig. 18–68**.

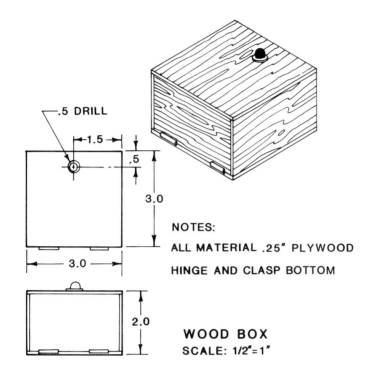

FIGURE 18–63 Draw the working drawings for the wood box container for a red flasher light and make the following ECOs:
1. Change the material to $3/8''$ plywood.
2. Change the height to 1.75".
3. Move the hole center for light from .5" to .65".
4. Specify the hinges and clasp.
5. Specify the assembly procedures.

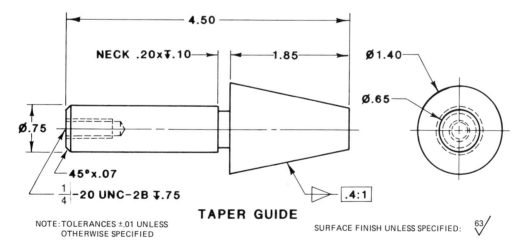

FIGURE 18–64 Draw the working drawings for the taper guide and make the following ECOs:
1. Change the taper length to 2.00".
2. Change the large diameter of taper to 1.50".
3. Change the small diameter of taper to .50".
4. Calculate a new taper: $\text{TAPER} = \dfrac{\Phi - \phi}{\text{LENGTH}}$.

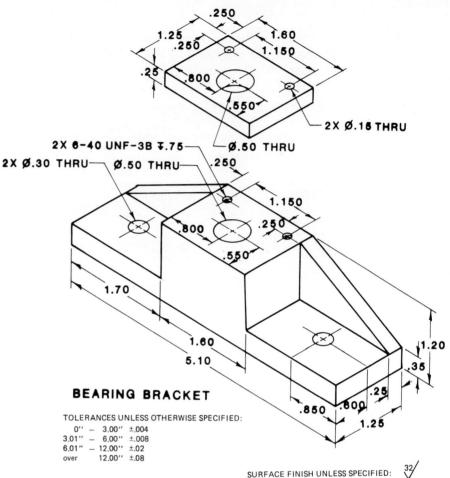

FIGURE 18-65 Draw the working drawings for the bearing bracket and make the following ECOs:
1. Add .05" radius fillets and rounds to the support webs.
2. Add a boss to both .30" diameter holes—.05" high, .50" diameter.
3. Change cap's thickness to .50".
4. Add .05" cork gasket seal.

BEARING BRACKET

TOLERANCES UNLESS OTHERWISE SPECIFIED:
0" – 3.00" ±.004
3.01" – 6.00" ±.008
6.01" – 12.00" ±.02
over 12.00" ±.08

SURFACE FINISH UNLESS SPECIFIED: 32

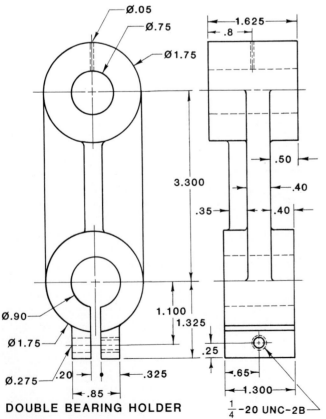

DOUBLE BEARING HOLDER

UNLESS OTHERWISE SPECIFIED:
TOLERANCES FOR MACHINING DIMENSIONS ARE ±.006
TOLERANCES FOR CASTING DIMENSIONS ARE ±.08

FIGURE 18-66 Draw the working drawings for the double bearing holder and make the following ECOs:
1. Change the bearing hole diameters .75" and .90" to .80" and 1.00" respectively.
2. Change the oil hole to .075" diameter.
3. Change the threaded hole to $3/16$" UNF, Class 3 fit.
4. Change the center distance of the bearing holes to 3.500".
5. Change the machining tolerance to plus or minus .008".

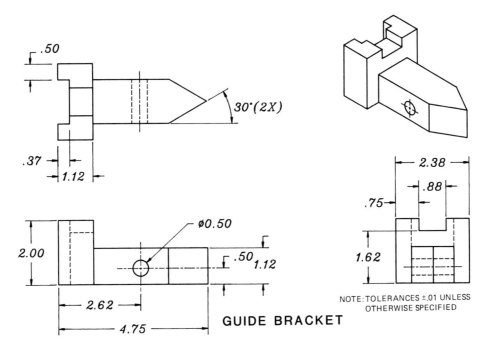

FIGURE 18-67 Draw the working drawings for the guide bracket and make the following ECOs:
1. Add UNC thread to .50" through.
2. Change the length to 5.00".
3. Change 30° point to 35°.
4. Change .88" slot to .90".

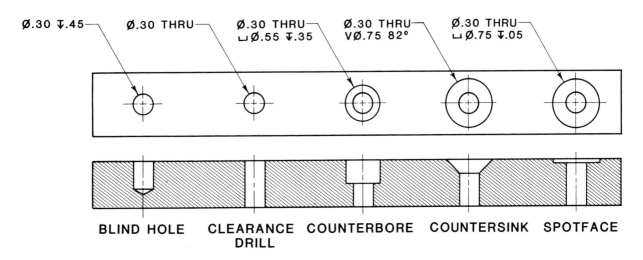

FIGURE 18-68 Design a plate cover for Fig. 18-65 using the five machine operations.

7. Redesign the bearing bracket in **Fig. 18–69**.
8. Design and draw the working drawings for:
    Kitchen cabinet door catch.
    Kitchen cabinet door and drawer pulls. Use standard spacing for pulls.
    • Childproof latch for medicine cabinet.
    • Hub cap for an automobile. Use standard dimensions for all diameters and stud locations.
    • Desktop holder for drafting supplies and equipment to be used while drafting.
    • Furniture to hold a CAD system's hardware.
    • Removable handle that can be used to lift heavy 2′ x 2′ x 3′ cardboard boxes.

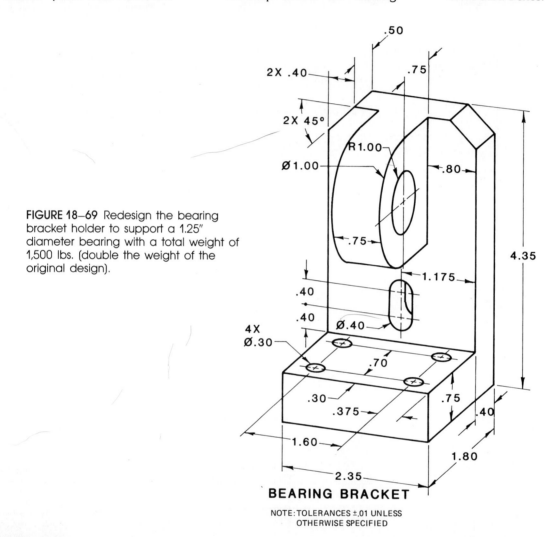

FIGURE 18–69 Redesign the bearing bracket holder to support a 1.25″ diameter bearing with a total weight of 1,500 lbs. (double the weight of the original design).

BEARING BRACKET
NOTE: TOLERANCES ±.01 UNLESS OTHERWISE SPECIFIED

KEY TERMS

Assembly drawing
Design assembly drawing
Detail drawing
Diagram assembly drawing
Erection assembly drawing

General assembly drawing
Installation assembly drawing
Layout assembly drawing
Operation assembly drawing
Outline assembly drawing

Pictorial assembly drawing
Subassembly drawing
Working assembly drawing
Working drawing

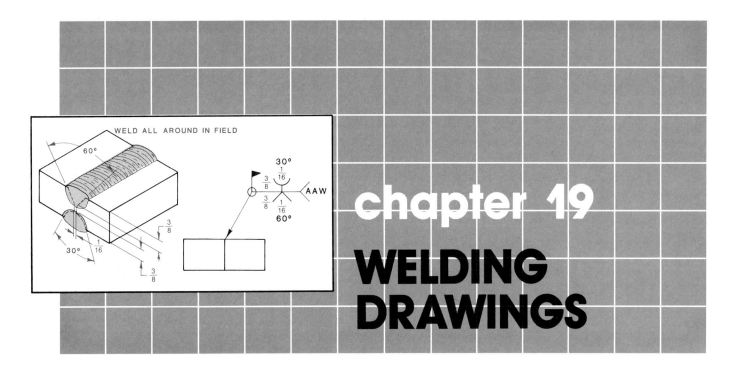

chapter 19
WELDING DRAWINGS

OBJECTIVES

The student will be able to:

- read welding abbreviations
- select the proper type of weld
- read welding symbols
- draw a weld reference symbol
- select the proper weld finish
- specify contour symbols
- specify the size and strength of a weld
- specify the length of a weld
- specify the pitch of a weld
- draw a weld symbol with a CAD system

INTRODUCTION

When designing products with multiple parts, designers have several options for joining parts together, including mechanical devices, adhesive bonding, and welding. The joining method depends on the material to be joined, the strength of the joint required, and the need for disassembly. Mechanical devices include nails, screws, bolts, rivets, keys, and splines. Adhesive bonding methods range from cement and epoxy glues, to plastic bonding solutions. Welding is a method of permanently joining metals and plastics through the application of heat. Welding is usually chosen when great strength, rigidity, permanence, or liquid-proofing is required.

Welding began over 4,000 years ago after the discovery of iron; but the science and technology of welding is relatively new. Early welders simply heated two pieces of iron in a forge, dipped them into a flux, then pounded the pieces together with a hammer. Later, oxygen and acetylene gas was added to increase the amount of heat produced. This enabled harder metals to be welded. Nevertheless, welding had been restricted to iron and iron alloys until the twentieth century. Today, hundreds of metallurgical processes are used to weld thousands of different metals and alloys. Vehicles, aircraft, ships, and machinery contain thousands of welds of different types and categories. These modern welds are usually stronger, lighter, more durable and economical than other methods of assembly **(Fig. 19–1)**.

FIGURE 19–1 Welding is a major procedure used to bond metals and plastics. (Courtesy of Arvin Industries Inc.)

363

WELDING PROCESSES

The specific metallurgic combination of heat, energy, and materials used to create a weld between two metals is called a **welding process.** There are hundreds of processes and combinations of processes now available. **Figure 19–2** shows some of the most commonly used processes organized under the major categories of welding. The abbreviations shown next to each process are used to specify the welding process on working drawings. All welding processes join metals together using either **solid-state welding, fusion welding, brazing,** or **soldering.**

Solid State Welding

These processes involve heating two workpieces and impacting them together until the hot material on the adjoining surfaces become mixed. Solid state welding processes include forge welding, friction welding, explosive welding, and roll welding. Thin, soft, malleable metals such as wrought iron, copper, and brass are well suited to solid-state welding processes.

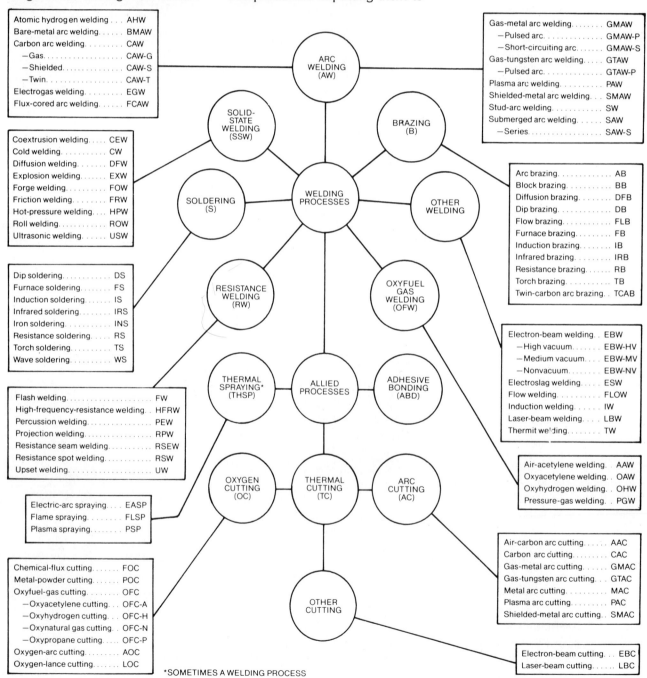

FIGURE 19–2 Categories of welding processes (Courtesy of The American Welding and Manufacturing Company)

Fusion Welding

In these processes, the common edge of two metal parts are melted to a molten state. The molten parts, combined with a filler material from a welding rod, form a molten puddle. When the puddle cools and solidifies, the separate pieces are permanently fused as one. The most common fusion welding processes include **arc welding, gas welding** and **resistance welding.**

Arc Welding. This process uses an electric arc struck between a workpiece and an electrode. The extreme heat from the arc fuses the metals together with material from the consumable electrode. Metals well suited to arc welding include copper, brass, soft iron, aluminum, low-carbon steel, cast steel, stainless steel and nickel alloys.

Gas Welding. The combination of gas (usually acetylene or hydrogen) with air (or oxygen) is known as **oxyacetylene welding.** Oxyacetylene welding can produce temperatures which will weld most metals, except very high carbon steel and chromium.

Resistance Welding. Both heat and pressure are used in resistance welding. A concentrated electrical current is passed through adjoining metals. The resistance to the electrical charge melts the metal while pressure is applied to the areas to be joined. When the metal cools and the pressure is removed, the separate pieces are permanently joined together. When the electrical charge is concentrated on a very small area, the type of welding is **spot welding.** When the charge is moved along the edge of a joint, the type of welding is **seam welding.** Resistance welding is widely used to spot or seam weld sheetmetal parts together.

Brazing and Soldering

These processes use a filler rod that melts at a lower temperature, leaving the workpiece metals solid. If the filler rod melts at less than 840°F (450°C), the process is called **soldering.** If the filler rod melts at a temperature higher than 840°F, but lower than the melting temperature of the base metal, the process is called **brazing.** When the molten filler rod cools and hardens, it adheres to both parts, forming a metallurgical bond. Since brazing or soldering temperatures are not hot enough to melt and mix the two workpiece metals, the joints are not as strong as those created by fusion or solid-state welding processes. Therefore, brazed or soldered joints are not used in areas of severe impact, high heat, shear, torsion, or tension stress.

WELDING SYMBOLS

When welders read working drawings, they need to know which welding process should be used. They also need to know the size and type

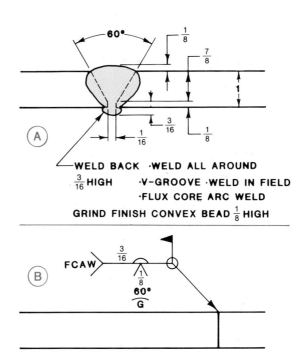

FIGURE 19–3 All the weld data in drawing A is included into the weld symbol in drawing B.

of each weld, the location, the type of joint, and the finish to be used. It is impossible or impractical to draw, dimension, or label all of this information on each weld location on complex working drawings. **Figure 19–3** shows a drawing of a weld joint fully dimensioned at A. A weld symbol containing the same information is shown at B. Notice how much less space the symbol information requires.

Anatomy of a Weld Symbol

Because of limited dimensioning space on most drawings, the actual appearance or dimensions of a weld are not drawn. The parts are drawn as they appear after welding without the weld bead or removed metal shown. For example, the material to be welded is shown at A in **Fig. 19–4** as being butted together. Examples at B and C are incorrect. A weld symbol (**Fig. 19–5**) substitutes for the drawing of the actual joint or weld. A weld symbol is a combination of graphics, text, and

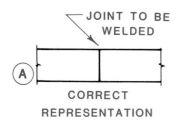

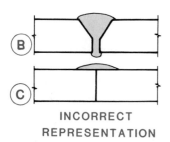

FIGURE 19-4 The actual welded area is not shown on the working drawing.

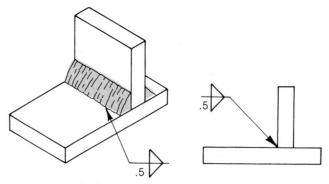

FIGURE 19-5 The configuration of the weld is not shown on the working drawing.

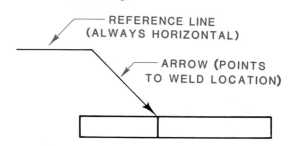

FIGURE 19-6 Welding reference line and arrow

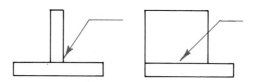

FIGURE 19-7 The arrow of the weld reference symbol touches the weld joint.

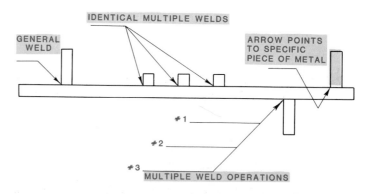

FIGURE 19-8 Variations of the weld reference symbol

numerical data. The following is a list of the common weld symbols:

Reference Line and Arrow. A welding reference line is a horizontal line upon which specifications for each weld are placed. An **arrow line** always extends at an angle from the reference line to the workpiece **(Fig. 19–6)**. The arrowhead at the end of the reference line touches the location of the weld **(Fig. 19–7)**. Variations of how a weld reference symbol may be used are shown in **Fig. 19–8**.

Weld Type Graphic Symbols. Weld type graphic symbols refer to the shape and configuration of the weld. A graphic symbol resembling the shape of the weld is used on, above, or under the reference line **(Fig. 19–9)**. **Figure 19–10** shows supplementary graphic symbols that are placed on, above, or below the reference line to convey additional information about the weld. These symbols will be covered in detail later in this chapter.

The position of the graphic symbol specifies the location of the weld **(Fig. 19–11)**. When the symbol is placed on the bottom side of the reference line, the joint is to be welded exactly where the arrowhead is placed. When the symbol is placed on the top side of the reference line, the weld is to be made on the opposite (other) side of the joint (Fig. 19–11, A and B). When the symbol is placed on both the arrow side and the other side of the reference line (Fig. 19–11, C), both sides of the joint are to be welded. The placement of all possible data in the weld reference symbol is shown in **Fig. 19–12**.

Tail Symbol. When specific information, such as detail specifications, procedures, welding processes, filler material, or preparation of the weld area is required, it is placed in the tail of the symbol **(Fig. 19–13)**. The abbreviations for the welding processes in Fig. 19–2 are placed in the

FIGURE 19-10 Supplementary graphic weld symbols

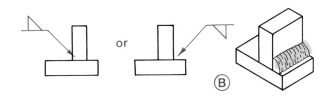

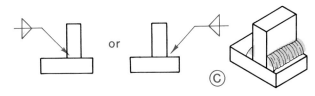

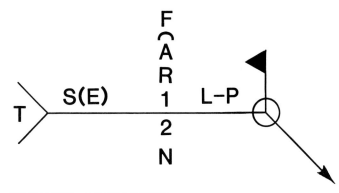

FIGURE 19-11 The position of the weld symbol (fillet) specifies the location of the weld.

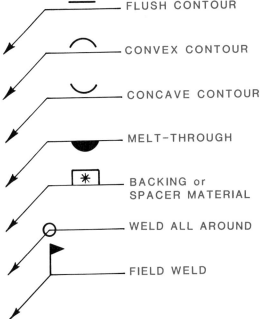

WELDING NOTATIONS:

F — FINISH SYMBOL
⌒ — CONTOUR SYMBOL
A — GROOVE ANGLE: INCLUDED ANGLE OF COUNTERSINK FOR PLUG WELDS
R — ROOT OPENING: DEPTH OF FILLING FOR PLUG AND SLOT WELDS
(E) — EFFECTIVE THROAT
T — SPECIFIC PROCESS OR REFERENCE
L — LENGTH OF WELD, LENGTH OF OVERLAP (BRAZED JOINTS)
S — DEPTH OF PREPARATION, SIZE OR STRENGTH FOR CERTAIN WELDS, HEIGHT OF WELD REINFORCEMENT, RADII OF FLARE-BEVEL GROOVES, RADII OF FLARE-V GROOVES, ANGLE OF JOINT (BRAZED WELDS)
P — PITCH OF WELDS (CENTER-TO-CENTER SPACING
1 — WELD SYMBOL ON THIS SIDE MEANS THE WELD MUST BE ON THE OPPOSITE SIDE OF THE MATERIAL THE ARROW IS TOUCHING
2 — WELD SYMBOL ON THIS SIDE MEANS THE WELD MUST BE ON THE SAME SIDE OF THE MATERIAL THE ARROW IS TOUCHING
(N) — THE NUMBER OF SPOT OR PROJECTION WELDS

⚑ — WELD DONE IN THE FIELD

⌀ — WELD ALL AROUND

FIGURE 19-12 Weld reference symbol with its data symbol information

FIGURE 19–13 When a specific welding notation is required a tail is added (oxyacetylene welding) as shown.

tail to indicate which process is to be used. If all welds use the same process, this can be covered in a general note, such as "all welds to be FCAW." When this is done, the tail information may be omitted, unless there are other specific instructions.

Finished Method Symbol. The method of finish for each weld is shown by a letter placed at the F position (Fig. 19–12). The following letters indicate the method of finish:

C—Chipping
G—Grinding
M—Machining
R—Rolling
H—Hammering

The exact degree of finish is not shown on a weld symbol since this is part of the finishing limits established for the entire surface on the working drawing. Surface finishes are explained in Chapter 9.

Contour Symbol. The shape of the finished surface of a weld is shown directly under the finishing method symbol (Fig. 19–12). A flat surface is indicated by a straight line. A rounded (concave) surface is shown by a curved line, like a smile, a depressed (convex) surface is shown by a curved line, like a frown **(Fig. 19–14)**.

Groove Angles. The letter directly below the contour symbol (position A, see Fig. 19–12) shows the groove angle if a groove joint is specified. Groove angles are expressed as included angles **(Fig. 19–15)**. This location on the reference weld symbol is also used to show the angle of any countersinks on plug welds.

SURFACE SHAPE	WELD	CONTOUR SYMBOL	REMEMBER
CONVEX		⌒	☹
CONCAVE		⌣	☺
FLUSH		—	😐

FIGURE 19–14 Contour symbols

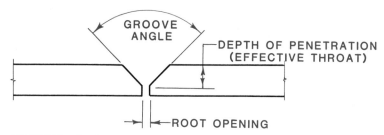

FIGURE 19–15 Groove angles

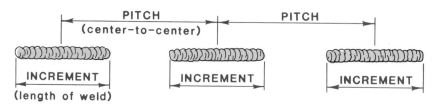

FIGURE 19–16 Intermittent welds (increments)

Root Opening. The gap between two workpieces before welding is known as the **root opening** or **root gap** (Fig. 19–15). The width of the root opening is located directly under the groove angle on the welding symbol (position R, Fig. 19–12). This location may also be used to show the filling depth for plug or slot welds.

Size and Strength. The size or strength of the weld is shown on or under the reference line directly to the left of the graphic weld symbol (position S, Fig. 19–12). Weld size can mean the height of the bead, radii of a flared bevel joint, or depth of penetration. The depth of penetration is sometimes called the **effective throat** (see position E, Fig. 19–12). For some welds, the minimum allowable strength of the joint in pounds is substituted for weld size in position S.

Length of Weld. The length of the weld is located on or below the reference line directly to the right of the weld type graphic symbol (position L, Fig. 19–12). If no length is indicated on the symbol, it is assumed that the weld extends the full length of the part. This location is also used to show the length of overlap on brazed joints.

Weld Pitch. When welds are intermittent over the length of a joint, the spacing of each weld and the length of unwelded spaces between welds must be shown. The center-to-center distance between intermittent welds is called the **pitch**. The distance between intermittent welds is called the **increment** (Fig. 19–16). The length of the pitch and increments is placed on or below the reference line (position P, Fig. 19–12). For example, a dimension of 6–10 on the symbol means 6″ of weld spaced 10″ apart from center-to-center.

Number of Spot or Projection Welds. When spot or projection welds are used, the spacing between welds is shown below the reference line (position N, Fig. 19–12).

Weld All Around. When a joint is to be welded around all of its edges, the weld all around symbol is used. This symbol is a circle drawn at the intersection of the reference line and the arrow line (Fig. 19–12). This is one of the supplementary symbols shown in Fig. 19–10.

Field Weld. If a part is to be welded on the construction site, a vertical line with a flag is drawn beginning at the intersection of the reference line and the arrow line (Fig. 19–12). This is also one of the supplementary symbols shown in Fig. 19–10.

Order of Information. Except for the tail symbol information, and the weld all around and field weld symbols, all data is located in the same horizontal order on the reference line, regardless of the position of the arrow. For example, **Fig. 19–17A** includes a left arrow line symbol and a right arrow line symbol containing the same information, read in the same sequence.

Vertically aligned data, such as the root opening, groove angle, contour symbol, and finish method symbol, are always in the same order from the reference line. Therefore, the order is reversed when these data are below the line. **Figure 19–17B** shows the order of the information applied to an arrow side symbol and an other side symbol.

DATA SELECTION

All data shown in Fig. 19–12 are not always used for each weld. Some information simply does not apply to some welds. For example, the groove angle degree does not apply to fillet welds, and the pitch distance does not apply to continuous welds.

When the data does apply to the weld, all information must be located on the symbol. If data is missing, the welder must determine the specifications of the weld without full knowledge of the design requirement. From the information supplied in **Fig. 19–18**, the welder knows only the location and size of a V-groove. Not known is the welding process, finish method, surface shape, groove angle, root opening, or length of weld. In comparison, **Fig. 19–19** shows a welding symbol containing complete data. From the information on this symbol, the welder knows exactly where a $\frac{1}{4}″$, 60°, V-groove weld with a $\frac{1}{8}″$ root opening is to be made. The welder also knows that the gas tungsten arc process is to be used in the field, and

FIGURE 19–17A The order of the weld data remains in the same position regardless of the arrow's direction.

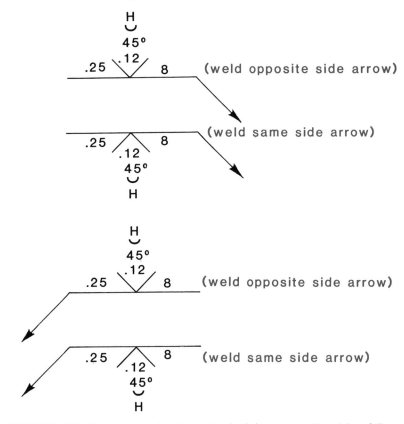

FIGURE 19–17B Placement of weld data depicts the location side of the weld.

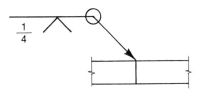

FIGURE 19–18 Incomplete weld information

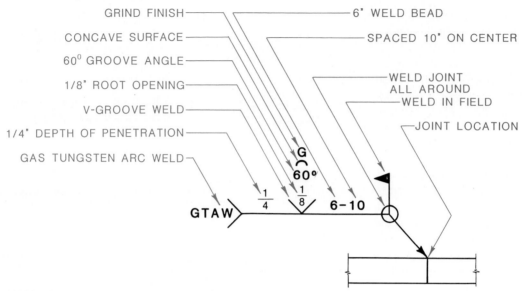

FIGURE 19-19 Fully noted weld instructions for a V-groove weld

that the weld is to be made all around and ground to a convex shape. The welder understands that this weld is intermittent with 6″ long welds spaced 10″ apart center-to-center.

Figure 19-20 shows another example of a complete symbol. It specifies an 8″ long, $^3/_{16}$″ gas-metal arc fillet weld to be made all around the joint and machined to a concave surface. Although this weld symbol contains only seven items compared to eleven in Fig. 19-19, it is complete since the other items are not required for this type of weld.

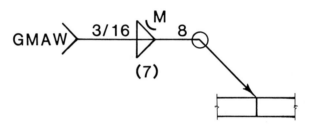

FIGURE 19-20 Fully noted weld instructions for a fillet weld

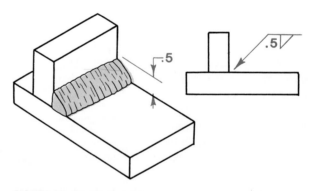

FIGURE 19-21 Fillet weld

WELDING TYPE SYMBOL APPLICATIONS

As shown in Fig. 19-9, the type of weld is represented by a graphic symbol over or under the reference line. A thorough knowledge of the relationship of the actual weld types to symbols is essential for the preparation of welding drawings. This understanding is also important because the selection of the welding type is directly related to and usually proceeds the establishment of other weld design specifications. Figures 19-21 through Fig. 19-42 show pictorial interpretations of the welding type symbols used for each major type of weld.

Fillet Welds

Fillet welds are used to weld two objects that join at right angles **(Fig. 19-21)**. Notice how the fillet weld symbol closely resembles the actual appearance of a fillet. The vertical line of the fillet weld symbol is always on the left and the slanted line on the right, regardless of the position of the weld. The size of a fillet weld is normally dimensioned as the height of the bead. If the height and width of the bead are different, both dimensions are given in parentheses with the height entered first **(Fig. 19-22)**.

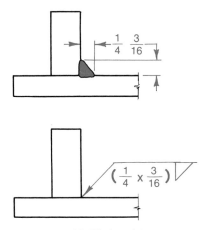

FIGURE 19-22 Fillet weld

Fillet weld contours are shown next to the slanted part of the fillet symbol **(Fig. 19–23)**.

Plug Welds

Plug welds substitute for rivets. A plug weld is a hole filled with a weld filler material. Five items are necessary on plug weld symbols **(Fig. 19–24)**. The graphic symbol for a plug weld is a rectangular box roughly resembling the cross section of a plug weld. The diameter of the hole to be filled is to the left of the rectangle, and the depth of the area to be filled is inside the symbol box. If there is more than one hole, the center-to-center spacing (pitch) is to the right of the box symbol. The symbol in Fig. 19–24 specifies $1/2''$ holes with 60° countersinks to be spaced 2" apart (center-to-center) and filled to a depth of $1/4''$.

Slot Welds

Since **slot welds** are basically elongated plug welds, the same rectangular box symbol is used for slot welds as for plug welds. The depth of slot, depth of fill, angle of countersink, size of hole, and length of slot must be specified. Because of the amount of data required, some or all of the dimensions for a slot weld are often shown on a detail drawing **(Fig. 19–25)**. The weld reference symbol for the slot weld can also be shown as in **Fig. 19–26**. In this case, a separate detail is needed to show the exact position of the slot on the part. This is also often true for dimensioning the position of a plug weld.

Spot Welds

Electrical resistance processes are usually used for spot welding thin sheets of metal, although fusion processes are sometimes used. In spot welding, the top sheet is melted into the bottom sheet. The graphic symbol for a spot weld is a circle **(Fig. 19–27)**. The position of the circle above or below the reference line indicates the direction of the source of heat. If the heat side has no significance in the design, the circle is placed directly on the line. When this is done, the circle is placed in the center of the reference line to avoid confusion with the weld all around symbol, which is also a circle drawn at the intersection of the reference line and the arrow line.

The size of a **spot weld** is the diameter of the spot and is located to the left of the circle symbol. A spot weld may also be classified by the minimum acceptable shear strength per spot. If this method is preferred,

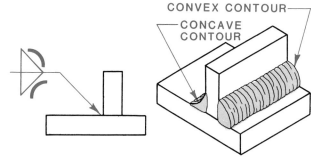

FIGURE 19-23 Contours for a fillet weld

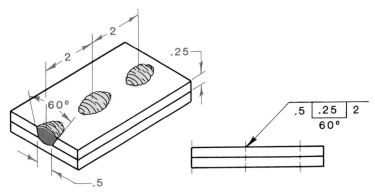

FIGURE 19-24 Plug welds

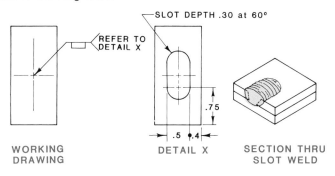

FIGURE 19-25 Detail drawing of slot weld

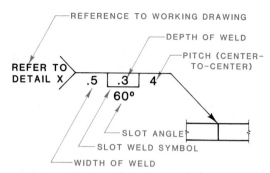

FIGURE 19-26 Slot weld reference symbol

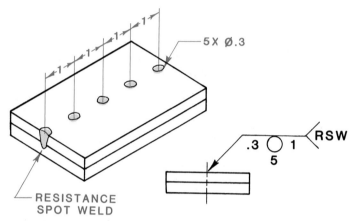

FIGURE 19-27 Spot welds

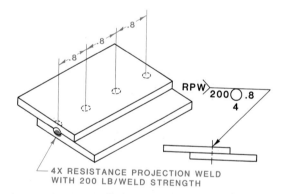

FIGURE 19-28 Projection welds

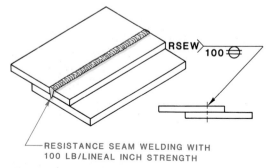

FIGURE 19-29 Seam weld

the strength in pounds per spot is located to the left of the circle. The pitch, located to the right of the circle, is the center-to-center spacing between each spot. Figure 19-27 specifies five .3" diameter resistance spot welds spaced 1" apart and located on the arrow side of the reference line.

Projection Welds

The only difference between a spot weld and a **projection weld** is the embossing of one of the metal sheets where the spot weld is to be made. The same graphic circle is used for both types of welds. The projection method is noted in the tail symbol data. The circle is located on the side of the reference line representing the side of the joint to be embossed. This means that the circle is placed below the reference line if the arrow side is to be embossed. Either the size or strength of spot welds may be noted to the left of the circle. The symbol in **Fig. 19-28** specifies four resistance projection welds with a joint strength of 200 lbs., spaced .8" apart on the arrow side of the joint.

Seam Welds

A continuous spot weld is called a **seam weld.** Seam welds are used instead of spot welds when a tight (storage of liquids or pressurized gas) or slightly stronger joint is required. The graphic symbol for a seam weld is a circle with two horizontal lines **(Fig. 19-29)**. This symbol can be placed above, below, or on the reference line to show the location of the heat source, as in spot weld symbols. All other spot weld dimensioning rules apply to seam welds, except the strength is expressed in pounds per linear inch. Figure 19-29 specifies a continuous resistance seam weld on the same side of the arrow, with a strength of 100 lbs. per linear inch.

Flange Welds

Welding flat surfaces of two sheets of metal bent to provide a surface for welding is called **flange welding.** There are two types of flange welds—edge flange welds and corner flange welds. Edge flange welds join sheets of metal that lie in the same plane **(Fig. 19–30)**. Corner flange welds join sheets of metal that intersect at right angles **(Fig. 19–31)**. The graphic symbol for each is different, but the rules for dimensioning are the same. Dimensions are shown on the reference line by using the bend radius, the distance to the point of tangency, and the height of the bead. The bend radius is located to the left of the graphic symbol, followed by a plus (+) sign and the distance from the edge of the metal to the point of tangency of the arc. The height of the bead is shown below this dual dimension. The symbol in Fig. 19–30 specifies a .10″ edge flange weld with a .20″ radius, and the point of tangency of the arc located .15″ from the edge of the metal. Figure 19–31 shows a $^3/_{16}$″ corner flange weld with a $^1/_8$″ radius, and a point of tangency located $^1/_8$″ from the edge of the metal.

Groove Welds

Often a groove or gap is made between two metal parts to be welded. During welding, this groove is filled with material from the welding rod. There are seven basic types of **groove welds:** the square groove, V-groove, bevel groove, U-groove, J-groove, flare V-groove, and flare bevel groove. The graphic symbol for each groove weld resembles the shape of the groove.

When two workpieces do not touch, the gap between the parts is called the root opening. The size of the root opening is located directly above or below the graphic symbol **(Fig. 19–32)**. The angle of the groove, if any, is placed directly above or below the root opening dimension, as in the V-groove symbol in **Fig. 19–33**. The symbol for a V-groove weld is always drawn at right angles, regardless of the actual angle of the groove. The symbol for a bevel groove weld is always drawn with one perpendicular line. The angle of the 45° bevel is shown in **Fig. 19–34**.

For all groove welds, the depth of penetration is always located to the left of the graphic symbol. The

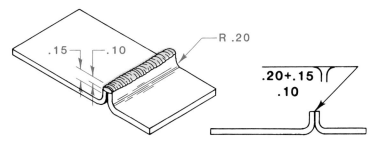

FIGURE 19–30 Edge flange weld

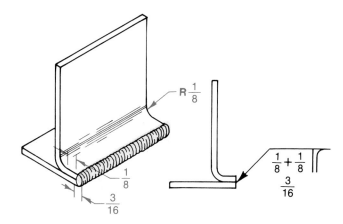

FIGURE 19–31 Corner flange weld

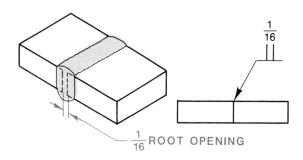

FIGURE 19–32 Root opening for a square groove weld

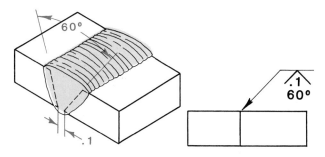

FIGURE 19–33 Angle opening for a V-groove weld

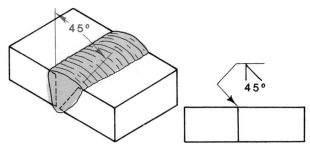

FIGURE 19-34 Opening for a bevel groove weld

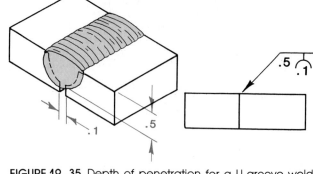

FIGURE 19-35 Depth of penetration for a U-groove weld

size of the U-groove weld, in **Fig. 19-35**, is .5" with a .1" root opening. The size of groove welds is the depth of penetration (effective throat) of the filler material. The effective throat of the J-groove weld, in **Fig. 19-36**, is .8" and the root opening is .2".

When two metal plates are bent to create flat adjacent surfaces for welding, the type of weld is called a flare groove weld. There are two types of flare groove welds—the flare V-groove and the flare bevel groove weld. The flare V-groove weld joins metal plates that lie in the same plane **(Fig. 19-37)**. The flare bevel groove weld joins two plates that intersect at an angle **(Fig. 19-38)**. The depth of penetration is the radius of the bevel for both types.

Backing and Melt-Through Welds

For maximum support, a weld bead is sometimes required on the back side of a joint in addition to the front. This is done in two separate operations. To specify a **backing weld**, a semicircle is drawn on the reference line directly opposite the graphic symbol. **Figure 19-39** shows this symbol applied to a V-groove weld. The depth of the backup bead is shown to the left of this semicircle.

Melt-through welds are made completely through the joint from one side with a backup bead. The semicircle symbol is made solid **(Fig. 19-40)**. It is similar to the backing weld, but is made with one welding operation.

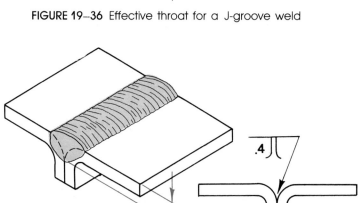

FIGURE 19-36 Effective throat for a J-groove weld

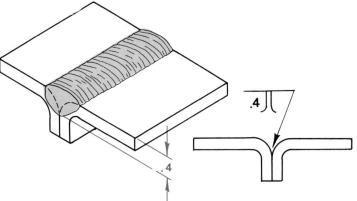

FIGURE 19-37 Flare V-groove weld

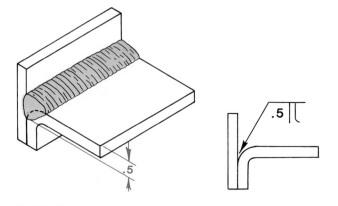

FIGURE 19-38 Flare bevel groove weld

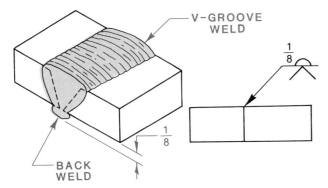

FIGURE 19-39 Back weld used with a V-groove weld

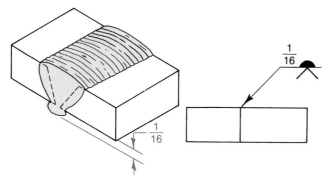

FIGURE 19-40 Melt-through weld

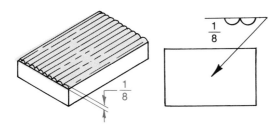

FIGURE 19-41 Surfacing weld

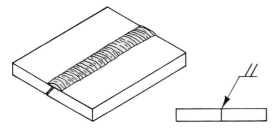

FIGURE 19-42 Scarf braze

Surfacing Weld

Surfaces are often made thicker with the application of filler material. This process is known as **surfacing** or **building up.** The graphic symbol for surfacing is two semicircles **(Fig. 19-41).** Since no joint is involved in surfacing welds, the minimum height of the filler material is the only dimension needed. However, the welding process and method of finishing the filled surface is very important. If only part of the build-up surface is to be finished, a separate detail drawing must be prepared showing the exact dimensions of the area to be finished. This information must be indexed in the welding tail symbol.

Brazing Symbols

Welding symbols are also used for brazing. Brazing occurs at lower temperatures, therefore the materials being joined are not melted. Usually, softer or thinner metals are involved **(Fig. 19-42).**

Combination Welds

For instructional purposes, the types of weld symbols shown thus far included only information about one type of weld. However, many welds contain combinations of different types of welds. **Figure 19-43** shows how a combination of different welding types and dimensions are incorporated into one symbol.

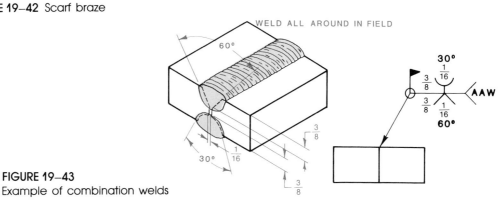

FIGURE 19-43
Example of combination welds

WELDED JOINTS

In addition to knowing which type of weld is needed, the designer must know which joint is appropriate to join metal parts by welding. The common joints used in welding are shown in **Fig. 19–44**. These include the butt, corner, lap, T-, and edge joints. The type of weld which can be used for each joint is also shown in Fig. 19–44. The final selection of the joint and weld type depends on the metal specified, and the structural and aesthetic requirements of the design.

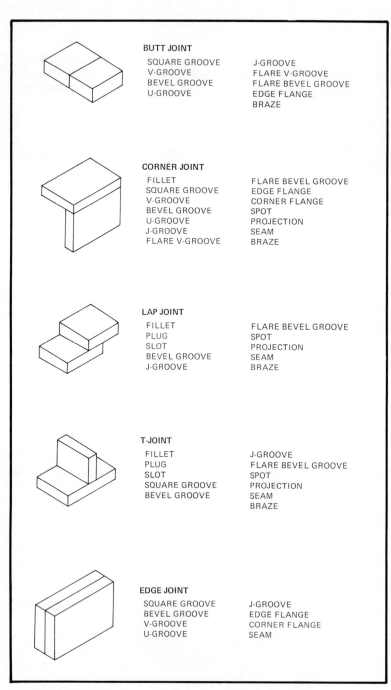

FIGURE 19–44 Weld joints and their recommended welds

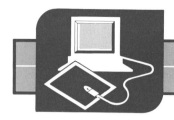

COMPUTER-AIDED DRAFTING & DESIGN

WELD SYMBOLS WITH CAD

The application of weld symbols on working drawings is an excellent task for a CAD system because it involves drawing repetitive details. The weld symbols can be drawn once, made into blocks, and inserted without redrawing each symbol. **Figure 19–45** shows a CAD monitor with standard weld symbols loaded along the right column of the screen. Standard symbols can be copied easily and rapidly on the drawing as many times as desired. Figures 19–45 through 19–48 show the following steps in laying out a typical working drawing with CAD welding symbols:

Step 1. Set up the drawing parameters to include ORTHO, SNAP and GRID. Load the weld symbols onto the screen menu (right column of screen) **(Fig. 19–45)**.

Step 2. Using the necessary entities, draw the required views of the weldment **(Fig. 19–46)**.

Step 3. Using the LEADER command, add the leaders for the weld symbols **(Fig. 19–47)**.

Step 4. Copy the standard weld symbols from the screen menu onto the drawing. In this example, the symbols that have been copied are the fillet, V-groove and all around symbols. Using the TEXT command, add the weld sizes **(Fig. 19–48)**.

Step 5. To complete the drawing, add dimensions, a parts list, and a title block.

Standard symbols may be inserted on CAD drawings in various ways depending on the system used. Another method is to use a graphics tablet with custom-made standard symbols **(Fig. 19–49)**. To use the graphics tablet, the CAD operator selects the desired symbol, moves it into place, and inserts it with a puck or mouse.

Using the graphics tablet is similar to using a screen menu. The advantage is that many symbols and drafting functions may be placed on one graphics tablet. With use, the CAD operator will gain a working knowledge of their locations, thereby greatly speeding up the CAD process.

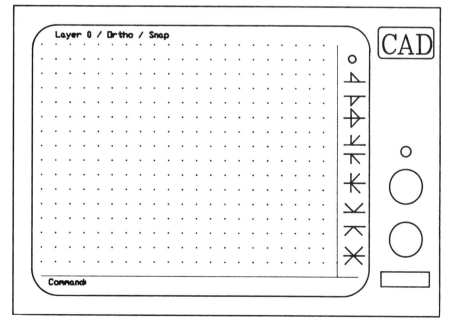

FIGURE 19–45 Set up the drawing parameters and load the weld symbols from the library.

378 Drafting in a Computer Age

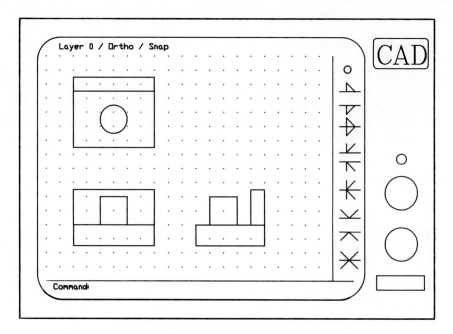

FIGURE 19-46 Draw or call up the weldment drawing from the CAD's storage system.

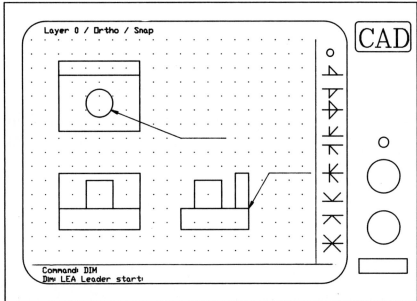

FIGURE 19-47 Call up the LEADER command and add the weld symbol leaders.

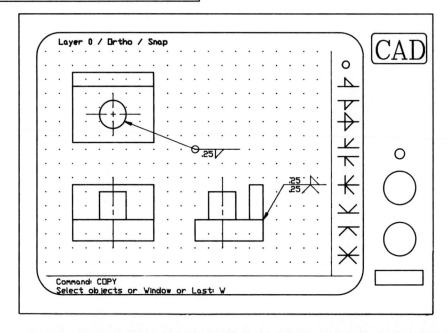

FIGURE 19-48 Place the required weld symbols of the drawing. Go to the TEXT command and add the weld sizes.

FIGURE 19–49 Adding graphic drawing symbols from a graphics tablet

1. Sketch the weld type, weld symbols, and a minimum of one additional piece of information for each of the following weld symbols: fillet, plug, slot, projection, seam, back, surfacing, edge flange, square groove, V-groove, bevel groove, U-groove, and J-groove. Print an explanation for each solution.
2. Sketch the symbols for two different weld types for each of the following joints: butt joint, inside corner joint, outside corner joint, T-joint, lap joint, and edge joint. Show all information and print the explanation for each.
3. Name and sketch the weld symbol for each weld in **Figs. 19–50** and **19–51**.

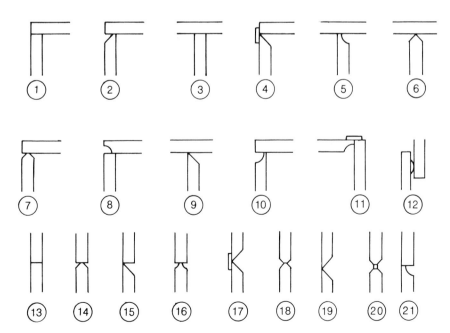

FIGURE 19–50 Sketch each example to be welded. Add the proper weld reference symbol and data required with another sketch as it will appear on a working drawing.

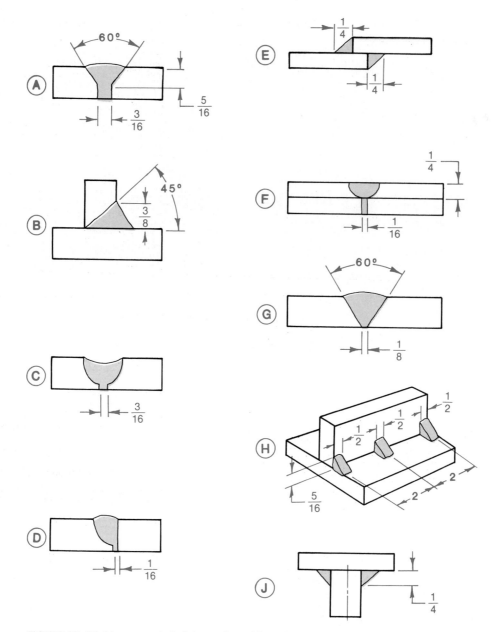

FIGURE 19-51 Name and sketch each weld symbol.

4. Draw the working drawings for **Figs. 19-52** through **19-58**, and add the appropriate welds and their weld symbols.
5. With a CAD system, lay out the working drawings, and the appropriate welds and their symbols for **Figs. 19-59** through **19-62**.

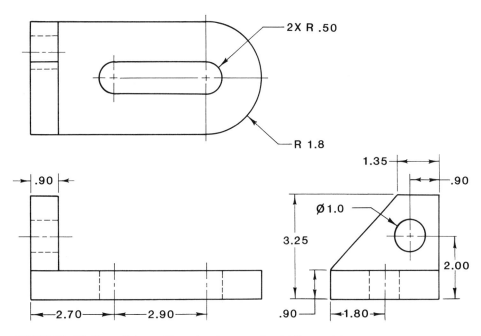

FIGURE 19-52 Draw the working drawings and add the weld notations as needed for the bearing plate.

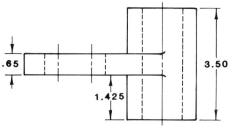

FIGURE 19-53 Draw the working drawings and add the weld notations as needed for the adjustable bearing locator.

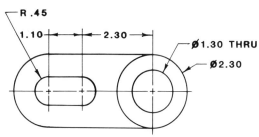

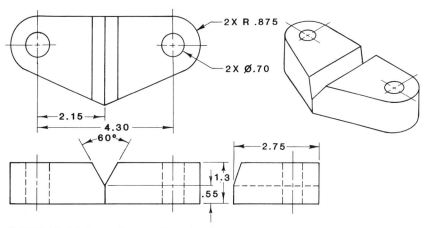

FIGURE 19-54 Draw the working drawings and add the weld notations as needed for the V-guide.

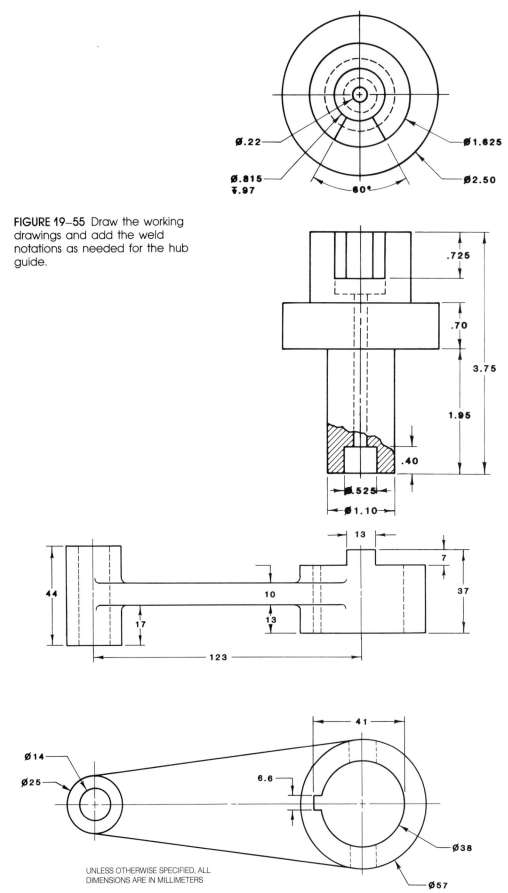

FIGURE 19-55 Draw the working drawings and add the weld notations as needed for the hub guide.

UNLESS OTHERWISE SPECIFIED, ALL DIMENSIONS ARE IN MILLIMETERS

FIGURE 19-56 Draw the working drawings and add the weld notations as needed for the web connector.

Chapter 19 Welding Drawings **383**

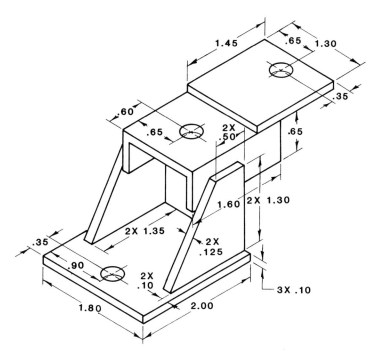

FIGURE 19–57 Draw the working drawings and add the weld notations as needed for the tool jig stand.

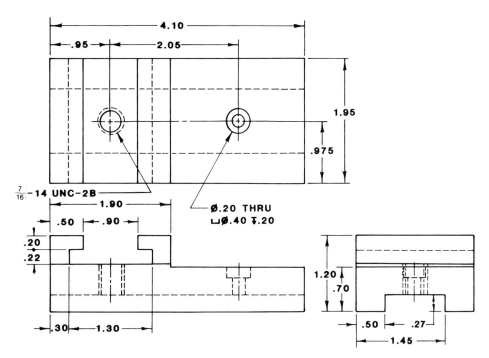

FIGURE 19–58 Draw the working drawings and add the weld notations as needed for the slotted fixture guide.

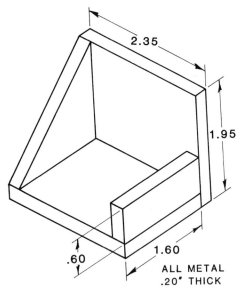

FIGURE 19-59 Draw the working drawings and add fillet welds with seam welds as needed.

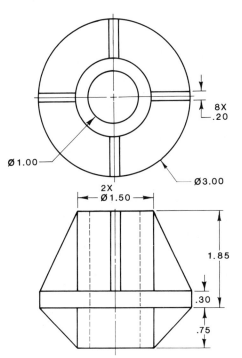

FIGURE 19-61 Draw the working drawings and add weld notations as needed for the cylinder and four webs with a CAD system.

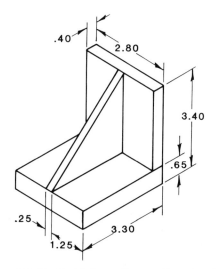

FIGURE 19-60 Draw the working drawings and add weld notations as needed for the bracket with a CAD system.

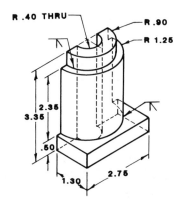

FIGURE 19-62 Draw working drawings of the semi-column support and complete the weld notations with a CAD system.

 **KEY TERMS**

Arc welding	Increment	Seam welds
Backing weld	Melt-through welds	Slot welds
Brazing	Oxyacetylene welding	Soldering
Effective throat	Pitch	Solid-state welding
Fillet welds	Plug welds	Spot welding
Flange welds	Projection welds	Spot welds
Fusion welding	Resistance welding	Surfacing (building up)
Gas welding	Root opening (root gap)	Welding process
Groove welds	Seam welding	Welding symbols

the world of CAD

OBJECT-ORIENTED PROGRAMMING

In traditional programming languages, the procedures and data are separated and the programmer must tell procedures to operate on data and report back the results. In Intergraph's object-oriented mechanical design package, I/EMS, a single statement can tell an object to perform an action and broadcast the results, using the intelligence inherent in the object.

In object-oriented programming, an object knows what it is, where it is, how it should act, and its proper relationship to its environment. Including all this intelligence within an object gives object-oriented programs a modular structure, simplifying maintenance and modification.

An object can be as generic as a line, an arc, or a circle, or as definitive as a telephone pole, a river, or a bicycle wheel. Unlike traditional programs in which the descriptive attributes reside in a database separate from the graphic image, object-based programs combine code describing the geometric characteristics of an element with a code defining the element's attributes. For example, an object representing a door would contain data describing the following.

- the physical attributes of the door—the shape of the door (a rectangular parallelepiped), height, width, depth, material, windows, etc.
- the functional attributes of a door—open and close, lock, etc.
- the relational attributes of the door—door/wall, door/molding, etc.
- messages to communicate requested changes to the door and messages to communicate resulting changes in its relationship with other objects—a change in door dimensions, for example, would affect walls, molding, and the hinges of the door itself.

A door created using the object-oriented method knows it is a door. It knows its dimensions, knows which wall it is a part of, knows it opens and closes, and knows its relationship to the wall and the room it is a part of. In a traditional program, a door would be created in graphics as a group of unrelated lines. It would then be identified as a single graphic element—a rectangular parallelepiped. All data relating to the graphic element's identity and function as a door would reside in a separate database. Separate messages to each attribute would be necessary whenever a change was necessary. For example, a change in the size of the door would necessitate messages to the graphic representation of the door, the graphic representation of wall parts contiguous to the door, the hinges, and so forth. However, a change to a door created with object-oriented programming requires a single message to the object door. From that point on, the object, which contains all the necessary intelligence, makes sure that all the relevant changes are made.

An object-oriented data structure is developed through the definition of classes and subclasses of objects. Individual objects contain information peculiar to them, plus they "inherit" all of the intelligence defined by each class above them in the hierarchy. Thanks to inheritance, when applications specialists create new objects as subcategories of existing objects, they automatically contain all of the intelligence inherent in the existing object. For example, if you ask a closet door to close and if you have defined doors as closing by swinging toward the wall on hinges, then the closet door already knows how to close. It has inherited the properties of the superclass doors. In addition, multiple inheritance allows classes to have more than one superclass.

To sum up, in an applications development environment, object-oriented programming offers three important benefits

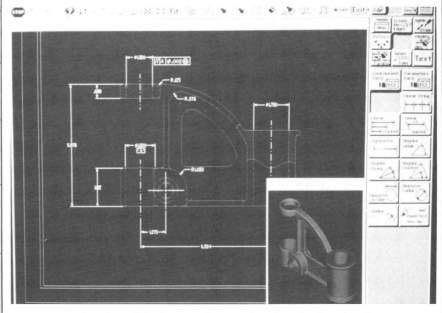

FIGURE 1 Solid model of a punch press bracket displayed with a drawing view

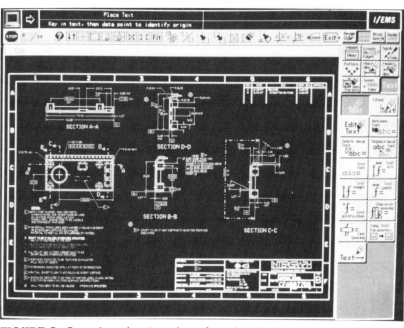

FIGURE 2 Complete drawing sheet for a key housing

1. Modular program design makes software maintenance easier, and allows programs to be modified and enhanced without affecting other parts of the program. Because modifications are only made to specific objects and the relationships contained within them, users need not be concerned about "breaking" other parts of the program.
2. The time required to develop new applications is reduced as libraries of objects are developed. Object-oriented programming allows a software engineer to take advantage of previously developed objects and all the intelligence inherent in them.
3. Knowledge-based systems can be developed by increasing the amount of intelligence inherent in the object, resulting in higher quality designs free of rule violations

Figures 1 and 2 are designs created by Intergraph's object-oriented mechanical design package, I/EMS.

(Courtesy of Eleanor Bowman, Intergraph Corporation, Huntsville, Alabama)

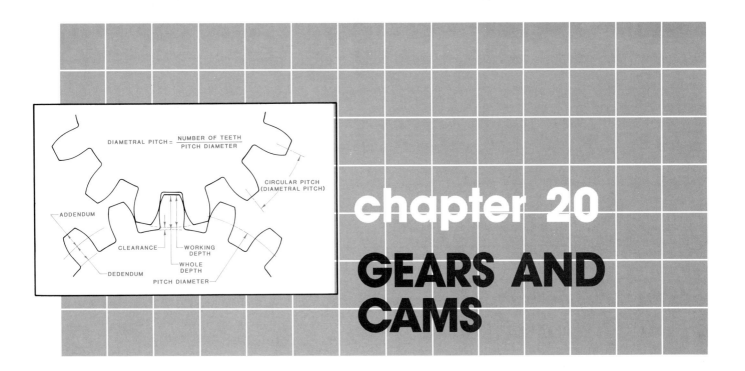

chapter 20
GEARS AND CAMS

OBJECTIVES

The student will be able to:

- draw a working drawing of a spur gear
- draw the teeth of a gear with a gear template
- draw a working drawing of a rack and pinion gear
- draw a working drawing of a bevel gear
- draw a working drawing of a worm gear
- draw a plate cam from displacement diagram data
- draw a drum cam
- lay out a displacement diagram from data in a plate cam profile drawing
- lay out a displacement diagram from data in a drum cam drawing
- recognize the various types of gears and cams
- prepare a gear drawing using a CAD system
- complete a cam drawing using a CAD system

INTRODUCTION

Gears, cams, and drives are mechanical devices used to control power. These devices can transmit motion from one device to another, or can change the speed and direction of motion. To design and draw gears, cams, and drives as integral parts of machine assemblies, a basic understanding of their function is vital.

GEARS AND DRIVES

Cylinders or cones used to transmit power and motion, and change the speed and direction of motion, are known as **gears.** Gears may be smooth (friction gears) or toothed. There are many types of gears designed to perform specific transmission, directional, and speed control functions. Gears are used to transmit rotary or reciprocating (back and forth) motion between machine parts.

Gears are divided into two general classifications based on the relative position of the gear shafts. **Parallel shaft gears** are connected with spur gears, helical gears, or herringbone gears. **Perpendicular shaft gears** are connected with bevel or miter gears. If intersecting gear shafts are neither parallel or perpendicular, connections are made with crossed helical gears or worm gears, called **angular gears.**

When one gear is rotated and brought into contact (meshed) with another gear, motion is transmitted. The second gear turns in the opposite direction of the first. If gears are equal in circumference, both gears will rotate one complete revolution in opposite directions. If the second gear is twice the circumference of the first, the second gear will rotate only one-half revolution, while the first gear rotates one full revolution. This results in the second gear rotating at one-half the speed of the first gear. The difference between the rotation of the two gears is known as the **gear ratio.** When gears are the same size and rotate at the same speed, the gear ratio is 1:1. When one gear rotates at twice the speed as the other, the ratio is 2:1. Three times rotation equals 3:1, and so forth.

The most common type of gears and drives are described below, and are illustrated in Figs. 20–1 through 20–9:

- **Spur gears** are one of the most common types of gears in which the teeth on the circumference are all parallel to the axis and perpendicular to the sides of the wheel (Fig. 20–1).
- **Rack and pinion gears** transmit reciprocating motion through a flat-

385

FIGURE 20–1 Spur gears (Courtesy of Boston Gear Division, Imo Corporation, Inc.)

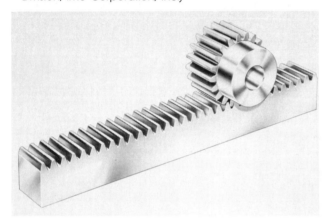

FIGURE 20–2 Rack and pinion gears (Courtesy of Boston Gear Division, Imo Corporation, Inc.)

toothed surface (rack) intersecting with a small cylindrical gear (pinion) **(Fig. 20–2)**. The pinion is the smaller gear in any gear system.
- **Bevel gears** transmit power between shafts that intersect at an angle **(Fig. 20–3)**. If the intersection between shafts is a right angle, they are often called **miter gears (Fig. 20–4)**.
- **Worm gears** connect shafts that are perpendicular with shaft centerlines that do not intersect. The small gear is the worm and drives the larger worm gear. The worm is shaped like a screw. The worm gear is round like a wheel. The worm includes at least one thread around the pitch surface. Worm gear teeth are cut on an angle to mesh with worm teeth **(Fig. 20–5)**.
- **Helical gears** operate on parallel shafts with gear teeth cut in a twisted, oblique angle to the shaft **(Fig. 20–6)**.
- **Herringbone gears** contain two sets of helical gear teeth placed side

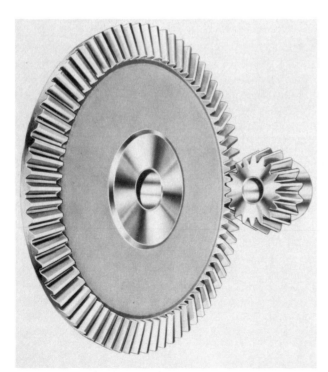

FIGURE 20–3 Bevel gears (Courtesy of Boston Gear Division, Imo Corporation, Inc.)

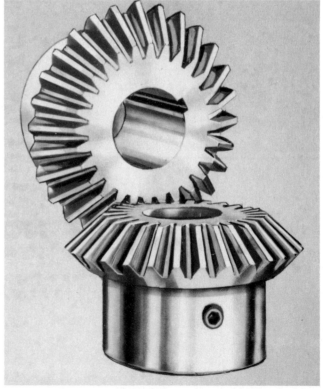

FIGURE 20–4 Miter gears (Courtesy of Boston Gear Division, Imo Corporation, Inc.)

Chapter 20 Gears and Cams **387**

FIGURE 20–5 Worm and worm gear (Courtesy of Boston Gear Division, Imo Corporation, Inc.)

FIGURE 20–6 Helical gears (Reprinted from MECHANICAL DRAFTING by Madsen, Shumaker, and Stewart © 1986, Delmar Publishers Inc.)

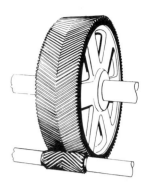

FIGURE 20–7 Herringbone gears (Reprinted from MECHANICAL DRAFTING by Madsen, Shumaker, and Stewart © 1986, Delmar Publishers Inc.)

FIGURE 20–8 Spiral bevel gears (Courtesy of Boston Gear Division, Imo Corporation, Inc.)

by side on an opposite slant on the same gear wheel **(Fig. 20–7)**.
- **Spiral bevel gears** contain oblique, curved teeth which provide more contact between the intersecting teeth **(Fig. 20–8)**.
- **Ring gears** are used primarily in automobile differentials. A propeller shaft pinion transmits power to a large ring gear that has teeth cut on the inside surface of the ring **(Fig. 20–9)**.
- **Chain and sprocket drives** are flexible power transmission devices. They consist of a continuous chain with links that mesh with toothed sprocket wheels attached to a drive shaft **(Fig. 20–10)**. Chain drives are used where ease of assembly and disassembly, and elasticity are required, or where shaft-to-shaft distances are extremely long.
- **Belt drives** transmit power between shafts with a belt wrapped around the outside circumference of a pulley mounted on each shaft. Belts which connect pulleys are either flat **(Fig. 20–11)**, grooved with longitudinally ribbed undersides, or v-belts. To connect perpendicular shafts, belts can be twisted by one-quarter. Belts can also be crossed (twisted by one-half) between pulleys to reverse the direction of the second pulley.
- **Friction wheels** are used to transmit power between shafts without using gears, chains, or belts. Friction wheels rely on heavy pressure to maintain sufficient contact to transfer power **(Fig. 20–12)**. Friction wheels engage easily but will slip when a predetermined amount of resistance is applied to the drive wheel.

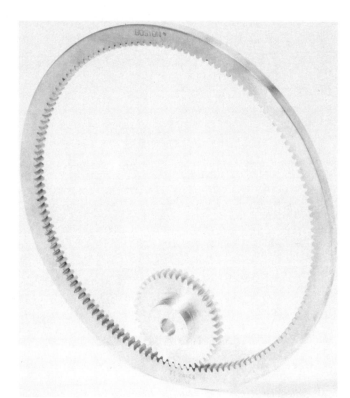

FIGURE 20-9 Ring gears (Courtesy of Boston Gear Division, Imo Corporation, Inc.)

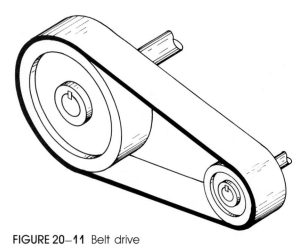

FIGURE 20-11 Belt drive

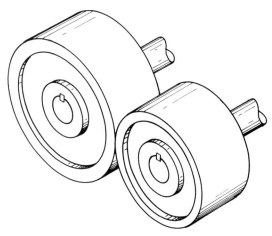

FIGURE 20-12 Friction drive

FIGURE 20-10 Chain and sprocket drive (Courtesy of Boston Gear Division, Imo Corporation, Inc.)

Gear Teeth

Two basic curves are used to define the shape of gear teeth—involute curves and cycloidal curves. An **involute curve** (Fig. 20-13) is produced by connecting the end points of a flexible and taut thread as it is unwound from a cylinder. A **cycloidal curve** (Fig. 20-14) is formed by

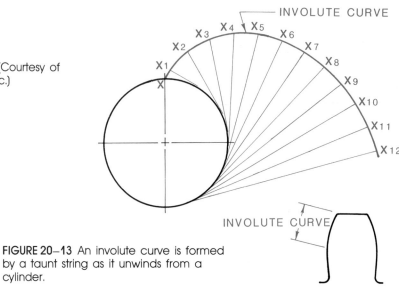

FIGURE 20-13 An involute curve is formed by a taunt string as it unwinds from a cylinder.

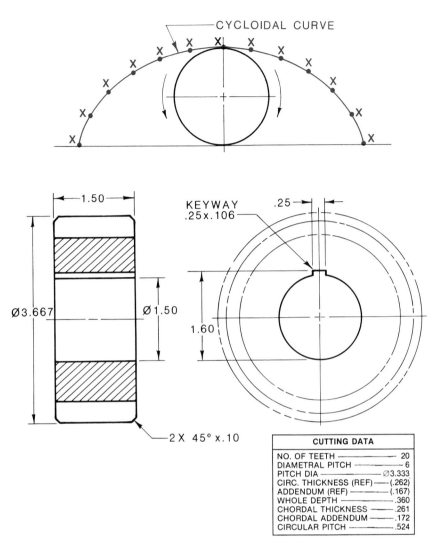

connecting the position of a point at evenly spaced intervals on the circumference of a circle as that circle is rolled on a flat surface.

The actual shape (profile) of gear teeth is rarely used on working drawings. Simplified symbols are used to show the outline of gear teeth; however, complete cutting data notations are always necessary **(Fig. 20–15)**. When it is necessary to show gear teeth on detail drawings, gear tooth templates of different types and sizes are used **(Figs. 20–16 and 20–17)**.

Gear Nomenclature

Before preparing working drawings involving gears, it is important to understand the following parts of a gear which define its size and shape **(Figs. 20–18 and 20–19)**:

- **Gear axis** is the central axis that the gear revolves around.
- **Pitch circle** is an imaginary circle concentric with the gear axis and passing through the pitch points.
- **Pitch point** is the point of intersection between the tooth face and pitch circle.

FIGURE 20–15 A working drawing for a gear seldom shows the actual outline of each gear tooth.

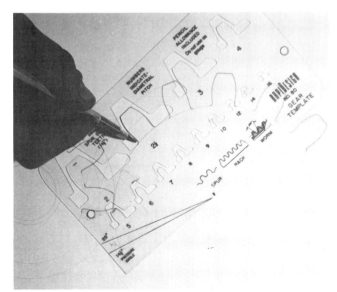

FIGURE 20–16 Involute rack and spur gear teeth drawing template

FIGURE 20–17 The diametral pitch determines the number and size of gear teeth on a gear. (Reprinted from MECHANICAL DRAFTING by Madsen, Shumaker, and Stewart © 1986, Delmar Publishers Inc.)

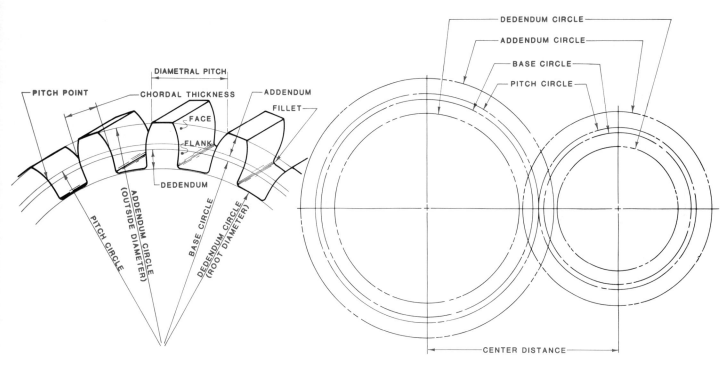

FIGURE 20-18 Gear terminology

- **Tooth face** is the surface of the gear tooth that applies pressure on the adjacent gear's tooth.
- **Addendum** is the radial distance from the top of the tooth to the pitch circle.
- **Addendum circle (outside diameter)** is a circle connecting the outer edge of the gear teeth.
- **Chordal thickness** is the thickness of a tooth measured on the pitch circle.
- **Pitch diameter** is the diameter of the pitch circle.
- **Diametral pitch** is the ratio equal to the number of teeth on the gear per inch of pitch diameter.
- **Dedendum** is the radial distance from the pitch circle to the root diameter.
- **Dedendum circle** is the diameter of the pitch diameter, minus the dedendum.
- **Root diameter** is the same as the dedendum circle. It is the smallest diameter for the base of a gear tooth.
- **Base circle** is the circle from which the involute tooth profile begins. When drawing gear teeth with a template, it is not necessary to lay out and draw the involute curve.
- **Fillet** is the concave intersection between the root and the side of a gear tooth.
- **Flank** is the lower side of a gear tooth between the base circle and root.
- **Center distance** is the distance between the centerlines of two parallel gear shafts.
- **Number of teeth** is the total number of teeth on the circumference of a gear.
- **Whole depth** is the total height of a gear tooth and is equal to the addendum, plus the dedendum.
- **Working depth** is the depth a tooth extends into the mating space of the opposite tooth.
- **Clearance** is the distance between the top of one tooth and bottom of the opposite mating tooth. It is equal to the dedendum minus the addendum.

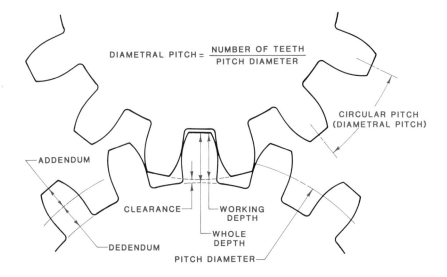

FIGURE 20-19 Gear terminology

- **Circular pitch** is the distance measured along the pitch circle from a point on one tooth to the corresponding point on the adjacent tooth. It includes the space and the tooth.

Gear Machining Formulas

Before gears can be drawn, accurate and complete specifications must be established. Once the basic data is gathered, formulas are used to determine the numerical values for the dimensions needed for gear machining. **Figure 20–20** shows the formulas used for tooth cutting of spur and pinion gears. **Figure 20–21** shows how the final data is used on a spur gear detail drawing.

Cutting data and formulas for rack and pinion gears is basically the same as for spur gears. Since the rack is a straight surface, all circular dimensions become linear, with the addendum, dedendum, whole depth, and tooth thickness identical to the mating pinion **(Fig. 20–22)**. Figure 20–23 shows a working drawing of a rack accompanied with the data notations needed for machining.

Information for cutting bevel gear teeth is based on the same involute curve as spur gear teeth; however, bevel gear teeth are tapered in a cone shape. Bevel gear teeth are designed in pairs and are not interchangeable. Dimensions are determined by the same formulas used for spur gears, with some modifications. The pitch diameter of bevel gears is the diameter of the base of the cone. The circular pitch

	REQUIRED CUTTING DATA		
ITEM	TO FIND:	HAVING	FORMULA
1	Number of teeth (N)	D & P	D x P
		DO & P	(DO x P) – 2
2	Diametral pitch (P)	p	$\frac{3.1416}{p}$
		D & N	$\frac{N}{D}$
		DO & N	$\frac{N+2}{DO}$
3	Pressure angle (∅)	–	20° STANDARD 14°-30' OLD STANDARD
4	Pitch diameter (D)	N & P	$\frac{N}{P}$
		N & DO	$\frac{N \times DO}{N+2}$
		DO & P	$DO - \frac{2}{P}$
5	Whole depth (ht)	a & b	a + b
		P	$\frac{2.157}{P}$
6	Outside diameter (DO)	N & P	$\frac{N+2}{P}$
		D & P	$D + \frac{2}{P}$
		D & N	$\frac{(N+2) \times D}{N}$
7	Addendum (a)	P	$\frac{1}{P}$
8	Working depth (hk)	a	2 x a
		P	$\frac{2}{P}$
9	Circular thickness (t)	P	$\frac{3.1416}{2 \times P}$
10	Chordal thickness	N, D & a	$\sin\left(\frac{90°}{N}\right) \times D$
11	Chordal addendum	N, D & a	$\left[1 - \cos\left(\frac{90°}{N}\right)\right] \times \frac{D}{2} + a$
12	Dedendum (b)	P	$\frac{1.157}{P}$

FIGURE 20–20 Required cutting data and formulas for spur gears (Reprinted from TECHNICAL DRAWING AND DESIGN by Goetsch and Nelson © 1986, Delmar Publishers Inc.)

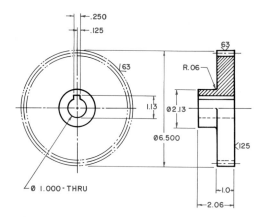

FIGURE 20-21 Example of a spur gear detail working drawing (Reprinted from TECHNICAL DRAWING AND DESIGN by Goetsch and Nelson © 1986, Delmar Publishers Inc.)

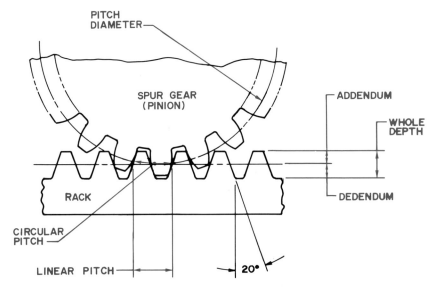

FIGURE 20-22 Example of rack gear terminology (Reprinted from MECHANICAL DRAFTING by Madsen, Shumaker, and Stewart © 1986, Delmar Publishers Inc.)

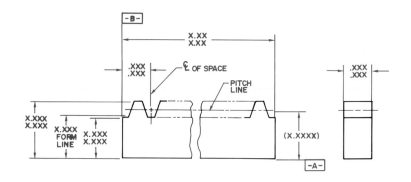

FIGURE 20-23 Partial example of a rack gear working drawing (Reprinted from MECHANICAL DRAFTING by Madsen, Shumaker, and Stewart © 1986, Delmar Publishers Inc.)

is measured on this base circle **(Figs. 20–24** and **20–25)**. Working drawings of bevel gears include only the outline of the gear blank. This is normally drawn as a single sectional view **(Fig. 20–26)** with all related gear tooth data provided in tabular notations.

Worm gears consist of a worm (screw form) and a worm gear (wheel form). The worm is basically a screw with single or multiple threads. The teeth on the worm gear are curved to conform to the shape of the worm teeth. Large reductions and high mechanical advantages result from the high ratios possible between worm and worm gear (up to 50:1). One revolution of the worm may result in a lead of only one tooth width on the worm gear. The lead of the worm is the distance the worm gear will travel in one revolution of the worm. Normally, only a one-view drawing of a worm gear assembly is necessary when cutting data is supplied. **Figure 20–27** shows cutting data for a worm. **Figure 20–28** shows cutting data for a worm gear. When a second view is required, the relationship of gear teeth is shown with radial centerlines on the side view, and interlocking profiles sectioned on the front view **(Fig. 20–29)**.

			REQUIRED CUTTING DATA	
ITEM	TO FIND:	HAVING	FORMULA	
			SPUR	PINION
1	Number of teeth (N)	—	AS REQ'D.	
2	Diametral pitch (P)	p	$\frac{3.1416}{p}$	
3	Pressure angle (∅)	—	20° STANDARD 14°-30' OLD STANDARD	
4	Cone distance (A)	D & d	$\sin d \overline{)\frac{D}{2}}$	
5	Pitch distance (D)	p	$\frac{N}{P}$	
6	Circular thickness (t)	p	$\frac{1.5708}{p}$	
7	Pitch angle (d)	N & d (of pinion)	90°-d(pinion)	$\tan d \frac{N \text{ pinion}}{N \text{ gear}}$
8	Root angle (⋎R)	d & ƀ	d − ƀ	
9	Addendum (a)	p	$\frac{1}{P}$	
10	Whole depth (ht)	p	$\frac{2.188}{P} + .002$	
11	Chordal thickness (C)	D & d	$\frac{1}{2}\left(\frac{D}{\cos d}\right) 1-\cos\frac{90°}{\frac{N}{\cos d}} + a$	
12	Chordal addendum (aC)	d	$\sin\left(\frac{\frac{90°}{N}}{\cos d}\right)$	
13	Dedendum (bC)	P	$\frac{2.188}{P} - a(\text{pinion})$	$\frac{2.188}{P} - a(\text{gear})$
14	Outside diameter (DO)	D, a & d	D + (2 x a) x cos. d	
15	Face	A	1/3 A (max.)	
16	Circular pitch (p)	p & N	$\frac{3.1416 \times p}{N}$	
17	Ratio	N gear & N pinion	$\frac{N \text{ gear}}{N \text{ pinion}}$	
18	Back angle (⋎O)	—	SAME AS PITCH ANGLE	
19	Angle of shafts	—	90°	
20	Part number of mating gear	—	AS REQ'D.	
21	Dedendum angle ƀ	A & b	$\frac{b}{A} = \tan ƀ$	

FIGURE 20–24 Required cutting data and formulas for bevel gears (Reprinted from TECHNICAL DRAWING AND DESIGN by Goetsch and Nelson © 1986, Delmar Publishers Inc.)

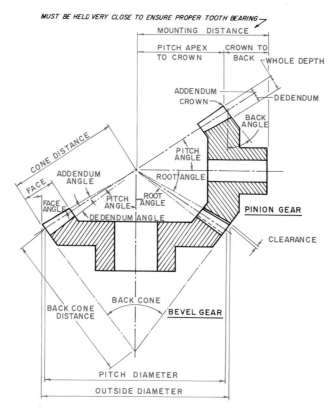

FIGURE 20-25 Example of bevel gear terminology (Reprinted from TECHNICAL DRAWING AND DESIGN by Goetsch and Nelson © 1986, Delmar Publishers Inc.)

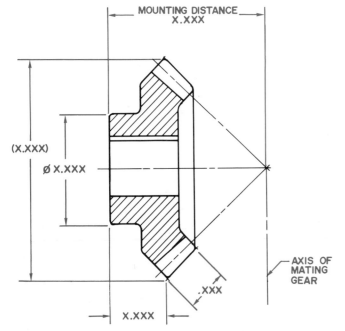

GEAR TOOTH DATA	
NUMBER OF TEETH	XX
DIAMETRAL PITCH	XX
PRESSURE ANGLE	XX.XX°
CONE DISTANCE	X.XXX
PITCH DIAMETER	X.XXX
CIRCULAR THICKNESS (REF)	.XXXX
PITCH ANGLE	XX.XX°
ROOT ANGLE	XX.XX°
ADDENDUM	.XXX
WHOLE DEPTH	.XXX
CHORDAL ADDENDUM	.XXX
CHORDAL THICKNESS	.XXX

1. INTERPRET TOOTH DATA PER ANSI Y14.7.1

NOTES:

FIGURE 20-26 Example of a bevel gear working drawing (Reprinted from MECHANICAL DRAFTING by Madsen, Shumaker, and Stewart © 1986, Delmar Publishers Inc.)

REQUIRED CUTTING DATA			
ITEM	TO FIND	HAVING:	FORMULA
1	Number of teeth (N)	P	$\dfrac{3.1416}{P}$
2	Pitch diameter (D)	Pa	$(2.4 \times Pa) + 1.1$
		DO & a	$DO - (2 \times a)$
3	Axial pitch (Pa)	—	Distance from a point on one tooth to same point on next tooth
4	Lead (L) Right or Left	p & N	$p \times N$
5	Lead angle (La)	L & D	$\dfrac{L}{3.1416 \times D} = \tan La$
6	Pressure angle (∅)	—	20° STANDARD 14°-30' OLD STANDARD
7	Addendum (a)	p	$p \times .3183$
		P	$\dfrac{1}{P}$
8	Whole depth (ht)	Pa	$.686 \times Pa$
9	Chordal thickness	N, D & a	$\left[1 - \cos\left(\dfrac{90°}{N}\right)\right] \times \dfrac{D}{2} + a$
10	Chordal addendum	N, D & a	$\sin\left(\dfrac{90°}{N}\right) \times D$
11	Outside diameter (DO)	D & a	$D + (2 \times a)$
12	Worm gear part no.	—	AS REQ'D.

Axial pitch (Pa) must be same as worm gear circular pitch (p)

FIGURE 20-27 Required cutting data and formulas for a worm (Reprinted from TECHNICAL DRAWING AND DESIGN by Goetsch and Nelson © 1986, Delmar Publishers Inc.)

REQUIRED CUTTING DATA			
ITEM	TO FIND:	HAVING	FORMULA
1	Number of teeth (N)	—	AS REQ'D.
2	Pitch diameter (D)	N & p	$\dfrac{N \times p}{3.1416}$
3	Addendum (a)	p	$p \times .3181$
		P	$\dfrac{1}{P}$
4	Whole depth (ht)	p	$p \times .6866$
		P	$\dfrac{2.157}{P}$
5	Lead (L) Right-Left	p & N	$p \times N$
6	Worm part no.	—	AS REQ'D.
7	Pressure angle ∅	—	20° STANDARD 14°-30' OLD STANDARD
8	Outside diameter (DO)	Dt & Pa	$Dt + .4775 \times Pa$
9	*Circular pitch (p)	P	$\dfrac{3.1416}{P}$
		L & N	$\dfrac{L}{N}$
10	Diametral pitch (P)	p	$\dfrac{3.1416}{p}$
11	Throat diameter (Dt)	D & Pa	$D + .636 \times Pa$
12	Ratio of worm/worm gear	N worm & N worm gear	$\dfrac{N\ \text{worm gear}}{N\ \text{gear}}$
13	Center to center distance between worm & worm gear.	D worm & D worm gear	$\dfrac{D\ \text{worm} + D\ \text{worm gear}}{2}$

Circular pitch (p) must be same as worm Axial pitch (Pa)

FIGURE 20-28 Cutting data and formulas for a worm gear (Reprinted from TECHNICAL DRAWING AND DESIGN by Goetsch and Nelson © 1986, Delmar Publishers Inc.)

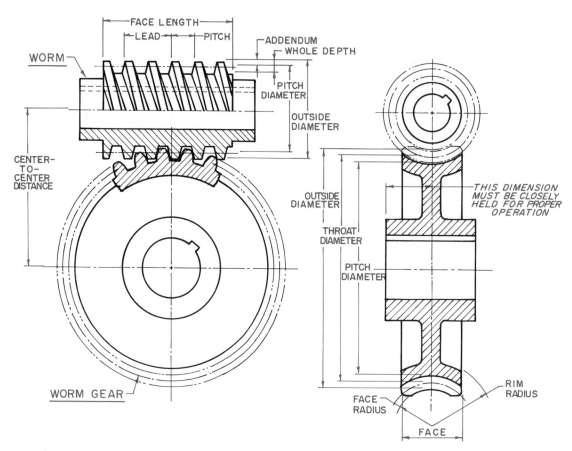

FIGURE 20-29 Example of worm and worm gear terminology (Reprinted from TECHNICAL DRAWING AND DESIGN by Goetsch and Nelson © 1986, Delmar Publishers Inc.)

Pictorial Representation

When multiple gears are connected with shafts that lie on many different planes, the positioning and relationship of gears is often difficult to visualize. This is especially true when the position of gears change as they do in an automobile transmission. Pictorial drawings (usually isometric) are used in those cases as an aid to interpretation **(Fig. 20-30)**. Pictorial drawings, unless prepared to a very large scale, do not include teeth details, since these drawings are not used for manufacturing.

Figure 20-31 is a pictorial assembly drawing designed to show the relationship of gears and other mechanical components. **Figure 20-32** is a pictorial cutaway view of a gear assembly revealing the internal configuration and relationship of gears and gear assemblies.

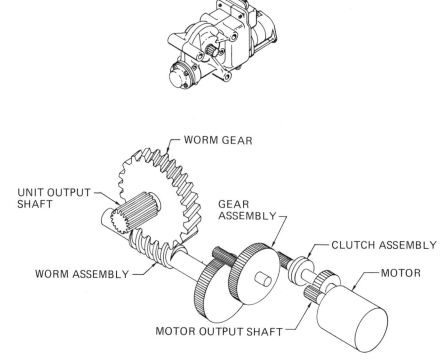

FIGURE 20-30 Pictorial drawing of a motor-driven gear assembly

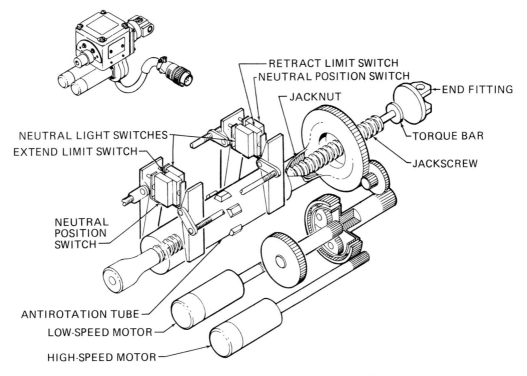

FIGURE 20-31 Pictorial drawing of a motor-driven gear assembly

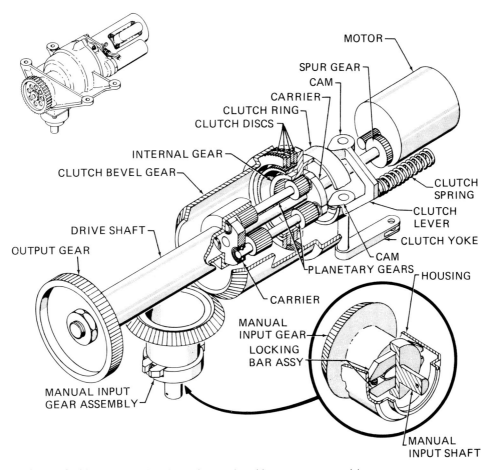

FIGURE 20-32 Pictorial drawing of a motor-driven gear assembly

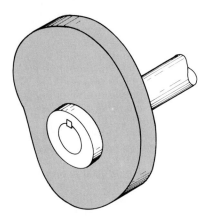

FIGURE 20-33 Plate cam

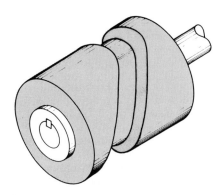

FIGURE 20-34 Drum cam

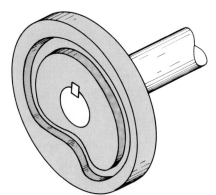

FIGURE 20-35 Face cam

CAM MECHANISMS

It is often necessary to convert the rotary motion of a motor into linear motion for many mechanical operations. It is also often necessary to convert this motion back again to rotary motion for successive manufacturing procedures. Gears and screws can only change or maintain motions that are evenly spaced at regular intervals. **Cam** mechanisms can convert regular motion to irregular or unusual motion. They also can convert one type of motion into another such as converting rotary motion to linear oscillating motion. Cam mechanisms consist of a cam, follower, and camshaft.

Cams are divided into three types—**plate cams (Fig. 20-33)**, **drum cams (Fig. 20-34)**, and **face cams (Fig. 20-35)**. Cam shapes are determined by the desired motion of a follower. A plate cam follower moves vertically as the cam rotates, converting rotary motion into vertical linear (displacement) reciprocating movement **(Fig. 20-36)**. Drum cams convert circular motion to linear horizontal reciprocating movement by guiding the follower through a groove **(Fig. 20-37)**.

Cam Followers

There are four types of cam followers—**roller, pointed, round and flat**

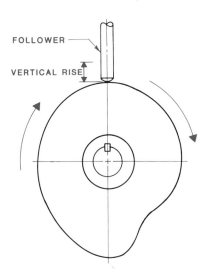

FIGURE 20-36 The circular motion of the plate cam is converted to a vertical linear movement by the follower.

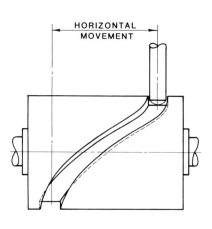

FIGURE 20-37 The circular motion of the drum cam is converted to a horizontal linear movement by the follower.

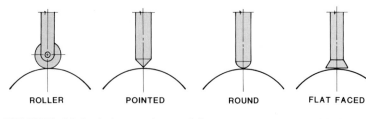

FIGURE 20-38 Basic types of cam followers

faced **(Fig. 20-38)**. Roller followers are used the most since they operate better at higher speeds with less friction and wear. The exact shape of a cam determines the exact timing and motion limits of the follower during one cam revolution. Think of a cam as a simple inclined plane with uneven surface contours. Although **Fig. 20-39** shows the follower position at only four points, it moves smoothly from one position to another. **Figure 20-40** summarizes the nomenclature used in the design and drawing of a plate cam.

Displacement Diagram

The first step in designing and drawing a cam shape is the development

Chapter 20 Gears and Cams **399**

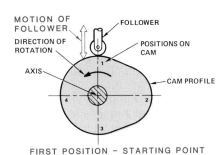

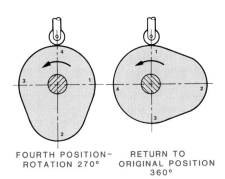

FIGURE 20-39 Movement of the cam's follower

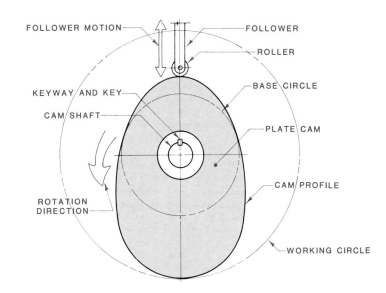

FIGURE 20-40 Plate cam terminology

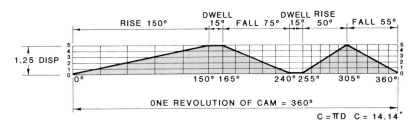

FIGURE 20-41 Displacement diagram for a cam

of a **displacement diagram.** A displacement diagram is a curve representing the desired position (displacement) of the follower at successive intervals during one revolution of the **camshaft (Fig. 20–41)**. Geometric line shapes used in the development of displacement diagrams include

1. straight angular lines, which produce uniform motion,
2. parabolic curves, which produce constant acceleration or deceleration,
3. harmonic curves, which produce a harmonic motion, and
4. straight horizontal lines, which hold the follower at a fixed (**dwell** or **rest**) position.

All nonhorizontal lines result in an upward movement (**rise**) of the follower or in a downward movement (**fall**) of the follower. The displacement diagram in Fig. 20–41 illustrates two uniform rise and fall curves with two dwell (rest) periods. In Fig. 20–41, the horizontal axis represents 360° or one revolution of the cam. The vertical grid represents the maximum distances the follower will move during one revolution of the cam. The profile line identifies the exact position of the follower required at specific degrees of rotation.

Displacement diagrams are flat while cams are basically round. The flat profile intersections or the displacement diagram must be converted to radial distances to create a cam shape **(Fig. 20–42)**. In Fig. 20–42, all horizontal distances are converted to degrees on concentric circles. These are spaced evenly to represent the movement range of the follower. Intersecting points are then plotted to coincide with the intersections on the displacement diagram. Connecting these points create the outline shape of the cam that will produce the follower action plotted in the displacement diagram.

CAM Motions

Care must be taken to provide the correct cam shape, since that controls the movement of the follower. The follower must move smoothly, without jerks, at the maximum planned speed of rotation. The following displacement curves are designed to produce specific results:

Uniform Motion. A displacement diagram and resulting cam shape designed to produce a uniform motion at a constant speed is shown in

400 Drafting in a Computer Age

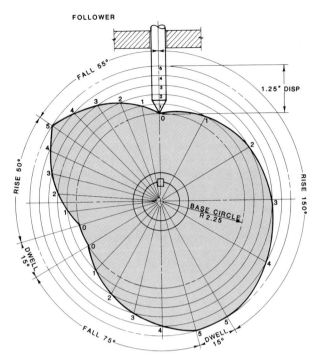

FIGURE 20-42 Layout of a CAM drawing made from the displacement diagram in Fig. 20-41

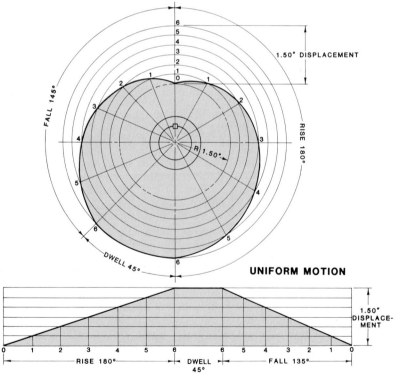

FIGURE 20-43 The uniform cam motion gives a jerk to the follower at the start and end of each phase.

Fig. 20-43. However, the changes in direction from this curve are abrupt, and cause a jerky motion in the follower.

Modified Uniform Motion. A displacement diagram with the uniform motion modified to produce a smoother transition between rise and rest, and rest and fall is shown in **Fig. 20-44**. Here, straight line intersections are converted to small arc intersections. The radius of the arcs is equal to one-third the displacement rise or fall.

Harmonic Motion. When even greater smoothness is required, usually for high speeds, harmonic curves are plotted **(Fig. 20-45)**. Harmonic curves provide a smooth transition between rise, fall, and dwell. They also allow smoother start and stop movements of the follower. A harmonic curve is plotted by projecting segments of a circle onto the displacement diagram in place of evenly spaced distances.

Uniform Accelerated and Decelerated Motion. When the desired motion of a follower is to increase or decrease steadily and smoothly, a parabolic curve is plotted on the displacement diagram **(Fig. 20-46)**. In plotting a parabola, lay out the 360° line or time line horizontally as before. Divide the vertical follower distance line into displacement ratios of 1:3:5. Project these lines to intersect degree lines, and connect the intersections to produce a parabola.

Cycloidal Motion. A displacement diagram designed to produce cycloidal motion in a cam mechanism is shown in **Fig. 20-47**. Cycloidal curves produce extremely smooth curves by eliminating the dwell contour. Cycloidal curves are used in cams that operate at optimum efficiency at high speeds.

Chapter 20 Gears and Cams 401

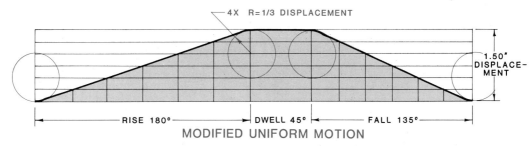

FIGURE 20–44 The modified uniform cam motion will smooth out the follower's jerky motion.

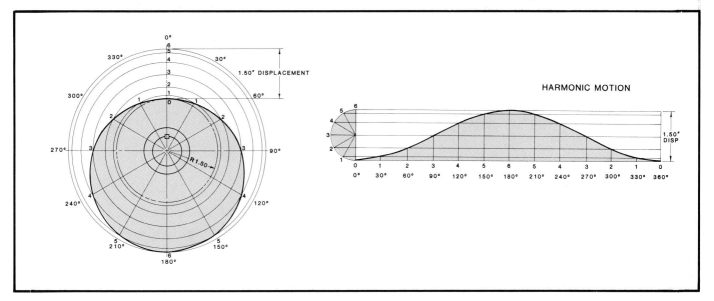

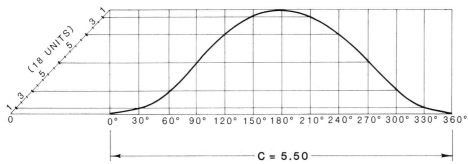

DISPLACEMENT DIAGRAM UNIFORM ACCELERATED MOTION

FIGURE 20–46 The uniform accelerated and decelerated motion produces a very smooth motion with a parabolic curve.

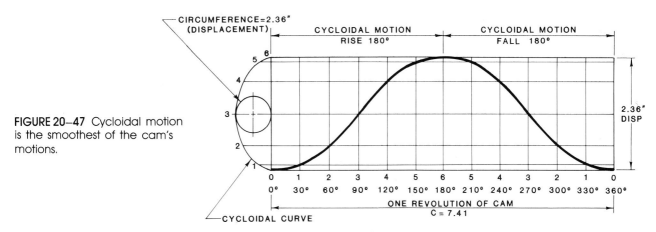

FIGURE 20–47 Cycloidal motion is the smoothest of the cam's motions.

DISPLACEMENT DIAGRAM – CYCLOIDAL MOTION

Design and Drawing Procedures

Figures 20–48 through 20–51B illustrate the specific steps used to design, draw, and dimension cams in detail.

A uniform motion displacement diagram **(Fig. 20–48)** is used to complete the cam profile shown in **Fig. 20–49**.

The procedures used to develop and draw a uniform motion plate cam is shown in **Fig. 20–50**. This involves laying out the base circle, followed by the segmenting of the circumference. The rise displacement circles are then added to represent the range of follower motion. Points are plotted to coincide with intersections on the displacement diagram. These points are connected to produce the final shape of the cam.

As in pattern drafting, the smoothness of the shape is directly

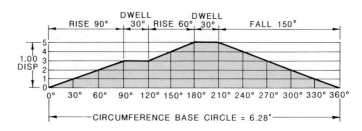

FIGURE 20–48 Displacement diagram designed with a uniform motion

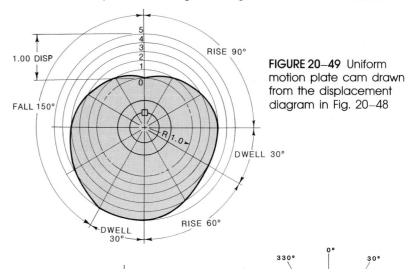

FIGURE 20–49 Uniform motion plate cam drawn from the displacement diagram in Fig. 20–48

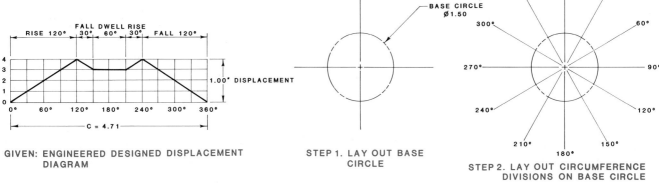

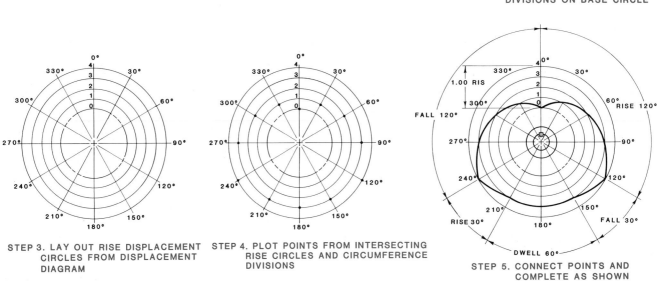

FIGURE 20–50 Steps to draw a uniform motion plate cam

related to the number of circle divisions used to develop the profile on the displacement diagram. If intersecting points are too far apart to produce a smooth curve, an irregular (French) curve template may be used to smooth out the contour.

Similar sequences used to develop a cam to produce a constant acceleration motion are shown in **Fig. 20–51A**. The only difference in this procedure is the use of an angular profile line to produce the fall-rise segments on the displacement diagram.

Since cam shapes are irregular, grid layouts may be used to establish the exact shape of the circumference. The shape of the cam also may be defined by locating the position of the follower at various degree increments during a 360° rotation. Again, more degree divisions produce a smoother contour on the cam circumference. All other sizes are dimensioned conventionally **(Fig. 20–51B)**. Figure 20–51B also shows geometric tolerancing limits which are extremely critical in cam machining.

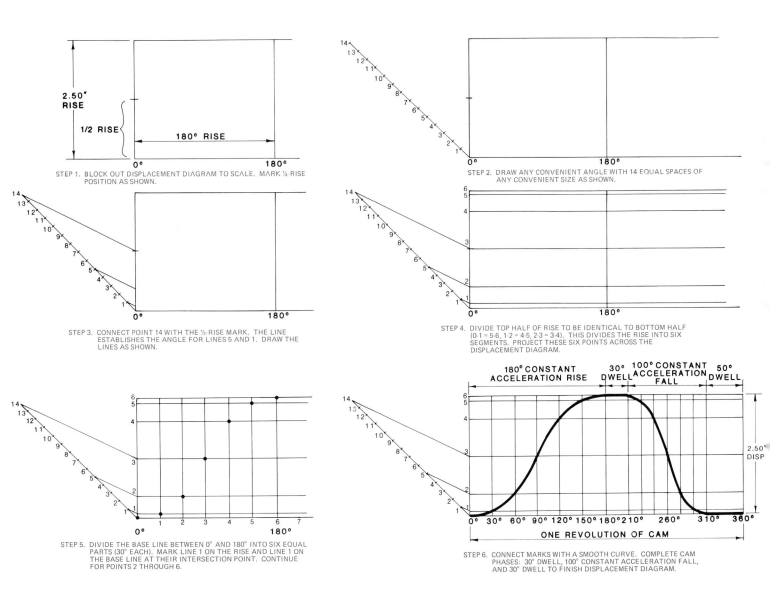

FIGURE 20–51A Steps to draw a constant acceleration cam with a rise of 180°, dwell 30°, fall 100°, and dwell of 50°

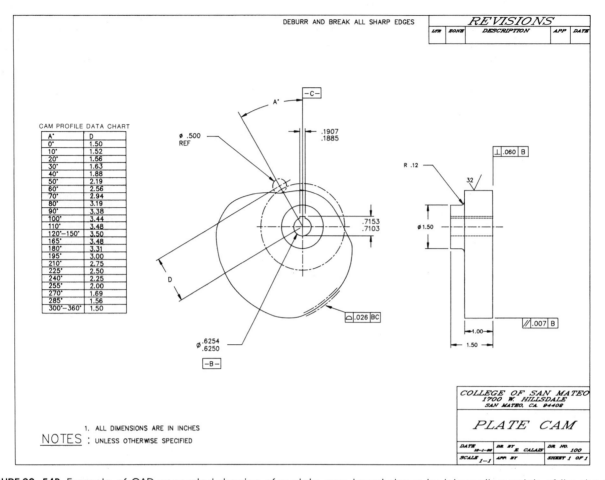

FIGURE 20–51B Example of CAD-generated drawing of a plate cam layout drawn by intersecting points of the rise (D) and the angular (A) locations (Courtesy of the College of San Mateo)

COMPUTER-AIDED DRAFTING & DESIGN

DRAWING GEARS WITH CAD

Gears can be accurately and rapidly drawn using a CAD system. The following information does not relate to specialized software for gear design and drawing, but uses normal CAD software which is universally available. A CAD command that is of great help in drawing gears is AR-RAY. The ARRAY command repeats a shape (such as a gear tooth) in a specified pattern **(Fig. 20–52)**. In the first step, centerlines and one circle have been drawn. In the second step the CAD operator has selected the ARRAY command, indicated the circle to be arrayed and specified the center of the circular array (box). The operator asked for six circles to be arrayed in 360° (step three).

The following sequence, used for drawing a spur gear, utilizes the ARRAY command to produce the individual gear teeth around the gear. The gear has a pitch diameter of 4.80″ and 36 teeth.

Step 1. The three main circles—pitch diameter, outside diameter, and root diameter are drawn **(Fig. 20–53)**. The gear calculations must be made prior to doing the drawing. The calculations may be

FIGURE 20–52 Using the ARRAY command

done with the aid of the computer, a hand calculator, or by student computations.

Step 2. The ARRAY command is used to array the center of each tooth **(Fig. 20–54)**. Since there are to be 36 teeth on the gear, the CAD operator enters 36 in the ARRAY command. The lines were automatically drawn.

Step 3. The ZOOM command is used to see the top center of the gear clearly. By zooming in, the operator can do fine detailed work and draw one of the gear teeth. One-half of a gear tooth is drawn using the results of the gear tooth calculations **(Fig. 20–55)**.

Step 4. The MIRROR command is used to provide a perfectly symmetrical gear tooth **(Fig. 20–56)**.

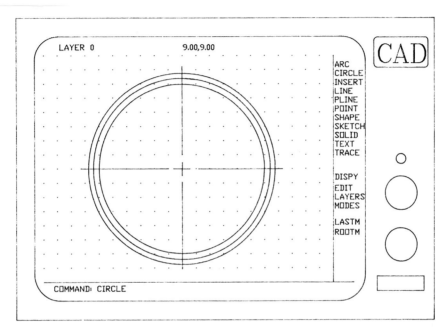

FIGURE 20–53 Lay out the outside diameter, pitch diameter, and root diameter after doing the gear calculations.

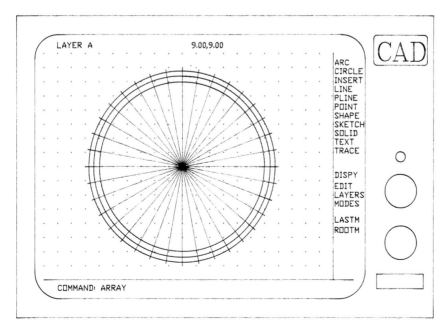

FIGURE 20–54 Gear construction with the ARRAY command to center 36 gear teeth

406 Drafting in a Computer Age

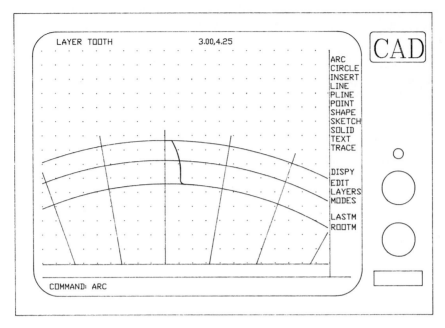

FIGURE 20–55 Use the ZOOM command to zoom in and draw half of the gear tooth profile from the gear calculations.

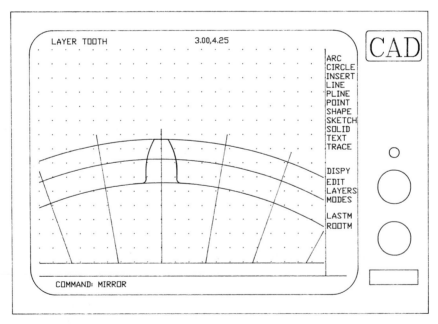

FIGURE 20–56 Use the MIRROR command to complete the symmetrical gear tooth.

Step 5. The ZOOM command again displays the entire gear, and the ARRAY command repeats the tooth 36 times around the gear **(Fig. 20–57)**.

Step 6. The gear profile is completed by erasing the construction lines, including the outside diameter, root diameter, and centerlines of the teeth **(Fig. 20–58)**.

Step 7. To complete the spur gear drawing, add the hub, center hole, side view, dimensions, and gear data **(Fig. 20–59)**.

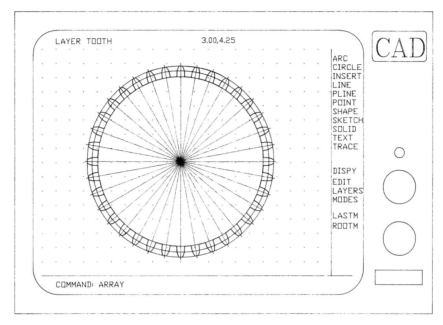

FIGURE 20–57 ZOOM out for the full drawing format and ARRAY the gear tooth 36 times.

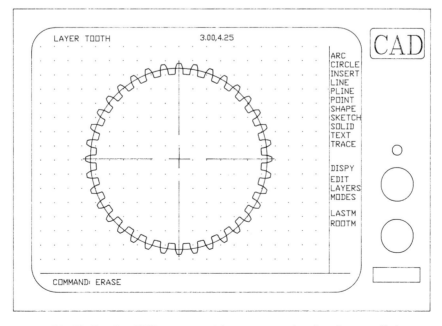

FIGURE 20–58 Use the EDIT command to erase construction lines, or if done on a separate layer, remove the layer.

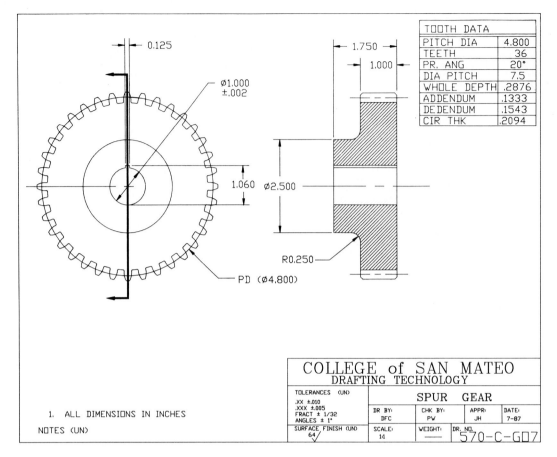

FIGURE 20-59 The completed CAD-generated spur gear drawing. (Courtesy of the College of San Mateo)

DRAWING CAMS WITH CAD

Although there are many types of cams, the following CAD procedures apply to a simple plate cam. These procedures can be applied easily to the design and drawing of other cam types on a CAD system. Smooth curved lines are a common feature of all cams.

When drawing cams manually, the drafter must use a French curve (flexible curve) to smooth out lines. CAD's counterpart to the French curve is a command called PLINE which stands for "polyline." The PLINE command has many features, one of which is to draw a smooth curve through a series of points. This is done in three steps **(Fig. 20–60)**. First, the points are drawn. Second, the points are connected using a polyline (PLINE command). The third step is to use a command called PEDIT, which

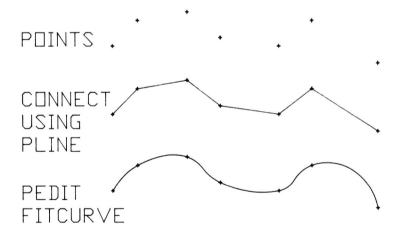

FIGURE 20-60 Use the PLINE and PEDIT commands for irregular curves.

stands for "polyline edit." One of the choices PEDIT gives the operator is **curve fit.** By selecting the PEDIT command and the curve fit option, and touching the PLINE command, the line automatically smooths out into a curve that goes through each of the given points. These commands will be used to construct both

the cam displacement diagram and the cam.

Figure 20–61 shows a CAD-generated displacement diagram. The curve was drawn using the PLINE command. This displacement diagram will be used to draw the profile of a plate cam (Figs. 20–62 through 20–65).

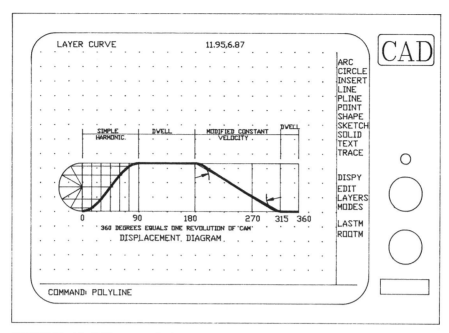

FIGURE 20-61 CAD-generated cam displacement diagram

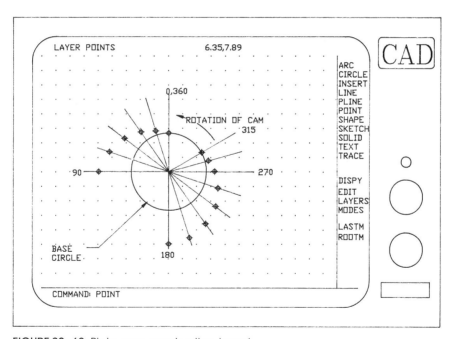

FIGURE 20-62 Plate cam construction layout

Step 1. Much of the layout design has been accomplished in **(Fig. 20-62)**. First, the base circle was drawn and the angles were labeled, indicating the counterclockwise rotation of the cam. The critical radial lines were then drawn representing the fall and rise of the cam. Note that no lines were drawn between 90° and 180°. This is because this cam dwells during this period. The distances of the radial lines were acquired from the displacement diagram in Fig. 20-61. The height of each point is transferred from the displacement diagram to the radial lines, starting from base circle.

Step 2. The points were connected using PLINE, and smoothed out using PEDIT **(Fig. 20-63)**. This provides for a smooth, accurate cam profile.

Step 3. The construction lines and points were erased and a profile drawn **(Fig. 20-64)**.

Step 4. Dimensions, a cam profile data chart, notes and a title block have been added to complete the drawing **(Fig. 20-65)**.

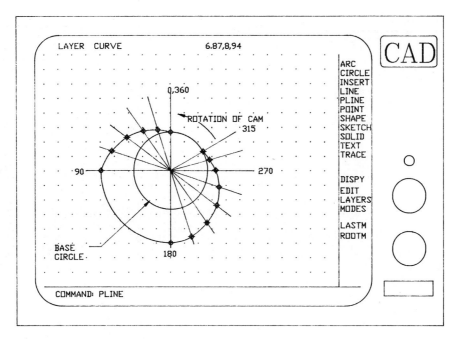

FIGURE 20-63 Plate cam outline drawn with the PLINE and PEDIT commands

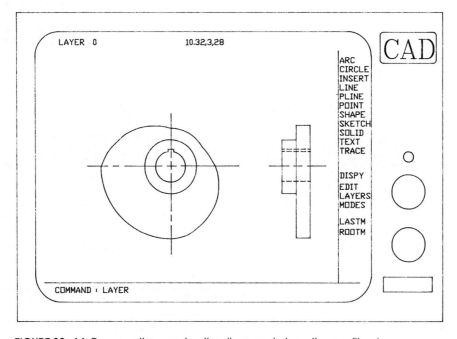

FIGURE 20-64 Remove the construction lines and draw the profile view.

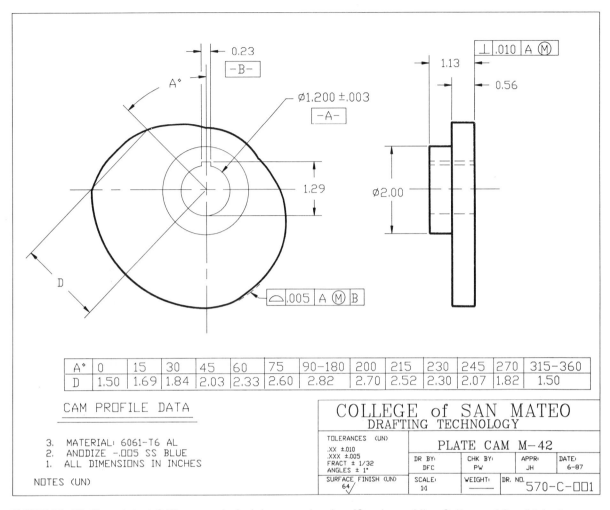

FIGURE 20-65 Completed CAD-generated plate cam drawing (Courtesy of the College of San Mateo)

EXERCISES

1. Redraw the spur gear in **Fig. 20-66** with drawing instruments and a gear template.
2. Redraw the helical gear **(Fig. 20-67)** with drawing instruments.
3. Redraw the two miter gears **(Fig. 20-68)** with drawing instruments.
4. Redraw the worm and worm gears **(Fig. 20-69)** with drawing instruments.
5. With a CAD system, draw the gears in Figs. 20-66 through 20-69.
6. Redraw the displacement diagrams and the cams for **Figs. 20-70** through **20-73**. Use drafting instruments or a CAD system.
7. Design a cam displacement diagram for a cam follower that rises with a harmonic motion 2.00″ in 180°, then has a dwell for 30°, then falls 2.00″ in 120°, and then dwells for 30°. The circumference of the base circle is 6.00″.

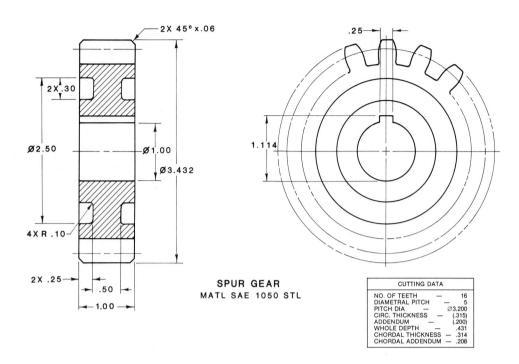

FIGURE 20-66 Draw the spur gear.

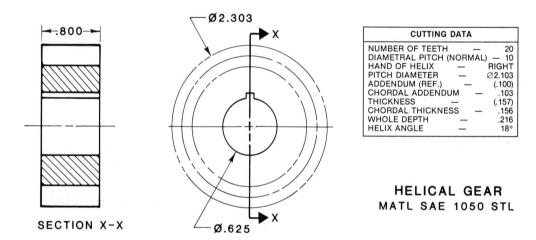

FIGURE 20-67 Draw the helical gear.

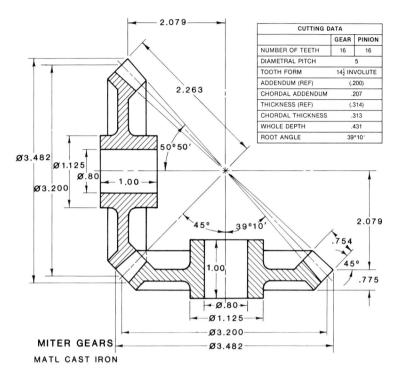

FIGURE 20-68 Redraw the two miter gears.

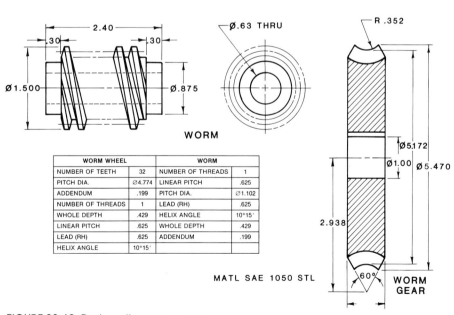

FIGURE 20-69 Redraw the worm gear.

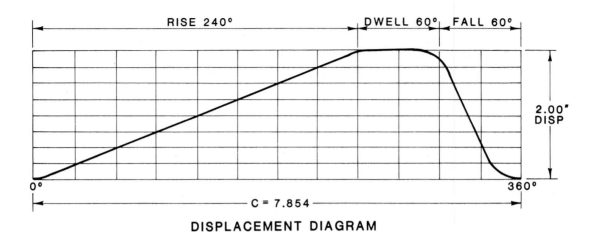

FIGURE 20–70 Draw the displacement diagram and the plate cam. To find the base circle's diameter, divide pi (π) into the circumference.

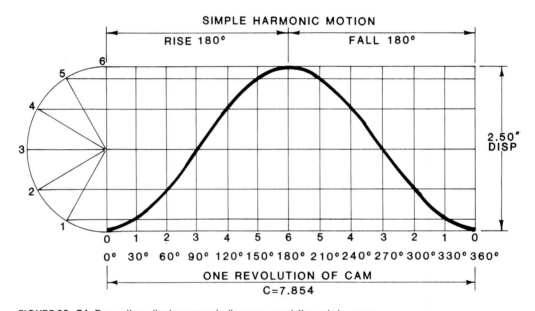

FIGURE 20–71 Draw the displacement diagram and the plate cam.

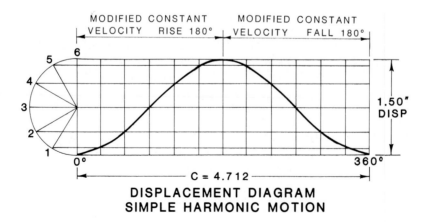

FIGURE 20–72 Draw the displacement diagram and the plate cam.

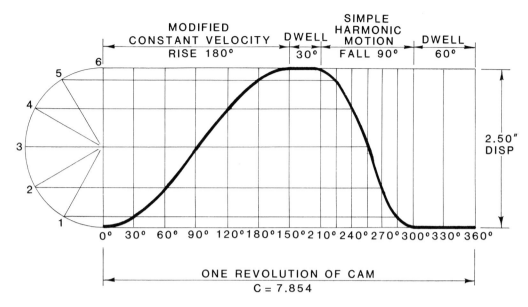

FIGURE 20-73 Draw the displacement diagram and the plate cam.

KEY TERMS

Addendum
Addendum circle
Angular gears
Base circle
Belt drives
Bevel gears
Cam
Camshaft
Center distance
Chain and sprocket drives
Chordal thickness
Circular pitch
Clearance
Curve fit
Cycloidal curve
Cycloidal motion
Dedendum
Dedendum circle
Diametral pitch
Displacement diagram

Drum cams
Dwell (rest)
Face cam
Fall
Fillet
Flank
Flat-faced followers
Friction wheels
Gear axis
Gear ratio
Gears
Harmonic motion
Helical gears
Herringbone gears
Involute curve
Miter gear
Modified uniform motion
Number of teeth
Parallel shaft gears
Perpendicular shaft gears

Pitch circle
Pitch diameter
Pitch point
Plate cams
Pointed followers
Rack and pinion gears
Ring gears
Rise
Roller followers
Root diameter
Round followers
Spiral bevel gears
Spur gears
Tooth face
Uniform accelerated motion
Uniform decelerated motion
Uniform motion
Whole depth
Working depth
Worm gears

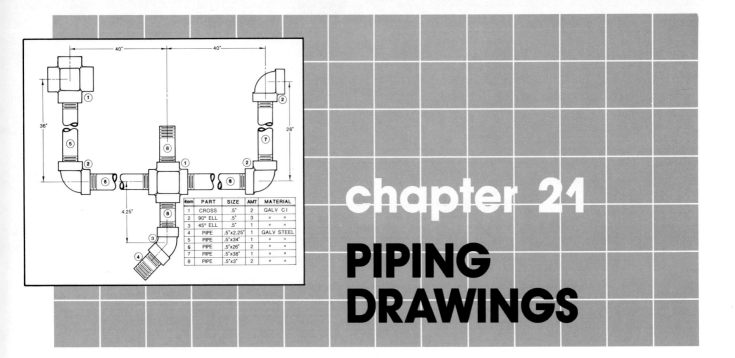

chapter 21

PIPING DRAWINGS

OBJECTIVES

The student will be able to:

- recognize pipe drafting symbols
- complete an orthographic single-line piping drawing
- complete an orthographic double-line piping drawing
- complete an isometric single-line piping drawing
- complete an isometric double-line piping drawing
- prepare a piping drawing on a CAD system

INTRODUCTION

Pipes are used in a wide range of applications, from small residences to large industrial plants **(Fig. 21–1A)**. Pipes carry all types of liquid, gases, and semisolids including water, oil, acid, liquid metal, nitrogen, oxygen, steam, granular materials, and exhausts. Pipes are also used structurally as support columns and for a wide variety of architectural applications, such as handrails and fence components.

Pipes are the arteries and veins of plumbing, heating, cooling, drainage, and manufacturing systems. Pipes are made of steel, wrought iron, cast iron, copper, brass, aluminum, plastic, clay, concrete, brass, glass, lead, rubber, wood, and other synthetic material combinations. Drawings which describe the pipe material, size, and location in a piping system are assembly drawings. These use orthographic or pictorial (isometric) projection **(Figs. 21–1B and 21–1C)**.

PIPE CONNECTIONS

Fittings are usually required to join pipes and attach them to other devices, such as pumps and tanks. Pipes are temporarily joined with threaded, flanged, or bell and spigot connections. Others are permanently attached by brazing, soldering, or welding. Some pipes can be bent to change direction.

FIGURE 21–1A Industrial plants use many pipes to transport liquids and gases. (Courtesy of FERRANTI Sciaky, Inc.)

416

Chapter 21 Piping Drawings **417**

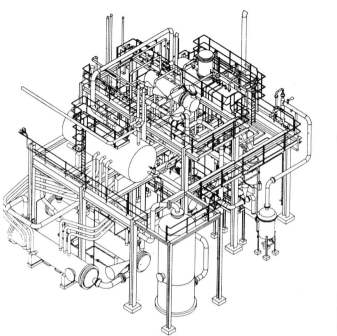

FIGURE 21–1B Piping drawing using an isometric drawing layout (Courtesy of Computervision Corporation)

FIGURE 21–1C Segment of an industrial plant's piping using an isometric drawing layout (Courtesy of Computervision Corporation)

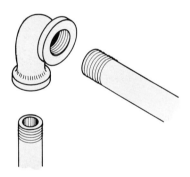

FIGURE 21–2 Threaded elbow pipe connection

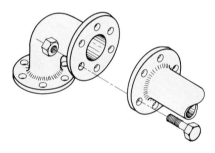

FIGURE 21–3 Bolted flange pipe connection

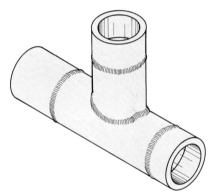

FIGURE 21–4 Welded pipe connections

Threaded Connections

Threaded connections are used primarily on pipes with 2″ diameters or less, and are available with straight or tapered threads. Threaded connectors, sometimes called **screwed fittings,** are available in many shapes—elbows **(Fig. 21–2)**, tees, Y bends, reducers, return bends, caps, and crosses. Couplings with inside threads are also used to connect straight pipes or nipples. Nipples are short pipe lengths with external threads on both ends.

Flange Connections

Flange connections are designed to enable a pipe joint to be disassembled by removing a series of bolts on the face of the flange. Flanges are manufactured in the form of elbows **(Fig. 21–3)**, return bends, Y bends, tees, taper reducers, and crosses.

Welded Connections

Since **welded connections** are virtually leakproof, welding is used to join pipes containing hot or volatile materials. Welded joints are also more economical to construct. Gas and arc welding methods are most commonly used to join pipe with butt, lap, tee, edge, and corner joints. **Figure 21–4** shows a welded tee shape. See Chapter 19 for welding joint details.

In the process of brazing and soldering, the base metal does not melt. Consequently brazing and soldering is restricted to joining small copper or brass pipe and tubing. Pipe joints are soldered with the use of

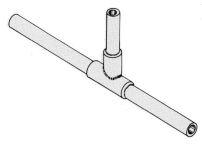

FIGURE 21-5 Soldered copper pipe connections

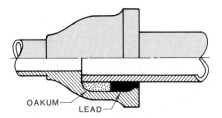

FIGURE 21-6 Bell and spigot pipe connection

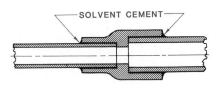

FIGURE 21-7 Cemented connection for plastic pipes

connectors **(Fig. 21-5)**. Connectors must be specified to fit tightly over the pipe to aid the sweat soldering process. Remember, solder has a low boiling point so extremely hot materials, especially under high pressure, must be avoided.

Bell and Spigot Connections

Bell and spigot connections are so named because one pipe end is shaped like a bell and inserted into a pipe with a spigot-like shape. These joints are fitted tightly with lead and oakum to make them waterproof **(Fig. 21-6)**.

Cementing

Many materials such as plastic, clay, concrete, and glass cannot be welded, soldered, or threaded effectively. Joining pipes made of these materials requires **cementing (Fig. 21-7)**.

PIPE STANDARDS

Pipes are specified by the type of material, sizes, and wall thickness. Pipe sizes relate to either inside or outside diameters. The **nominal pipe size** for steel pipe up to 12", is the inside diameter (ID). For pipes over 12", the nominal size is the outside diameter (OD). Pipe wall thicknesses are standardized by schedules of pipe weights. Pipe schedules range from 10 to 160 with wall thickness increasing as the schedule number increases **(Fig. 21-8)**.

STANDARDIZED PIPE SIZES—SCHEDULE 20		
Nominal Size (inches)	Outside Diameter (inches)	Inside Diameter (inches)
1/8	.405	.269
1/4	.540	.364
3/8	.675	.493
1/2	.840	.622
3/4	1.050	.824
1	1.315	1.049
1 1/4	1.660	1.380
1 1/2	1.900	1.610
2	2.375	2.067
2 1/2	2.875	2.469
3	3.500	3.068
3 1/2	4.000	3.548
4	4.500	4.026
5	5.563	5.047
6	6.625	6.065
8	8.625	7.981
10	10.750	10.020
12	12.750	11.938
14	14.000	13.126
16	16.000	15.000
18	18.000	16.876
20	20.000	18.814
24	24.000	22.626

FIGURE 21-8 Standardized pipe diameters for Schedule 20 pipes

PIPING DRAWINGS

There are two general types of piping drawings—orthographic and pictorial. Either type may be prepared using a **single line** to represent each pipe, or a **double line** to represent the outline of a pipe. Double-line drawings appear more realistic but require extensive drafting

Chapter 21 Piping Drawings 419

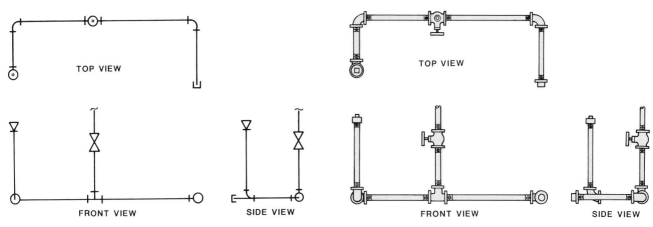

FIGURE 21-9 Single-line orthographic piping drawing

FIGURE 21-10 Double-line orthographic piping drawing

time. They are used sparingly, except for large pipes or for presentation and instructional purposes. Single-line drawings are used the majority of the time. In single-line drawings, a single thick line is drawn at the location of each pipe centerline, regardless of the pipe size.

When orthographic projection is used, there are usually three basic views. **Figure 21-9** is a single-line orthographic drawing of a simple piping layout. **Fig. 21-10** depicts a double-line piping drawing of Fig. 21-9. Orthographic piping drawings may involve only two views. In architectural work, plumbing systems are depicted in top and front views, and are labeled, *plan view* and *elevation view* respectively.

Orthographic drawings effectively show the layout of piping systems when pipes align on one or two planes. When pipes lie on many different planes and angles, orthographic drawings become cluttered and complex. For this reason pictorial drawings (isometric or oblique) are used to eliminate the confusion created by overlapping pipes, fittings, and fixtures. **Figure 21-11** shows a single-line isometric piping drawing, and **Fig. 21-12** shows a double-line isometric drawing of the same layout. Isometric piping drawings are especially effective for describing complex systems that involve a variety of pipe configurations, fittings, valves, fixtures, and devices **(Fig. 21-13)**.

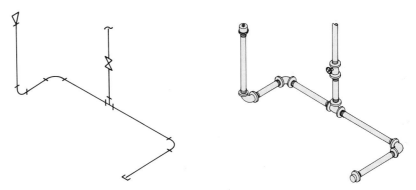

FIGURE 21-11 Single-line isometric piping drawing

FIGURE 21-12 Double-line isometric piping drawing

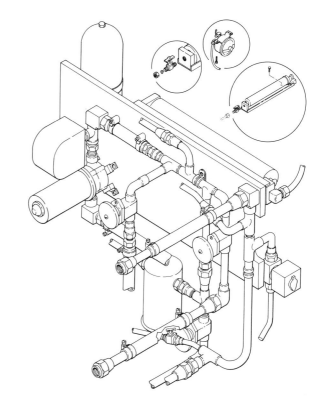

FIGURE 21-13 Double-line isometric piping drawing of a segment of a hydraulic system

If a single projection plane is involved, an orthographic drawing **(Fig. 21–14)** can be drawn to a more accurate scale. Piping systems are often broken into small segments which fall on one plane. Dimensioned orthographic drawings are prepared for these segments. Isometric drawings are used only to show the general assembly of the entire system or subsystem. Note the orthographic drawing shown in Fig. 21–14 represents the back of the hydraulic system shown in Fig. 21–13.

Piping Symbols

The symbol for a pipe is either a single or a double solid line. Pipes represent only one component in a complete piping system. Other components include valves, fittings, instruments, flanges, and other devices. Inserting an orthographic view of these components at hundreds of locations on a piping drawing is time-consuming and impractical.

Valves are complex devices which control the flow of pressurized liquids and gases. There are dozens of valve types, manufactured in hundreds of sizes and types of materials **(Figs. 21–15 through 21–18)**. This makes the preparation of orthographic drawings of valves extremely involved. Templates simply outline the drawing of valves and other components **(Fig. 21–19)**. Nevertheless, most valves, fittings, flanges, instruments, fixtures, special lines, and connection methods are represented by simplified symbols on piping drawings. Piping symbols are shown in **Fig. 21–20** in orthographic single- and double-line views, and in the isometric single-line view. The various types of pipe connection symbols must be apparent on the piping drawings to insure correct assembly. **Figure 21–21** shows the application of pipe connection symbols.

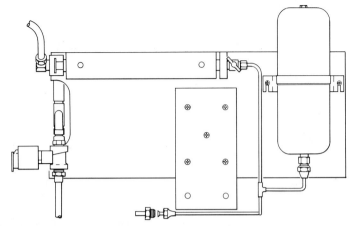

FIGURE 21–14 An orthographic view of the backside of a segment of the hydraulic system in Fig. 21–13

FIGURE 21–15 Gate valve (Courtesy of Crane Co.)

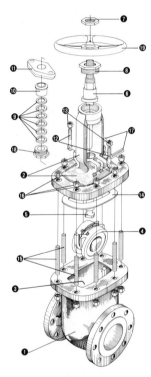

FIGURE 21–16 Exploded view of a gate valve (Courtesy of Crane Co.)

FIGURE 21–17 Ball valve (Courtesy of Crane Co.)

9
17
7
8
6
16
10
1
CRANE
5
12
4
13
14
18
3
2
15

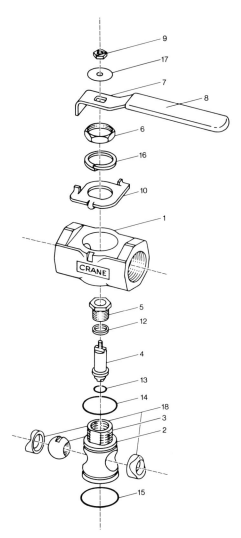

FIGURE 21–18 Exploded view of a ball valve (Courtesy of Crane Co.)

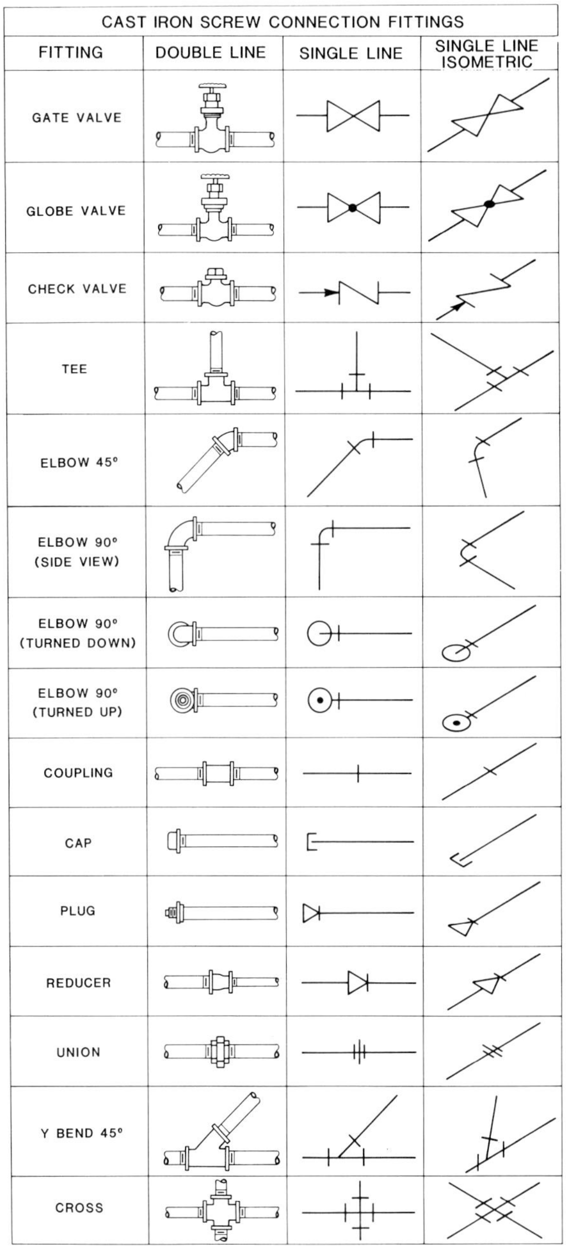

CAST IRON SCREW CONNECTION FITTINGS
FITTING
DOUBLE LINE
SINGLE LINE
SINGLE LINE ISOMETRIC
GATE VALVE
GLOBE VALVE
CHECK VALVE
TEE
ELBOW 45°
ELBOW 90° (SIDE VIEW)
ELBOW 90° (TURNED DOWN)
ELBOW 90° (TURNED UP)
COUPLING
CAP
PLUG
REDUCER
UNION
Y BEND 45°
CROSS

FIGURE 21–20 Screw pipe fitting symbols

PIPE O.D.
END BELL
FLANGE
FITTINGS
45°
45°
45°
PIPING TEMPLATE
VALVES

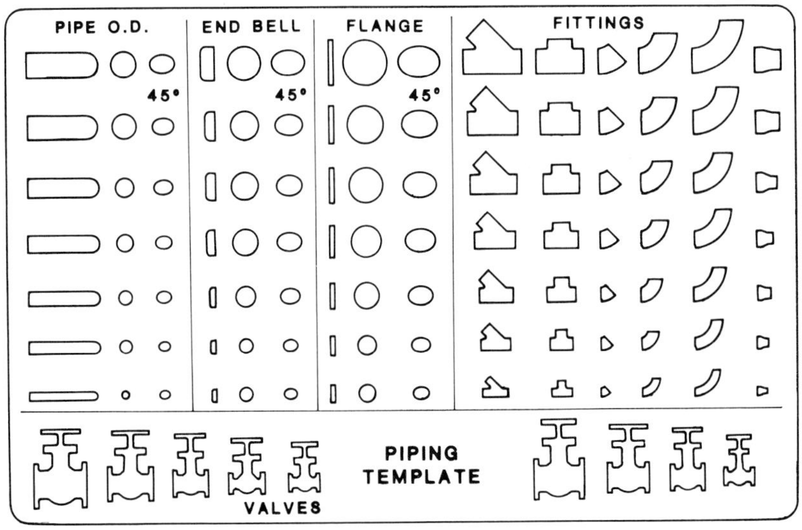

FIGURE 21–19 Piping templates help to speed piping working drawings.

TYPE OF PIPE CONNECTION	SYMBOL
SCREWED	
FLANGED	
BELL AND SPIGOT	
WELDED	
SOLDERED	

FIGURE 21–21 Pipe connection symbols

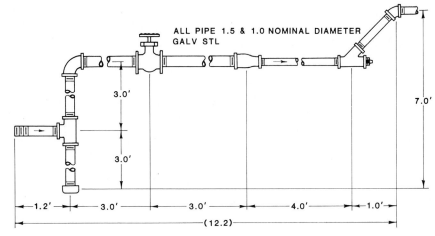

FIGURE 21–22 Example of a dimensioned pipe drawing

Dimensioning

Piping dimensions are similar to normal orthographic or isometric dimensioning standards with the following special considerations (Fig. 21–22):

1. Dimensions perpendicular to a pipe are located to the centerline of the pipe. Dimension lines are connected to the centerline in double-line drawings and to the pipe line in one-line drawings.
2. Dimensions parallel with pipe lengths are placed to the center of the pipe or flange.
3. Pipe sizes and pipe fitting sizes are provided in a general note if all pipes are equal. Dimensions and notes are connected to a pipe with a leader if pipe sizes vary. Notations include information on the type of material, size, and wall thickness.
4. When more than one length of pipe is required for a total run, only the total length of the run is dimensioned, not the length of the various pieces.
5. Keep dimensions and extension lines thinner than pipe object and symbol lines.
6. Do not chain-dimension. Dimension to a base datum to avoid accumulating errors.
7. Use arrows to show the direction of content flow.
8. Label the contents of each pipe subsystem.
9. Keep all dimensions in feet and inches if using the U.S. customary system, or in millimeters if using the metric system.
10. Use unilateral numerals for isometric dimensioning.

COMPUTER-AIDED DRAFTING & DESIGN

COMPUTER-AIDED PIPING DRAWINGS

Identical symbols appear in many different locations on a piping drawing. Repetition of these symbols on the same drawing is time-consuming when drawing with either instruments or templates. Since piping symbols are standardized, they are drawn only once when preparing a piping drawing on a CAD system.

There are several methods of inserting standard symbols onto a CAD drawing. The following are the three most common methods:

1. Draw the symbol once, then use the COPY and ROTATE commands to repeat the symbol on the drawing.
2. Use a graphics tablet overlay that contains piping symbols (Fig. 21–23). The overlay may be a commercial variety or may be created by the CAD operator.
3. Use a screen menu which may be inserted into the drawing. Figure 21–24 shows a screen menu containing double-line pipe symbols.

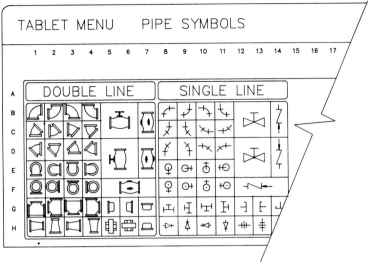

FIGURE 21–23 Piping symbols on the graphics tablet are stored in the CAD system's library.

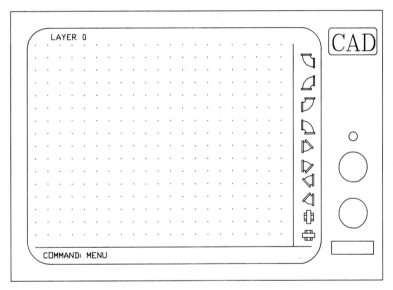

FIGURE 21–24 Pipe symbols can also be shown on the monitor.

Figures 21–23 through 21–28 show the following five steps in creating a double-line piping drawing, using any one of the systems for inserting standard piping symbols:

Step 1. Load the CAD piping program that includes the necessary symbols (Figs. 21–23 and 21–24).

Step 2. Draw the centerlines of the piping system to scale **(Fig. 21–25)**.

Step 3. Insert the piping symbols **(Fig. 21–26)**.

Step 4. Draw the pipes using a LINE command or a special PARALLEL LINE command **(Fig. 21–27)**.

Step 5. Add dimensions to complete the drawing **(Fig. 21–28)**.

Examples of a CAD-generated single-line isometric drawing and a double-line orthographic drawing of the same system (a hot tub heating

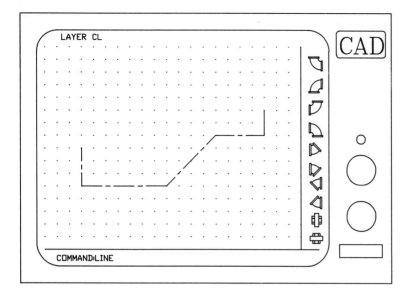

FIGURE 21–25 Lay out the centerlines of the pipes with LINE command.

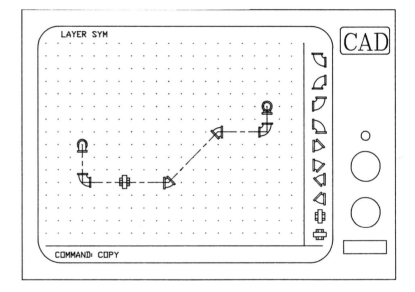

FIGURE 21–26 Call up and place the piping symbols from the LIBRARY.

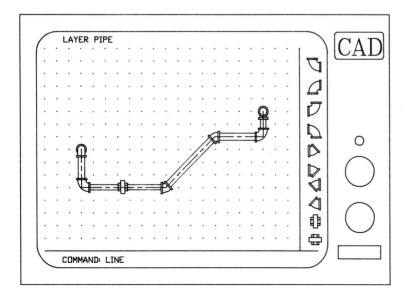

FIGURE 21–27 Draw lines for pipes with LINE command or PARALLEL LINE command.

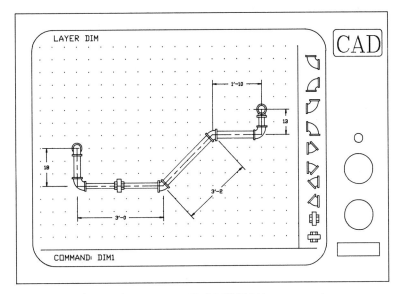

FIGURE 21–28 Add dimensions with DIMENSION command.

and filtering system) are shown in **Figs. 21–29** and **21–30. Figures 21–31** through **21–34** are further examples of computer-generated isometric and orthographic piping working drawings.

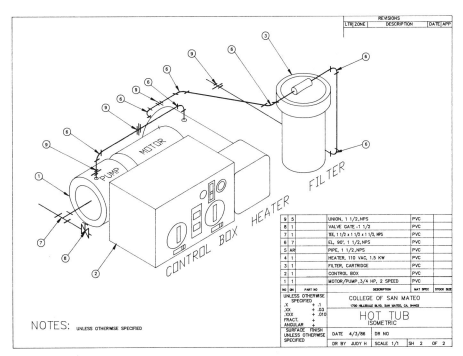

FIGURE 21–29 CAD-generated, single-line isometric piping drawing for a hot tub system (Courtesy of the College of San Mateo)

426 Drafting in a Computer Age

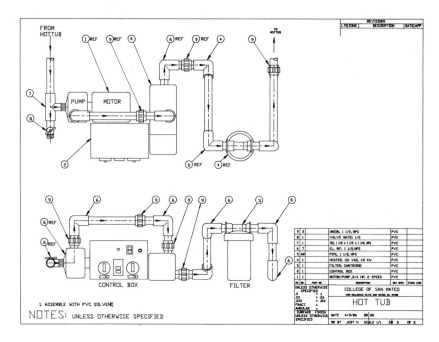

FIGURE 21-30 CAD-generated, double-line orthographic piping drawing for the hot tub system in Fig. 21-29 (Courtesy of the College of San Mateo)

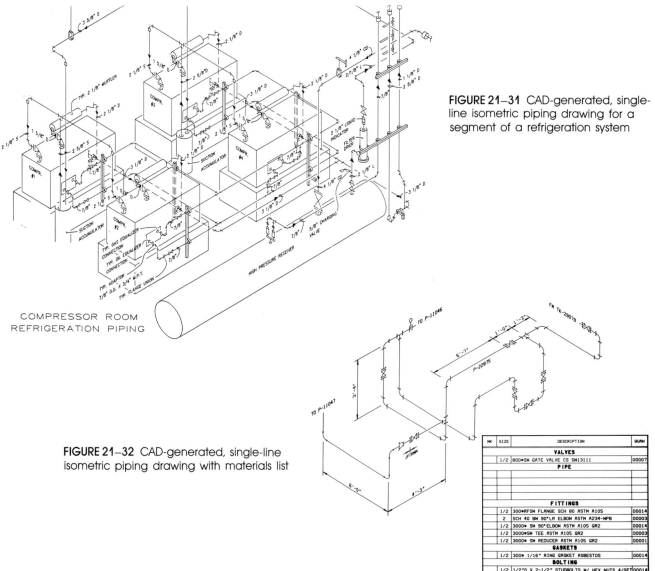

FIGURE 21-31 CAD-generated, single-line isometric piping drawing for a segment of a refrigeration system

FIGURE 21-32 CAD-generated, single-line isometric piping drawing with materials list

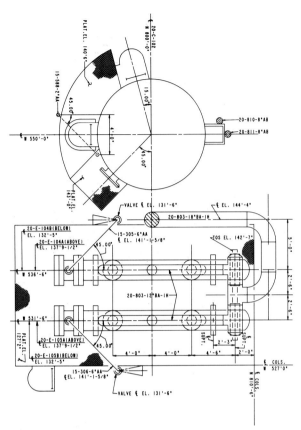

FIGURE 21-33 CAD-generated, double-line orthographic piping drawing for a segment of a steam system

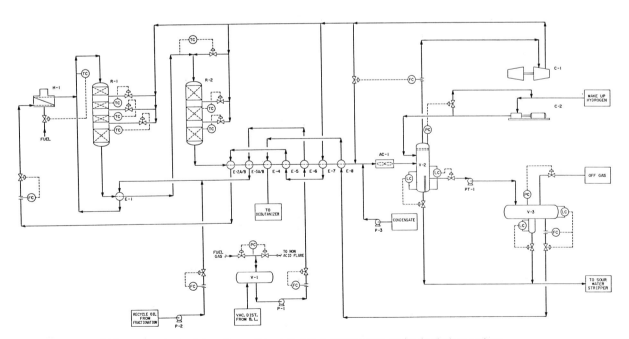

FIGURE 21-34 CAD-generated piping flow diagram for a petroleum manufacturing system

EXERCISES

1. Sketch a single-line orthographic drawing of the piping system shown in **Fig. 21–35**.
2. Sketch a double-line orthographic drawing of the piping system in Fig. 21–35.
3. Sketch a single-line isometric piping drawing of the piping system in Fig. 21–35.
4. Sketch a double-line isometric piping drawing of the piping system in Fig. 21–35.
5. Make the following ECOs to **Fig. 21–36**:
   a. Change the horizontal pipe diameter to .75".
   b. Lengthen pipe #4 to 5" and cap.
   c. Correct pipe #7 to 26" in the materials list and drawing.
   d. Correct pipes #6 (2) to 38" in the materials list.
   e. Shorten pipe #8 (top only) to 28".
   f. Update all changes on the parts list.
   g. List the ECOs on the drawing.

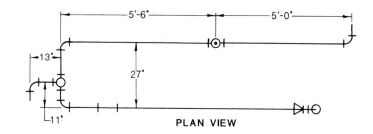
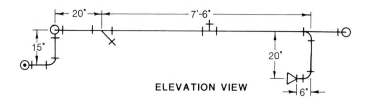

FIGURE 21–35 Draw the following, using .5" pipe and fittings throughout:
1. three-view, single-line orthographic piping drawing
2. single-line isometric piping drawing
3. double-line isometric piping drawing with materials list

FIGURE 21–36 Draw the double-line orthographic piping drawing. Make the ECOs listed in Exercise 5.

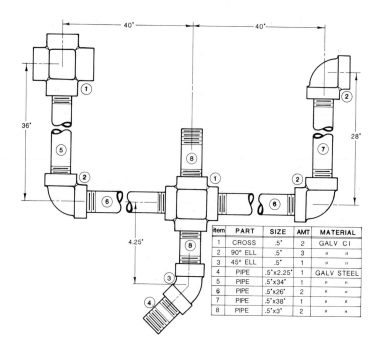

item	PART	SIZE	AMT	MATERIAL
1	CROSS	.5"	2	GALV CI
2	90° ELL	.5"	3	" "
3	45° ELL	.5"	1	" "
4	PIPE	.5"x2.25"	1	GALV STEEL
5	PIPE	.5"x34"	1	" "
6	PIPE	.5"x26"	2	" "
7	PIPE	.5"x38"	1	" "
8	PIPE	.5"x3"	2	" "

6. With instruments, draw an orthographic single- and double-line piping drawing of the sketched piping layout in **Fig. 21–37**.
7. With instruments, draw an isometric single- and double-line piping drawing of the piping layout in Fig. 21–37.
8. Using the cartesian coordinate system, lay out the pipe lines from the data table in **Fig. 21–38** on $1/4"$ grid paper. Add fittings as noted.
9. Using the cartesian coordinate system, complete the data table in **Fig. 21–39**.
10. With a CAD system, draw the 90° flanged pipe elbow in **Fig. 21–40**.
11. Design a single-line isometric pipe drawing plan for a water heater and dryer hook-up with a .5" gas pipeline for **Fig. 21–41**.

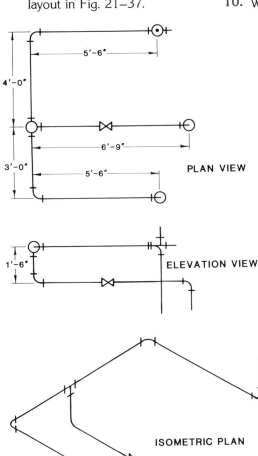

FIGURE 21–37 Draw the single-line piping drawings and their double-line counterparts with instruments or a CAD system.

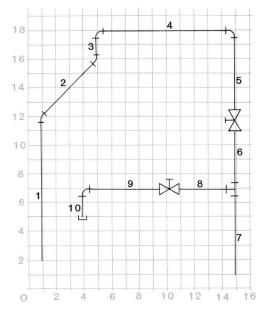

PIPE 1/2"	X	Y	go to	X	Y	FITT-ING
1	0	7	→	0	0	90° EL
2	0	0	→	9	0	45° EL
3	9	0	→	14	5	45° EL
4	14	5	→	14	7	90° EL
5	14	7	→	8	7	.5" CAP

FIGURE 21–38 Lay out the piping drawing on $1/4"$ grid paper with the coordinates from the data table. Add the fittings as noted.

1" PIPE	X	Y	go to	X	Y	FITT-ING
1	1	2	→	1	12	45° EL
2	1	12	→	5	16	45° EL
3	5	16	→	15	18	90° EL
4	15	18	→			
5						
6						

FIGURE 21–39 Complete the data table for the piping drawing.

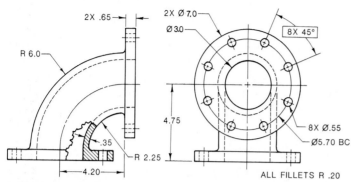

FIGURE 21-40 Draw the 90° flanged elbow with instruments or a CAD system.

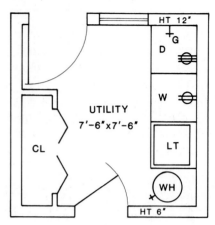

FIGURE 21-41 Design a single-line isometric piping drawing for the water heater hookup with a .5" gas pipe and dimension.

KEY TERMS

Bell and spigot connections
Cementing
Double-line piping drawing
Fittings

Flange connections
Nominal pipe size
Single-line piping drawing

Threaded connections (screwed fittings)
Valves
Welded connections

the world of CAD

Every day new horizons of understanding are opening up with the help of computers. The use of computers in problem solving is relatively new. As the art of computer science has progressed, many individuals have adapted their machines to find solutions for many situations. James Edward Fuller is one of these people who have used the computer as a problem solver.

Mr. Fuller first became involved with computers as a student of architecture at the University of Tennessee. He applied the computer to the problems of structural design. This computer was a time-share mainframe computer with a remote teletype terminal. Data storage was achieved with punch tapes. There was no monitor to look into; only a constant printout on paper fed from a roll behind the teletype. The use of the machine allowed a designer to experiment with different types of structural configurations. It spewed out data in an engineering format consisting of moments, tensile, compressive, and other force measurements pertaining to the structural members.

Several years later, Mr. Fuller established his architectural practice. His fascination with computers still intact, he began to explore ways to use these machines to solve problems in his office. With an engineering partner, a contract was secured to perform energy audits on large-scale buildings and complexes. A minicomputer was used for the extensive calculations required in work of this type.

As Fuller's practice grew, producing the plans and specifications for his designs became a bottleneck in the process of creating structures. The answer to the specifications was a word processing system. Specifications, the collection of written technical information on the individual components of a construction project, are bound into a booklet that may contain from 100 to 600 pages.

Now the problem of producing plans had to be solved. CAD, which had been around for many years, was intriguing to Mr. Fuller. However, it was only available on expensive mainframe systems and its capabilities were not extensive. Mr. Fuller needed a system that was cost efficient, flexible, and could be programmed for specific uses.

The computer industry was just introducing a new type of computer—the personal computer. These microcomputers were unique because, unlike mainframe and minicomputers, the entire computer was at the disposal of a single user. In addition, the cost was very attractive. Several companies had begun to write CAD programs for use on these machines. After researching the different options, Mr. Fuller decided on a system based on the IBM PC microcomputer, AutoCAD software for two-dimensional drawings, and MicroCADD software for three-dimensional drawings. This was in 1980 (long ago in computer history) and CAD systems for microcomputers were not packaged as they are

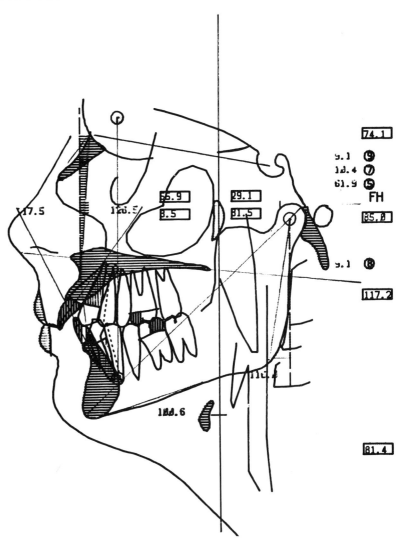

FIGURE 1 A CAD system was used to determine what section of bone a dental surgeon should remove in order to reposition a patient's "bite."

today.

Mr. Fuller set out to purchase the components necessary to assemble his systems. Starting with two systems, he ordered computers, disk drives, math coprocessor chips, a plotter, digitizing tablets, software, and a myriad of other necessities. When it all arrived, there were no cables available to connect the equipment. After much study, cables were made, and after that, made to work! It was time to draw.

The new way of drafting took some adjustment and quite a bit of setup time. The software was new and was not the immediate solution as was hoped. Mr. Fuller and his staff went to work in their off time to develop the system into a usable tool. Standards were set, symbol libraries were developed, techniques were honed. A cartridge disk subsystem was purchased to solve the problem of massive file storage. After many months of work, a system was in place to solve the problem of producing design and technical drawings.

In the following years, the system was refined, upgraded, and added to many times. The capabilities to solve many problems were now possible. The system was first used for its intended purpose; to produce architectural drawings. However, Mr. Fuller began to spend more time discovering ways to apply the computer to different types of problems. And as he found these ways, people with those problems found him and his computer.

A dental surgeon needed to chart a section of bone to remove from a patient's jaw bone so that his "bite" could be repositioned. The problem was solved by using a backlit digitizer pad to digitize (electronically "trace") an X-ray of the patient's jaw bone. A section was removed and the remaining bone was repositioned in the drawing to show the results. Careful measurements were used to determine the exact distances and angles to cut, Figure 1.

An electronics engineer was trying to create a device that would use a rotating cam to activate several switches and different points in the cam's rotation. He had constructed several models over a period of days. After each model was tested, a new problem was discovered. Mr. Fuller created a drawing of the cam and positioned switches around it. Rotating the cam in the drawing and repositioning the switches led to a new cam design that solved the problem. The time to discover this new design was not days or weeks, but only 20 minutes.

As a college professor, Mr. Fuller has directed students in a number of interesting projects. One of these involved developing a system of criminal identification. Drawings of noses, eyes, hair, and other face parts were used to create the face of a suspect. Another project was developed to manage the college facilities from CAD drawings and an internal data base. The data base maintained records that were used by the facilities management department.

After a few years, Mr. Fuller began to phase out his architectural practice and began to use computers to solve unique problems. He has written books on the use of computer-aided designs that are currently used in college and university classrooms all over the country. The books are written with a word processor, artwork is produced with CAD and scanned on a computerized optical scanner, the type is set with a computer publishing system, and the pages are printed by the computer on a laser printer. The use of computers to produce the book solves the problem of creating books in a timely fashion. He is currently involved in producing computer video training tapes and research in electronic computer imaging.

(Courtesy of James Edward Fuller, Morristown, Tennessee. Mr. Fuller is the author of successful books such as USING AUTOCAD®· 3rd Edition, Release 10 with 3-D and USING VERSACAD®· both published by Delmar Publishers Inc.)

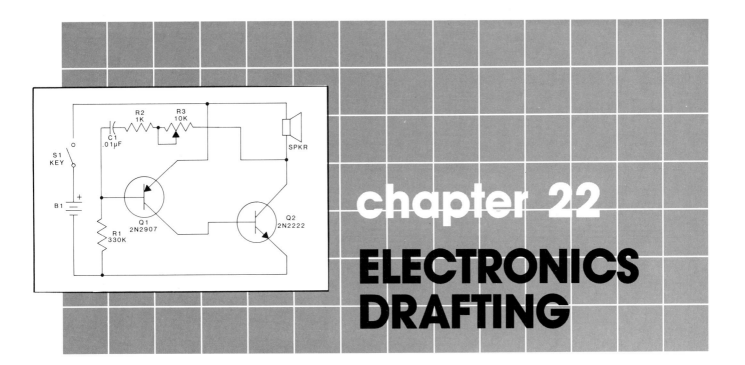

chapter 22
ELECTRONICS DRAFTING

OBJECTIVES

The student will be able to:

- identify the block diagrams, schematic diagrams, connection diagrams, logic diagrams, and cableform diagram
- differentiate between schematic and simple outline electronic symbols
- with a CAD system or manually, draw a block diagram, schematic diagram, connection diagram, logic diagram, cableform diagram, and a chassis pattern
- name the associations that set electronic standards
- understand how sophisticated CAD systems are used to create electronic drawings

INTRODUCTION

The electronics industry is in a virtual explosion of expansion and advancement in research, product development, and manufacturing. Extremely rapid growth has occurred in such high-tech areas as computers, telecommunication, aerospace, automotive, consumer products, and industrial instrumentation **(Figs. 22–1A** through 22–1D). The electronics industry touches almost every facet of our daily lives.

This expanding electronics industry is creating an increasing demand for qualified electronics drafters to produce drawings manually and with computer-aided drafting systems. Before electronics products are produced, drawings are needed to direct the manufacture, assembly, and packaging of all component parts. In addition, drawings provide necessary information for the installation, control, repair, and maintenance of every product. Electronics drafters are a vital communications link in this process.

ELECTRONICS INDUSTRY STANDARDS

Electronics drafting uses a broad combination of governmental and professional association standards in the development of drawings and diagrams. This standardization im-

FIGURE 22–1A Desktop calculator (Courtesy of Monroe Systems for Business)

432 Drafting in a Computer Age

FIGURE 22–1B Aerial targets can simulate enemy aircraft, sea-skimming cruise missiles, and antiship missiles, and are used to train U.S. Navy personnel. (Courtesy of Northrup Corporation)

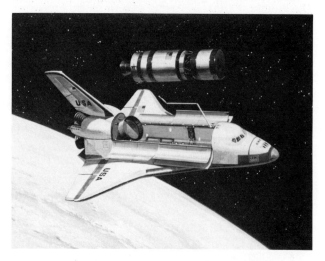

FIGURE 22–1C Space shuttle (Courtesy of NASA)

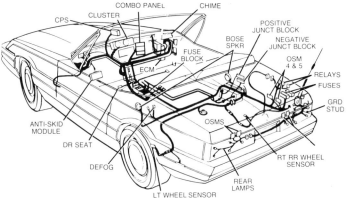

FIGURE 22–1D Pictorial drawing of an automobile's electrical system

proves the efficiency and minimizes the potential errors in the interpretation of drawings. Standards cover the uniform selection, formation, symbolism, and reference notations used for all electronics and electrical circuits, equipment, and devices. Standardization for electronics and electrical drawings are prepared by the following organizations:

- **ANSI**—American National Standard Institute
- **ASA**—American Standards Association
- **IEC**—International Electrotechnical Commission
- **EIA**—Electronics Industries Association
- **IEEE**—Institute of Electrical and Electronics Engineers
- **MIL STD**—Military Standards

ELECTRONICS DRAWING SYMBOLS

Graphic symbols (electronics shorthand) play a major role in preparing **electronics drawings.** The use of symbols is a much faster drafting method than using orthographic views, pictorials, or notations. Because of the complexity of electronic circuits and devices, the use of graphic symbols is also the most accurate method of preparing electronics and electrical drawings. The most universally accepted set of electronics symbols is the ANSI Y32.2 standard symbology, as shown in Figs. 22–12 and 22–13, page 439. In using electronic symbols to represent electronic circuit devices and components, use the following guidelines:

1. Layout circuits should read from left to right. Complex circuits should read from upper left to lower right.
2. Symbols can be drawn at any size, but should be approximately the same size and proportional to the size of the drawing.
3. Symbol directions have no significance and component symbols may be rotated to any direction on the drawing.

4. Avoid line (circuit path) crossovers wherever possible.
5. Keep the length of circuit lines and distance between components to a minimum.
6. Different and discrete parts of a diagram need not adhere to the same scale.
7. Although angular lines have no significance, avoid wherever possible.
8. If additional data is needed to describe a symbol, standard abbreviations should be used next to the symbol.
9. Lay out circuit paths with symbols placed in logical order as the circuit functions, from power source to loads.
10. Use vertical or horizontal lines wherever possible. Avoid using angular lines.
11. If lines are interrupted and progress to another circuit or drawing, identify the lines and indicate where they are connected on the original and receiving drawing.

Most electronic drawings are prepared with pencil lines. However, since many drawings are used in technical manuals, ink drawings are used to produce high-quality lines for this type for reproduction. See Chapter 3 for inking information.

TYPES OF ELECTRONICS DRAWINGS

There are many types of electronics drawings. Most start as sketches or written instructions from an electronics engineer or designer. The electronics drafter interprets these instructions and graphically transforms them into working drawings. Each type of electronic drawings serves a specific purpose in describing the design, construction, fabrication, maintenance, or repair of a product. It is the responsibility of the electronics drafter to select the best type or combination of drawings to represent each design.

Block Diagrams

A **block diagram** is a general presentation drawing used to simplify and show the interrelationship of the various parts of an entire electronics system. Block diagrams are often used in the first planning and design stages because they are a fast and uncomplicated method of recording general design ideas. Circuit details are omitted from block diagrams.

Blocks in the shape of rectangles, squares, or triangles are joined by a single line with arrows. Arrows show the direction of current flow. The direction of block diagram design progresses from left to right and the function of each block is lettered within the block. **Figure 22–2A** shows a simple block diagram for a crystal radio. Examples of a block diagram for a television and a radio are shown in **Figs. 22–2B and 22–2C.**

Schematic Diagram

Schematic diagrams are the most frequently used drawings in the electronics industry. They show the connection and function of each circuit in a simplified form **(Fig. 22–3A)**. Nevertheless, **schematic diagrams** show more detail than block dia-

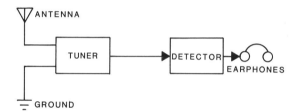

FIGURE 22–2A Block diagram of a crystal set

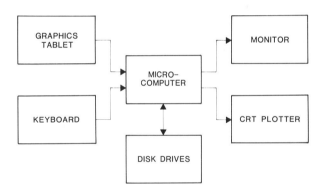

FIGURE 22–2B Block diagram of a microcomputer CAD system

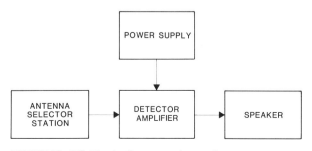

FIGURE 22–2C Block diagram of a radio

434 Drafting in a Computer Age

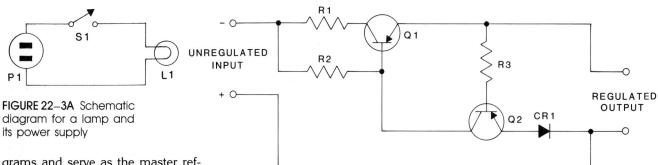

FIGURE 22–3A Schematic diagram for a lamp and its power supply

FIGURE 22–3B Schematic for a regulated power supply

grams and serve as the master reference for the preparation of production drawings, parts lists, and component specifications. They contain graphic symbols, connections, and noted functions to describe each specific circuit arrangement. Schematic diagrams also show the sequence in which components are connected to complete a circuit. Each component is represented by its own special schematic symbol, and is drawn to show its relationship to other components **(Fig. 22–3B)**. Schematic diagrams are sometimes called **elementary diagrams**. The following are general rules for drawing schematic diagrams:

1. Current flows from left to right, or upper left to lower right.
2. Allow space between symbols and connectors to avoid crowding.
3. Reference callouts are positioned in the same location, next to all symbols.
4. Keep the crossing of connector lines to a minimum.
5. If scale is a factor, use a grid overlay to avoid excessive dimensioning.

Connection Diagrams

Connection or **wiring diagrams** are electronic assembly drawings which contain details needed to make or trace all electronic connections.

This type of diagram shows precisely where components are located in a circuit and how wires are connected. The components are drawn pictorially or as single orthographic views. Connection diagrams are supplementary drawings used with a schematic diagram for installing and repairing electronic equipment. Types of connection diagrams include point-to-point, highway or trunkline, baseline or airline, and cableform diagrams.

Point-to-Point Diagrams. **Point-to-point diagrams** show components, drawn as simple outlines, with connecting lines and terminals. These drawings are used by technicians for the assembly of electrical systems. An example of a point-to-point diagram is the motorcycle wiring diagram shown in **Fig. 22–4**. Automotive manuals frequently contain point-to-point diagrams for troubleshooting purposes.

Highway Diagrams. Highway diagrams are preferred when the addi-

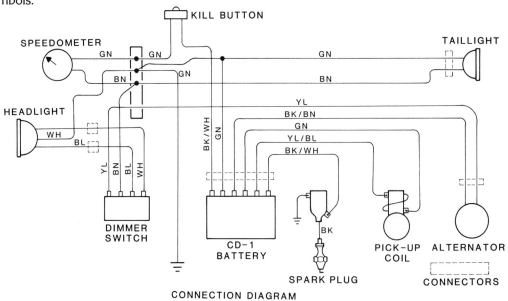

FIGURE 22–4 Point-to-point connection diagram for a dirt bike

tion of a large number of point-to-point interconnecting lines would be excessive. **Highway** or **trunkline diagrams** are similar to point-to-point diagrams, except all wires are joined to a thick line called a highway or trunkline. Wires are separated where they reach the component to which the wire is connected. The long, thick line is called a **path** or highway, and shorter lines are called **feeders (Fig. 22–5A)**. To show where connections are made, wire data is labeled on feeder lines. **Figure 22–5B** shows a highway diagram for a meter box with the appropriate wire data. This wire data includes the wire destination (hookup), wire color, and wire size. EIA color code standards are as follows:

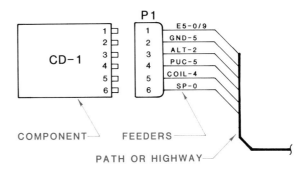

FIGURE 22–5A Highway diagram's feeders and paths

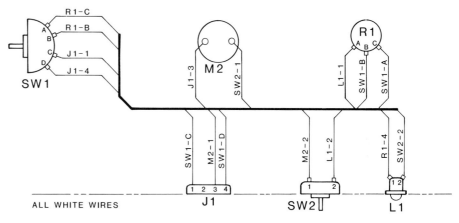

FIGURE 22–5B Highway diagram for a meter box with wire destinations

Wire Number	Color
0	Black
1	Brown
2	Red
3	Orange
4	Yellow
5	Green
6	Blue
7	Violet
8	Grey
9	White

Baseline Diagrams. **Baseline** or **airline diagrams** are similar to highway or trunkline diagrams. The connection lines are theoretical and do not depict the true size and form of the connectors. The baseline diagram is a drawing shortcut procedure to allow the repositioning of components so the connectors can be shortened and drawn conveniently. All the feed lines meet at a right angle to the baseline which is always vertical or horizontal **(Fig. 22–6)**. The use of a wire data chart will list the connection points for each wire connector and their sizes and colors.

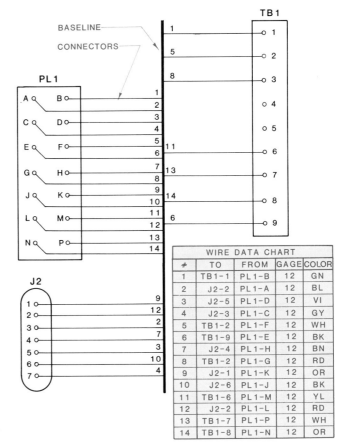

FIGURE 22–6 A baseline diagram with a wire data chart

Cableform Diagrams. A **cable** is a group of wires bound together **(Fig. 22-7)**. A **cableform diagram** shows where each wire leaves the cable harness to make a connection **(Fig. 22-8)**. Line corners are drawn rounded to simulate cables and wires. Information such as color, gage, length, and type of connectors are also included on cableform diagrams.

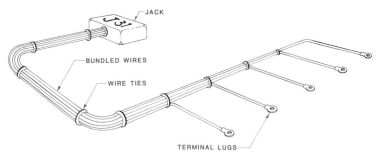

FIGURE 22-7 Cable (bundled wires) with jack, ties, and terminal lugs

Logic Diagrams

Logic diagrams are used to design the microcircuitry used in many contemporary devices such as digital watches, calculators, and computers. A logic diagram is a specialized type of block diagram showing circuitry flow details. Blocks of specific shapes are used to represent different logic functions **(Fig. 22-9)**.

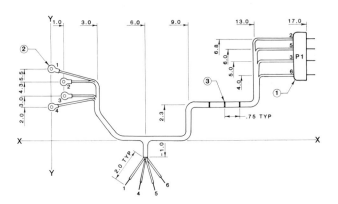

FIGURE 22-8 Cable form diagram

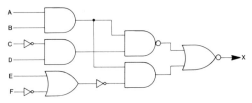

FIGURE 22-9 Simple logic diagram

Chassis Drawings

Chassis drawings are pattern or development drawings of cabinets which house electronic circuits and equipment. Chassis drawings include overall dimensions and stretchout views of a chassis. They also include the location and size of all holes and other openings needed to mount components and provide openings for wires **(Figs. 22-10A and 22-10B)**.

Printed Circuits

Printed circuits are contemporary replacements for wired circuits. One component is electrically connected to another through printed copper conductors on an insulated board, called a **printed circuit board (PCB)**. Printed circuits eliminate wiring faults and errors, reduce production costs, and enable complex circuits to be miniaturized.

Printed circuit drawings are prepared at full scale or larger, and represent the exact layout of a circuit.

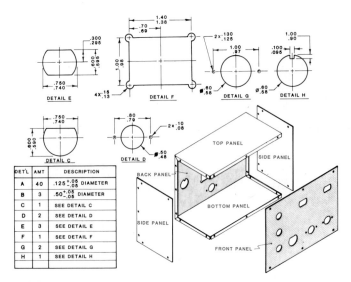

FIGURE 22-10A Cutout dimensions for a chassis pattern

Chapter 22 Electronics Drafting

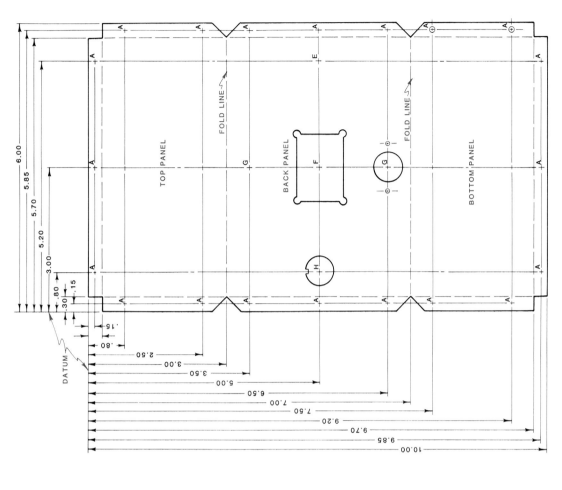

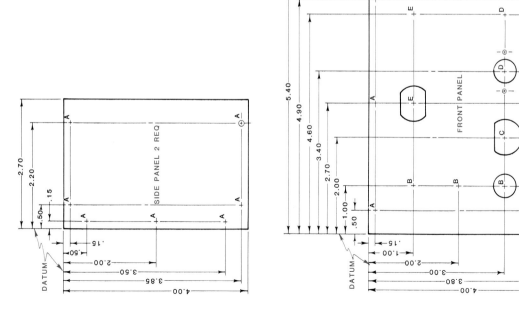

FIGURE 22-10B A chassis pattern's working drawing using rectangular coordinate dimensioning

After the drawings are photographically reproduced and reduced in size, they are used in the etching process to create printed circuit boards. Many drawings are necessary to design and create the documentation necessary to produce a printed circuit board **(Figs. 22–11A through 22–11E)**.

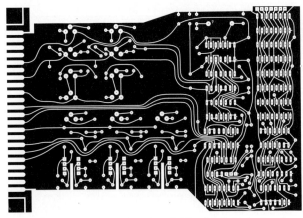

FIGURE 22–11A Example of a finished printed circuit board layout

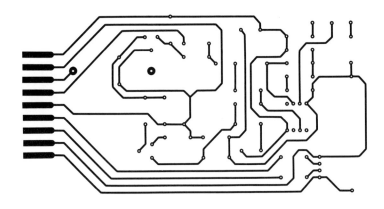

FIGURE 22–11B Example of a tape-up layout for a printed circuit board

FIGURE 22–11C Both sides of a printed circuit board with components installed

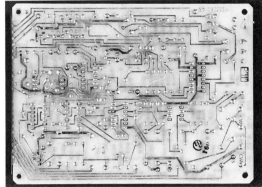

FIGURE 22–11D Solder side of a printed circuit board

FIGURE 22–11E Printed circuit board with components for a digital display

STANDARD SYMBOLS CHARTS

There are two types of standard symbols for electronics components on connecting diagrams—**schematic symbols** and **simple outline symbols (pictorial) (Fig. 22–12)**. Different symbols are used to represent logic diagrams **(Fig. 22–13)**.

DRAWING SCHEMATIC DIAGRAMS

Schematic diagrams contain schematic symbols, connecting lines, **reference designations** (part numbers) and **value designations.** Procedures for drawing a simple schematic diagram are shown in Figs. 22–14A, through 22–14D, and include the following steps:

COMPONENT	REFER. DESG	SCHEMATIC SYMBOL ANSI Y32.2	PICTORIAL	COMPONENT	REFER. DESG	SCHEMATIC SYMBOL ANSI Y32.2	PICTORIAL
ANTENNA	E			FUSE	F		
BATTERY	BT			SWITCH, OPEN CONTACT	S		
RESISTOR	R			CONNECTOR	P/J		
RESISTOR, VARIABLE	R			JACK, MALE	J		
POTENTIOMETER	R			JACK, FEMALE	J		
CAPACITOR, GENERAL	C			SPEAKER	LS		
CAPACITOR, VARIABLE	C			TRANSFORMER, MAGNETIC CORE	T		
DIODE, GENERAL	CR			HEADSET, DOUBLE	T		
DIODE, BRIDGE	CR			MOTOR	B		
TRANSISTOR PNP	Q			GROUND, EARTH	—		
TRANSISTOR NPN	Q			GROUND, CHASSIS	—		

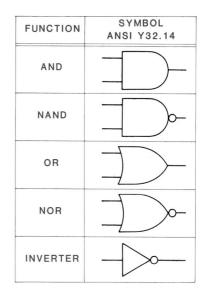

FIGURE 22-13 Logic diagram symbols

FIGURE 22-12 Schematic diagram symbols with their pictorial representations

Step 1. Place power input to the left and power output to the right **(Fig. 22-14A)**.

FIGURE 22-14A Locate the power input on the left side and power output on the right side.

Step 2. Place major components from left to right across the diagram **(Fig. 22-14B)**.

Step 3. Arrange minor components to eliminate crossed lines and keep the flow correct **(Fig. 22-14C)**.

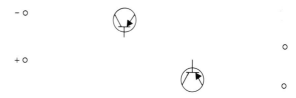

FIGURE 22-14B Locate the major components.

Step 4. Place the reference designations and value designations **(Fig. 22-14D)**.

Electronics drawing templates greatly speed the lay out and drawing of schematic diagrams **(Fig. 22-15)**.

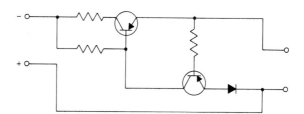

FIGURE 22-14C Locate the minor components and connecting lines (also called wires or traces).

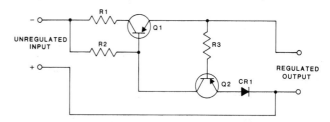

FIGURE 22-14D Place reference and value designations to complete the schematic diagram.

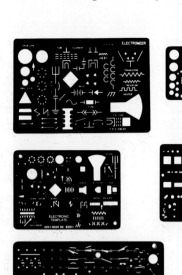

FIGURE 22–15 Examples of templates used by electronics drafters (Courtesy of Koh-I-Noor)

FIGURE 22–16A Draw a chassis outline.

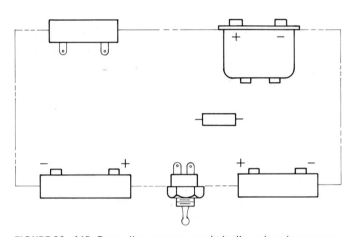

FIGURE 22–16B Draw the components in the chassis as per location.

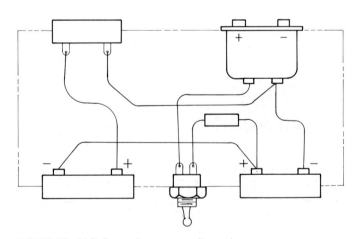

FIGURE 22–16C Draw the connecting wires.

DRAWING CONNECTION DIAGRAMS

In a connection diagram, the placement of components (simple outlines) are determined by their relative location on the chassis, rather than by electron flow. Lines which connect components represent wires. It is common to draw the wires with corner radii and color labels. Procedures for drawing a point-to-point connection diagram for a charging system tester are shown in Figs. 22–16A through 22–16D, and include the following steps:

Step 1. Draw the outline of the chassis **(Fig. 22–16A)**.

Step 2. Place the components. Arrange them as they should be located in the chassis **(Fig. 22–16B)**.

Step 3. Draw the connecting wires **(Fig. 22–16C)**.

Step 4. Label the reference designations on the components and label wire colors **(Fig. 22–16D)**.

DRAWING LOGIC DIAGRAMS

A logic diagram is a special type of block diagram. It consists of logic symbols, connecting lines, and input/output letters. Procedures are shown in Figs. 22–17A through 22–17C, and include the following steps:

Step 1. Arrange logic symbols so they are evenly spaced and the system flows smoothly from left to right **(Fig. 22–17A)**.

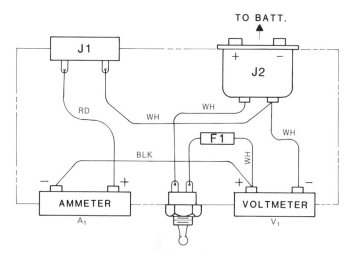

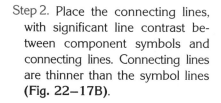

FIGURE 22–16D Complete the connection diagram with labels and designations.

Step 2. Place the connecting lines, with significant line contrast between component symbols and connecting lines. Connecting lines are thinner than the symbol lines (Fig. 22–17B).

Step 3. Place input/output letters on the diagram (Fig. 22–17C).

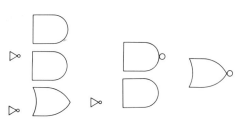

FIGURE 22–17A Lay out and draw the logic symbols.

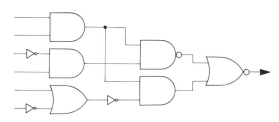

FIGURE 22–17B Draw the connecting lines.

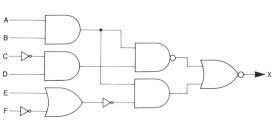

FIGURE 22–17C Place input and output letters on the drawing.

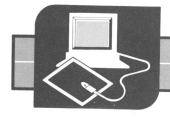

COMPUTER-AIDED DRAFTING & DESIGN

COMPUTER-AIDED ELECTRONICS DRAFTING

Electronics drafting was one of the first drafting fields to use the computer. This is primarily because early designers of CAD software were familiar with electronics and electronic diagrams. In addition, electronics diagrams contained many standard symbols and repeated details, and did not require complex dimensions or accurate geometric data bases. Computer-aided electronics drafting has understandably experienced more growth than any other field of drafting.

CAD-system software, now available for all types of computers, includes extensive design capabilities, especially in printed circuit board design. As a CAD operator develops schematic or logic diagrams on a computer, circuit information is stored in the computer's data base. This information is processed by printed circuit board software to locate the components on the board and to automatically complete the circuit routing.

Operators may use standard editing commands, such as MOVE, COPY, and ROTATE to modify designs as desired. The computer then checks the circuit to verify if it matches the schematic diagram. If the circuit does not match the diagram, the operator may ask the computer to review the board to match the schematic, or may revise the schematic to match the board.

Some of the major features of sophisticated electronics drawing CAD software are:

- automatic placement of components,
- automatic routing of circuit paths,
- use of multiple colored layers,
- use of multiple PCB trace widths and pad sizes,
- automatic circuit and continuity checking capabilities,
- creation of artwork for the manufacture of the PCB,
- parts list printouts, and
- creation of assembly drawings.

On a simpler level, electronics drawings can easily be created with any CAD system. Since electronic symbols are standardized, they can be drawn and made into blocks or tablet menu icons. Prepared electronics symbols may be purchased as software, saving the time needed to make your own symbols and file them in the library. Grids on a monitor also aid in the placement of components and text. Figures 22–18 through 22–22 illustrate the following sequence of creating schematic diagrams using standard CAD software:

Step 1. Create and draw standardized symbols **(Fig. 22–18)** and save them as blocks, or use purchased software icons with a tablet.

Step 2. Turn on GRID to aid in the placement of symbols **(Fig. 22–19)**.

Step 3. Locate the symbols **(Fig. 22–20)**. This placement need not be perfect at the start because it is easy to move symbols at any time with a CAD system to conform with the design.

Step 4. Place the connectors (wires) and move the symbols as necessary **(Fig. 22–21)**.

Step 5. Use the TEXT command to label the diagram **(Fig. 22–22)**.

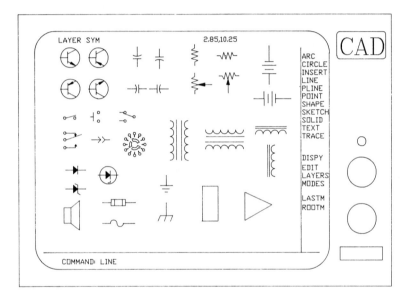

FIGURE 22–18 Create or purchase a standard electronics symbol library for your CAD system.

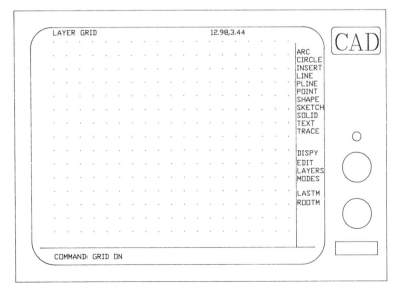

FIGURE 22–19 Select the GRID command and set grids for symbol placement.

Chapter 22 Electronics Drafting   443

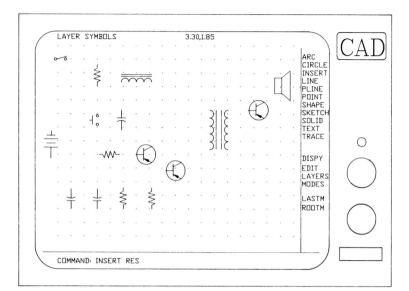

FIGURE 22-20 Place and position the symbols.

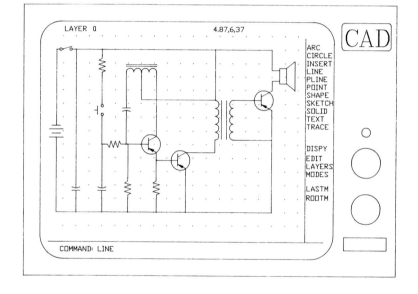

FIGURE 22-21 Place the connectors. Reposition the symbols, if necessary.

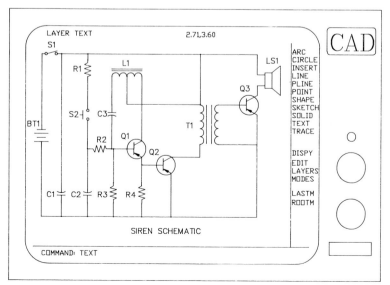

FIGURE 22-22 Add text.

EXERCISES

1. Sketch the electronics symbols in **Fig. 22–23**.
2. With drafting instruments, draw the electronics symbols in Fig. 22–23.
3. With a CAD system, draw the electronics symbols in Fig. 22–23.
4. With drafting instruments or a CAD system, draw the schematic diagrams in **Figs. 22–24** through **22–31**.
5. With drafting instruments or a CAD system, draw the block diagrams in Figs. 22–2A, 22–2B, and 22–2C (page 433).
6. With drafting instruments or a CAD system, draw the connection diagrams in Figs. 22–4 (page 434), 22–6 (page 435), and 22–16D (page 441).
7. With drafting instruments or a CAD system, draw the logic diagrams in Figs. 22–9 (page 436) and 22–17C (page 441).
8. With drafting instruments or a CAD system, draw the cable-form diagram in Fig. 22–8 (page 436).

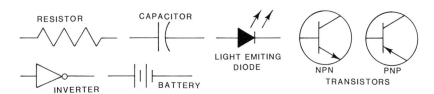

FIGURE 22–23 Practice drawing the electronics symbols.

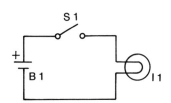

FIGURE 22–24 Draw the schematic diagram for the flashlight.

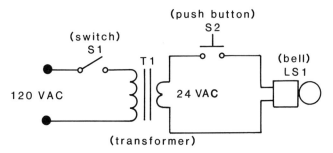

FIGURE 22–26 Draw the schematic diagram for the residential doorbell.

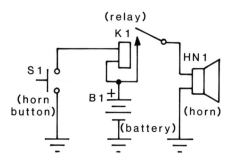

FIGURE 22–25 Draw the schematic diagram for the automobile horn.

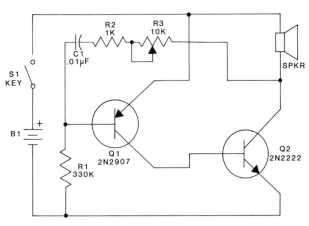

FIGURE 22–27 Draw the schematic diagram for the code practice oscillator.

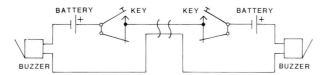

FIGURE 22–28 Draw the schematics diagram for the two-station telegraph.

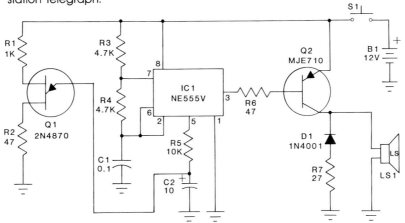

NOTE: RESISTANCE IN OHMS, CAPACITANCE IN MICROFARADS

FIGURE 22–29 Draw the schematic diagram for the electronic siren.

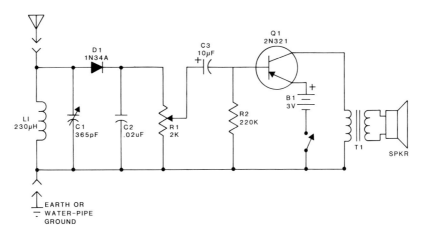

FIGURE 22–30 Draw the schematic diagram for the crystal set with audio amplifier and speaker.

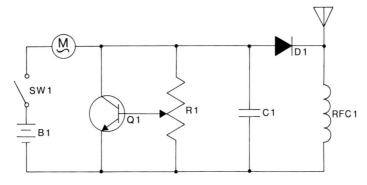

FIGURE 22–31 Draw the schematic diagram for the sensitive field-strength meter.

9. With drafting instruments or a CAD system, draw the chassis pattern in Fig. 22–10B (page 437).

10. With the data table information in Fig. 22–32, draw the figure on ¼" grid paper.

11. Complete the data table for the horn symbol in Fig. 22–33.

12. Complete the data table for the resistor symbol in Fig. 22–34.

COORDINATE DATA TABLE					
LN	X	Y	go to	X	Y
1	1	3	→	4	3
2	4	3	→	4	4
3	4	4	→	6	3
4	6	3	→	4	2
5	4	2	→	4	3
			new start		
6	6	4	→	6	2
			new start		
7	6	3	→	10	3
			FINISH		

FIGURE 22–32 Draw the illustration on ¼" grid paper from the coordinates in the data table.

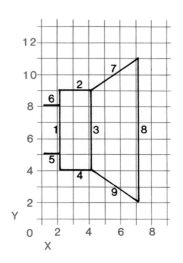

COORDINATE DATA TABLE					
LN	X	Y	go to	X	Y
1	2	4	→	2	9
2	2	9	→	4	9
3	4	9	→		4
4			→		
5					

FIGURE 22–33 Complete the data table for the horn symbol.

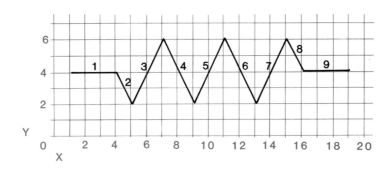

COORDINATE DATA TABLE					
LN	X	Y	go to	X	Y
1	1	4	→	4	4
2	4	4	→	5	2
3	5	2	→		
4			→		
5					
6					

FIGURE 22–34 Complete the data table for the resistor symbol.

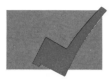

KEY TERMS

Airline diagrams (baseline diagrams)
ANSI (American National Standards Institute)
ASA (American Standards Association)
Block diagrams
Cable
Cableform diagrams
Chassis drawings
Connection diagrams (wiring diagrams)
Electronics drawings
EIA (Electronics Industries Association)
Feeder
Highway diagrams (trunkline diagrams)
IEEE (Institute of Electrical and Electronics Engineers)
IEC (International Electrotechnical Commission)
Logic diagrams
MIL STD (Military standards)
Path
Point-to-point diagrams
Printed circuit board (PCB)
Reference designations
Schematic diagrams (elementary diagrams)
Schematic symbols
Simple outline symbols
Value designations

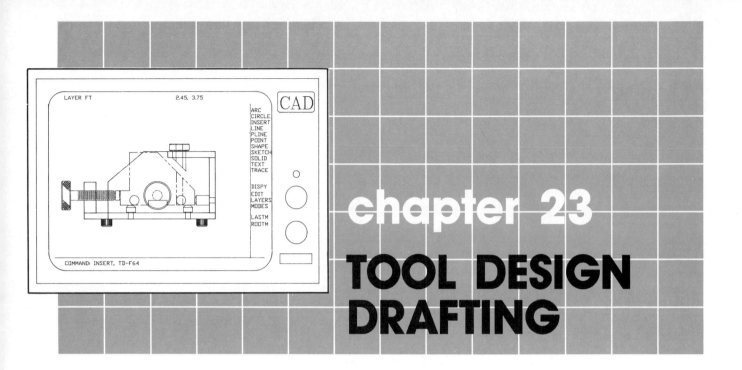

chapter 23
TOOL DESIGN DRAFTING

OBJECTIVES

The student will be able to:

- read jig and fixture working drawings
- design a simple jig
- design a simple fixture
- complete a working drawing for a simple jig
- complete a working drawing for a simple fixture

INTRODUCTION

To compete successfully in today's marketplace, industrial products must be accurately and economically produced. This requires effective tool engineering and design. **Tool engineering** is the branch of mechanical engineering responsible for the design and development of tools and processes needed in the mass production of products. This includes the planning, designing, and detail drawing of all tools and devices used in the manufacturing process.

Tool designers must possess a good working knowledge of all machine tool operations. Since tool designers must communicate ideas and describe tools in great detail, a thorough understanding of working drawings and tolerancing is also mandatory. Tool engineers, tool designers, and tool design drafters are all involved at different levels of the machine tool operations and have the following responsibilities:

1. Control production costs by designing tools for rapid production.
2. Design special tools for the manufacture of complete parts.
3. Design tools that aid in the manufacture of the highest quality product within the established limits.
4. Select the proper materials for the manufacture of products.
5. Design tools that require minimum maintenance.
6. Simplify manufacturing and assembly processes.
7. Design tools that are safe under all manufacturing conditions.
8. Design tools that are fault-free and reduce the potential for human error.
9. Design tools that produce accurate interchangeable parts within acceptable tolerances.

To help fulfill these responsibilities, many devices are used to securely hold, guide, move, and position workpieces through a wide variety of manufacturing processes. Most are classified as either jigs or fixtures.

JIGS AND FIXTURES

Jigs are used in the machining and assembly of products (Fig. 23–1). A **jig** is a device designed to position, hold, and support a workpiece in a fixed position during a machining operation. In addition, a jig holds or guides cutting tools through prescribed paths. A jig may be fastened to a worktable or can be hand held. **Figure 23–2** shows a jig used to hold a workpiece during a drill press operation.

Jigs used in product assembly are designed to hold separate components in fixed positions while final connections are completed. This may involve the use of all types of fasteners, weldments, or adhesives.

Fixtures, like jigs, hold and support workpieces. However, **fixtures** are fixed to a worktable or a machine surface to hold a workpiece rigidly. Fixtures do not hold or guide cutting tools. They are generally larger than jigs and are used in milling, grinding, and honing operations. **Figure 23–3** shows a typical numerically controlled machining center fixture.

Chapter 23 Tool Design Drafting 449

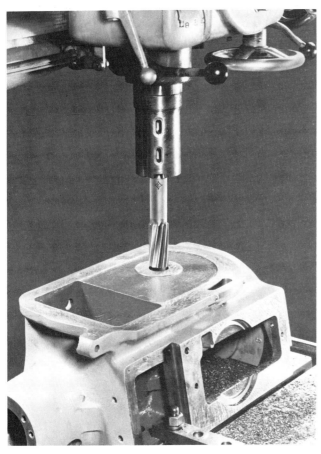

FIGURE 23-1 Tool design use in industry (Courtesy of Cleveland Twist Drill Company)

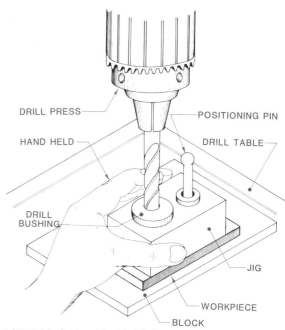

FIGURE 23-2 Hand-held drill jig

FIGURE 23-3 A typical numerically controlled machining center fixture (Courtesy of Cincinnati Milacron Inc.)

STANDARD PARTS

Most jigs and fixtures use standard components. Only when a tool design problem cannot be solved with a standard component will a tool designer use a unique type, size, and configuration of materials. Standard parts in the design and fabrication of jigs and fixtures include locators, feet and rest buttons, clamps, drill bushings, pins, screws, V-blocks, plates, and aluminum or steel shapes. When cast shapes are used, all burrs and sharp edges are removed to protect machine operators.

Locators

Locators are devices, such as pins, dowels, buttons, and pads, used to position a workpiece on a machine surface or worktable. **Figure 23–4** shows the design and size options for standard spherical radius locator buttons. This type of locator is normally positioned with the workpiece touching the tip of the button **(Fig. 23–5)**. This locator also can be used with the workpiece designed to touching the flattened side of the locator **(Fig. 23–6)**. Note the location and direction of forces in relation to the locators in Fig. 23–6.

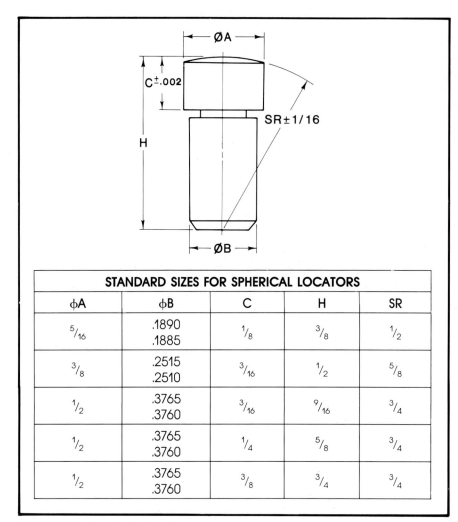

FIGURE 23-4 Standard spherical locators

STANDARD SIZES FOR SPHERICAL LOCATORS				
ɸA	ɸB	C	H	SR
5/16	.1890 .1885	1/8	3/8	1/2
3/8	.2515 .2510	3/16	1/2	5/8
1/2	.3765 .3760	3/16	9/16	3/4
1/2	.3765 .3760	1/4	5/8	3/4
1/2	.3765 .3760	3/8	3/4	3/4

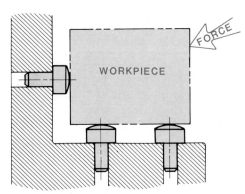

FIGURE 23-5 Positioning workpiece against ends of locators

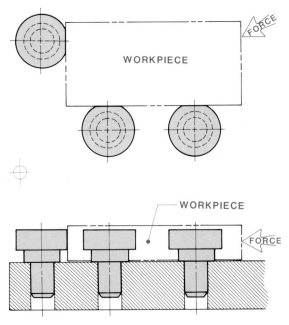

FIGURE 23-6 Positioning workpiece against flattened sides of locators

If a workpiece contains holes, locators should be positioned to align with hole centers. **Figure 23–7A** shows the alignment of bullet nose locators with the holes in a workpiece. Figure 23–7A also contains standard dimensions for bullet nose locators. **Figure 23–7B** shows standard diamond pin locator dimensions.

Feet and Rest Buttons

Feet are used to elevate jigs or fixtures from a machine surface. **Feet** and **rest buttons** are designed to

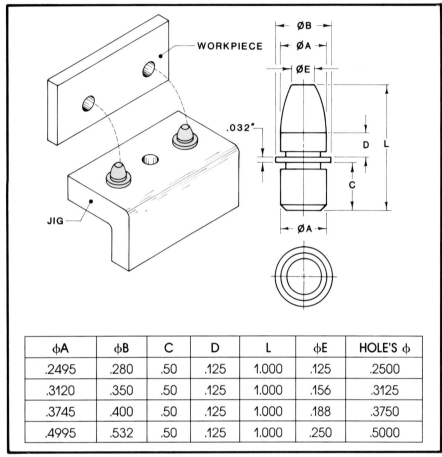

ɸA	ɸB	C	D	L	ɸE	HOLE'S ɸ
.2495	.280	.50	.125	1.000	.125	.2500
.3120	.350	.50	.125	1.000	.156	.3125
.3745	.400	.50	.125	1.000	.188	.3750
.4995	.532	.50	.125	1.000	.250	.5000

FIGURE 23–7A Bullet nose locators

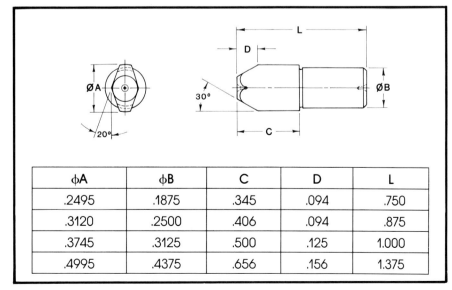

ɸA	ɸB	C	D	L
.2495	.1875	.345	.094	.750
.3120	.2500	.406	.094	.875
.3745	.3125	.500	.125	1.000
.4995	.4375	.656	.156	1.375

FIGURE 23–7B Diamond pin locators

provide flexibility in the leveling a jig and fixture to precise surface angles. Standard dimensions for jig feet and rest buttons are shown in **Fig. 23–8A**. **Figure 23–8B** shows standard dimensions for hexhead screw type jig feet and rest buttons. Socket head cap screws are used for the same purpose. Socket head dimensions are shown in **Fig. 23–9**.

Clamps

Clamps are used to hold workpieces firmly against locators and rest buttons without distorting the workpiece. There are many types of clamps for jig and fixture design. The three major categories of clamps are **toggle head clamps, cam clamps,** and **screw clamps.**

Toggle clamps, such as the push-pull toggle clamp shown in **Fig. 23–10,** are used where maximum surface contact with the workpiece is required. Cam clamps **(Fig. 23–11)** are used where ease of changing the workpieces is needed to speedup production. Screw clamps **(Fig. 23–12)** reduce the effect of vibration on jigs and workpieces. Using screw clamps often slows down production unless swing-out features are used for loading and unloading workpieces.

Appropriate application of clamps allow clamping from multiple directions **(Fig. 23–13)**. Generally, cutting actions should not be directed towards a clamp, but towards a rigid locator. **Figure 23–14** also shows an example of a cutting motion directed toward rigid locators, and away from a screw clamp.

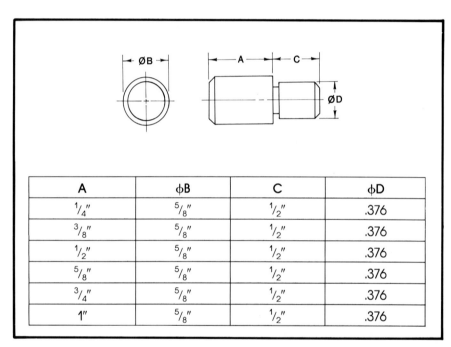

FIGURE 23–8A Jig feet (rest buttons)

A	ϕB	C	ϕD
1/4"	5/8"	1/2"	.376
3/8"	5/8"	1/2"	.376
1/2"	5/8"	1/2"	.376
5/8"	5/8"	1/2"	.376
3/4"	5/8"	1/2"	.376
1"	5/8"	1/2"	.376

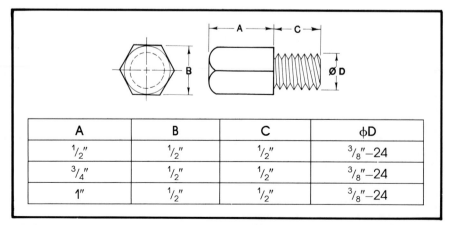

FIGURE 23–8B Hexhead screw-type jig feet (rest buttons)

A	B	C	ϕD
1/2"	1/2"	1/2"	3/8"–24
3/4"	1/2"	1/2"	3/8"–24
1"	1/2"	1/2"	3/8"–24

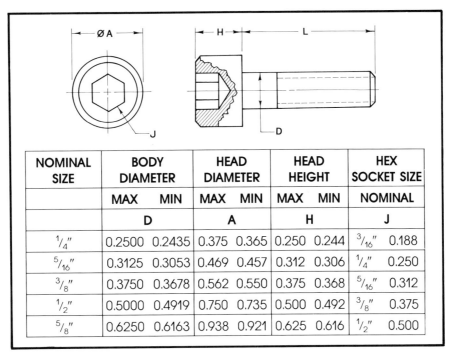

FIGURE 23-9 Socket head cap screws

NOMINAL SIZE	BODY DIAMETER		HEAD DIAMETER		HEAD HEIGHT		HEX SOCKET SIZE	
	MAX	MIN	MAX	MIN	MAX	MIN	NOMINAL	
	D		A		H		J	
1/4"	0.2500	0.2435	0.375	0.365	0.250	0.244	3/16"	0.188
5/16"	0.3125	0.3053	0.469	0.457	0.312	0.306	1/4"	0.250
3/8"	0.3750	0.3678	0.562	0.550	0.375	0.368	5/16"	0.312
1/2"	0.5000	0.4919	0.750	0.735	0.500	0.492	3/8"	0.375
5/8"	0.6250	0.6163	0.938	0.921	0.625	0.616	1/2"	0.500

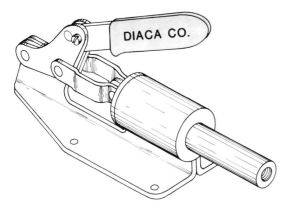

FIGURE 23-10 Push-pull toggle clamp

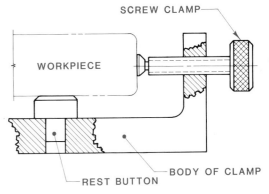

FIGURE 23-12 Knurled head swivel pad screw clamp

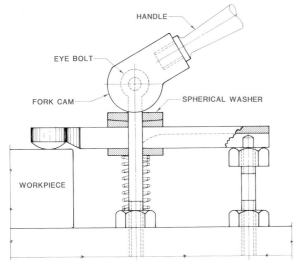

FIGURE 23-11 Cam clamp assembly

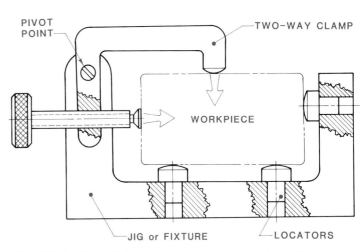

FIGURE 23-13 Multiple swing-out screw clamp. Note the clamping action is towards the locators.

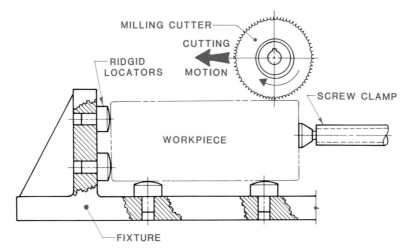

FIGURE 23-14 Cutting motion is away from the clamp (towards the rigid locators).

Drill Bushings

Drill bushings align, hold, and guide drill bits to an exact location, and at a specific angle for drilling **(Fig. 23-15)**. Drill bushings are made of a harder steel than the drill bit. This prevents the drill from damaging the bushings. Drill bushing sizes correspond to standard drill sizes.

Bushings may be either fixed or replaceable. **Fixed drill bushings** are force-fitted into jig bodies. Fixed drill bushings are relatively inexpensive, but are difficult or impossible to replace. They are used for short production runs that do not require a bushing change. There are two kinds—headed and headless. **Headed fixed drill bushings** can withstand greater pressures than headless bushings because they have shoulders to support their position. **Headless fixed drill bushings** can be placed closer together and allow jig surfaces to remain flat **(Fig. 23-16)**.

Replaceable drill bushings are inserted into a liner that has been force-fitted into a jig body **(Fig. 23-17)**. Once the replaceable bushing is inserted, it is held in place with a lock screw. These bushings can be easily replaced by removing the lock screw.

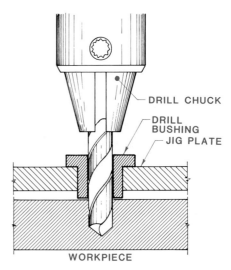

FIGURE 23-15 Drill bushing guides drills to exact location on workpiece.

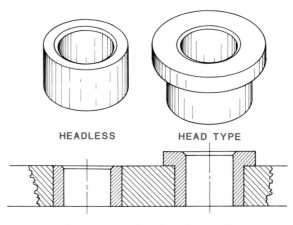

FIGURE 23-16 Fixed drill bushings (press fit)

Chapter 23 Tool Design Drafting 455

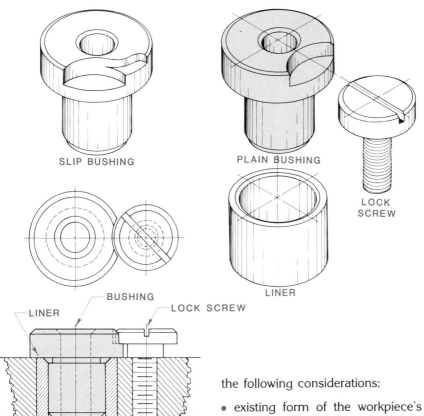

FIGURE 23–17 Renewable drill bushings

Bodies

Jig and fixture bodies are made in three general forms—**weld bodies, cast bodies,** and **built-up bodies.** Figure 23–18 shows the same jig body produced by these three methods.

The most commonly used form is the built-up body. When attaching plates using the built-up body method, dowel pins and socket head cap screws are used **(Fig. 23–19).** Dowel pins are used to align the plates in an exact position without slippage, and the cap screws apply pressure to hold the plates rigid. When this standard fastening technique is used, both dowels and camp screws are flush with the plates.

JIG AND FIXTURE DESIGN

Designing the most appropriate jig for a specific operation depends on the following considerations:

- existing form of the workpiece's raw material
- dimensional and geometric tolerance requirements in the finished product
- type of machine operator needed
- number of repetitive operations needed
- number of duplicate parts to be made
- safety considerations, such as chip clearance, lubricant flow, operator distance, visibility, loading and unloading clearances, and weights.

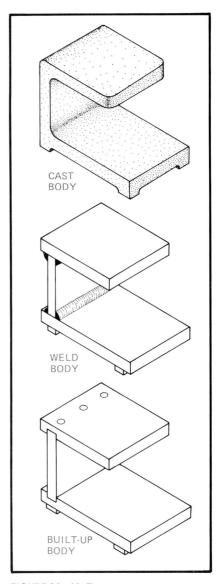

FIGURE 23–18 Three common forms used to produce bodies for jigs and fixtures

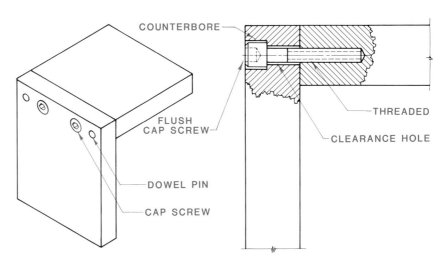

FIGURE 23–19 Built-up body connections

With these considerations in mind, established practices and design sequences can be used to arrive at a sound jig or fixture design. The proper sequence for jig or fixture design is shown in Figs. 23–20 through 23–27, and include the following steps:

Step 1. To draw the workpiece shown in **Fig. 23–20**, first draw the outline with red phantom lines **(Fig. 23–21)**.

Step 2. Since the placement of the locators is critical, study the workpiece drawing carefully before drawing the locators on the workpiece datums **(Fig. 23–22)**.

Step 3. Draw the rest buttons so they support the workpiece while the machining operation (drilling) is performed **(Fig. 23–23)**.

Step 4. Draw the bushings so they are placed above the holes to be drilled (fixed head type here). The distance from the bottom of the bushing to the top of the workpiece should equal one-half the drill's diameter, in this case, .188″ **(Fig. 23–24)**.

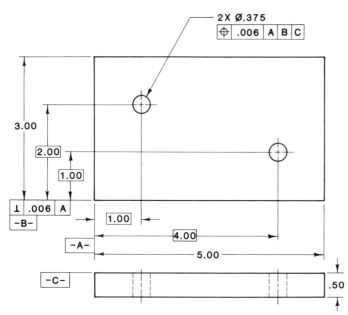

FIGURE 23–20 Drawing of workpiece with two holes to be drilled with a drill jig

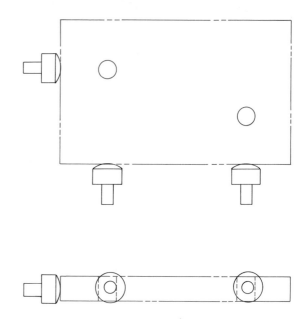

FIGURE 23–22 Add the locators.

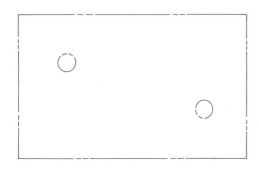

FIGURE 23–21 Draw an outline of the workpiece with a colored pencil (red preferred).

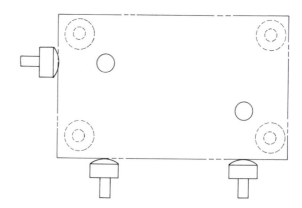

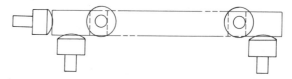

FIGURE 23–23 Add the rest buttons.

Step 5. Select and draw all clamps so they force the workpiece against the locators and hold it firmly. In this example, a double-acting clamp **(Fig. 23–25)** forces the workpiece against the locators to the left and bottom.

Step 6. Design the body of the jig **(Fig. 23–26)**. The jig body holds all jig components together, and simultaneously allows the workpiece to be inserted and removed easily.

Step 7. Draw the jig feet in position. In this case, socket head cap screws are used. Finish the design by adding fasteners to hold the body plates together and with clamps in place **(Fig. 23–27)**.

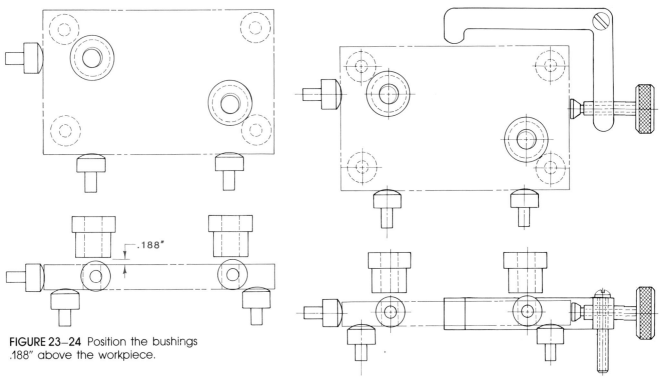

FIGURE 23–24 Position the bushings .188" above the workpiece.

FIGURE 23–25 Locate the clamp's position.

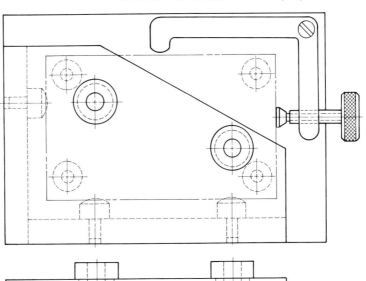

FIGURE 23–26 Design the body to support all holding components

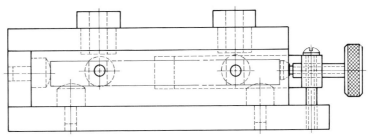

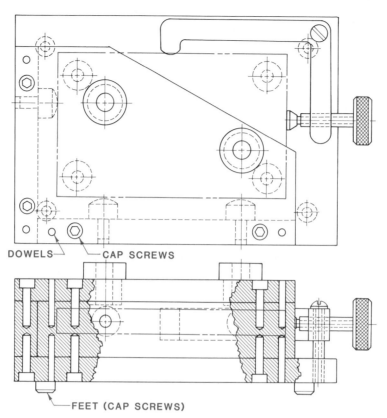

FIGURE 23-27 Add fasteners and feet to the jig body.

Step 8. Dimension the jig assembly. Jig and fixture dimensioning is completed using the dimensioning and tolerancing practices outlined in chapters 9 and 10. Since most jigs and fixtures are made of many standard parts, few dimensions are needed because it is not necessary to specify standard parts sizes. Some size dimensions are necessary for the body, but the majority of dimensions on a jig or fixture drawing are location dimensions.

COMPUTER-AIDED DRAFTING & DESIGN

COMPUTER-AIDED JIG AND FIXTURE DESIGN

Because jigs and fixtures are fabricated from standard parts, designing them on a CAD system is simple and fast. Designers need to understand design procedures, and know how to select and position standard locators, rest buttons, bushings, clamps, feet, and bodies.

Drafters must also be able to modify standard parts. This modification may be extensive or simple, such as shortening a locator, trimming a bushing, grinding a rest button, or modifying a clamp. Editing on a CAD system makes it easy to modify standard components and to move parts to their proper position on a drawing.

In designing jig and fixtures, a CAD-system operator first creates, draws, and saves a library of the standard component drawings. This process is time-consuming at first, but enables a CAD system to function efficiently and rapidly later. The top and side views are drawn and saved as standard components. **Figure 23-28** is an example of a drill bushing drawing that has been drawn and stored in the library. This bushing is a standard headed press fit bushing. In this case, the CAD-system operator has assigned a number (#TD-B1) to this bushing. Each subsequent standard component is drawn and saved in this manner.

Figures 23-29 and 23-30 show how a CAD system is used to modify a standard part. In these examples,

Chapter 23 Tool Design Drafting 459

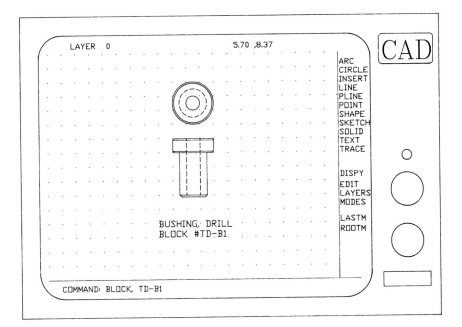

FIGURE 23-28 A library of jig and fixture components is a necessity. Components may be drawn and saved, or purchased as software.

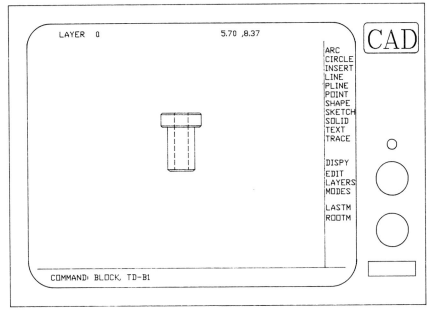

FIGURE 23-29 Drill bushing #TD-B1 is called up from the library. The top view is erased.

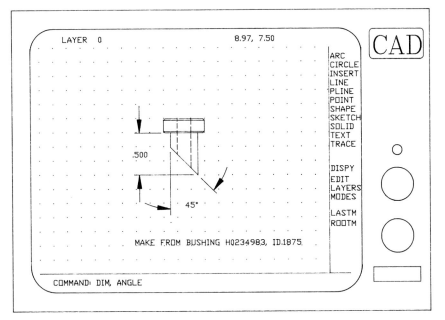

FIGURE 23-30 Modify the bushing to fit the specific machining requirements.

the part to be modified is drill bushing #TD–B1.

Step 1. The operator inserts the block number TD–B1 and erases the unnecessary top view **(Fig. 23–29)**.

Step 2. The modifications are made on the drawing **(Fig. 23–30)**. Note the bushing has been cut on a 45° angle. Because the bushing is a standard part and its size dimensions are known, the only dimensions needed are the ones describing the modification.

Figures 23–31 through 23–37 illustrate how to create a jig design using a CAD system. The workpiece to be drilled is shown in **Fig. 23–31**. The hole to be drilled is on a slanted surface. If drilled without a jig to support the drill, the drill bit would bend and slide on the inclined surface. A modified drill bushing is needed to

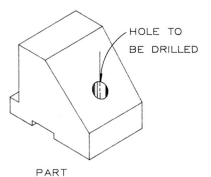

FIGURE 23–31 The part and the required machining operation (hole)

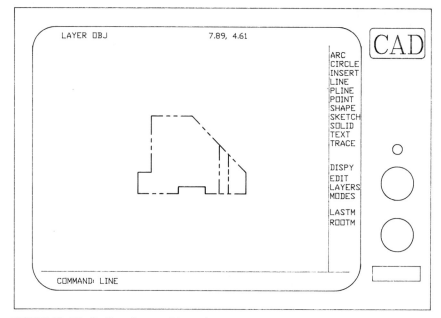

FIGURE 23–32 Outline the part in color.

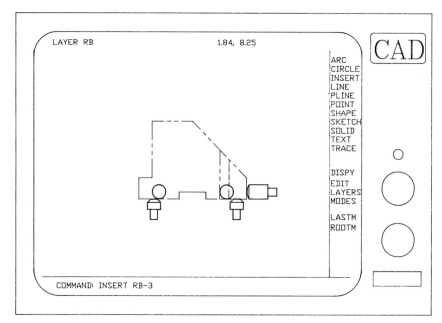

FIGURE 23–33 Position the locators and the rest buttons.

solve this problem. The following steps are used in designing a jig using a CAD system:

Step 1. Draw the part with phantom lines using a red layer **(Fig. 23–32)**. In this case, the operator has called the layer OBJ for object.

Step 2. Insert standard locators and rest buttons **(Fig. 23–33)**. Once inserted, they may be moved to any location.

Step 3. Insert the modified drill bushing directly above the hole to be drilled **(Fig. 23–34)**.

Step 4. Insert the two thumb-screw clamps **(Fig. 23–35)**.

Step 5. Design a body to hold the standard jig components, using normal drawing commands **(Fig. 23–36)**. If necessary, any standard part may be easily relocated using the MOVE command.

Step 6. Using CAD DRAW and EDIT commands, complete the front view of the jig **(Fig. 23–37)**. Hidden lines are changed and broken-out sections are added for clarity. This step also includes the addition of fasteners which hold the body plates.

Step 7. To complete a jig working drawing, a top view, the removed sections, details, and tolerance dimensions are necessary. These tasks are covered in chapters 8 through 11.

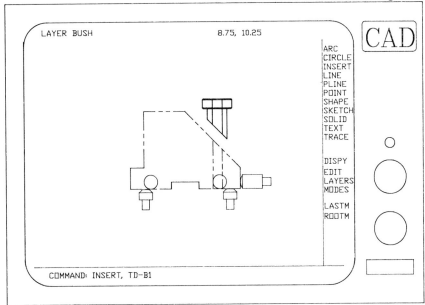

FIGURE 23–34 Position the modified drill bushing.

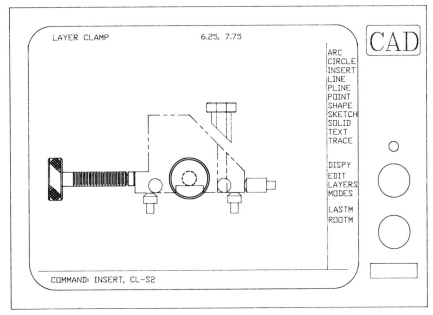

FIGURE 23–35 Position the clamps.

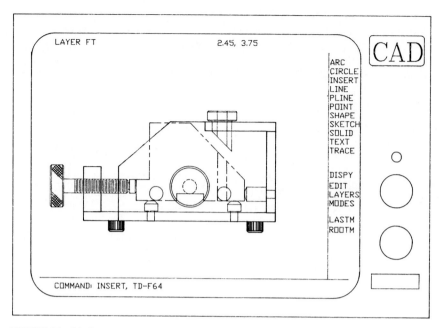

FIGURE 23-36 Design a body to support the standard jig components.

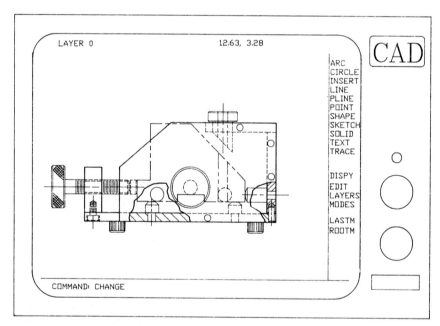

FIGURE 23-37 Complete the working drawing by adding fasteners, feet, and the dimensions.

EXERCISES

1. Design a jig or fixture to hold a plate for drilling **(Fig. 23–38)**.
2. Design a jig or fixture to hold a plate for milling the slot **(Fig. 23–39)**.
3. Design a jig or fixture to hold a U-plate for surface grinding of the top only **(Fig. 23–40)**.
4. Design a jig or fixture to hold a triangular plate for drilling and top surface grinding operations **(Fig. 23–41)**.
5. Design a jig or fixture to hold a channel block for milling and drilling operations **(Fig. 23–42)**.
6. Design a jig or fixture to hold a pipe for 4" cutoffs **(Fig. 23–43)**.
7. Design a jig or fixture to hold a guide block for milling and drilling operations **(Fig. 23–44)**.
8. Design a jig or fixture to hold a milling guide block for milling, drilling, and grinding operations **(Fig. 23–45)**.

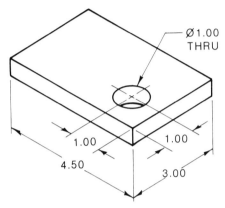

FIGURE 23–38 Design a jig or fixture to hold the plate for drilling.

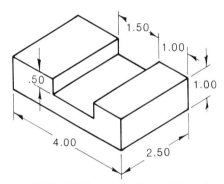

FIGURE 23–39 Design a jig or fixture to hold the plate for milling the slot.

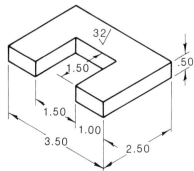

FIGURE 23–40 Design a jig or fixture to hold the U-plate for surface grinding (top surface only).

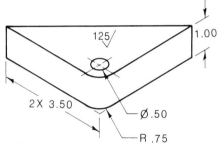

FIGURE 23–41 Design a jig or fixture to hold the triangular plate for a drilling and grinding operation (top surface only).

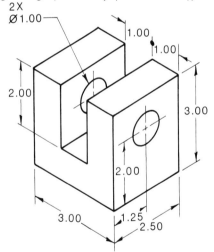

FIGURE 23–42 Design a jig or fixture to hold the channel block for a milling (U shape) and drilling operation.

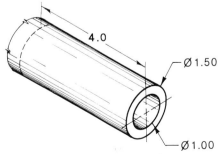

FIGURE 23–43 Design a jig or fixture to hold the pipe for 4" cutoffs.

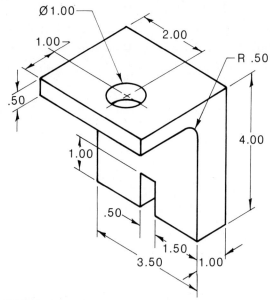

FIGURE 23–44 Design a jig or fixture to hold the guide block for a slot milling and drilling operation.

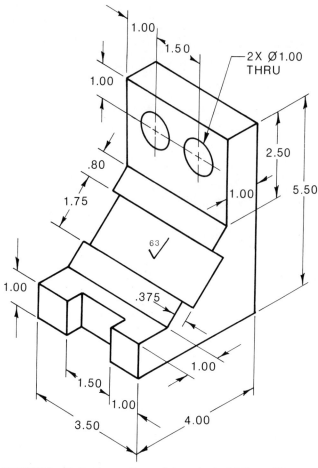

FIGURE 23–45 Design a jig or fixture to hold the milling guide block for milling two slots (1.50" and 1.75"), drilling operation for two holes, and a grinding operation for the 1.75" slot.

KEY TERMS

Built-up bodies
Cam clamps
Cast bodies
Clamps
Drill bushings
Feet

Fixed drill bushings
Fixtures
Headed fixed drill bushings
Headless fixed drill bushings
Jigs
Locators

Replaceable drill bushings
Rest buttons
Screw clamps
Toggle head clamps
Tool engineering
Weld bodies

the world of CAD

The Ocean Drilling Program (ODP) is an international research program of scientific ocean drilling. The main objectives of the program are to study the origin, composition and evolution of the earth's crust and to develop new tools and technology for deep ocean exploration and drilling. Both the engineering and art departments use CAD systems in their day-to-day operations.

The engineering department uses CAD for drafting new drill parts and to illustrate new drilling techniques, Figures 1 and 2. The use of CAD saves the draftsman countless hours of work. The actual execution of the initial drawing takes the same amount of time as it did when drafted on the board, however the time is saved during the revision cycle. An engineer may revise a design 4 to 6 times before approving the final version. Before the use of CAD, this often resulted in redrafting the same part over and over. CAD allows the draftsman to make changes without redrafting and to plot a new vellum original. When drafting on the board there is always the danger that with each revision the drawing may be off by just a pencil width. This could create a tolerance problem that may result in parts not fitting together properly. CAD insures that the drawing will not have tolerance problems because it allows the draftsman to accurately work with unlimited tolerance.

The use of CAD has improved communications for ODP's engineering department. It is now possible for the engineers onshore to electronically transfer hard copies of a design to the ship via satellite. An electronic hard copy is much more accurate than a copy from a FAX machine, and when you are onsite, drilling accuracy is critical. Some of the engineers are also using CAD to make preliminary sketches that can then be transferred to the draftsman to clean-up. It will eventually be pos-

sible to transfer electronic hard copies via floppy disks and magnetic tapes to machine shop vendors so that they can manufacture a part from the disk or tape.

For each cruise the Ocean Drilling Program publishes two books detailing the scientific and engineering objectives and results. Onboard the ship, scientists complete a series of experiments on samples from the core. These analyses generate numerous statistical graphs and diagramatic drawings. It is the responsibility of ODP's art department to produce final illustrations for publication from these graphs and charts.

Currently the art department uses CAD to complete most of the artwork generated on each cruise. A CAD package flexible in creating and manipulating illustrations was chosen. The art department has created complicated cross-hatch patterns and

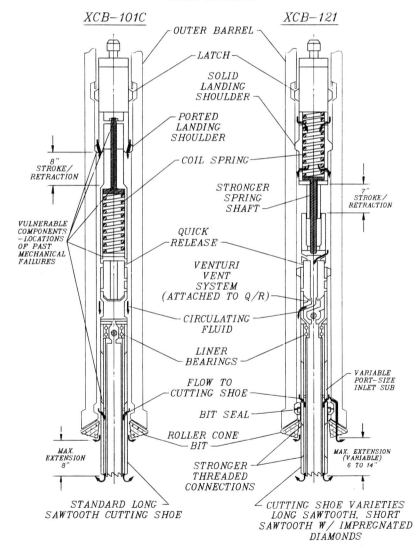

FIGURE 1 CAD is used to draft new drill parts. (Courtesy of Eric Schulte, Draftsman, Engineering Dept., ODP)

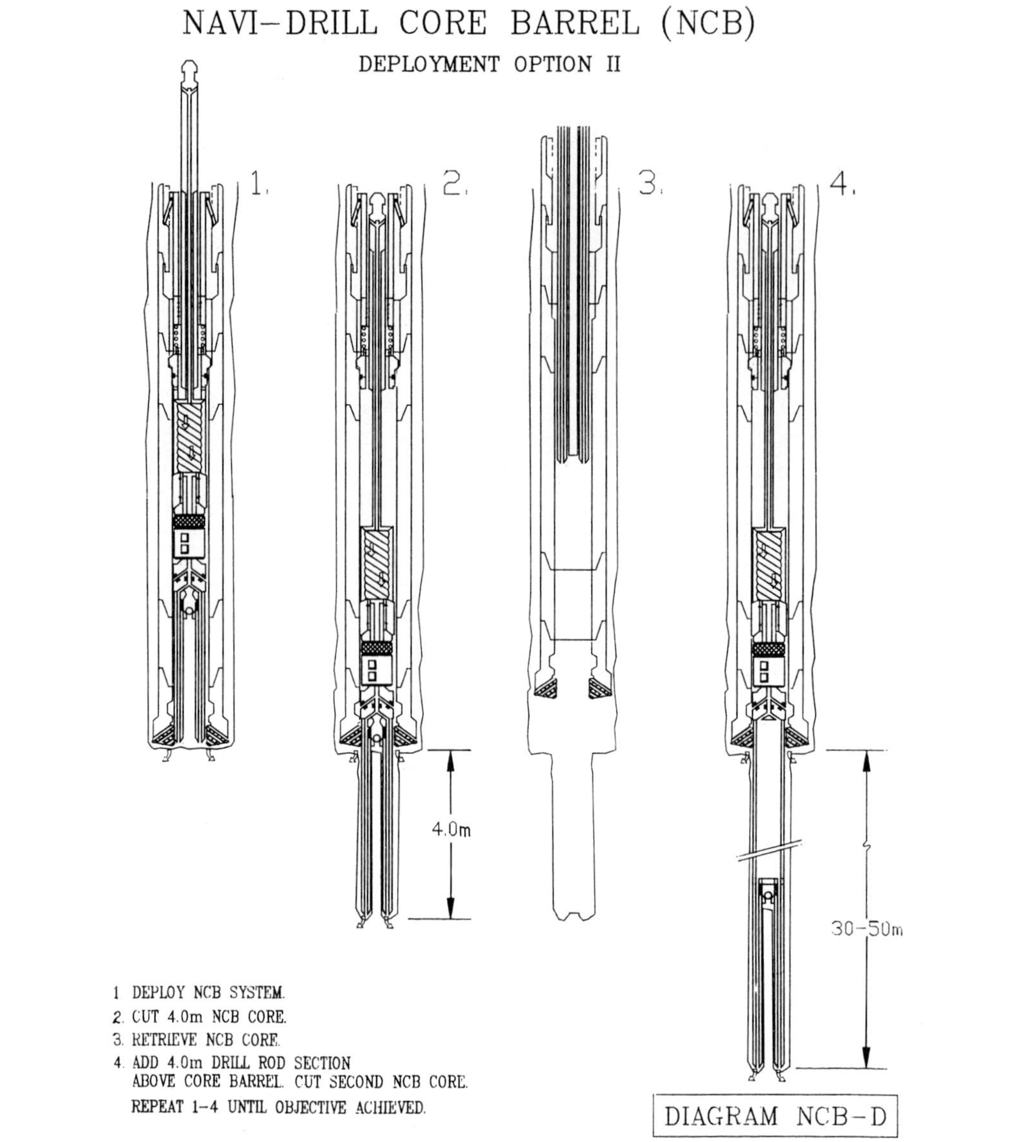

NAVI-DRILL CORE BARREL (NCB)
DEPLOYMENT OPTION II

1. DEPLOY NCB SYSTEM.
2. CUT 4.0m NCB CORE.
3. RETRIEVE NCB CORE.
4. ADD 4.0m DRILL ROD SECTION ABOVE CORE BARREL. CUT SECOND NCB CORE.
   REPEAT 1-4 UNTIL OBJECTIVE ACHIEVED.

DIAGRAM NCB-D

FIGURE 2 CAD is used to illustrate new drilling techniques. (Courtesy of Eric Schulte, Draftsman, Engineering Dept., ODP)

structure symbols that are stored in a system library to be used at any time. This CAD package also interfaces with a scientific graphics program used onboard the ship; this gives the illustrator an electronic hard copy of the graph. Using the scaling and editing tools of CAD, the illustrator can easily make the editorial and size changes necessary for final publication. When a computer file is not available for a graph or diagram, the illustrator will use a digitizer to trace the graph. This produces an electronic version of the illustration that can be easily modified and printed, Figures 3 and 4.

Using CAD saves the art department time in the initial drafting of an illustration and during the revision cycle. When an illustration is created on the board using the traditional methods of pen, ink, and paste-up, it is very difficult for the illustrator to make changes to the final illustration. However using CAD, corrections and even extensive changes can be done quickly and less expensively.

The majority of the illustrations for each cruise are barrel sheets. Barrel sheets are graphic representations of the summary description of each core. Today an illustrator can complete 10 to 15 barrel sheets on a CAD system in the time it takes to complete 2 to 3 barrel sheets on the drafting board. This represents a significant amount of time and money saved using CAD. Figure 4 is an example of a barrel sheet drafted using CAD.

CAD has increased the ability of the illustrator and draftsman to prepare scientific and engineering presentations on short notice. An electronic version of a drawing is always on file, and multiple copies can be made plotted on paper vellum or overhead transparency material.

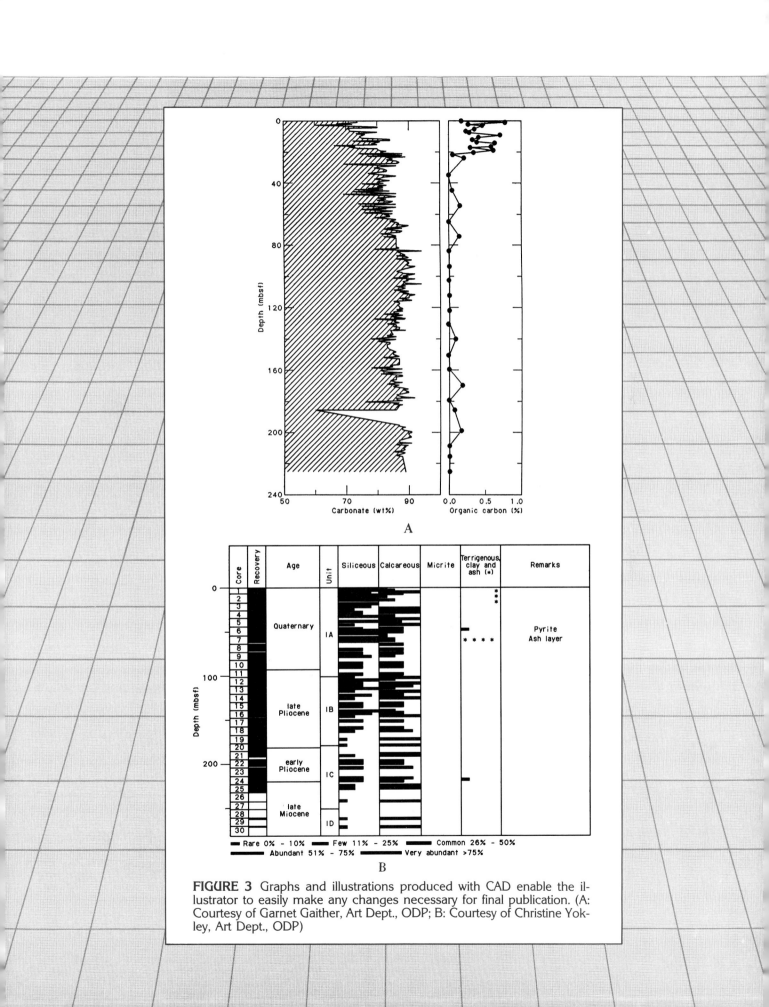

FIGURE 3 Graphs and illustrations produced with CAD enable the illustrator to easily make any changes necessary for final publication. (A: Courtesy of Garnet Gaither, Art Dept., ODP; B: Courtesy of Christine Yokley, Art Dept., ODP)

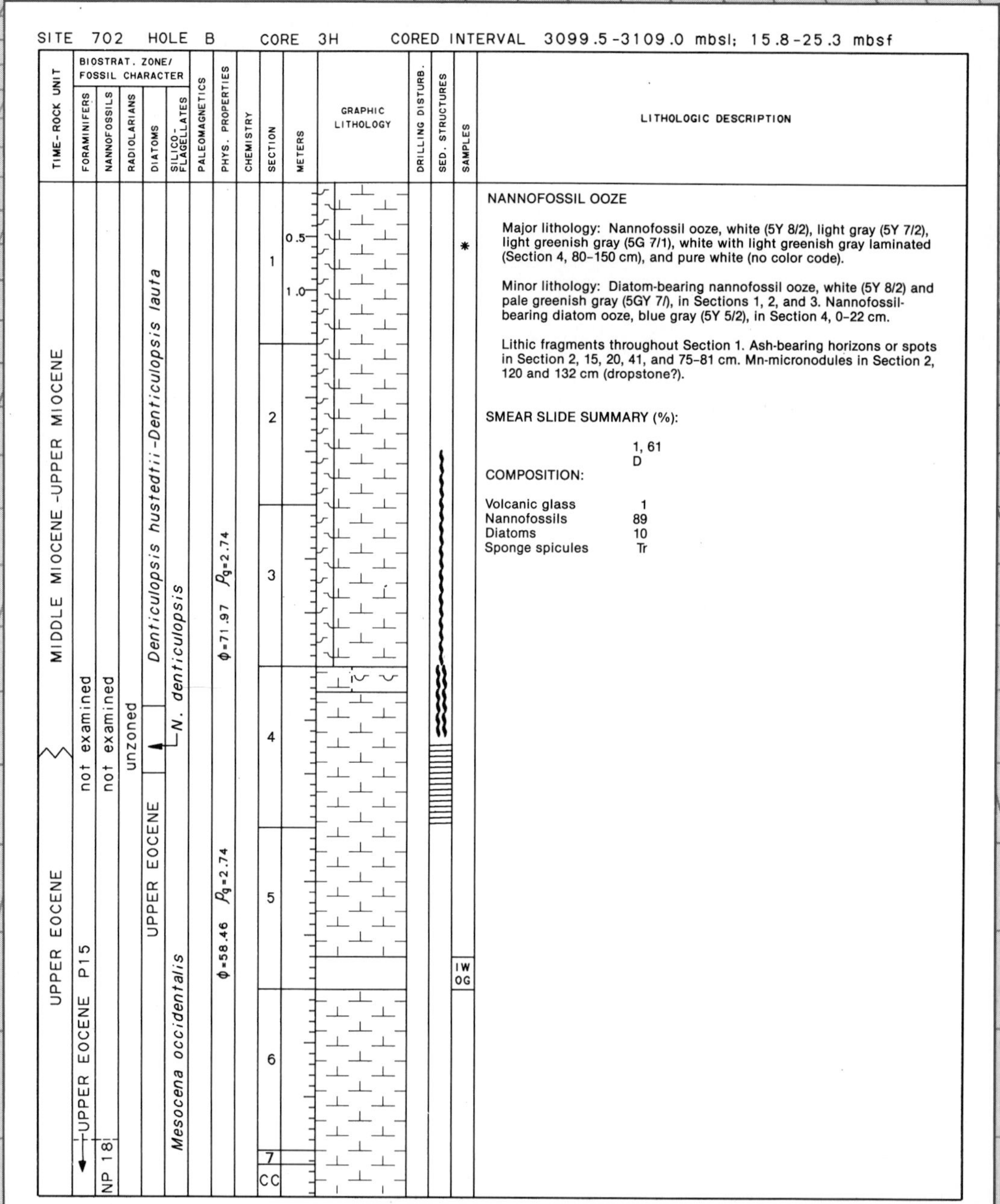

FIGURE 4 An illustrator can complete about 5 times as many barrel sheets using a CAD system compared to hand drafting in the same amount of time. (Courtesy of Jaime Gracia, Art Dept., ODP)

CAD has provided the Ocean Drilling Program with a powerful array of tools to produce high-quality engineering drawings and scientific illustrations faster and more efficiently. ODP is continuing to find new uses for their CAD systems to save both time and money.

(Courtesy of Karen Benson, Chief Illustrator for the Ocean Drilling Program, Texas A & M University, College Station, TX)

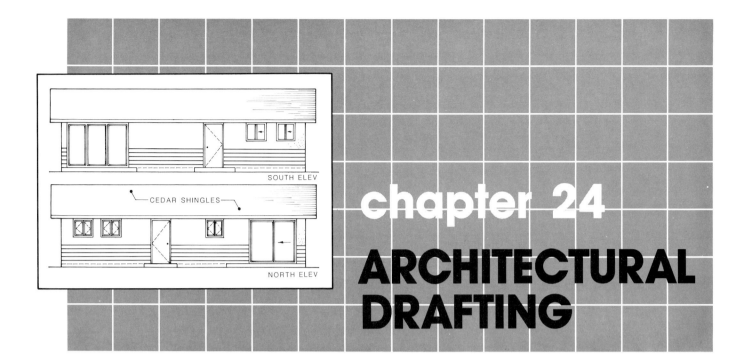

chapter 24
ARCHITECTURAL DRAFTING

OBJECTIVES

The student will be able to:

- read architectural drawings
- prepare an exterior two-point perspective drawing
- draw a plot plan
- draw a floor plan
- draw a foundation plan
- draw four exterior elevations
- draw an interior elevation
- draw a sectional construction detail
- design a small structure to meet specific objectives

INTRODUCTION

The design of any architectural structure follows similar procedures. Only the complexity of the design and drafting process, and the levels of structural engineering, change as the size of a structure increases. In designing a structure, whether a small residence or a high-rise building, the functions of the structure must first be determined and carefully organized. Only then can steps be taken to create a functional, aesthetically pleasing, and structurally sound design. Because of size and familiarity, residential applications are used in this chapter. However, the principles and practices apply to the design of any structure, regardless of size or function.

GATHERING INFORMATION

Structures are built primarily for people. Detailed information concerning the goals, habits, and tastes of all building inhabitants must be matched with the conditions of the site, building code limitations, and available funds or financing. The first step in designing a residence **(Fig. 24–1)** is to gather the following information:

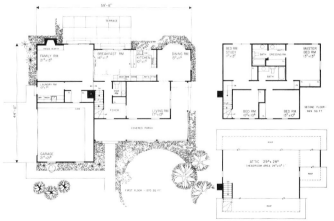

FIGURE 24–1 Architectural floor-plan drawings for a family residence (Courtesy of Home Planners, Inc.)

1. Number of occupants, present and future, and their age and sex
2. Price range and monthly mortgage limits
3. Site conditions which affect the type and orientation of structures

(Fig. 24–2A). This includes a study of potential views, solar orientation, prevailing winds, slope angles, vegetation, lot shape, street access, and soil conditions.

The **plat plan** (Fig. 24–2B) is a map of a large number of properties and surrounding streets. The following data is included:

- property-line dimensions
- building setbacks
- compass orientation
- property lines with compass orientation
- easements
- street widths
- street names
- block numbers
- lot numbers
- scale

The **site plan** (Fig. 24–2C) is a drawing describing the physical elements of a single piece of property. The following data is included:

- property corner elevations
- contour lines with elevation heights
- compass orientation
- property-line dimensions
- property lines with compass orientation
- landscaping
- legal descriptions
- waterways
- recontoured lines (dotted)
- utility lines

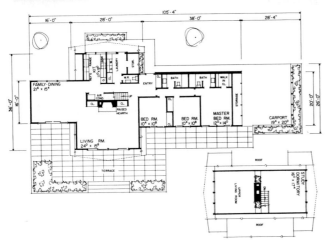

FIGURE 24–2A The conditions of the building site will affect the design of the house. (Courtesy of Home Planners, Inc.)

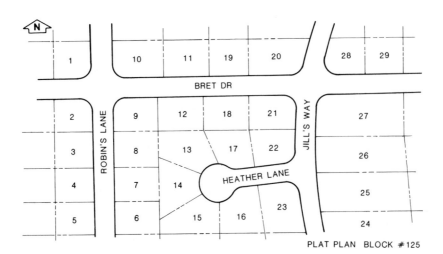

FIGURE 24–2B Example of a simplified plat plan

Chapter 24 Architectural Drafting **467**

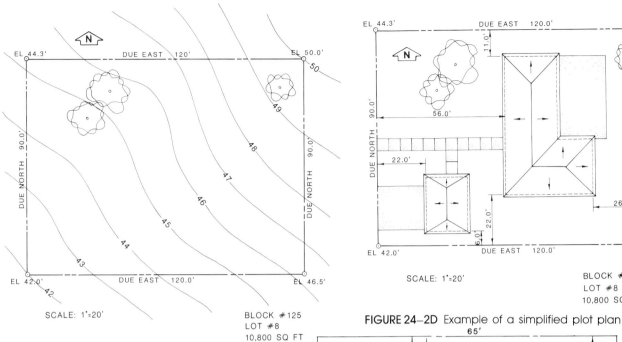

FIGURE 24–2C Example of a simplified site plan

FIGURE 24–2D Example of a simplified plot plan

The **plot plan (Fig. 24–2D)** is a drawing of the structure oriented on the property. The following data is included:

- structure setbacks from property lines
- outline of house and roof
- outline of auxiliary buildings
- property-line dimensions
- property lines with compass orientation
- property corner elevations
- walks, patios
- landscaping
- legal description

4. Local zoning laws and building codes which affect and control minimum lot size, maximum building height, areas of structures, structural types, and setbacks. **Setback** requirements dictate the minimum distance a structure can be located from property lines. The zoning code setbacks set up the boundaries for the buildable area. These dimensions are not shown. Only the actual setbacks are drawn on the plot plan (**Fig. 24-3**).

FIGURE 24–3 Example of typical setback minimums for a structure on a building site

5. Architectural style preference. In addition to personal preference, some local codes restrict or require the use of specific architectural styles and materials. Architectural styles fall into several broad categories; European (traditional), Early American, Later American, and Contemporary.

RESIDENTIAL STYLES

Since most early settlers imigrated from Europe, the first American houses **(Fig. 24–4A)** were based on the characteristics of northern European houses. However, the lack of some building materials and constructural constraints resulted in the simplification of many northern European styles.

Early American Style

One of the first American adaptations was the one-and-one-half story, gable roof, Cape Cod style **(Fig. 24–4B)**. Cold New England winters contributed to the addition of shutters, large fireplaces, and small window areas. Another early American adaptation was the Dutch colonial **(Fig. 24–4C)**, which included dormers and a gambrel (double pitch) roof. Other early American styles added individual and shed dormers to create more second story space **(Fig. 24–4D)**. The early American saltbox offset the roof angles to create a full second story on the front of the house **(Fig. 24–4E)**.

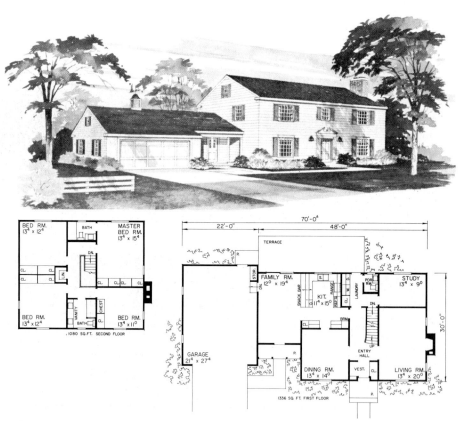

FIGURE 24–4A An example of contemporary early American colonial architecture (Courtesy of Home Planners, Inc.)

Chapter 24 Architectural Drafting 469

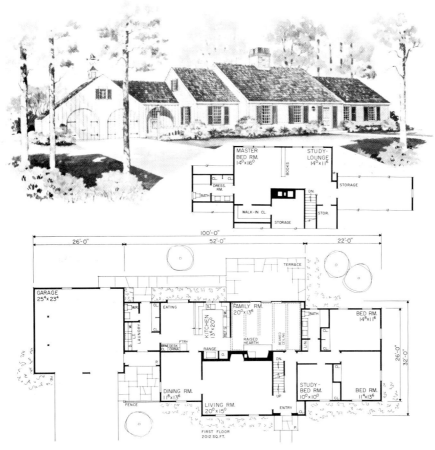

FIGURE 24–4B An example of contemporary Cape Cod architecture (Courtesy of Home Planners, Inc.)

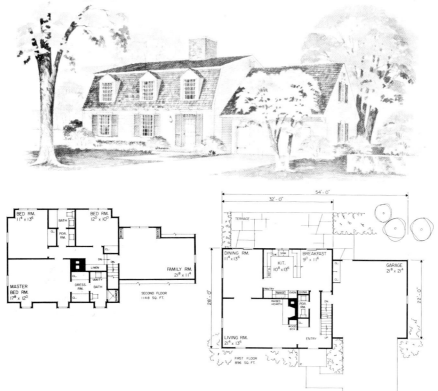

FIGURE 24–4C An example of contemporary Dutch colonial architecture (Courtesy of Home Planners, Inc.)

470 Drafting in a Computer Age

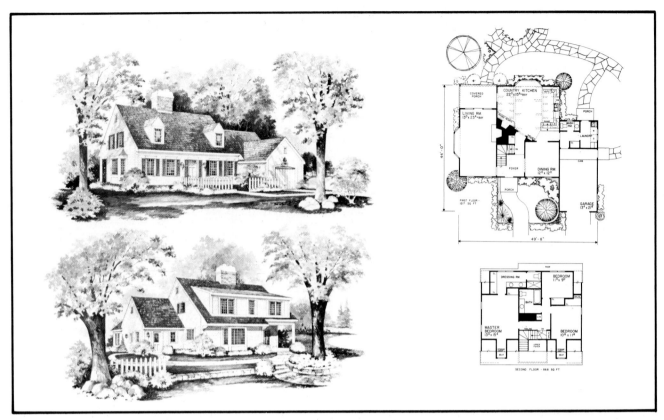

FIGURE 24-4D An example of contemporary early American architecture with gable and shed roof dormers (Courtesy of Home Planners, Inc.)

FIGURE 24-4E An example of contemporary early American saltbox architecture (Courtesy of Home Planners, Inc.)

A full second level, which cantilevered (overhung) the first, was added to create the garrison style **(Fig. 24–4F)**. The garrison was built mainly in the north, therefore maintained small windows and had little outdoor living space. By contrast, southern colonial homes **(Fig. 24–4G)** also included a full second floor but also expanded the size of windows and added verandas (large porches) for southern outdoor living.

Later American Styles

After the colonial period, architectural styles were influenced by climatic conditions and the southern European styles as part of the classical revival movement. Styles developed during this period were strongly affected by French, Spanish, and classical Roman and Greek architecture. Styles of this period include federal, Victorian, western ranch, monterey, and French mansard **(Fig. 24–4H)**. This period was also marked by the return to some traditional European styling. This became possible as building materials became more abundant, building skills improved and affluence grew. A popular style of this period was the English Tudor, characterized by high pitched gable roofs, diamond-leaded windows, and brick

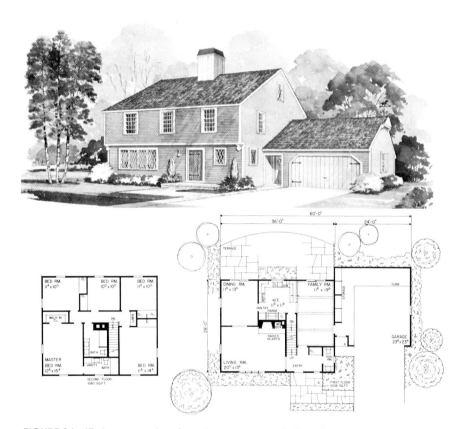

FIGURE 24–4F An example of contemporary early American garrison architecture (Courtesy of Home Planners, Inc.)

472 Drafting in a Computer Age

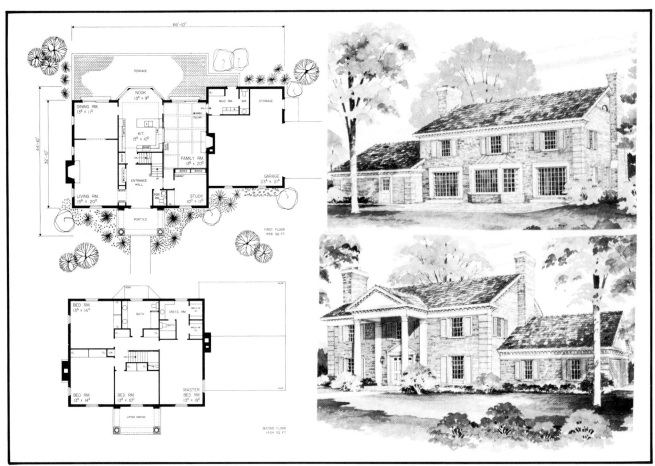

FIGURE 24-4G An example of contemporary southern colonial architecture (Courtesy of Home Planners, Inc.)

FIGURE 24-4H An example of contemporary French mansard architecture (Courtesy of Home Planners, Inc.)

and stucco half-timber walls **(Fig. 24–4I)**. Style changes often lead to eclectic designs. These designs combined the features of many styles without regard for the principles and elements of design. **Figure 24–4J** shows an example of an eclectic architectural style.

Contemporary Styles

As stronger, lighter, and more portable building materials were developed, and as construction methods became more sophisticated, architectural styles were effected. Designs became more functional, broader distances spanned, larger glass areas manufactured and supported, greater heights reached, and difficult site conditions mastered.

These developments led to the use of larger and more unusual shapes. Circles, spheres, pyramids, and triangles became viable building shapes, breaking the rectangular restrictions of the past. For example, the A-frame structure **(Fig. 24–4K)** is based on an ancient building type. However, not until the development of laminated arches and beams did the design become practical for large structures. Concurrently, contemporary styles also became simpler **(Fig. 24–4L)**.

DESIGN CONSIDERATIONS

In the design process, many architectural factors must be considered and compared to the personal goals and requirements of the occupants. The following considerations provide guidelines for the formulation and development of a basic architectural design.

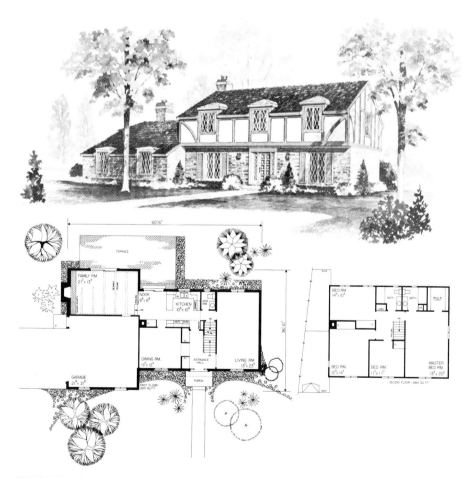

FIGURE 24–4I An example of contemporary English Tudor architecture (Courtesy of Home Planners, Inc.)

474 Drafting in a Computer Age

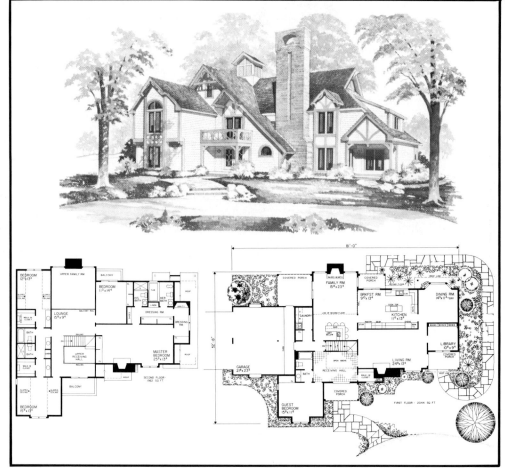

FIGURE 24–4J An example of mixed architectural styles called eclectic (Courtesy of Home Planners, Inc.)

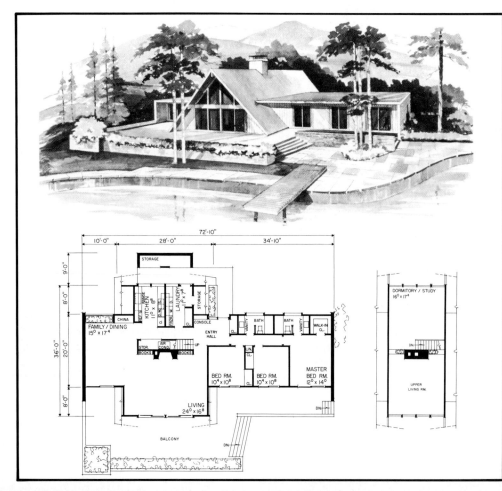

FIGURE 24–4K An example of an A-frame architectural style (Courtesy of Home Planners, Inc.)

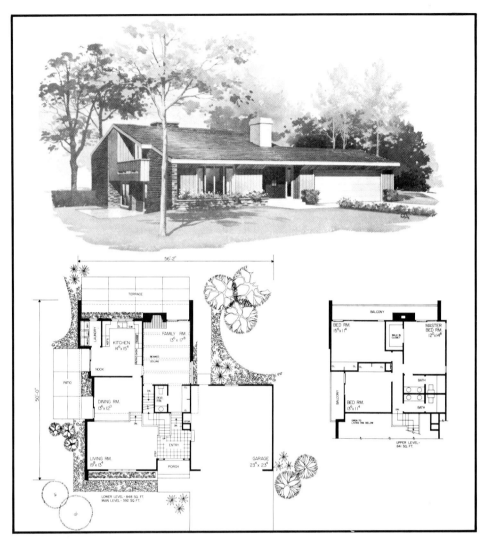

FIGURE 24–4L An example of contemporary style architecture (Courtesy of Home Planners, Inc.)

Area Design

For design purposes, each structure is divided into functional areas. In an educational facility these design areas include classroom areas, administration area, athletic activity areas, technology labs, science labs, food service areas, assembly areas, lavatory facilities, and traffic areas. In an office or industrial building the areas are quite different.

In a residence the design areas include traffic (entry) areas, living area, sleeping area, and service area. The traffic area includes all entrances, foyers, and halls. The living area includes the living room, family room or recreation room, den, study, lavatory, and dining room. Sleeping areas include bedrooms, dressing rooms, baths, and other rooms requiring a quiet surrounding. The service area includes kitchen, utility room, work areas, bath, and garage. A typical breakdown of these areas in a residence is shown in **Fig. 24–5**.

Room Size

Room sizes are determined by the function of each room and the total amount of space available. Furniture, appliances, fixtures, equipment, storage, door and window areas, and traffic requirements must be considered in selecting room sizes. **Figure 24–6** shows typical room sizes for small, average, and large rooms for a residence.

Traffic Flow

Traffic patterns are critical to functional residential design. External traffic must be adequately planned around the main (front) entry, service entry, and special purpose entries **(Fig. 24–7)**. External traffic also includes the effective design of driveways and walkways. Internal traffic planning includes determining the size and location of foyers, halls, stairs, and circulation through each room.

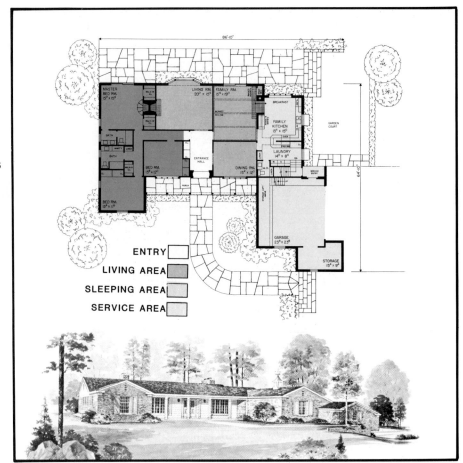

FIGURE 24–5 Architectural designers must carefully plan the relationship of the entry, living, sleeping, and service areas. (Courtesy of Home Planners, Inc.)

A GENERAL GUIDE TO ROOM SIZES

BASIC ROOMS	TYPICAL ROOM SIZES (FEET)		
	SMALL	AVERAGE	LARGE
LIVING ROOM	12 x 18	16 x 20	22 x 28
DINING ROOM	10 x 12	12 x 15	15 x 18
KITCHEN	5 x 10	10 x 16	12 x 20
UTILITY ROOM	6 x 7	6 x 10	8 x 12
BEDROOM	10 x 10	12 x 12	14 x 16
BATHROOM	5 x 7	7 x 9	9 x 12
HALLS	3' WIDE	3'–6" WIDE	3'–9" WIDE
GARAGE	10 x 20	20 x 20	22 x 25
STORAGE WALL	6" DEEP	12" DEEP	18" DEEP
DEN	8 x 10	10 x 12	12 x 16
FAMILY ROOM	12 x 15	15 x 18	15 x 22
WARDROBE CLOSET	2 x 4	2 x 8	2 x 15
ONE ROD WALK-IN CLOSET	4 x 3	4 x 6	4 x 8
TWO ROD WALK-IN CLOSET	6 x 4	6 x 6	6 x 8
PORCH	6 x 8	8 x 12	12 x 20
ENTRY	6 x 6	8 x 10	8 x 15

FIGURE 24–6 A general guide to room sizes for primary residences

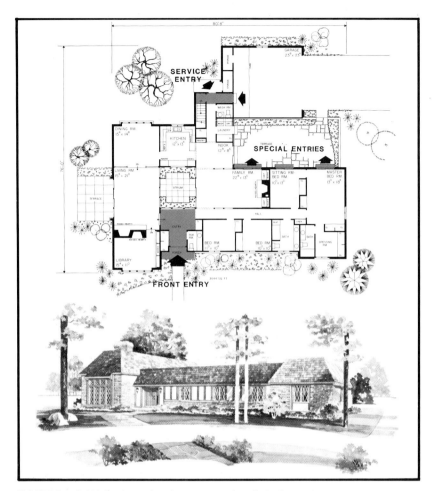

FIGURE 24-7 Well-placed entrances and exits help make an efficient traffic pattern. (Courtesy of Home Planners, Inc.)

Lighting

Architectural lighting is grouped into four types—natural lighting, general lighting, specific (local) lighting, and decorative lighting. General lighting provides diffused lighting throughout an entire room. Local (task) lighting provides light for a specific task such as reading, drawing, or cooking. Decorative lighting helps create moods by highlighting architectural features. The lighting in each room must include adequate outlets and switches at each entrance **(Fig. 24-8)**.

Storage

Storage areas are an integral part of every functional design. Adequate

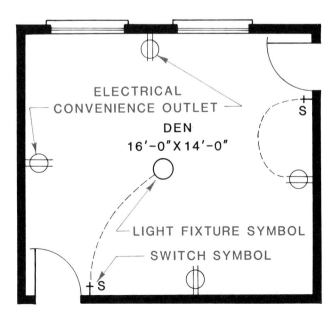

FIGURE 24-8 Light switches must be accessible at the opening of each door and conform to local building codes for their placement.

storage is planned into a residence through the design of closets, built-in storage facilities, and furniture and general storage areas, such as attics, basements, and garages. **Figure 24–9** shows standard sizes for residential storage facilities.

Doors

Doors are of three basic types—swinging, folding, and sliding. Swinging doors are the most popular since they move with the person opening the door. To conserve wall space, swinging doors should swing into rooms toward an adjacent wall, with the hinge as close to the wall as possible. All types of doors are available in a wide range of panels, materials, and styles **(Fig. 24–10)**. Figure 24–10 shows the elevation and floor plan symbols for common door types.

Windows

Windows admit light, air, and to provide a view. The effectiveness of a window depends on its size, location, compass orientation, roof overhang, and the amount of exterior and interior reflected light. **Figure 24–11** shows elevation and floor plan symbols used to represent common window types.

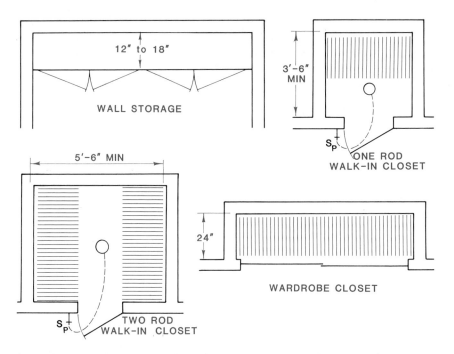

FIGURE 24–9 Storage and closet areas must be adequate and easily accessible.

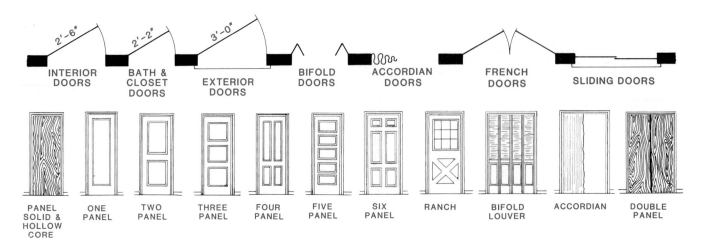

FIGURE 24–10 Door styles and their symbols

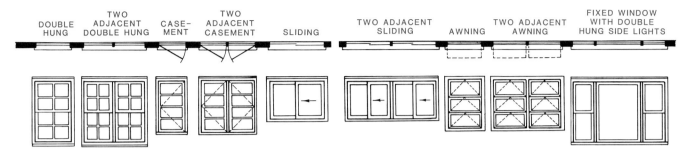

FIGURE 24–11 Window styles and their symbols

Heating, Ventilation, and Air Conditioning (HVAC)

Comfort levels must be designed into each structure. This involves the maintenance of a desired temperature, humidity, and the movement of clean air. To accomplish this, heating, ventilation, and air conditioning (**HVAC**), and humidifying systems are used in combination with insulation and ventilation control through vents and windows.

DESIGN SEQUENCE

After gathering information about the occupants and the site, the next step in the design process is to develop a set of objectives for each area. A floor plan for each room should be sketched containing the features that will help meet these objectives. Keep the personal data in mind while completing **rough sketches** of each area. Keep refining and resketching the plan as many times as needed to arrive at a final **composite sketch.** Professional designers may sketch dozens of revisions before arriving at a final composite sketch.

The architectural symbols shown in **Figs. 24–12A** through **24–12G** should be used in sketching the final plan, and in the development of the final working drawings.

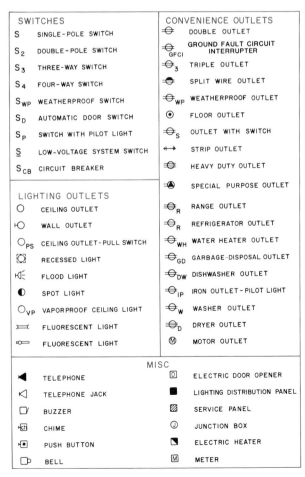

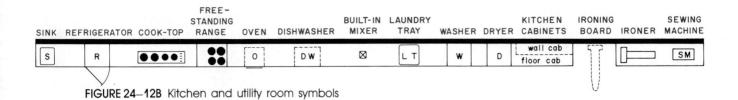

FIGURE 24-12B Kitchen and utility room symbols

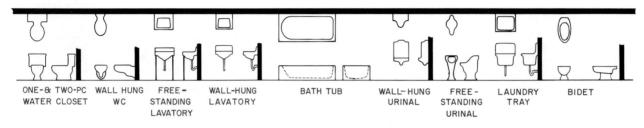

FIGURE 24-12C Bathroom fixture floor plan and elevation symbols

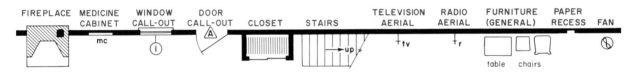

FIGURE 24-12D Built-in component symbols

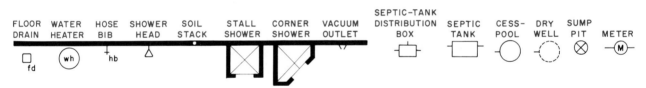

FIGURE 24-12E Wastewater facility symbols

FIGURE 24-12F HVAC (heating, ventilation, and air conditioning) symbols

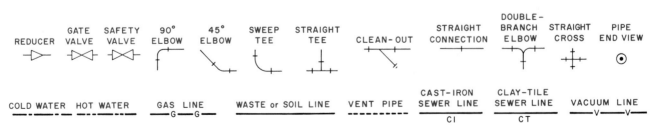

FIGURE 24-12G Plumbing line symbols

WORKING DRAWINGS

The final composite sketch is used as a guide in preparing a set of working drawings. A full set of architectural working drawings includes pictorials, floor plans, elevations, and construction detail drawings. A set of working drawings also includes supporting documents such as schedules, specifications, and may include budgets and contracts.

Pictorial Drawing

Architectural pictorial drawings show the general appearance of a building. They are rarely used as working drawings. Because of the large size of buildings, perspective drawings **(Fig. 24–13)** are used for most architectural pictorials. Where structural details are shown pictorially, isometric drawings are usually used. Refer to Chapter 15 for pictorial drawing information.

Presentation Floor Plans

Presentation or **abbreviated floor plans** show the general layout of a floor plan. They do not contain extensive detailing. These plans are not used as working drawings but are prepared primarily for sales and general reference purposes. **Figure 24–14** shows a typical presentation floor plan. Note the lack of detail dimensions and the use of abbreviated symbols.

FIGURE 24–13 Two point exterior perspective drawing (Courtesy of Home Planners, Inc.)

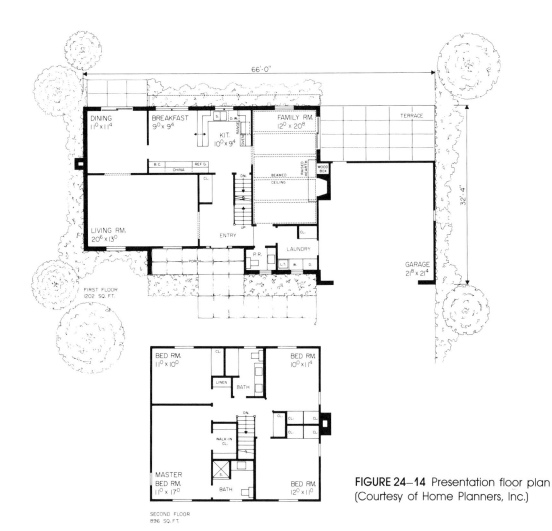

FIGURE 24–14 Presentation floor plan (Courtesy of Home Planners, Inc.)

Plot Plans, Survey Plans, and Site Plans

Before finalizing a floor plan, an outline of the plan must be located on a layout of the lot for which it was designed. A plan of a lot which includes a dimensioned building outline, setback dimensions, driveways, walks, and street locations, is a **plot plan (Fig. 24–15)**.

When site contours and utility line locations are added, the plan is known as a **survey plan.** A plan which includes all of these features plus landscaping details is known as a **site plan.**

Floor Plan

Floor plans are horizontal sections of a building projected from a cutting plane line placed approximately 4 ft. above the floor level. Working drawing floor plans contain symbols (see Figs. 24–12A through G on pages 479–480) which show the top view and position of fixtures, stairs, walls, windows, doors, electrical outlets and switches, plumbing, and HVAC equipment. Floor plans are usually drawn at a scale of $1/4'' = 1'-0''$ or $1/8'' = 1'-0''$ depending on the size of the building.

In drawing floor plans such as the one in **Fig. 24–16,** first draw the outline of all interior and exterior walls. Next, add the door, window, stair, and fireplace symbols. Finish the plan by adding the fixtures, dimensions, and labeling.

Second-floor plans are drawn on tracing paper placed over the first-floor plan. The position of outside walls, chimneys, and stairs must align with their first-floor counterparts. **Figure 24–17** shows a second-floor plan that can be used as a working drawing. Successive level floor plans and foundation plans are prepared in the same manner.

Foundation plans are floor plans that show the outline of footings and sections through foundation walls. In addition, a foundation plan shows the location and size of all structural support members such as columns and beams.

Elevations

The orthographic projection of the front, rear, and side views of a building are known as **elevations.** A rectangular building has four sides, or elevations—front, rear, left, and right. These are labeled by compass direction—north, south, east, and west. Horizontal distances on each elevation are projected from related points on each exterior floor-plan wall. This is similar to the way orthographic front, rear, and side views are projected from a top view of an object.

Vertical distances are measured and drawn from an established **datum plane,** such as a grade line or first-floor line. **Figures 24–18A** through **24–18D** show the elevation drawings projected from the floor plans shown in Figs. 24–16 and 24–17. Note that elevation symbols resemble the appearance of building materials and architectural features more closely than do floor-plan symbols.

Interior Elevations

An orthographic view of an interior wall is an **interior elevation.** Interior elevations show the materials, windows, doors, openings, fixtures, and built-in details to be constructed or installed on a wall. Interior elevations **(Figs. 24–19A** and **B)** are identified by room title, compass direction, or detail reference number. For example, an elevation drawing could be labeled kitchen or west wall elevation. The drawing also could be identified with a number or letter corresponding to an index number and arrow on the floor plan. Interior elevations are most commonly prepared for bath and kitchen walls, and walls which contain intricate details such as built-in bookcases, panel-

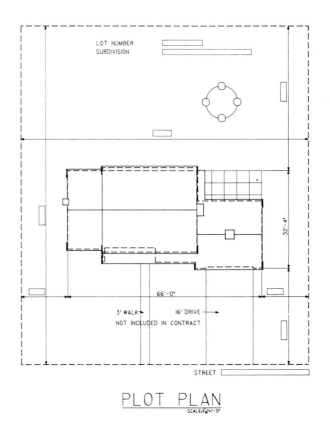

FIGURE 24–15 Plot plan (Courtesy of Home Planners, Inc.)

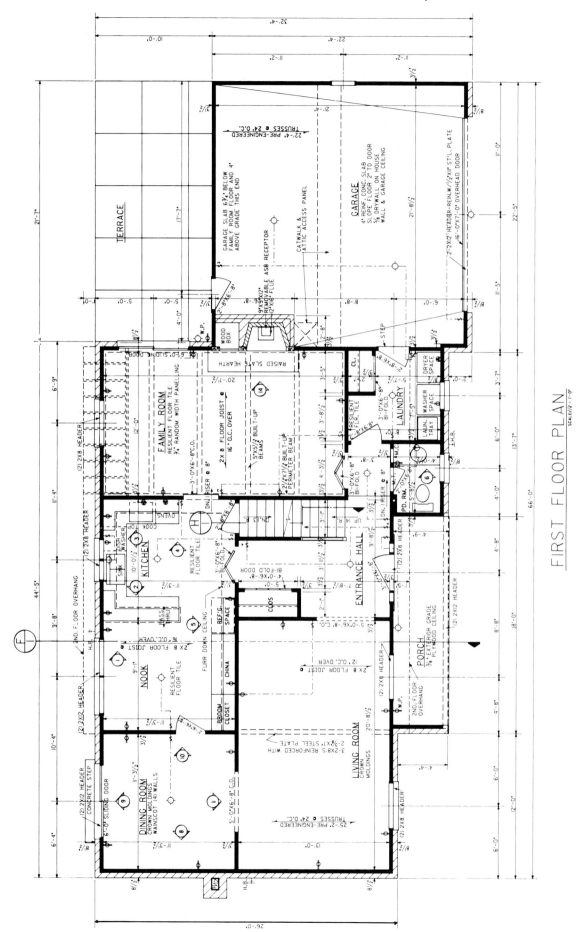

FIGURE 24–16 Floor plan (Courtesy of Home Planners, Inc.)

484 Drafting in a Computer Age

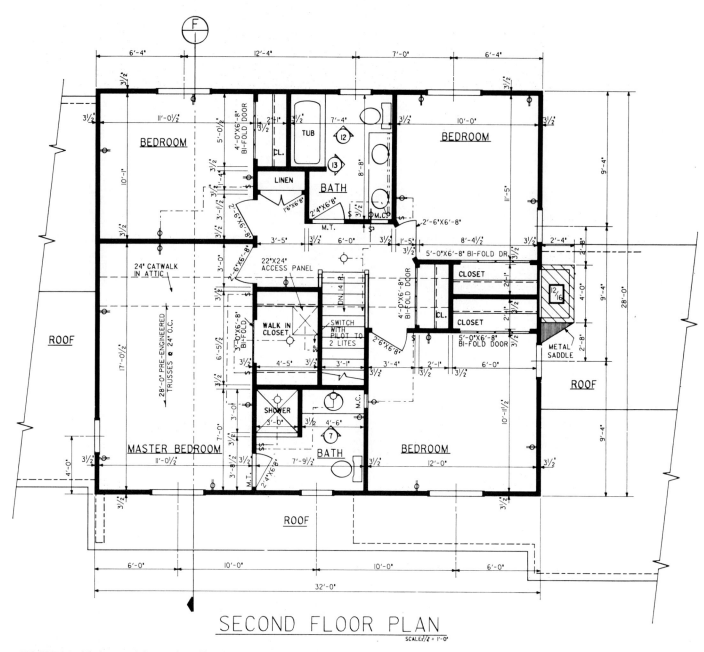

FIGURE 24–17 Second floor plan (Courtesy of Home Planners, Inc.)

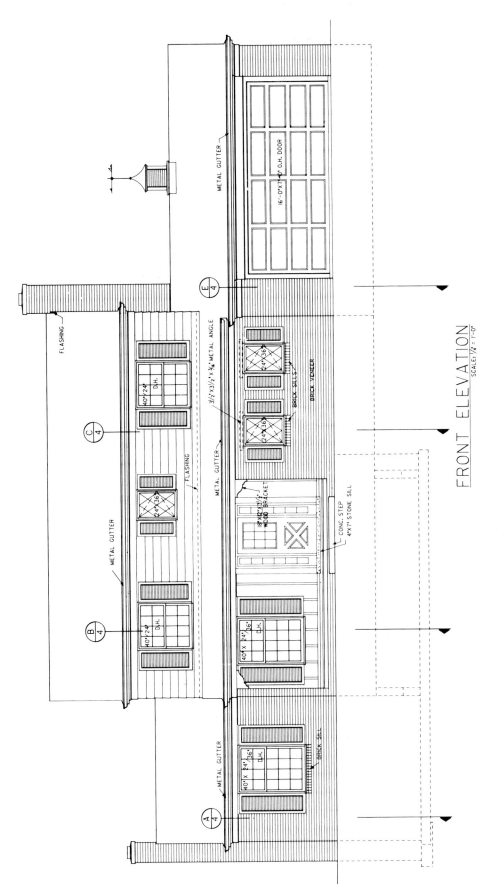

FIGURE 24-18A Front elevation (Courtesy of Home Planners, Inc.)

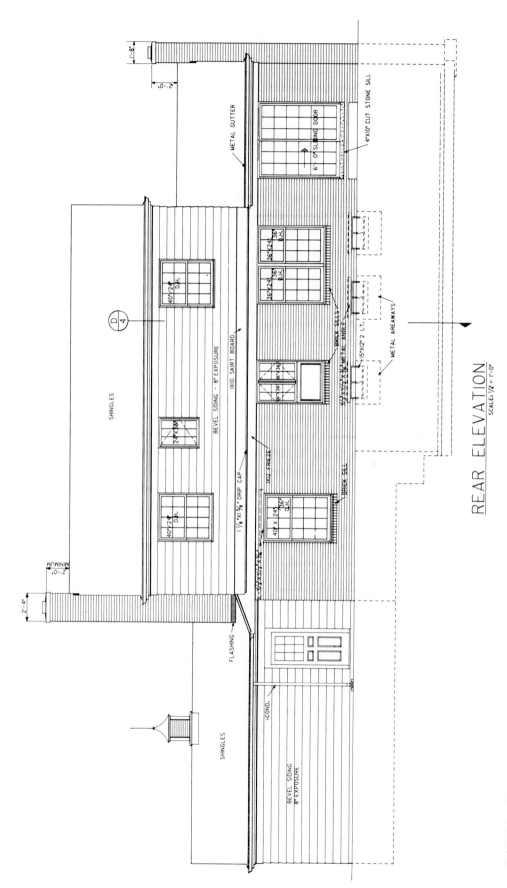

FIGURE 24-18B Rear elevation (Courtesy of Home Planners, Inc.)

FIGURE 24–18C Right-side elevation (Courtesy of Home Planners, Inc.)

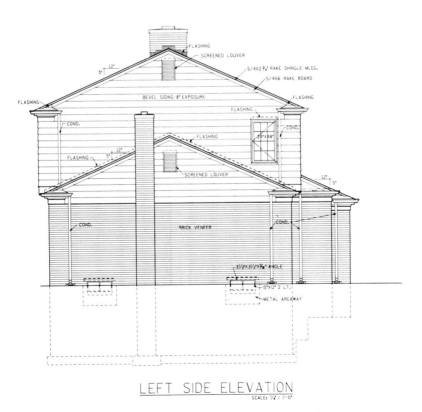

FIGURE 24–18D Left-side elevation (Courtesy of Home Planners, Inc.)

488 Drafting in a Computer Age

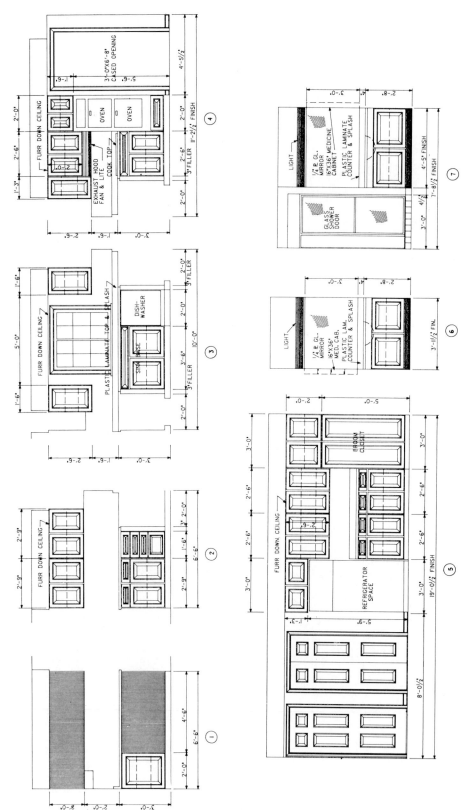

FIGURE 24-19A Interior elevations (Courtesy of Home Planners, Inc.)

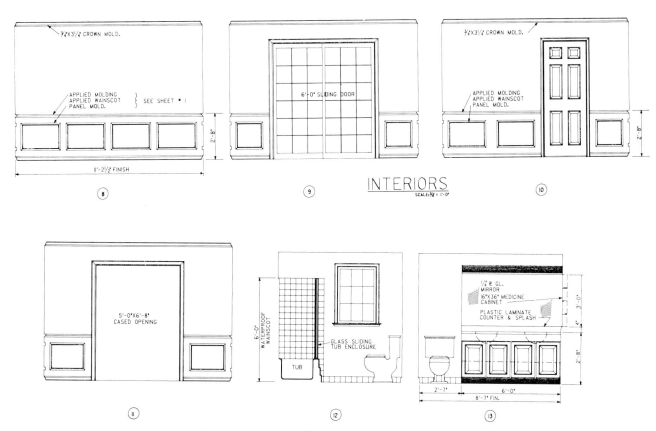

FIGURE 24–19B Interior elevations (Courtesy of Home Planners, Inc.)

ing, music centers, or fireplace walls **(Fig. 24–20)**. If interior elevations are drawn to the same scale as the floor plan, horizontal distances can be projected from the floor-plan wall.

Sectional Drawings

Architectural sectional drawings include detail sections and full sections. Detail sections are either **plan** (horizontal) **sections** or **elevation** (vertical) **sections.** Detail sections are drawn to a large scale and show the details of a small area of a structure. Detail sections **(Figs. 24–21A, B, and C)** are designed to show the exact size and relationship of construction members and building components.

Full sections are designed to show the type of construction used for the entire structure. Full sections are drawn to a small scale and do not reveal small member details and intersections **(Fig. 24–22)**. A full section through the shorter dimension of a building is known as a **transverse section.** A section drawn through the longer dimension of a building is known as a **longitudinal section.**

Cutting plane lines are used in the same manner as described in Chapter 11. Because of the size of architectural drawings and the diversity of building materials, a wide variety of section-lining symbols are used in architectural sectional drawings.

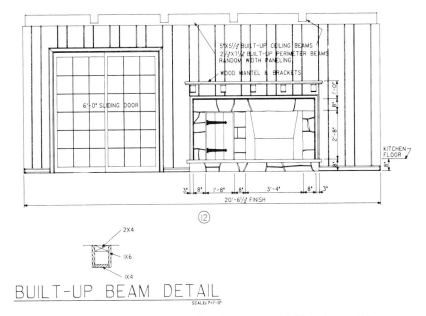

FIGURE 24–20 Interior elevation of the fireplace wall (Courtesy of Home Planners, Inc.)

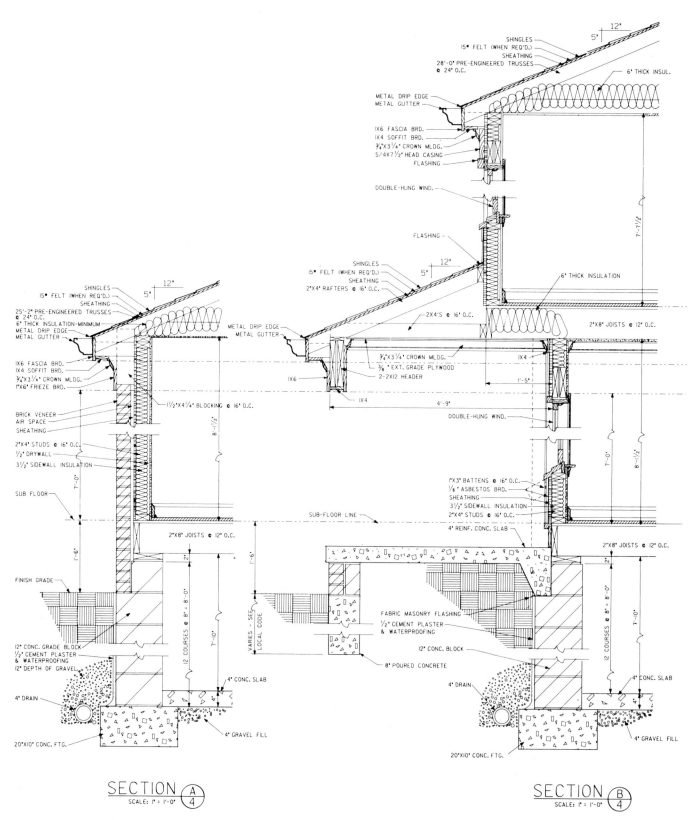

FIGURE 24-21A Detailed house sections A and B (Courtesy of Home Planners, Inc.)

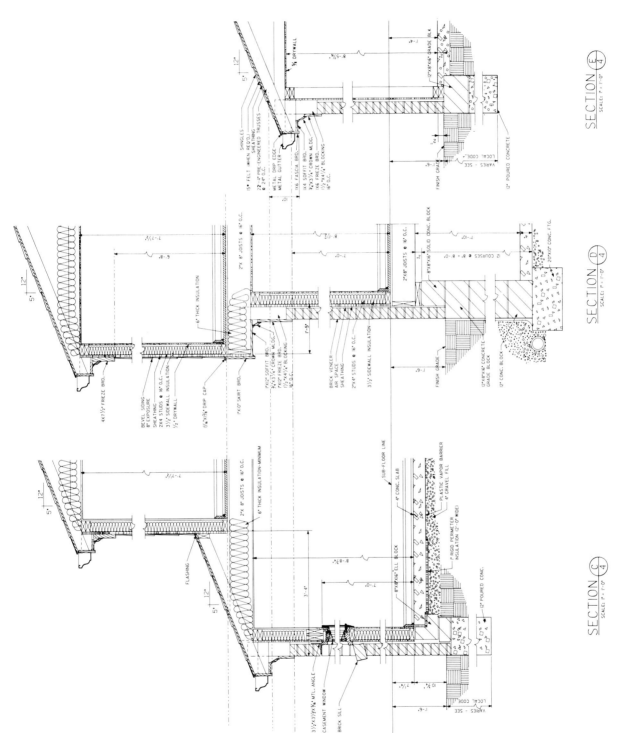

FIGURE 24-21B Contruction details C, D, and E (Courtesy of Home Planners, Inc.)

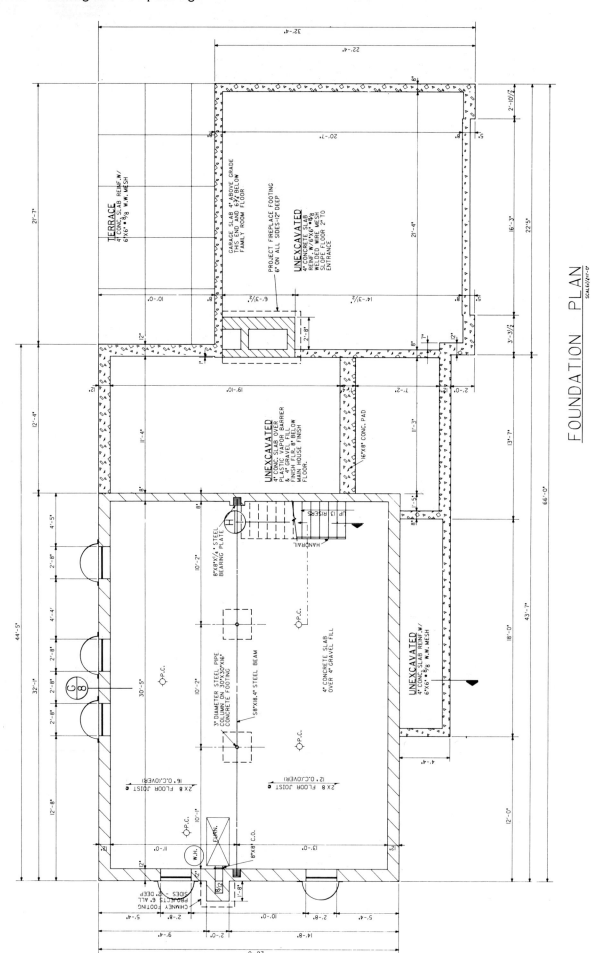

FIGURE 24-21C Foundation plan (Courtesy of Home Planners, Inc.)

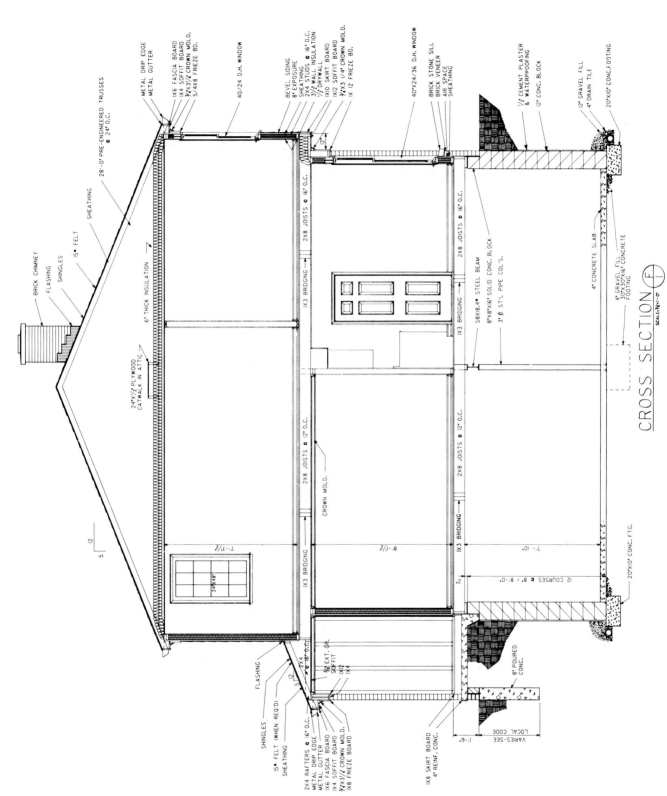

FIGURE 24-22 Full cross section through house (Courtesy of Home Planners, Inc.)

Stair assemblies are often complex and the intersections with floors difficult to describe on floor plans or interior elevation drawings. Sectional drawings are prepared to reveal stair details, material sizes, and dimensional relationships with floors and walls. **Figure 24–23** shows a typical stair detail drawing which covers three floor levels.

The internal structure of a fireplace and chimney system can only be described on a sectional drawing. Fireplace sections **(Fig. 24–24)**, describe the size and position of all masonry materials. They also show the methods of intersection with slabs, floors, walls, and roofs.

When a designer wants to specify an exact size and shape of a detail, a full-size sectional drawing may be used. For example, the details shown in **Fig. 24–25** are prepared so that the builder knows to use these specific trim shapes. Otherwise, a builder is free to use any molding with these overall dimensions.

SUPPORT DOCUMENTS

In addition to the drawings prepared to describe a building, many other documents are needed to insure a building is constructed exactly as designed. These include schedules, specifications, budgets, contracts, and bids. Explanations relating to the set of plans or structural limitations **(Fig. 24–26)** may also be included to emphasize a design concern.

Schedules

Schedules are listings of materials and components used in the construction of a building. Schedules contain information, in tabular form that is too extensive or detailed to include on a drawing. This includes such data as detail dimensions, materials, colors, style numbers, and manufacturers. Schedules are commonly prepared for doors, windows, finishing methods and materials, fixtures, appliances, and furniture.

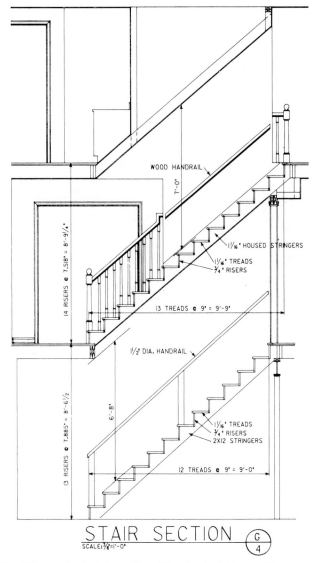

FIGURE 24–23 Detail of stair section (Courtesy of Home Planners, Inc.)

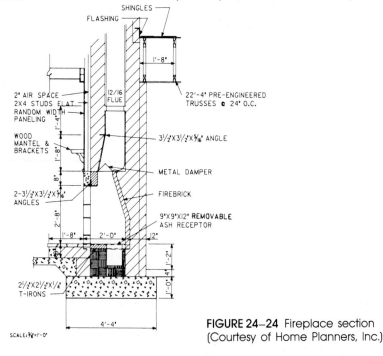

FIGURE 24–24 Fireplace section (Courtesy of Home Planners, Inc.)

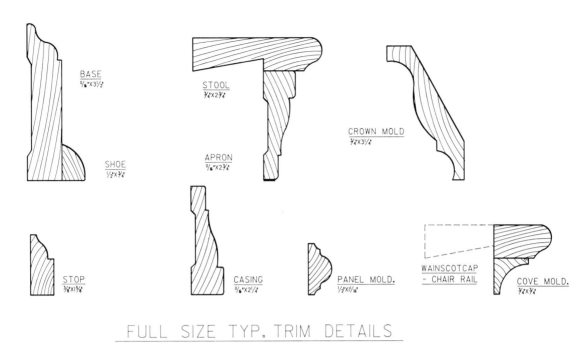

FIGURE 24–25 Trim details (Courtesy of Home Planners, Inc.)

FIGURE 24–26 Floor area, abbreviations, and design loads (Courtesy of Home Planners, Inc.)

Specifications

Specifications are lists of materials (Figs. 24–27A and B) that describe the size, type, and quantity of the building materials and components to be used in the construction of a building. It also includes a list for all building procedures not covered in the working drawings. Specifications are organized by building-material type as outlined by the **Construction Specification Institute (CSI)**. The CSI has established a format that subdivides building areas into sixteen headings, and subdivides each heading into the various building trades. This results in better construction documentation which means safer, more efficient, and more profitable construction for builders.

LINE NO.	QUANTITY REQUIRED	MATERIAL DESCRIPTION DESIGN NO. 1285 COLUMN NO. 2	UNIT COST	TOTAL COST
1	ROOFING & SHEET METAL			
2	3/4 SQUARES	ASHPHALT RIDGE SHINGLES		
3	23 SQUARES	ASHPHALT SELF-SEALING SHINGLES		
4	328 LIN.FT.	METAL DRIP EDGE		
5	70 LBS.	ROOFING NAILS		
5	126 PCS.	5"X7" METAL STEP FLASHING		
7	16 LIN.FT.	CHIMNEY COUNTER		
8	28 LIN.FT.	CORNICE RETURN		
9	44 LIN.FT.	WALL TO ROOF		
10	14 LIN.FT.	WINDOW HEAD		
11	214 LIN.FT.	12" GIRTH "K" SECTION SHEET METAL GUTTER		
12	82 LIN.FT.	2"X3" SHEET METAL CONDUCTOR PIPE & FITTINGS		
13				
14	EXTERIOR FINISH			
15		WINDOWS		
16		ALL WINDOWS WITH BRICK MOLD CASING		
17	4 SINGLE	24"X36" CASEMENT WINDOWS CUT 3-WIDE 3-HIGH		
18	1 SINGLE	24"X48" CASEMENT WINDOWS CUT 3-WIDE 4-HIGH		
19	1 DOUBLE	16"X36" CASEMENT WINDOWS CUT 2-WIDE 3-HIGH		
20	4 SINGLE	40"X24/24" DOUBLE HUNG WINDOW CUT 4-WIDE 4-HIGH		
21	3 SINGLE	40"X24/36" DOUBLE HUNG WINDOW CUT 4-WIDE 5-HIGH		
22	1 DOUBLE	32"X24/36" DOUBLE HUNG WINDOW CUT 4-WIDE 5 HIGH		
23	10 LIN.FT.	1"X4" OGEE PANEL MOLD		
24	1 PC.	2'X4'X1/8" ASBESTOS CEMENT WALLBOARD WINDOW PANEL		
25	3 SINGLE	24"X36" CASEMENT WINDOWS CUT DIAMOND PATTERN		
26				
27				
28		DOOR FRAMES		
29	1 REAR	2'-8"X6'-8" RABBETED 1-3/4"X 5/4"X4" CASING		
30	2 TERRACE	6'-0"X6'-8" SLIDING DOOR, FRAME, TRACK & HDW.		
31		BRICK MOLD CASING		
32	1 GARAGE	16'-0"X7'-0" 2X6" OVERHEAD DOOR FRAME,		
33		5/4"X6" JAMBS, 5/4"X10" AND 5/4"X4" HEAD		
34		CASING WITH 3/4"X3 1/4" CROWN MOLD AND 1/2" ROUND		
35	1 UNIT	5'-4"X6'-8" RABBETED FOR 3'-0X6'-8" X 1 3/4"		
36		DOOR AND WITH TWO 10"X76" SIDE-		
37		LIGHTS CUT 1-WIDE 6-HIGH BRICK MOLD CASING		
38				
39		CORNICE TRIM		
40	214 LIN.FT.	1"X6" FASCIA BOARD		
41	214 LIN.FT.	1"X4" SOFFIT BOARD		
42	292 LIN.FT.	3/4"X3 1/4" CROWN MOLD		
43	80 LIN.FT.	1"X8" FRIEZE BOARD		
44	104 LIN.FT.	5/4"X8" FRIEZE BOARD		
45	68 LIN.FT.	1"X12" SOFFIT & FRIEZE BOARD		
46	18 LIN.FT.	1"X4" FRIEZE BOARD		
47	92 SQ.FT.	3/8" EXTERIOR GRADE PLYWOOD SOFFIT BOARD		
48	24 LIN.FT.	1X4" BEAM SOFFIT BOARD-CUT TO FIT		
49	32 LIN.FT.	3/4"X4 1/4" WINDOW HEAD SOFFIT FILLER		
50	116 LIN.FT.	5/4"X6" RAKE BOARD		
51	116 LIN.FT.	5/4"X2-3/4" RAKE SHINGLE MOLD		
52	2 UNITS	48" RIGHT ANGLE 5/12 PITCH-TRIANGLE		
53		GABLE LOUVERS-SCREENED		
54	2 UNITS	16"X24" GABLE LOUVERS-SCREENED BRICK MOLD CASING		
55	1 UNIT	60" TRIANGLE SCREENED GABLE LOUVER		
56				
57		PORCH COLUMN TRIM		
58	2 PCS.	18" X11 1/2"X3-5/8 " ORNAMENTAL WOOD BRACKETS		
59				
60		SIDING & WATER TABLE		
61	1636 LIN.FT.	5/8"X8 3/4" BEVEL SIDING		
62	40 LIN.FT.	1"X10" SKIRT BAORD		
63	40 LIN.FT.	1 1/16"X1 3/8" DRIP CAP MOLD		
64				
65	170 SQ.FT.	1/8" CEMENT ASBESTOS WALLBOARD		
66	118 LIN.FT.	1"X3" BATTEN STRIPS		
67				
68		CUPOLAS		
69	1 UNIT	30"X30"X48" LOUVERED CUPOLA COMPLETE		
70		WITH SHEET METAL ROOF AND WEATHERVANE		
71				
72		SHUTTERS		
73	3 PAIRS	16"X42" LOUVERED SHUTTERS		
74	2 PAIRS	20"X54" LOUVERED SHUTTERS		
75	2 PAIRS	20"X66" LOUVERED SHUTTERS		
76				
77	INTERIOR FINISH			
78		EXTERIOR DOORS		
79	1 REAR	2'-8"X6'-8"X1 3/4" 4-PANEL GRADE WITH		
80		22"X36" GLAZED PANEL		
81	1 GARAGE	16'-0"X7'-0"X1 3/4" OVERHEAD DOOR, TRACK		
82		AND HARDWARE		
83	1 FRONT	3'-0"X6'-8"X1 3/4" DIAMOND PANEL COLONIAL-		
84		GLAZED 26"X36" CUT 3-WIDE 3 HIGH		
85				
86		INSULATION		
87	1306 SQ.FT.	6" THICK CEILING INSULATION		
88	1672 SQ.FT.	3 1/2" SIDEWALL INSULATION		
89				
90		DRYWALL WALLBOARD		
91	613 SQ.FT.	5/8" GYPSUM DRYWALL WALLBOARD		
92	6965 SQ.FT.	1/2" GYPSUM DRYWALL WALLBOARD		
93	2526 LIN.FT.	PERFORATED DRYWALL JOINT TAPE		
94	420 LBS.	DRYWALL JOINT COMPOUND		
95	60 LBS.	HOLDTITE DRYWALL NAILS		
96	420 LIN.FT.	METAL DRYWALL CORNERBEADS		
97				
98		TILEWORK		
99	10 SQ.FT.	3/8" VERMONT SLATE HEARTH		
100	3 PCS.	MARBLE THRESHOLD		
101	94 SQ.FT.	UNGLAZED CERAMIC TILE FLOOR		
102	124 SQ.FT.	GLAZED CERAMIC TILE WALLS		
103	3 PCS.	PAPER HOLDERS		
104	2 PCS.	SOAP & GRAB BARS		
105	6 PCS.	TOWEL BARS		
106				
107		FINISH FLOORS		
108	526 SQ.FT.	5/8" PLYWOOD FLOORING		
109	482 SQ.FT.	RESILIENT FLOOR TILE		
110	1783 BD.MEAS.	HARDWOOD FLOORING		
111	1700 SQ.FT.	FLOORING FELT		
112	70 LBS.	8D FLOORING NAILS		
113				
114		STAIRS		
115	2 PCS.	2"X12"X13'-0" STRINGERS		
116	2 PCS.	5/4"X12"X14'-0" HOUSED STRINGERS		
117	20 PCS.	5/4"X10"X3'-0" TREADS		
118	5 PCS.	5/4"X10"X3'-4" TREADS		
119	21 PCS.	1"X8"X3'-0" RISERS		
120	6 PCS.	1"X8"X3'-4" RISERS		
121	11 LIN.FT.	5/4"X3 1/2" STARTER NOSING		
122	84 LIN.FT.	3/4" COVE MOLD		
123	2 PCS.	1 1/2"DIA. X13'-0" HANDRAIL		
124	2 PCS.	3 1/4"X3 1/4"X42" TURNED NEWEL POSTS		
125	11 PCS.	1 1/16"X1 1/16"X32" TURNED BALUSTERS		
126	6 LIN.FT.	2 1/8"X2 7/8" HANDRAIL		
127	48 PCS.	HARDWOOD WEDGES		
128				
129				
130				
131				
132				
133				
134				
135				
136				

FIGURE 24–27A Specification list for building materials

LINE NO.	QUANTITY REQUIRED	MATERIAL DESCRIPTION DESIGN NO. 1285 COLUMN NO. 1	UNIT COST	TOTAL COST
1				
2	**MASONRY**			
3		FOOTINGS		
4	88 LIN.FT.	20X10"		
5	45 LIN.FT.	16X8"		
5	35 LIN.FT.	12X42"		
7	104 LIN.FT.	8X42"		
8	11 LIN.FT.	10X42"		
9	2 PADS	30"X30"X16"		
10	1 PAD	3'-0"X3'-8"X12"		
11	1 PAD	4'-0"X6'-8"X12"		
12	22 CU.YDS.	2500# CONCRETE		
13	124 PCS.	4" DRAIN TILE		
14	8 PCS.	4" DRAIN ELS		
15	6 PCS.	4" DRAIN TEES		
16	6 PCS.	4" VITRIFIED CROCKS		
17	1 ROLL	15# 4" STRIP FELT		
18	5 CU.YDS.	PEA GRAVEL		
19	128 SQ.FT.	1" FIBERGLASS PERIMETER INSULATION		
20				
21		MASONRY BLOCK WALLS		
22	68 PCS.	12X8X16" MASONRY GRADE BLOCKS		
23	30 PCS.	12X8X16" MASONRY CORNER BLOCKS		
24	7 PCS.	12X8X8" MASONRY CORNER BLOCKS		
25	24 PCS.	8X8X16" MASONRY CORNER BLOCKS		
26	9 PCS.	8X8X8" MASONRY CORNER BLOCKS		
27	700 PCS.	12X8X16" MASONRY REGULAR BLOCKS		
28	198 PCS.	8X8X16" MASONRY REGULAR BLOCKS		
29	10 PCS.	10X8X16" MASONRY SOLID BLOCKS		
30	210 PCS.	8X8X16" MASONRY SOLID BLOCKS		
31	20 PCS.	4X8X12" MASONRY SOLID BLOCKS		
32	10 PCS.	4X8X16" MASONRY SOLID BLOCKS		
33	12 CU.YDS.	50/50 MASON SAND		
34	63 SACKS	MORTAR		
35	20 SACKS	CEMENT		
36	20 GALS.	ASPHALT FOUNDATION COATING		
37	2 PCS.	8"X8" CLEANOUT DOORS		
38	1 PC.	8" DIAM. STEEL FURNACE THIMBLE		
39	1 PC.	6" DIAM. STEEL WATER HEATER THIMBLE		
40	4 PCS.	8"X12" FLUE LININGS		
41				
42		BASEMENT STEEL & SASH		
43	2 PCS.	8"X8"X1/4" STEEL BEARING PLATES		
44	2 PCS.	3" DIAM. STEEL PIPE COLUMNS		
45	1 PC.	S8X18.4#X31'-2" STEEL BEAM		
46	5 UNITS	2-LIGHT 15"X12" STEEL SASH		
47	5 PCS.	37" DIAM.X24" CORRUGATED STEEL AREAWAY		
48				
49		VENEER & FIREPLACE		
50	1 UNIT	42" DOME DAMPER		
51	1 UNIT	9"X9"X12" REMOVABLE ASH RECEPTOR		
52	3 PCS.	3 1/2"X 3 1/2"X5/16"X4'-8" FIREPLACE LINTEL ANGLES		
53	6 PCS.	2 1/2"X2 1/2"X1/4"X22" HEARTH T-IRON S		
54	110 PCS.	FIRE BRICKS		
55	50 LBS.,	FIRE CLAY		
56	10 PCS.	12"X16" FLUE LINING		
57	8 PCS.	8"X12" FLUE LININGS		
58	3655 PCS.	COMMON BRICKS		
59	7700 PCS.	FACE BRICKS		
60	12 CU.YDS	50/50 MASON SAND		
61	85 SACKS	MORTAR		
62	750 PCS.	METAL WALL TIES		
63	5 PCS.	3 1/2"X3 1/2"X5/16" X3'-4" STEEL LINTEL ANGLES		
64	5 PCS.	2 1/2"X5"X32" FLAGSTONE SILL		
65	2 PCS.	4"X10"X6'-4" CUT STONE DOOR SILL		
66	1PC.	4"X7"X6'-4" CUT STONE DOOR SILL		
67	840 PCS.	FACE BRICKS		
68	2 PCS.	3 1/2"X3 1/2"X3/16 "X3'-4" STEEL LINTEL ANGLES		
69				
70				
71				
72		CONCRETE SLABS		
73		GARAGE & PORCHES (SIDEWALKS AND DRIVEWAY NOT INCLUDED)		
74	600 SQ. FT.	6"X6" #8/8 WELDED WIRE REINFORCING MESH		
75	3 CU. YDS.	FILL GRAVEL		
76	8 CU. YDS.	3000# CONCRETE		
77	6 GALS.	LIQUID CEMENT HARDENER		
78				
79		HOUSE FLOOR		
80	375 SQ.FT.	PLASTIC MEMBRANE VAPOR BARRIER		
81	16 CU. YDS.	FILL GRAVEL		
82	14 CU.YDS	3000# CONCRETE		
83	43 PCS.	4" VITRIFIED CROCKS		
84	5 PCS.	4" VITRIFIED 45° BENDS		
85	5 PCS.	4" VITRIFIED WYES		
86				
87	**FRAMING LUMBER**	(LUMBER SIZES BASED ON DESIGN STRESS, F=1500		
88		PLATFORMS		
89	116 LIN.FT.	2"X8" SILL PLATE		
90	32 LIN.FT.	2"X4" BEAM PLATE		
91	164 LIN.FT.	2"X8" JOIST TRIMMERS		
92	32 PCS.	2"X8"X12'-0" JOISTS		
93	12 PCS.	2"X8"X10'-0" JOISTS		
94	76 PCS.	2"X8"X14'-0" JOISTS		
95	16 PCS.	2"X8"X16'-0" JOISTS		
96	290 LIN.FT.	1"X3" BRIDGING (OPTIONAL)		
97	1800 SQ.FT.	SUBFLOORING		
98	1 UNIT	3 PCS. 2"X8"X13'-6" REINFORCED WITH 2 PCS.		
99		1/2"X7"X13'-6" STEEL PLATE CEILING BEAM		
100	100 LIN.FT.	2"X4" ATTIC CATWALK RUNNERS		
101				
102		EXTERIOR WALLS		
103	78 PCS.	2"X4"X9'-0" STUDS		
104	15 PCS.	2"X4"X12'-0" GABLE STUDS		
105	5 PCS.	2"X4"X10'-0" GABLE STUDS		
106	300 PCS.	2"X4"X8'-0" STUDS		
107	1382 LIN.FT.	2"X4" PLATES		
108	2 PCS.	2"X12"16'-6" REINFORCED WITH ONE 1/2"X11"X16'-6"		
109		STEEL PLATE-GARAGE DOOR HEADER		
110	1 PC.	2"X12"X12'-0" HEADERS		
111	3 PCS.	2"X12"X14'-0" HEADERS		
112	3 PCS.	2X8X8'-0" HEADERS		
113	80 LIN.FT.	2"X6" HEADERS		
114	2 PCS.	2"X12"X18'-0" PORCH HEADERS		
115	2800 SQ.FT.	EXTERIOR WALL SHEATHING		
116	3100 SQ.FT.	15# FELT (OPTIONAL)		
117				
118		INTERIOR PARTITIONS		
119	45 PCS.	2"X4"X9'-0" STUDS		
120	210 PCS.	2"X4"X8'-0" STUDS		
121	1000 LIN.FT.	2"X4" PLATES		
122	174 LIN.FT.	2"X6" HEADERS		
123				
124		ROOFING & FRAMING		
125	5 UNITS	25'-2" 5/12 PITCH PRE-ENGINEERED TRUSSES		
126	15 UNITS	28'-0" 5/12 PITCH PRE-ENGINEERED TRUSSES		
127	11 UNITS	22'-4" 5/12 PITCH PRE-ENGINEERED TRUSSES		
128	8 PCS.	2"X4"X16'-0" NAILER RAFTERS & CEILING JOISTS		
129	28 PCS.	2"X4"X8'-0" PORCH CEILING JOIST & RAFTERS		
130	80 LIN.FT.	1 1/2"X4 1/4" CORNICE BLOCKING		
131	5 PCS.	2"X4"X12'-0" OVERHANG RAFTERS & SOFFIT FRAMING		
132	2140 SQ.FT.	ROOF SHEATHING		
133	3300 SQ.FT.	15# FELT		
134				
135				
136				

LINE NO.	QUANTITY REQUIRED	MATERIAL DESCRIPTION DESIGN NO. 1285 COLUMN NO. 3	UNIT COST	TOTAL COST
1		PANELING & FALSE BEAMS		
2	414 SQ.FT.	3/4" RANDOM WIDTH, V-CUT, T&G PANELING		
3	118 LIN.FT.	3/4"X3 1/2" CROWN MOLD		
4	32 LIN.FT.	3/4"X2 3/4" CUT DOWN WINDOW STOOL CHAIR RAIL		
5	32 LIN.FT.	3/4"X3/4" COVE MOLD		
5	108 LIN.FT.	1/2"X1 1/16" PANEL MOLD		
7	118 LIN.FT.	2"X4" BEAM BLOCKING		
8	70 LIN.FT.	1"X2" BEAM BOTTOM		
9	48 LIN.FT.	1"X4" BEAM BOTTOM		
10	70 LIN.FT.	1"X8" BEAM SIDES		
11	96 LIN.FT.	1"X6" BEAM SIDES		
12				
13		FIREPLACE TRIM		
14	16 LIN.FT.	1 1/2"X3 1/2" MANTEL SURROUND TRIM		
15	1 PC.	5/4"X7 1/2"X8'-0" MANTEL SHELF		
16	5 PCS.	5 1/2"X4-3/4"X4 1/2" ORNAMENTAL WOOD BRACKET		
17	8 LIN.FT.	3/4"X1 1/4" COVE MOLD		
18				
19				
20		INTERIOR DOORS		
21	2 INTERIOR	1'-6"X6'-8"X1 3/8" FLUSH		
22	1 INTERIOR	2'-0"X6'-8"X1 3/8" FLUSH		
23	2 INTERIOR	2'-4"X6'-8"X1 3/8" FLUSH		
24	7 INTERIOR	2'-6"X6'-8"X1 3/8" FLUSH		
25	1 INTERIOR	2'-8"X6'-8"X1 3/8" FLUSH		
26	1 INTERIOR	2'-8"X6'-8"X1 3/4" MINERAL CORE FIRE DOOR		
27				
28		ALL BIFOLD DOORS COMPLETE WITH TRACK & HARDWARE		
29	5 INTERIOR	3'-0"X6'-8"X1 3/8" 2- SECTION BIFOLD		
30	3 INTERIOR	4'-0"X6'-8"X1 3/8" 2-SECTION BIFOLD		
31	2 INTERIOR	5'-0"X6'-8"X1 3/4" 2-SECTION BIFOLD		
32				
33		DOOR FRAMES		
34	1 SET	2'-0"X6'-8"X4 5/8" JAMBS WITH STOPS		
35	2 SET	2'-4"X6'-8"X4 5/8" JAMBS WITH STOPS		
36	6 SET	2'-6"X6'X-8"X4 5/8" JAMBS WITH STOPS		
37	2 SETS	2'-8"X6'-8"X4 5/8" JAMBS WITH STOPS		
38	4 SETS	3'-0"X6'-8"X4 5/8" JAMBS WITH STOPS		
39	2 SETS	5'-0"X6'-8"X4 5/8" JAMBS NO STOPS		
40				
41		DOOR TRIM		
42	2 SIDES	2 SIDES 2'-0"X6'-8" CASING		
43	4 SIDES	2'-4"X6'-8" CASING		
44	12 SIDES	2'-6"X6'-8" CASING		
45	5 SIDES	2'-8"X6'-8" CASING		
46	9 SIDES	3'-0"X6'-8" CASING		
47	4 SIDES	5'-0"X6'-8" CASING		
48	1 SIDE	6'-0"X6'-8" CASING		
49	1 SIDE	6'-6"X6'-8" CASING		
50				
51		WINDOW TRIM		
52	67 LIN.FT.	3/4"X2 3/4" WINDOW STOOL		
53	320 LIN.FT.	1 1/16"X2 1/4" CASING & APRON		
54				
55		RUNNING TRIM		
56	566 LIN.FT.	1 1/16"X3 1/2" BASE		
57	588 LIN.FT.	1/2"X3/4" SHOE		
58	42 LIN.FT.	1"X12" SHELVING		
59	58 LIN.FT.	1"X3" HOOK STRIP		
60	96 LIN.FT.	1"X2" SHELF CLEAT		
61	4 PCS.	4'-0"X8'-0"X1/2" PLYWOOD SHELVING		
62				
63		CABINETS		
64		BATHS		
65	1 BASE	47 1/2"X32"X21" LAVATORY CABINET		
66	1 BASE	53"X32"X21" LAVATORY CABINET		
67	1 BASE	72"X32"X21" LAVATORY CABINET		
68	1 PC.	47 1/2"X21" PLASTIC LAMINATE COUNTER & SPLASH		
69	1 PC.	53"X21" PLASTIC LAMINATE COUNTER & SPLASH		
70	1 PC.	72"X21" PLASTIC LAMINATE COUNTER & SPLASH		
71	3 UNITS	16"X36" MEDICINL CABINETS		
72	1 UNIT	60"X60" ALUMINUM & GLASS TUB ENCLOSURE		
73	1 UNIT	36"X72" ALUMINUM & GLASS SHOWER ENCLOSURE		
74				
75		KITCHEN		
76	1 BASE	24"X36"X24" DISHWASHER		
77	1 BASE	18"X36"X24" 4-DRAWER STORAGE CUPBOARD		
78	1 BASE	33"X36"X24" SINK AND STORAGE CUPBOARD		
79	1 BASE	42"X36"X24" SINK AND STORAGE CUPBOARD		
80	2 BASE	30"X36"X24" STORAGE CUPBOARD		
81	1BASE	30"X36"X24" COOKING TOP & STORAGE CUPBOARD		
82	1 BASE	24"X60"X24"X OVEN AND STORAGE CUPBOARD		
83	1 BASE	36"X60"X24"X BROOM STORAGE CUPBOARD		
84	2 UPPER	30"X30"X12" 3-SHELF STORAGE CUPBOARD		
85	1 UPPER	36"X24"X12"X 2-SHELF STORAGE CUPBOARD		
86	2 UPPER	33"X30"X12" 3-SHELF STORAGE CUPBOARD		
87	2 UPPER	18"X30"X12"X 3-SHELF STORAGE CUPBOARD		
88	1 UPPER	15"X30"X12" 3-SHELF STORAGE CUPBOARD		
89	1 UPPER	24"X18"X12" 1-SHELF STORAGE CUPBOARD		
90	1 UPPER	30"X24"X12"X 2-SHELF STORAGE CUPBOARD		
91	1 UPPER	36"X15"X12" 1-SHELF STORAGE CUPBOARD		
92	1 PC.	U-SHAPED 78"X26"+70"X25"+57"X25"		
93		PLASTIC LAMINATED COUNTER & SPLASH		
94	1 PC.	60"X25" PLASTIC LAMINATE COUNTER & SPLASH		
95	1 UNIT	30" RANGE HOOD, EXHAUST FAN & DUCT		
96				
97	**FINISH HARDWARE**			
98	1 FRONT	DOOR LOCK SET		
99	2 REAR	DOOR LOCK SETS		
100	1 SET	SAFETY CHAIN		
101	1 SET	MAIL SLOT HARDWARE		
102	2 SETS	MORTISE DEAD BOLT		
103	1 SET	AUTOMATIC FIRE DOOR CLOSER		
104	3 PAIRS	4"X4" VUTTS		
105	3 INTERIOR	PRIVACY LOCK SETS		
106	8 INTERIOR	PASSAGE LATCH SETS		
107	2 INTERIOR	PASSAGE KNOB SETS		
108	2 PCS.	3/4" BULLET CATCHES		
109	13 PAIRS	3 1/2"X3 1/2" BUTTS		
110	16 PCS.	DOOR STOPS		
111	4 SETS	HANDRAIL BRACKETS		
112	8 PCS.	METAL ADJUSTABLE CLOSET POLES		
113	12 SETS	DOUBLE HUNG SASH LOCKS		
114	24 SETS	DOUBLE HUNG SASH LIFTS		
115	1 PINT	GLUE		
116	1 REAM	SAND PAPER		
117				
118				
119				
120	**ROUGH HARDWARE**			
121	10 LBS	8D MASONRY NAILS		
122	50 lbs.	16D COMMON NAILS		
123	250 lbs.	8D COMMON NAILS		
124	25 lbs.	ROOFING		
125	40 lbs.	8D CASING		
126	30 lbs.	6D CASING		
127	20 lbs.	8D FINISH		
128	10 lbs.	6D FINIAH NAILS		
129	2 lbs.	4D FINISH NAILS		
130	36 pcs.	8" JOIST ANGLES		
131	25 LBS.	8D ALUMINUM SINKER HEAD SIDING NAILS		
132				
133				
134				
135				
136				

FIGURE 24–27B Specification list for building materials (Courtesy of Home Planners, Inc.)

TYPES OF BUILDINGS

This chapter has referred to residential buildings. It should be understood that there are several types of architecturally designed structures. The following is a general outline of these categories:

1. A single-family dwelling is a residence for one family only.
2. Multiple-family dwellings are multiple-living residences, such as a duplex, triplex, or fourplex. The structure is usually two or three stories high.
3. High-rise dwellings are multiple-living residences of four stories or more. The height of the building is limited by the zoning codes and the structural engineering.
4. Industrial plants are structures that house the production operations of our industries. They may range from small buildings to huge structures covering a vast amount of land.
5. Commercial buildings are structures that range from small business offices to huge skyscrapers. These structures house such operations as business offices and wholesale/retail stores. An example of a partial set of working drawings for a 7–ELEVEN convenience store is shown in **Figs. 24–28** through **24–32**.
6. Other types of structures that must be designed and drawn by architects, engineers, and drafters are government buildings, military structures, educational facilities, medical buildings, entertainment centers, transportation centers, bridges, tunnels, and highways.

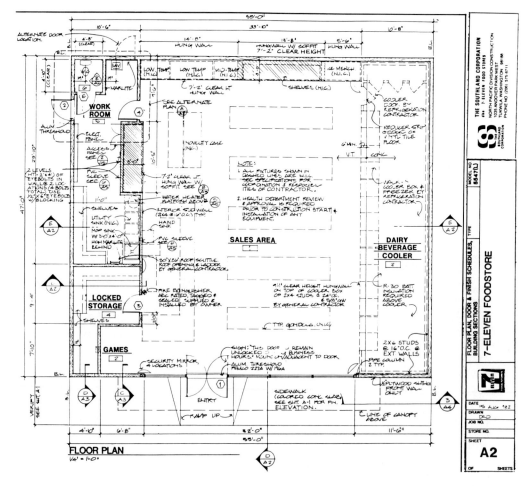

FIGURE 24–28 Commercial floor plan for a 7-ELEVEN convenience store (Courtesy of The Southland Corporation, Washington Division)

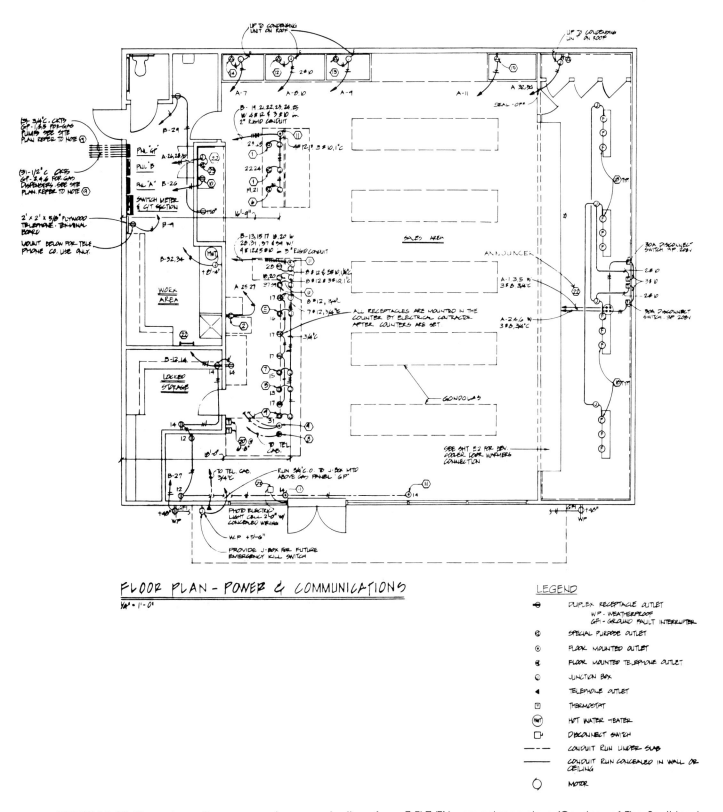

FIGURE 24–29 Floor plan with power and communications for a 7-ELEVEN convenience store (Courtesy of The Southland Corporation, Washington Division)

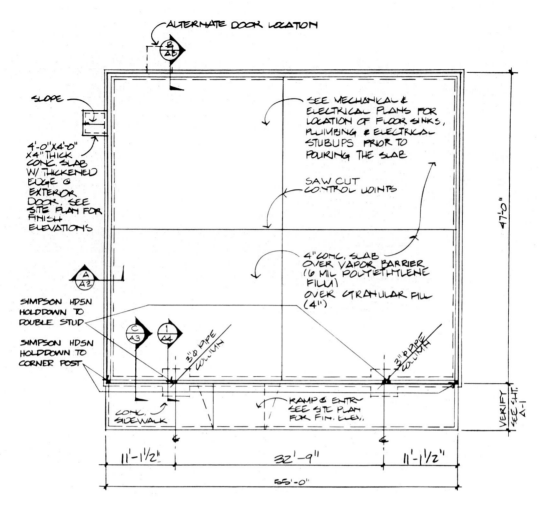

FIGURE 24–30 Foundation plan for a 7-ELEVEN convenience store. (Courtesy of The Southland Corporation, Washington Division)

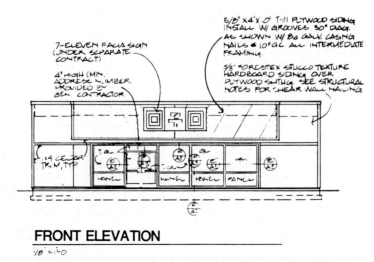

FIGURE 24–31 Front elevation for a 7-ELEVEN convenience store. (Courtesy of The Southland Corporation, Washington Division)

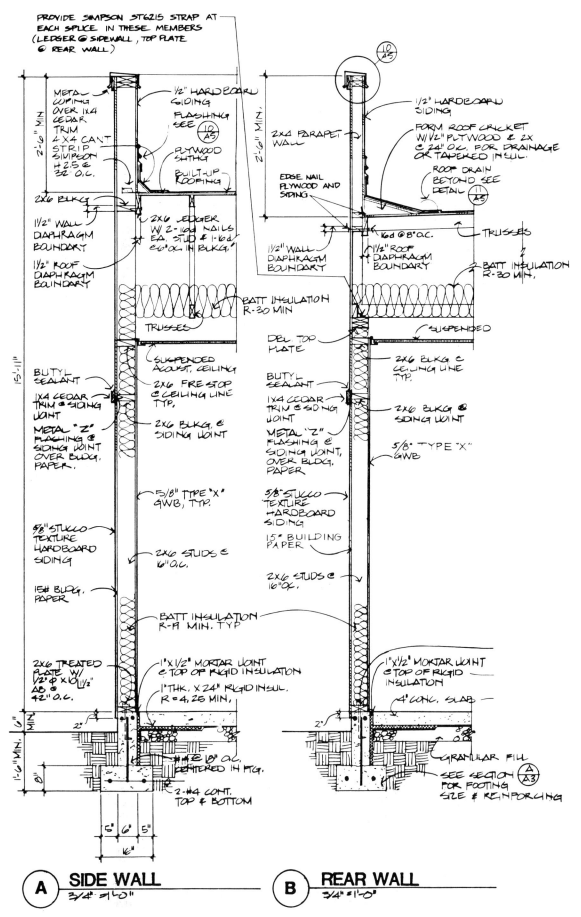

FIGURE 24-32 Construction details for a 7-ELEVEN convenience store. (Courtesy of The Southland Corporation, Washington Division)

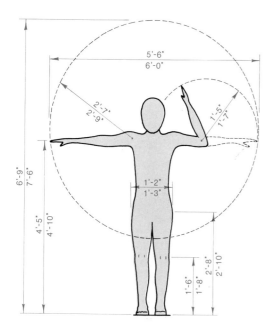

FIGURE 24-33 Typical human dimensions for women and men

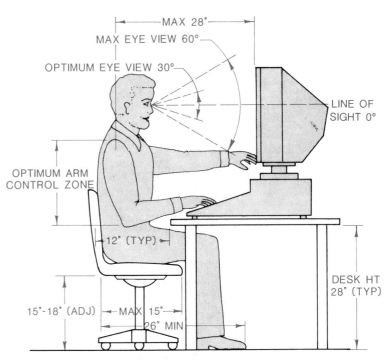

FIGURE 24-34 Human design data for desk work

ERGONOMICS

To enable people to do their best, the areas where they live, work, and play must be carefully designed to match their physical size, shape, reach, and mobility. This is the science of **ergonomics.** In order to meet certain design requirements, the human figure must be considered in all phases of the design **(Figs. 24-33 and 24-34).** Careful planning and an awareness of the specific needs of people is essential for good design.

The Handicapped

Part of ergonomic planning is to create an environment sensitive to the needs of individuals confined to a wheelchair, or to those who are sight impaired or hearing impaired. Wherever there is a properly designed environment for the handicapped, an international handicap symbol must be posted nearby **(Fig. 24-35).**

The following factors will affect a design for persons confined to a wheelchair **(Figs. 24-36** through **24-39):**

1. Eye level is between 43" and 51".
2. Lap level is about 27", therefore clearance lavatories must be 30".
3. Maximum reaching height is 54".
4. Maximum slope of ramps is 1:12 (8.3% slope or 4°-46′).
5. Minimum width of ramps and halls is 36".
6. Sidewalks at corners should have gradient ramps to the street.
7. Adjacent to the automobile, the loading zone must be 4′ x 20′.
8. Minimum width of doors is 32" (36" preferred).
9. Doors should open 90°.
10. Minimum 5′ turning area.
11. Water closet (toilet) height is 18".
12. Area for water closet stall is 36" x 60".
13. Area adjacent to bath tub is 36" x 60".
14. Area adjacent to lavatory is 30" x 48".
15. Low mirror installation.

The following items must be included in a design for persons with a hearing impairment:

FIGURE 24-35 This sign is posted for public areas that are ergonomically designed for the handicapped.

1. Eye level signs
2. High-frequency alarms
3. Video warning signals
4. Flashing signal lights
5. Telephone amplifier attachment

The following are factors that will affect a design for persons with a sight impairment:

1. All floors should have nonslip surfaces.
2. Braille signs should be posted.
3. All stairs should have uniform tread widths and riser heights.

4. There should be no projections from floors or walls.
5. There should be no protruding walls or corners.
6. There should be no swinging doors (use sliding doors).
7. There should be guardrails on walls.
8. Noise alarms should be installed.

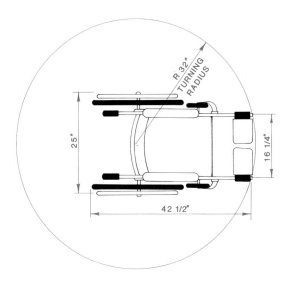 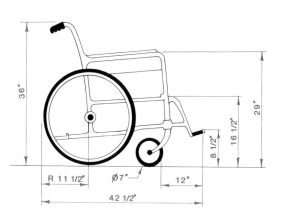

FIGURE 24-36 Standard size wheel with turning radius and vertical heights needed for clearance

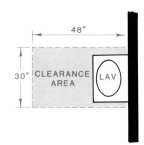

FIGURE 24-37 The maneuverable area required for a wheelchair under a lavatory

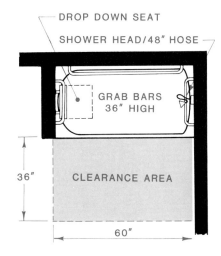

FIGURE 24-38 The maneuverable area required for a wheelchair next to a bathtub

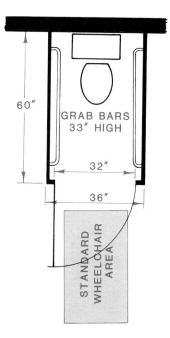

FIGURE 24-39 The maneuverable area required for a wheelchair in a water closet stall

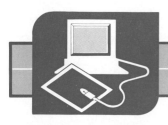

COMPUTER-AIDED DRAFTING & DESIGN

COMPUTER-AIDED ARCHITECTURAL DRAFTING

The standardization of architectural symbols makes the preparation of floor plans, elevations, and details on a CAD system extremely fast, accurate, and convenient. Specialized drawings, such as electrical, plumbing, and HVAC plans, are also easy to prepare using the LAYERING command. In addition, computer spreadsheeting and special software programs greatly improve speed and accuracy in preparing schedules, specifications, budgets, contracts, and bids.

There are many types of architectural drafting software available for specific drawing needs, such as an architectural menu tablet overlay **(Fig. 24–40)**. This tablet contains all common drawing and editing commands, such as LINE, CIRCLE, ERASE, MOVE, COPY, TEXT, and DIMENSIONING. It also contains an array of architectural commands and

FIGURE 24–40 Architectural menu tablet for a CAD system (Courtesy of AutoCAD. AutoCAD is a trademark of Autodesk, Inc.)

symbols. These commands are divided into the following five areas:

1. WALLS—This command area allows the drafter to draw walls to a scaled thickness as easily as drawing a single line. It also allows for the insertion of insulation into walls.
2. DIMENSIONING—In addition to the normal dimensioning commands in the lower-left corner of the tablet, this command area provides automatic dimensions in an architectural style. This includes optional tick marks in place of arrows, measurements in feet and inches, and the placement of dimensions above dimension lines.
3. DOORS and WINDOWS—This command area consists of standard floor-plan symbols for doors and windows that can be inserted into previously drawn walls.
4. TAGS—This command area includes common symbols for removed sections and callouts on architectural drawings.
5. SHAPES—Standard symbols for plumbing fixtures, electrical and lighting components, furniture, and other commonly used architectural symbols are located in this command area.

The architectural menu overlay shown in Fig. 24–40 is used in the following sequence to complete a floor plan. The same steps can be followed to prepare elevations and structural detail drawings.

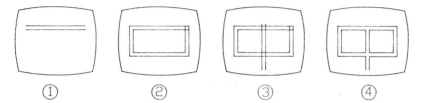

FIGURE 24–41 Drawing walls with the WALL command

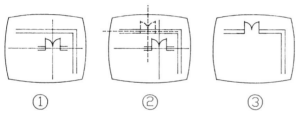

FIGURE 24–42 Inserting symbols from the library onto the floor plan

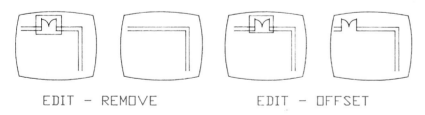

FIGURE 24–43 Editing the symbols on the floor plan with the EDIT command

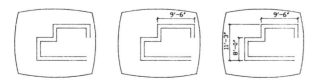

FIGURE 24–44 Dimensioning the floor plan with the architectural DIMENSIONING command

Step 1. Draw parallel lines using the WALL command **(Fig. 24–41)**.

Step 2. Select the distance between lines. After blocking in walls, automatically trim the corners using one of the three selections under the WALL command (Fig. 24–41).

Step 3. Select the desired door or window (single-swing, double-leaf door in **Fig. 24–42**) with the puck. The symbol is inserted into the wall. Note the wall has been automatically trimmed without any editing, although editing is extremely easy with the use of an architectural menu. The two examples shown in **Fig. 24–43** illustrate how easily symbols may be erased or offset (moved).

Step 4. The DIMENSION commands provide proper horizontal and vertical dimensions for architectural drawings **(Fig. 24–44)**.

Figures 24–45 through 24–49 illustrate the following steps used in constructing a simple floor plan for a building addition using common architectural commands and symbols on a CAD system:

Step 1. Block-in and trim all the walls **(Fig. 24–45)**.

Step 2. Insert all of the doors and windows **(Fig. 24–46)**.

Step 3. Insert the electrical symbols representing wall outlets, switches and overhead lights **(Fig. 24–47)**.

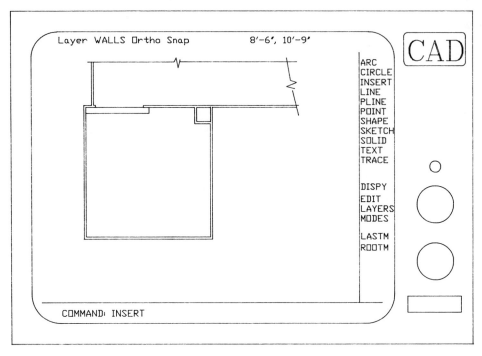

FIGURE 24–45 Block in walls.

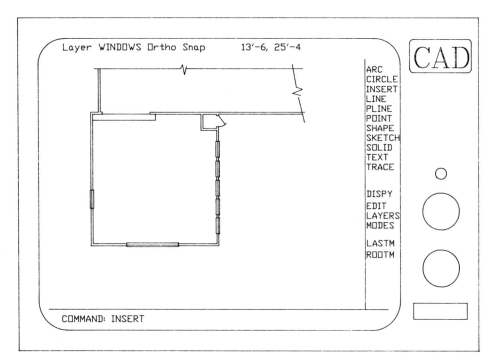

FIGURE 24–46 Insert window and door symbols.

Step 4. Label the drawing using the TEXT command **(Fig. 24–48)**.

Step 5. The architectural dimensioning command is used for inserting dimensions using the architectural dimensioning style **(Fig. 24–49)**.

Once an architectural drawing is completed on a CAD system, the drawing can be generated on paper with a plotter or a laser printer **(Fig. 24–50)**.

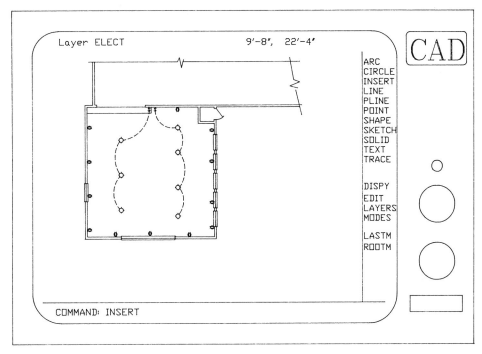

FIGURE 24–47 Insert electrical symbols.

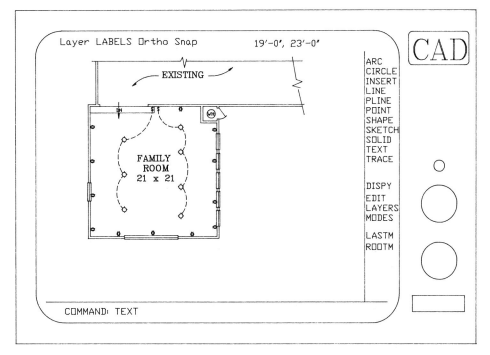

FIGURE 24–48 Insert text.

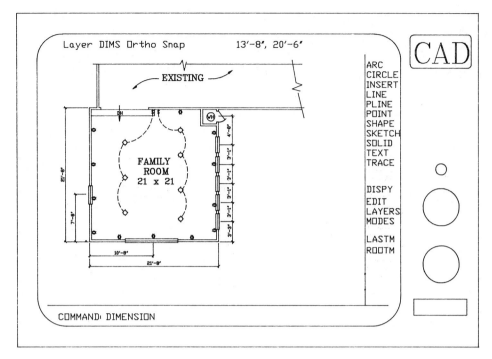

FIGURE 24-49 Place dimensions.

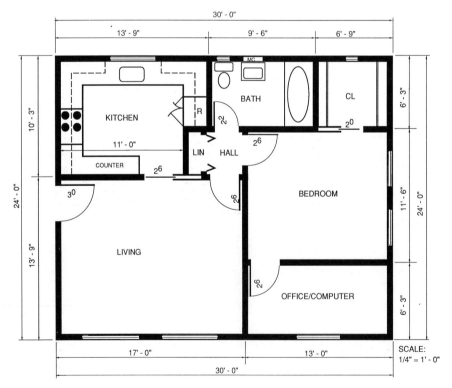

FIGURE 24-50 Small floor plan designed with a CAD system and printed with a laser printer

EXERCISES

1. Redraw the architectural symbols in Figs. 24–12A through G (pages 479–480) with drawing instruments. Keep the proportions of each symbol the same, and size them to be easily read on $1/4'' = 1'-0''$ scaled drawings.
2. With a CAD system, draw the architectural symbols in Figs. 24–12A through G (pages 479–480), and file them in a CAD-system library.
3. Draw one of the floor plans shown in Figs. 24–4A through L (pages 468–475). Use a scale of $1/4'' = 1'-0''$. Use drawing instruments or a CAD system.
4. With drafting instruments or a CAD system, prepare the following architectural drawings in **Fig. 24–51**:
   - floor plan (scale: $1/4'' = 1'-0''$) Redesign and correct dimensional changes
   - exterior elevations (scale: $1/4'' = 1'-0''$)
   - roof plan (scale: $1/8'' = 1'-0''$)
5. Add the following drawings to the plan set for Fig. 24–51:
   - plot plan (scale: $1/8'' = 1'-0''$)
   - foundation plan (scale: $1/4'' = 1'-0''$)
   - window schedule
   - door schedule
   - electrical plan (scale: $1/4'' = 1'-0''$)
6. Remodel the plan in Fig. 24–51 to include the following:
   - four bedrooms
   - larger living area
   - fireplace
   - attached two-car garage
7. Design a house to fit the needs of your family. If possible, find an actual building site to use for your design. Draw the floor plan and plot plan.
8. Copy the set of working drawings in this chapter to scale as noted.
9. Obtain or design a site plan. Design a house and a set of architectural working drawings for the site.
10. Design a one-bedroom cabin on a CAD system. Draw the floor plan, plot plan, and four eleva-

FIGURE 24–51 Follow the drawing instructions in Exercises 4, 5, and 6.

tions with a CAD system.

11. Design a vacation weekend cabin for a family of five. Children include two boys, ages 10 and 13, and one girl, age 7. The site **(Fig. 24-52)**, is located adjacent to a large lake suitable for swimming and nonpower boating. The zoning data includes the lot size is 6600 sq. ft. with an allowable building area of 2600 sq. ft. Local codes restrict the structural coverage to a 40% maximum.

    The area of the plan must be 1000 sq. ft. or less. Indicate the square footage on the plan and include two overall dimensions, plus the size of each room. Draw the symbols and location of all fixtures, appliances, windows, and doors. Add the location and symbols for electrical switches, lights, and outlets. Remember to show the compass orientation on the plan as an aid to correct orientation. Be careful to avoid the following:

    - Rooms too small
    - Miscalculations of square footage
    - Offset rooms
    - Too much hall space
    - Lack of storage or poorly designed storage space
    - Disregarding the design criteria and parameters given

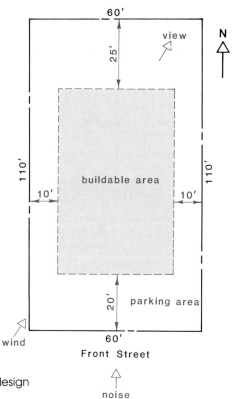

FIGURE 24-52 Follow the design instructions in Exercise 11.

KEY TERMS

Architectural pictorial drawings
Composite sketch
Construction Specifications Institute (CSI)
Datum plane
Elevation section
Elevations
Ergonomics
Floor plan

Foundations plan
Heating, Ventilation, and Air Conditioning (HVAC)
Interior elevation
Longitudinal section
Plan section
Plat plan
Plot plan

Presentation (abbreviated) floor plan
Rough sketch
Schedules
Setback
Site plan
Specifications
Survey plan
Transverse section

APPENDICES

TABLES

Table 1 Inches to Millimeters 512
Table 2 Millimeters to Inches 512
Table 3 Inch/Metric—Equivalents 513
Table 4 Inch/Metric—Conversion 514
Table 5 Rules Relative to the Circle 515
Table 6 Dimensioning Symbols 516
Table 7 ASTM and SAE Grade Markings for Steel Bolts and Screws 517
Table 8 Unified Standard Screw Thread Series 518
Table 9 Dimensions of Hex Cap Screws (Finished Hex Bolts) 519
Table 10 Dimensions of Square Bolts 520
Table 11 Dimensions of Hex Nuts and Hex Jam Nuts 521
Table 12 Dimensions of Square Nuts 522
Table 13 Dimensions of Slotted Flat Countersunk Head Machine Screws 523
Table 14 Dimensions of Cap Screw Heads 524
Table 15 Dimensions of Slotted Oval Countersunk Head Machine Screws 525
Table 16 Dimensions of Hexagon and Spline Socket Flat Countersunk Head Cap Screws 526
Table 17 Dimensions of Type 1 Cross Recessed Flat Countersunk Head Machine Screws 527
Table 18 Dimensions of Slotted Pan Head Machine Screws 528
Table 19 Dimensions of Hexagon and Spline Socket Head Cap Screws (1960 Series) 529
Table 20 Dimensions of Square Head and Headless Set Screws 530
Table 21 Dimensions of Hexagon and Spline Socket Set Screws 531
Table 22 Dimensions of Preferred Sizes of Type A Plain Washers 532
Table 23 Dimensions of Regular Helical Spring Lock Washers 533
Table 24 American National Standard Large Rivets 534
Table 25 Drill and Tap Sizes 535
Table 26 Spur/Pinion Gear Tooth Parts 536
Table 27 Woodruff Key Sizes for Different Shaft Diameters 536
Table 28 Key and Keyway Sizes 537
Table 29 Key Dimensions and Tolerances 538
Table 30 Gib Head Nominal Dimensions 539
Table 31 Sheet Metal and Wire Gage Designation 540
Table 32 American National Standards of Interest to Designers, Architects, and Drafters 541

ABBREVIATIONS 542

GLOSSARY 546

TABLES

TABLE 1 INCHES TO MILLIMETERS

in.	mm	in.	mm	in.	mm	in.	mm
1	25.4	26	660.4	51	1295.4	76	1930.4
2	50.8	27	685.8	52	1320.8	77	1955.8
3	76.2	28	711.2	53	1346.2	78	1981.2
4	101.6	29	736.6	54	1371.6	79	2006.6
5	127.0	30	762.0	55	1397.0	80	2032.0
6	152.4	31	787.4	56	1422.4	81	2057.4
7	177.8	32	812.8	57	1447.8	82	2082.8
8	203.2	33	838.2	58	1473.2	83	2108.2
9	228.6	34	863.6	59	1498.6	84	2133.6
10	254.0	35	889.0	60	1524.0	85	2159.0
11	279.4	36	914.4	61	1549.4	86	2184.4
12	304.8	37	939.8	62	1574.8	87	2209.8
13	330.2	38	965.2	63	1600.2	88	2235.2
14	355.6	39	990.6	64	1625.6	89	2260.6
15	381.0	40	1016.0	65	1651.0	90	2286.0
16	406.4	41	1041.4	66	1676.4	91	2311.4
17	431.8	42	1066.8	67	1701.8	92	2336.8
18	457.2	43	1092.2	68	1727.2	93	2362.2
19	482.6	44	1117.6	69	1752.6	94	2387.6
20	508.0	45	1143.0	70	1778.0	95	2413.0
21	533.4	46	1168.4	71	1803.4	96	2438.4
22	558.8	47	1193.8	72	1828.8	97	2463.8
23	584.2	48	1219.2	73	1854.2	98	2489.2
24	609.6	49	1244.6	74	1879.6	99	2514.6
25	635.0	50	1270.0	75	1905.0	100	2540.0

The above table is exact on the basis: 1 in. = 25.4 mm

TABLE 2 MILLIMETERS TO INCHES

mm	in.	mm	in.	mm	in.	mm	in.
1	0.039370	26	1.023622	51	2.007874	76	2.992126
2	0.078740	27	1.062992	52	2.047244	77	3.031496
3	0.118110	28	1.102362	53	2.086614	78	3.070866
4	0.157480	29	1.141732	54	2.125984	79	3.110236
5	0.196850	30	1.181102	55	2.165354	80	3.149606
6	0.236220	31	1.220472	56	2.204724	81	3.188976
7	0.275591	32	1.259843	57	2.244094	82	3.228346
8	0.314961	33	1.299213	58	2.283465	83	3.267717
9	0.354331	34	1.338583	59	2.322835	84	3.307087
10	0.393701	35	1.377953	60	2.362205	85	3.346457
11	0.433071	36	1.417323	61	2.401575	86	3.385827
12	0.472441	37	1.456693	62	2.440945	87	3.425197
13	0.511811	38	1.496063	63	2.480315	88	3.464567
14	0.551181	39	1.535433	64	2.519685	89	3.503937
15	0.590551	40	1.574803	65	2.559055	90	3.543307
16	0.629921	41	1.614173	66	2.598425	91	3.582677
17	0.669291	42	1.653543	67	2.637795	92	3.622047
18	0.708661	43	1.692913	68	2.677165	93	3.661417
19	0.748031	44	1.732283	69	2.716535	94	3.700787
20	0.787402	45	1.771654	70	2.755906	95	3.740157
21	0.826772	46	1.811024	71	2.795276	96	3.779528
22	0.866142	47	1.850394	72	2.834646	97	3.818898
23	0.905512	48	1.889764	73	2.874016	98	3.858268
24	0.944882	49	1.929134	74	2.913386	99	3.897638
25	0.984252	50	1.968504	75	2.952756	100	3.937008

The above table is approximate on the basis: 1 in. = 25.4 mm, 1/25.4 = 0.039370078740+

(Reprinted from TECHNICAL DRAWING AND DESIGN by Goetsch and Nelson, © 1986 Delmar Publishers Inc.)

TABLE 3 INCH/METRIC—EQUIVALENTS

Fraction	Decimal Equivalent Customary (in.)	Metric (mm)	Fraction	Decimal Equivalent Customary (in.)	Metric (mm)
1/64	.015625	0.3969	33/64	.515625	13.0969
1/32	.03125	0.7938	17/32	.53125	13.4938
3/64	.046875	1.1906	35/64	.546875	13.8906
1/16	.0625	1.5875	9/16	.5625	14.2875
5/64	.078125	1.9844	37/64	.578125	14.6844
3/32	.09375	2.3813	19/32	.59375	15.0813
7/64	.109375	2.7781	39/64	.609375	15.4781
1/8	.1250	3.1750	5/8	.6250	15.8750
9/64	.140625	3.5719	41/64	.640625	16.2719
5/32	.15625	3.9688	21/32	.65625	16.6688
11/64	.171875	4.3656	43/64	.671875	17.0656
3/16	.1875	4.7625	11/16	.6875	17.4625
13/64	.203125	5.1594	45/64	.703125	17.8594
7/32	.21875	5.5563	23/32	.71875	18.2563
15/64	.234375	5.9531	47/64	.734375	18.6531
1/4	.250	6.3500	3/4	.750	19.0500
17/64	.265625	6.7469	49/64	.765625	19.4469
9/32	.28125	7.1438	25/32	.78125	19.8438
19/64	.296875	7.5406	51/64	.796875	20.2406
5/16	.3125	7.9375	13/16	.8125	20.6375
21/64	.328125	8.3384	53/64	.828125	21.0344
11/32	.34375	8.7313	27/32	.84375	21.4313
23/64	.359375	9.1281	55/64	.859375	21.8281
3/8	.3750	9.5250	7/8	.8750	22.2250
25/64	.390625	9.9219	57/64	.890625	22.6219
13/32	.40625	10.3188	29/32	.90625	23.0188
27/64	.421875	10.7156	59/64	.921875	23.4156
7/16	.4375	11.1125	15/16	.9375	23.8125
29/64	.453125	11.5094	61/64	.953125	24.2094
15/32	.46875	11.9063	31/32	.96875	24.6063
31/64	.484375	12.3031	63/64	.984375	25.0031
1/2	.500	12.7000	1	1.000	25.4000

(Reprinted from DRAFTING FOR TRADES AND INDUSTRY—BASIC SKILLS by Nelson, © 1979 Delmar Publishers Inc.)

TABLE 4 INCH/METRIC—CONVERSION

Measures of Length

1 millimeter (mm) = 0.03937 inch
1 centimeter (cm) = 0.39370 inch
1 meter (m) = 39.37008 inches
    = 3.2808 feet
    = 1.0936 yards
1 kilometer (km) = 0.6214 mile
1 inch = 25.4 millimeters (mm)
    = 2.54 centimeters (cm)
1 foot = 304.8 millimeters (mm)
    = 0.3048 meter (m)
1 yard = 0.9144 meter (m)
1 mile = 1.609 kilometers (km)

Measures of Area

1 square millimeter = 0.00155 square inch
1 square centimeter = 0.155 square inch
1 square meter = 10.764 square feet
    = 1.196 square yards
1 square kilometer = 0.3861 square mile
1 square inch = 645.2 square millimeters
    = 6.452 square centimeters
1 square foot = 929 square centimeters
    = 0.0929 square meter
1 square yard = 0.836 square meter
1 square mile = 2.5899 square kilometers

Measures of Capacity (Dry)

1 cubic centimeter (cm^3) = 0.061 cubic inch
1 liter = 0.0353 cubic foot
    = 61.023 cubic inches
1 cubic meter (m^3) = 35.315 cubic feet
    = 1.308 cubic yards
1 cubic inch = 16.38706 cubic centimeters (cm^3)
1 cubic foot = 0.02832 cubic meter (m^3)
    = 28.317 liters
1 cubic yard = 0.7646 cubic meter (m^3)

Measures of Capacity (Liquid)

1 liter = 1.0567 U.S. quarts
    = 0.2642 U.S. gallon
    = 0.2200 Imperial gallon
1 cubic meter (m^3) = 264.2 U.S. gallons
    = 219.969 Imperial gallons
1 U.S. quart = 0.946 liter
1 Imperial quart = 1.136 liters
1 U.S. gallon = 3.785 liters
1 Imperial gallon = 4.546 liters

Measures of Weight

1 gram (g) = 15.432 grains
    = 0.03215 ounce troy
    = 0.03527 ounce avoirdupois
1 kilogram (kg) = 35.274 ounces avoirdupois
    = 2.2046 pounds
1000 kilograms (kg) = 1 metric ton (t)
    = 1.1023 tons of 2000 pounds
    = 0.9842 ton of 2240 pounds
1 ounce avoirdupois = 28.35 grams (g)
1 ounce troy = 31.103 grams (g)
1 pound = 453.6 grams
    = 0.4536 kilogram (kg)
1 ton of 2240 pounds = 1016 kilograms (kg)
    = 1.016 metric tons
1 grain = 0.0648 gram (g)
1 metric ton = 0.9842 ton of 2240 pounds
    = 2204.6 pounds

TABLE 5 RULES RELATIVE TO THE CIRCLE

To Find Circumference—
Multiply diameter by 3.1416 .... Or divide diameter by 0.3183

To Find Diameter—
Multiply circumference by 0.3183 .... Or divide circumference by 3.1416

To Find Radius—
Multiply circumference by 0.15915 .... Or divide circumference by 6.28318

To Find Side of an Inscribed Square—
Multiply diameter by 0.7071
Or multiply circumference by 0.2251 .... Or divide circumference by 4.4428

To Find Side of an Equal Square—
Multiply diameter by 0.8862 .... Or divide diameter by 1.1284
Or multiply circumference by 0.2821 .... Or divide circumference by 3.545

Square—
A side multiplied by 1.4142 equals diameter of its circumscribing circle.
A side multiplied by 4.443 equals circumference of its circumscribing circle.
A side multiplied by 1.128 equals diameter of an equal circle.
A side multiplied by 3.547 equals circumference of an equal circle.

To Find the Area of a Circle—
Multiply circumference by one-quarter of the diameter.
Or multiply the square of diameter by 0.7854
Or multiply the square of circumference by .07958
Or multiply the square of $1/2$ diameter by 3.1416

To Find the Surface of a Sphere or Globe—
Multiply the diameter by the circumference.
Or multiply the square of diameter by 3.1416
Or multiply four times the square of radius by 3.1416

TABLE 6 DIMENSIONING SYMBOLS

SYMBOL FOR:	ANSI Y14.5	ISO
DIMENSION ORIGIN	⌀→	NONE
FEATURE CONTROL FRAME	⊕ Ø0.5Ⓜ A B C	⊕ Ø0.5Ⓜ A B C
CONICAL TAPER	▷	▷
SLOPE	◁	◁
COUNTERBORE/SPOTFACE	⊔	NONE
COUNTERSINK	∨	NONE
DEPTH/DEEP	↧	NONE
SQUARE (SHAPE)	□	□
DIMENSION NOT TO SCALE	<u>15</u>	<u>15</u>
NUMBER OF TIMES/PLACES	8X	8X
ARC LENGTH	⌒105	NONE
RADIUS	R	R
SPHERICAL RADIUS	SR	NONE
SPHERICAL DIAMETER	SØ	NONE

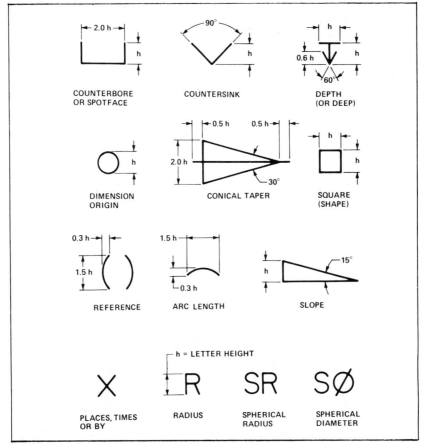

(Reprinted from The American Society of Mechanical Engineers—ANSI Y14.5M—1982)

TABLE 7 ASTM AND SAE MARKINGS FOR STEEL BOLTS AND SCREWS

Grade Marking	Specification	Material
NO MARK	SAE—Grade 1	Low or Medium Carbon Steel
NO MARK	ASTM—A307	Low Carbon Steel
NO MARK	SAE—Grade 2	Low or Medium Carbon Steel
(3 radial lines)	SAE—Grade 5	Medium Carbon Steel, Quenched and Tempered
(3 radial lines)	ASTM—A 449	Medium Carbon Steel, Quenched and Tempered
(1 line with T)	SAE—Grade 5.2	Low Carbon Martensite Steel, Quenched and Tempered
A 325	ASTM—A 325 Type 1	Medium Carbon Steel, Quenched and Tempered Radial dashes optional
A 325	ASTM—A 325 Type 2	Low Carbon Martensite Steel, Quenched and Tempered
A 325	ASTM—A 325 Type 3	Atmospheric Corrosion (Weathering) Steel, Quenched and Tempered
BC	ASTM—A 354 Grade BC	Alloy Steel, Quenched and Tempered
(5 radial lines)	SAE—Grade 7	Medium Carbon Alloy Steel, Quenched and Tempered, Roll Threaded After Heat Treatment
(6 radial lines)	SAE—Grade 8	Medium Carbon Alloy Steel, Quenched and Tempered
(6 radial lines)	ASTM—A 354 Grade BD	Alloy Steel, Quenched and Tempered
(rays)	SAE—Grade 8.2	Low Carbon Martensite Steel, Quenched and Tempered
A 490	ASTM—A 490 Type 1	Alloy Steel, Quenched and Tempered
A 490	ASTM—A 490 Type 3	Atmospheric Corrosion (Weathering) Steel, Quenched and Tempered

(Reprinted from The American Society of Mechanical Engineers—ANSI B18.2.1–1981)

TABLE 8 UNIFIED STANDARD SCREW THREAD SERIES

Sizes		Basic Major Diameter	THREADS PER INCH											Sizes
			Series with graded pitches			Series with constant pitches								
Primary	Secondary		Coarse UNC	Fine UNF	Extra fine UNEF	4UN	6UN	8UN	12UN	16UN	20UN	28UN	32UN	
0		0.0600	—	80	—	—	—	—	—	—	—	—	—	0
	1	0.0730	64	72	—	—	—	—	—	—	—	—	—	1
2		0.0860	56	64	—	—	—	—	—	—	—	—	—	2
	3	0.0990	48	56	—	—	—	—	—	—	—	—	—	3
4		0.1120	40	48	—	—	—	—	—	—	—	—	—	4
5		0.1250	40	44	—	—	—	—	—	—	—	—	—	5
6		0.1380	32	40	—	—	—	—	—	—	—	—	UNC	6
8		0.1640	32	36	—	—	—	—	—	—	—	—	UNC	8
10		0.1900	24	32	—	—	—	—	—	—	—	—	UNF	10
	12	0.2160	24	28	32	—	—	—	—	—	—	UNF	UNEF	12
1/4		0.2500	20	28	32	—	—	—	—	—	UNC	UNF	UNEF	1/4
5/16		0.3125	18	24	32	—	—	—	—	—	20	28	UNEF	5/16
3/8		0.3750	16	24	32	—	—	—	—	UNC	20	28	UNEF	3/8
7/16		0.4375	14	20	28	—	—	—	—	16	UNF	UNEF	32	7/16
1/2		0.5000	13	20	28	—	—	—	—	16	UNF	UNEF	32	1/2
9/16		0.5625	12	18	24	—	—	—	UNC	16	20	28	32	9/16
5/8		0.6250	11	18	24	—	—	—	12	16	20	28	32	5/8
	11/16	0.6875	—	—	24	—	—	—	12	16	20	28	32	11/16
3/4		0.7500	10	16	20	—	—	—	12	UNF	UNEF	28	32	3/4
	13/16	0.8125	—	—	20	—	—	—	12	16	UNEF	28	32	13/16
7/8		0.8750	9	14	20	—	—	—	12	16	UNEF	28	32	7/8
	15/16	0.9375	—	—	20	—	—	—	12	16	UNEF	28	32	15/16
1		1.0000	8	12	20	—	—	UNC	UNF	16	UNEF	28	32	1
	1 1/16	1.0625	—	—	18	—	—	8	12	16	20	28	—	1 1/16
1 1/8		1.1250	7	12	18	—	—	8	UNF	16	20	28	—	1 1/8
	1 3/16	1.1875	—	—	18	—	—	8	12	16	20	28	—	1 3/16
1 1/4		1.2500	7	12	18	—	—	8	UNF	16	20	28	—	1 1/4
	1 5/16	1.3125	—	—	18	—	—	8	12	16	20	28	—	1 5/16
1 3/8		1.3750	6	12	18	—	UNC	8	UNF	16	20	28	—	1 3/8
	1 7/16	1.4375	—	—	18	—	6	8	12	16	20	28	—	1 7/16
1 1/2		1.5000	6	12	18	—	UNC	8	UNF	16	20	28	—	1 1/2
	1 9/16	1.5625	—	—	18	—	6	8	12	16	20	—	—	1 9/16
1 5/8		1.6250	—	—	18	—	6	8	12	16	20	—	—	1 5/8
	1 11/16	1.6875	—	—	18	—	6	8	12	16	20	—	—	1 11/16
1 3/4		1.7500	5	—	—	—	6	8	12	16	20	—	—	1 3/4
	1 13/16	1.8125	—	—	—	—	6	8	12	16	20	—	—	1 13/16
1 7/8		1.8750	—	—	—	—	6	8	12	16	20	—	—	1 7/8
	1 15/16	1.9375	—	—	—	—	6	8	12	16	20	—	—	1 15/16
2		2.0000	4 1/2	—	—	—	6	8	12	16	20	—	—	2
	2 1/8	2.1250	—	—	—	—	6	8	12	16	20	—	—	2 1/8
2 1/4		2.2500	4 1/2	—	—	—	6	8	12	16	20	—	—	2 1/4
	2 3/8	2.3750	—	—	—	—	6	8	12	16	20	—	—	2 3/8
2 1/2		2.5000	4	—	—	UNC	6	8	12	16	20	—	—	2 1/2
	2 5/8	2.6250	—	—	—	4	6	8	12	16	20	—	—	2 5/8
2 3/4		2.7500	4	—	—	UNC	6	8	12	16	20	—	—	2 3/4
	2 7/8	2.8750	—	—	—	4	6	8	12	16	20	—	—	2 7/8
3		3.0000	4	—	—	UNC	6	8	12	16	20	—	—	3
	3 1/8	3.1250	—	—	—	4	6	8	12	16	—	—	—	3 1/8
3 1/4		3.2500	4	—	—	UNC	6	8	12	16	—	—	—	3 1/4
	3 3/8	3.3750	—	—	—	4	6	8	12	16	—	—	—	3 3/8
3 1/2		3.5000	4	—	—	UNC	6	8	12	16	—	—	—	3 1/2
	3 5/8	3.6250	—	—	—	4	6	8	12	16	—	—	—	3 5/8
3 3/4		3.7500	4	—	—	UNC	6	8	12	16	—	—	—	3 3/4
	3 7/8	3.8750	—	—	—	4	6	8	12	16	—	—	—	3 7/8
4		4.0000	4	—	—	UNC	6	8	12	16	—	—	—	4
	4 1/8	4.1250	—	—	—	4	6	8	12	16	—	—	—	4 1/8
4 1/4		4.2500	—	—	—	4	6	8	12	16	—	—	—	4 1/4
	4 3/8	4.3750	—	—	—	4	6	8	12	16	—	—	—	4 3/8
4 1/2		4.5000	—	—	—	4	6	8	12	16	—	—	—	4 1/2
	4 5/8	4.6250	—	—	—	4	6	8	12	16	—	—	—	4 5/8
4 3/4		4.7500	—	—	—	4	6	8	12	16	—	—	—	4 3/4
	4 7/8	4.8750	—	—	—	4	6	8	12	16	—	—	—	4 7/8
5		5.0000	—	—	—	4	6	8	12	16	—	—	—	5
	5 1/8	5.1250	—	—	—	4	6	8	12	16	—	—	—	5 1/8
5 1/4		5.2500	—	—	—	4	6	8	12	16	—	—	—	5 1/4
	5 3/8	5.3750	—	—	—	4	6	8	12	16	—	—	—	5 3/8
5 1/2		5.5000	—	—	—	4	6	8	12	16	—	—	—	5 1/2
	5 5/8	5.6250	—	—	—	4	6	8	12	16	—	—	—	5 5/8
5 3/4		5.7500	—	—	—	4	6	8	12	16	—	—	—	5 3/4
	5 7/8	5.8750	—	—	—	4	6	8	12	16	—	—	—	5 7/8
6		6.0000	—	—	—	4	6	8	12	16	—	—	—	6

(Reprinted from TECHNICAL DRAWING AND DESIGN by Goetsch and Nelson, © 1986 Delmar Publishers Inc.)

AMERICAN NATIONAL STANDARD
SQUARE AND HEX BOLTS AND SCREWS — INCH SERIES

ANSI B18.2.1–1981

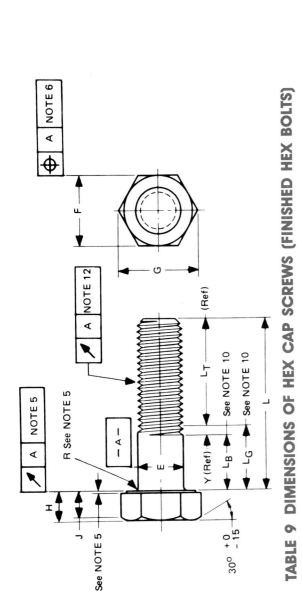

TABLE 9 DIMENSIONS OF HEX CAP SCREWS (FINISHED HEX BOLTS)

Nominal Size or Basic Product Dia (18)	E Body Dia (8)		F Width Across Flats			G Width Across Corners (4)		H Height			J Wrenching Height (4)	L_T Thread Length For Screw Lengths (10)		Y Transition Thread Length (10)	Runout of Bearing Surface FIM (5)
	Max	Min	Basic	Max	Min	Max	Min	Basic	Max	Min	Min	6 in. and Shorter Basic	Over 6 in. Basic	Max	Max
1/4 0.2500	0.2500	0.2450	7/16	0.438	0.428	0.505	0.488	5/32	0.163	0.150	0.106	0.750	1.000	0.250	0.010
5/16 0.3125	0.3125	0.3065	1/2	0.500	0.489	0.577	0.557	13/64	0.211	0.195	0.140	0.875	1.125	0.278	0.011
3/8 0.3750	0.3750	0.3690	9/16	0.562	0.551	0.650	0.628	15/64	0.243	0.226	0.160	1.000	1.250	0.312	0.012
7/16 0.4375	0.4375	0.4305	5/8	0.625	0.612	0.722	0.698	9/32	0.291	0.272	0.195	1.125	1.375	0.357	0.013
1/2 0.5000	0.5000	0.4930	3/4	0.750	0.736	0.866	0.840	5/16	0.323	0.302	0.215	1.250	1.500	0.385	0.014
9/16 0.5625	0.5625	0.5545	13/16	0.812	0.798	0.938	0.910	23/64	0.371	0.348	0.250	1.375	1.625	0.417	0.015
5/8 0.6250	0.6250	0.6170	15/16	0.938	0.922	1.083	1.051	25/64	0.403	0.378	0.269	1.500	1.750	0.455	0.017
3/4 0.7500	0.7500	0.7410	1 1/8	1.125	1.100	1.299	1.254	15/32	0.483	0.455	0.324	1.750	2.000	0.500	0.020
7/8 0.8750	0.8750	0.8660	1 5/16	1.312	1.285	1.516	1.465	35/64	0.563	0.531	0.378	2.000	2.250	0.556	0.023
1 1.0000	1.0000	0.9900	1 1/2	1.500	1.469	1.732	1.675	39/64	0.627	0.591	0.416	2.250	2.500	0.625	0.026
1 1/8 1.1250	1.1250	1.1140	1 11/16	1.688	1.631	1.949	1.859	11/16	0.718	0.658	0.461	2.500	2.750	0.714	0.029
1 1/4 1.2500	1.2500	1.2390	1 7/8	1.875	1.812	2.165	2.066	25/32	0.813	0.749	0.530	2.750	3.000	0.714	0.033
1 3/8 1.3750	1.3750	1.3630	2 1/16	2.062	1.994	2.382	2.273	27/32	0.878	0.810	0.569	3.000	3.250	0.833	0.036
1 1/2 1.5000	1.5000	1.4880	2 1/4	2.230	2.175	2.598	2.480	15/16	0.974	0.902	0.640	3.250	3.500	0.833	0.039
1 3/4 1.7500	1.7500	1.7380	2 5/8	2.625	2.538	3.031	2.893	1 3/32	1.134	1.054	0.748	3.750	4.000	1.000	0.046
2 2.0000	2.0000	1.9880	3	3.000	2.900	3.464	3.306	1 7/32	1.263	1.175	0.825	4.250	4.500	1.111	0.052
2 1/4 2.2500	2.2500	2.2380	3 3/8	3.375	3.262	3.897	3.719	1 3/8	1.423	1.327	0.933	4.750	5.000	1.111	0.059
2 1/2 2.5000	2.5000	2.4880	3 3/4	3.750	3.625	4.330	4.133	1 17/32	1.583	1.479	1.042	5.250	5.500	1.250	0.065
2 3/4 2.7500	2.7500	2.7380	4 1/8	4.125	3.988	4.763	4.546	1 11/16	1.744	1.632	1.151	5.750	6.000	1.250	0.072
3 3.0000	3.0000	2.9880	4 1/2	4.500	4.350	5.196	4.959	1 7/8	1.935	1.815	1.290	6.250	6.500	1.250	0.079

(Reprinted from The American Society of Mechanical Engineers—ANSI B18.2.1–1981)

AMERICAN NATIONAL STANDARD
SQUARE AND HEX BOLTS AND SCREWS – INCH SERIES

ANSI B18.2.1–1981

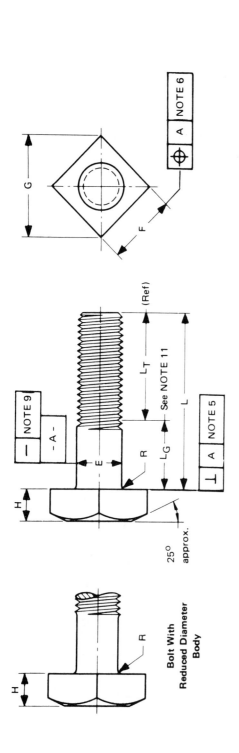

TABLE 10 DIMENSIONS OF SQUARE BOLTS

Nominal Size or Basic Product Dia (17)	E Body Dia (7), (14) Max	F Width Across Flats (4) Basic	F Max	F Min	G Width Across Corners Max	G Min	H Height Basic	H Max	H Min	R Radius of Fillet Max	R Min	L_T Thread Length For Bolt Lengths (11) 6 in. and shorter Basic	L_T over 6 in. Basic
1/4 0.2500	0.260	3/8	0.375	0.362	0.530	0.498	11/64	0.188	0.156	0.03	0.01	0.750	1.000
5/16 0.3125	0.324	1/2	0.500	0.484	0.707	0.665	13/64	0.220	0.186	0.03	0.01	0.875	1.125
3/8 0.3750	0.388	9/16	0.562	0.544	0.795	0.747	1/4	0.268	0.232	0.03	0.01	1.000	1.250
7/16 0.4375	0.452	5/8	0.625	0.603	0.884	0.828	19/64	0.316	0.278	0.03	0.01	1.125	1.375
1/2 0.5000	0.515	3/4	0.750	0.725	1.061	0.995	21/64	0.348	0.308	0.03	0.01	1.250	1.500
5/8 0.6250	0.642	15/16	0.938	0.906	1.326	1.244	27/64	0.444	0.400	0.06	0.02	1.500	1.750
3/4 0.7500	0.768	1 1/8	1.125	1.088	1.591	1.494	1/2	0.524	0.476	0.06	0.02	1.750	2.000
7/8 0.8750	0.895	1 5/16	1.312	1.269	1.856	1.742	19/32	0.620	0.568	0.06	0.02	2.000	2.250
1 1.0000	1.022	1 1/2	1.500	1.450	2.121	1.991	21/32	0.684	0.628	0.09	0.03	2.250	2.500
1 1/8 1.1250	1.149	1 11/16	1.688	1.631	2.386	2.239	3/4	0.780	0.720	0.09	0.03	2.500	2.750
1 1/4 1.2500	1.277	1 7/8	1.875	1.812	2.652	2.489	27/32	0.876	0.812	0.09	0.03	2.750	3.000
1 3/8 1.3750	1.404	2 1/16	2.062	1.994	2.917	2.738	29/32	0.940	0.872	0.09	0.03	3.000	3.250
1 1/2 1.5000	1.531	2 1/4	2.250	2.175	3.182	2.986	1	1.036	0.964	0.09	0.03	3.250	3.500

(Reprinted from The American Society of Mechanical Engineers—ANSI B18.2.1–1981)

**AMERICAN NATIONAL STANDARD
SQUARE AND HEX NUTS**

ANSI B18.2.2-1972

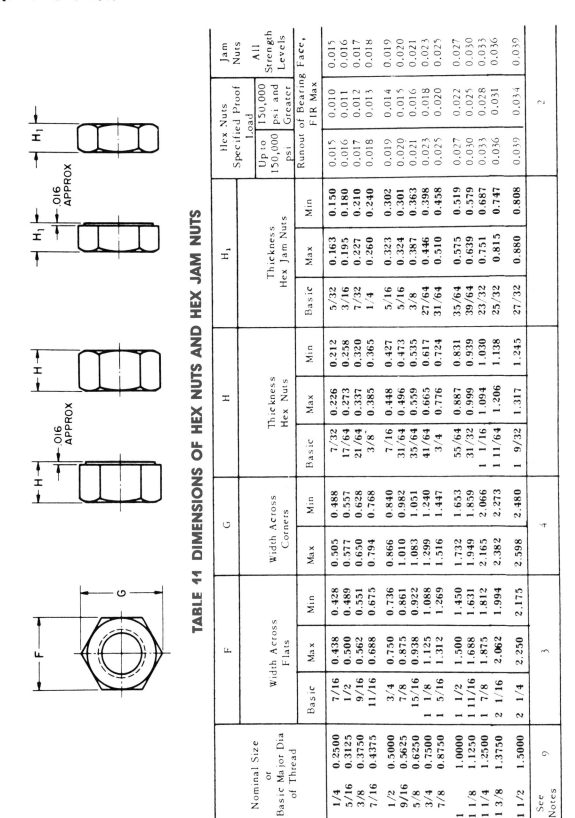

TABLE 11 DIMENSIONS OF HEX NUTS AND HEX JAM NUTS

Nominal Size or Basic Major Dia of Thread		F Width Across Flats			G Width Across Corners		H Thickness Hex Nuts			H_1 Thickness Hex Jam Nuts			Hex Nuts Specified Proof Load		Jam Nuts All Strength Levels
		Basic	Max	Min	Max	Min	Basic	Max	Min	Basic	Max	Min	Up to 150,000 psi	150,000 psi and Greater	
													Runout of Bearing Face, FIR Max		
1/4	0.2500	7/16	0.438	0.428	0.505	0.488	7/32	0.226	0.212	5/32	0.163	0.150	0.015	0.010	0.015
5/16	0.3125	1/2	0.500	0.489	0.577	0.557	17/64	0.273	0.258	3/16	0.195	0.180	0.016	0.011	0.016
3/8	0.3750	9/16	0.562	0.551	0.650	0.628	21/64	0.337	0.320	7/32	0.227	0.210	0.017	0.012	0.017
7/16	0.4375	11/16	0.688	0.675	0.794	0.768	3/8	0.385	0.365	1/4	0.260	0.240	0.018	0.013	0.018
1/2	0.5000	3/4	0.750	0.736	0.866	0.840	7/16	0.448	0.427	5/16	0.323	0.302	0.019	0.014	0.019
9/16	0.5625	7/8	0.875	0.861	1.010	0.982	31/64	0.496	0.473	5/16	0.324	0.301	0.020	0.015	0.020
5/8	0.6250	15/16	0.938	0.922	1.083	1.051	35/64	0.559	0.535	3/8	0.387	0.363	0.021	0.016	0.021
3/4	0.7500	1 1/8	1.125	1.088	1.299	1.240	41/64	0.665	0.617	27/64	0.446	0.398	0.023	0.018	0.023
7/8	0.8750	1 5/16	1.312	1.269	1.516	1.447	3/4	0.776	0.724	31/64	0.510	0.458	0.025	0.020	0.025
1	1.0000	1 1/2	1.500	1.450	1.732	1.653	55/64	0.887	0.831	35/64	0.575	0.519	0.027	0.022	0.027
1 1/8	1.1250	1 11/16	1.688	1.631	1.949	1.859	31/32	0.999	0.939	39/64	0.639	0.579	0.030	0.025	0.030
1 1/4	1.2500	1 7/8	1.875	1.812	2.165	2.066	1 1/16	1.094	1.030	23/32	0.751	0.687	0.033	0.028	0.033
1 3/8	1.3750	2 1/16	2.062	1.994	2.382	2.273	1 11/64	1.206	1.138	25/32	0.815	0.747	0.036	0.031	0.036
1 1/2	1.5000	2 1/4	2.250	2.175	2.598	2.480	1 9/32	1.317	1.245	27/32	0.880	0.808	0.039	0.034	0.039
See Notes	9		3		4									2	

(Reprinted from The American Society of Mechanical Engineers—ANSI B18.2.2-1972)

**AMERICAN NATIONAL STANDARD
SQUARE AND HEX NUTS**

ANSI B18.2.2-1972

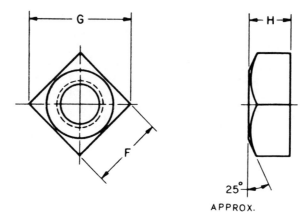

TABLE 12 DIMENSIONS OF SQUARE NUTS

Nominal Size or Basic Major Dia of Thread		F Width Across Flats			G Width Across Corners		H Thickness		
		Basic	Max	Min	Max	Min	Basic	Max	Min
1/4	0.2500	7/16	0.438	0.425	0.619	0.584	7/32	0.235	0.203
5/16	0.3125	9/16	0.562	0.547	0.795	0.751	17/64	0.283	0.249
3/8	0.3750	5/8	0.625	0.606	0.884	0.832	21/64	0.346	0.310
7/16	0.4375	3/4	0.750	0.728	1.061	1.000	3/8	0.394	0.356
1/2	0.5000	13/16	0.812	0.788	1.149	1.082	7/16	0.458	0.418
5/8	0.6250	1	1.000	0.969	1.414	1.330	35/64	0.569	0.525
3/4	0.7500	1 1/8	1.125	1.088	1.591	1.494	21/32	0.680	0.632
7/8	0.8750	1 5/16	1.312	1.269	1.856	1.742	49/64	0.792	0.740
1	1.0000	1 1/2	1.500	1.450	2.121	1.991	7/8	0.903	0.847
1 1/8	1.1250	1 11/16	1.688	1.631	2.386	2.239	1	1.030	0.970
1 1/4	1.2500	1 7/8	1.875	1.812	2.652	2.489	1 3/32	1.126	1.062
1 3/8	1.3750	2 1/16	2.062	1.994	2.917	2.738	1 13/64	1.237	1.169
1 1/2	1.5000	2 1/4	2.250	2.175	3.182	2.986	1 5/16	1.348	1.276
See Notes		8	3						

(Reprinted from The American Society of Mechanical Engineers—ANSI B18.2.2–1972)

Tables 523

SLOTTED FLAT
Type of Head

AMERICAN NATIONAL STANDARD
MACHINE SCREWS AND MACHINE SCREW NUTS

ANSI B18.6.3—1972

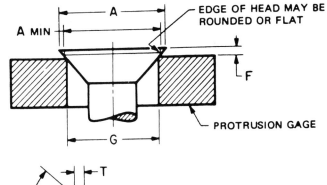
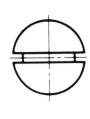
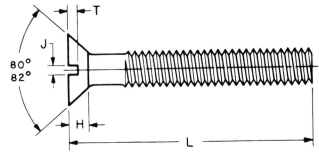

TABLE 13 DIMENSIONS OF SLOTTED FLAT COUNTERSUNK HEAD MACHINE SCREWS

Nominal Size[1] or Basic Screw Diameter		L^2 These Lengths or Shorter are Undercut	A Head Diameter		H^3 Head Height	J Slot Width		T Slot Depth		F^4 Protrusion Above Gaging Diameter		G^4 Gaging Diameter
			Max, Edge Sharp	Min, Edge Rounded or Flat	Ref	Max	Min	Max	Min	Max	Min	
0000	0.0210	—	0.043	0.037	0.011	0.008	0.004	0.007	0.003	*	*	*
000	0.0340	—	0.064	0.058	0.016	0.011	0.007	0.009	0.005	*	*	*
00	0.0470	—	0.093	0.085	0.028	0.017	0.010	0.014	0.009	*	*	*
0	0.0600	1/8	0.119	0.099	0.035	0.023	0.016	0.015	0.010	0.026	0.016	0.078
1	0.0730	1/8	0.146	0.123	0.043	0.026	0.019	0.019	0.012	0.028	0.016	0.101
2	0.0860	1/8	0.172	0.147	0.051	0.031	0.023	0.023	0.015	0.029	0.017	0.124
3	0.0990	1/8	0.199	0.171	0.059	0.035	0.027	0.027	0.017	0.031	0.018	0.148
4	0.1120	3/16	0.225	0.195	0.067	0.039	0.031	0.030	0.020	0.032	0.019	0.172
5	0.1250	3/16	0.252	0.220	0.075	0.043	0.035	0.034	0.022	0.034	0.020	0.196
6	0.1380	3/16	0.279	0.244	0.083	0.048	0.039	0.038	0.024	0.036	0.021	0.220
8	0.1640	1/4	0.332	0.292	0.100	0.054	0.045	0.045	0.029	0.039	0.023	0.267
10	0.1900	5/16	0.385	0.340	0.116	0.060	0.050	0.053	0.034	0.042	0.025	0.313
12	0.2160	3/8	0.438	0.389	0.132	0.067	0.056	0.060	0.039	0.045	0.027	0.362
1/4	0.2500	7/16	0.507	0.452	0.153	0.075	0.064	0.070	0.046	0.050	0.029	0.424
5/16	0.3125	1/2	0.635	0.568	0.191	0.084	0.072	0.088	0.058	0.057	0.034	0.539
3/8	0.3750	9/16	0.762	0.685	0.230	0.094	0.081	0.106	0.070	0.065	0.039	0.653
7/16	0.4375	5/8	0.812	0.723	0.223	0.094	0.081	0.103	0.066	0.073	0.044	0.690
1/2	0.5000	3/4	0.875	0.775	0.223	0.106	0.091	0.103	0.065	0.081	0.049	0.739
9/16	0.5625	—	1.000	0.889	0.260	0.118	0.102	0.120	0.077	0.089	0.053	0.851
5/8	0.6250	—	1.125	1.002	0.298	0.133	0.116	0.137	0.088	0.097	0.058	0.962
3/4	0.7500	—	1.375	1.230	0.372	0.149	0.131	0.171	0.111	0.112	0.067	1.186

[1] Where specifying nominal size in decimals, zeros preceding decimal and in the fourth decimal place shall be omitted.
[2] Screws of these lengths and shorter shall have undercut heads as shown in Table 5.
[3] Tabulated values determined from formula for maximum H, Appendix V.
[4] No tolerance for gaging diameter is given. If the gaging diameter of the gage used differs from tabulated value, the protrusion will be affected accordingly and the proper protrusion values must be recalculated using the formulas shown in **Appendix I**.
*Not practical to gage.
(Reprinted from The American Society of Mechanical Engineers—ANSI B18.6.3-1972)

TABLE 14 DIMENSIONS OF CAP SCREW HEADS

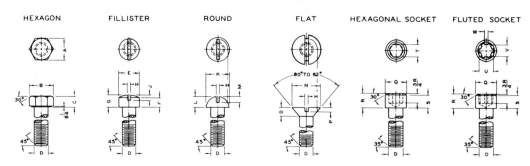

Diam.	A	B	C	E	F	G	H	J	K	L	M	N	O	P	Q	R	S	T	U	V	W	
2-.086															.140 .136	.086 .083	.079	1/16	.074 .073	.064 .063	.016 .015	Max. Min.
3-.099															.161 .157	.099 .096	.091	5/64	.098 .097	.082 .080	.022 .021	Max. Min.
4-.112															.183 .178	.112 .109	.103	5/64	.098 .097	.082 .080	.022 .021	Max. Min.
5-.125															.205 .200	.125 .122	.115	3/32	.115 .113	.098 .096	.025 .023	Max. Min.
6-.138															.226 .221	.138 .134	.127	3/32	.115 .113	.098 .096	.025 .023	Max. Min.
8-.164															.270 .265	.164 .160	.150	1/8	.149 .147	.128 .126	.032 .030	Max. Min.
10-.190															5/16	.190 .185	.174	5/32	.188 .186	.163 .161	.039 .037	Max. Min.
12-.216															11/32	.216 .211	.198	5/32	.188 .186	.163 .161	.039 .037	Max. Min.
1/4	7/16	.505	3/16	3/8	.172 .157	.216 .195	.075 .064	.097 .077	7/16	.191 .175	.117 .097	1/2	.140	.069 .046	3/8	1/4	.229	3/16	.221 .219	.190 .188	.050 .048	Max. Min.
5/16	1/2	.577	15/64	7/16	.203 .186	.253 .230	.084 .072	.115 .090	9/16	.246 .226	.151 .126	5/8	.176	.086 .057	7/16	5/16	.286	7/32	.256 .254	.221 .219	.060 .058	Max. Min.
3/8	9/16	.649	9/32	9/16	.250 .229	.314 .285	.094 .081	.143 .113	5/8	.273 .252	.168 .138	3/4	.210	.103 .069	9/16	3/8	.344	5/16	.380 .377	.319 .316	.092 .089	Max. Min.
7/16	5/8	.722	21/64	5/8	.297 .274	.368 .337	.094 .081	.168 .133	3/4	.328 .302	.202 .167		.210	.103 .069	5/8	7/16	.401	5/16	.380 .377	.319 .316	.092 .089	Max. Min.
1/2	3/4	.866	3/8	3/4	.328 .301	.412 .376	.106 .091	.188 .148	13/16	.355 .328	.219 .179	7/8	.210	.103 .069	3/4	1/2	.458	3/8	.463 .460	.386 .383	.112 .109	Max. Min.
9/16	13/16	.938	27/64	13/16	.375 .347	.466 .428	.118 .102	.214 .169	15/16	.410 .379	.253 .208	1	.245	.120 .080	13/16	9/16	.516	3/8	.463 .460	.386 .383	.112 .109	Max. Min.
5/8	7/8	1.010	15/32	7/8	.422 .392	.521 .480	.133 .116	.240 .190	1	.438 .405	.270 .220	1 1/8	.281	.137 .092	7/8	5/8	.573	1/2	.604 .601	.509 .506	.138 .134	Max. Min.
3/4	1	1.155	9/16	1	.500 .466	.612 .566	.149 .131	.283 .233	1 1/4	.547 .506	.337 .277	1 3/8	.352	.171 .115	1	3/4	.688	9/16	.631 .627	.535 .531	.149 .145	Max. Min.
7/8	1 1/8	1.299	21/32	1 1/8	.594 .556	.720 .669	.167 .147	.334 .264				1 5/8	.423	.206 .139	1 1/8	7/8	.802	9/16	.709 .705	.604 .600	.168 .164	Max. Min.
1	1 5/16	1.516	3/4	1 5/16	.656 .613	.802 .744	.188 .166	.372 .292				1 7/8	.494	.240 .162	1 5/16	1	.917	5/8	.801 .797	.685 .681	.189 .185	Max. Min.
1 1/8	1 1/2	1.732	27/32												1 1/2	1 1/8	1.031	3/4	.970 .966	.828 .824	.231 .227	Max. Min.
1 1/4	1 11/16	1.949	15/16												1 3/4	1 1/4	1.146	3/4	.970 .966	.828 .824	.231 .227	Max. Min.
1 3/8															1 7/8	1 3/8	1.260	3/4	.970 .966	.828 .824	.231 .227	Max. Min.
1 1/2															2	1 1/2	1.375	1	1.275 1.271	1.007 1.003	.298 .294	Max. Min.

Length of Thread on Hexagon, Fillister, Round and Flat = 2 D + $\frac{1}{4}$".
Length of Thread on Hex. and Fluted Socket, For NC Thread = 2D + $\frac{1}{2}$", or $\frac{1}{2}$ Bolt Length
For NF Thread = $1\frac{1}{2}$D + $\frac{1}{2}$, or $\frac{5}{8}$ Bolt Length   Whichever is Greater

(Reprinted from BASIC DRAFTING TECHNOLOGY by Rotmans, Horton, and Good, © 1980 Delmar Publishers Inc.)

AMERICAN NATIONAL STANDARD
MACHINE SCREWS AND MACHINE SCREW NUTS   ANSI B18.6.3-1972

SLOTTED OVAL
Type of Head

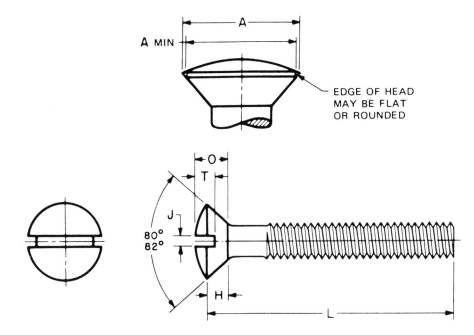

TABLE 15 DIMENSIONS OF SLOTTED OVAL COUNTERSUNK HEAD MACHINE SCREWS

Nominal Size[1] or Basic Screw Diameter		L^2 These Lengths or Shorter are Undercut	A Head Diameter		H^3 Head Side Height	O Total Head Height		J Slot Width		T Slot Depth	
			Max, Edge Sharp	Min, Edge Rounded or Flat	Ref	Max	Min	Max	Min	Max	Min
00	0.0470	—	0.093	0.085	0.028	0.042	0.034	0.017	0.010	0.023	0.016
0	0.0600	1/8	0.119	0.099	0.035	0.056	0.041	0.023	0.016	0.030	0.025
1	0.0730	1/8	0.146	0.123	0.043	0.068	0.052	0.026	0.019	0.038	0.031
2	0.0860	1/8	0.172	0.147	0.051	0.080	0.063	0.031	0.023	0.045	0.037
3	0.0990	1/8	0.199	0.171	0.059	0.092	0.073	0.035	0.027	0.052	0.043
4	0.1120	3/16	0.225	0.195	0.067	0.104	0.084	0.039	0.031	0.059	0.049
5	0.1250	3/16	0.252	0.220	0.075	0.116	0.095	0.043	0.035	0.067	0.055
6	0.1380	3/16	0.279	0.244	0.083	0.128	0.105	0.048	0.039	0.074	0.060
8	0.1640	1/4	0.332	0.292	0.100	0.152	0.126	0.054	0.045	0.088	0.072
10	0.1900	5/16	0.385	0.340	0.116	0.176	0.148	0.060	0.050	0.103	0.084
12	0.2160	3/8	0.438	0.389	0.132	0.200	0.169	0.067	0.056	0.117	0.096
1/4	0.2500	7/16	0.507	0.452	0.153	0.232	0.197	0.075	0.064	0.136	0.112
5/16	0.3125	1/2	0.635	0.568	0.191	0.290	0.249	0.084	0.072	0.171	0.141
3/8	0.3750	9/16	0.762	0.685	0.230	0.347	0.300	0.094	0.081	0.206	0.170
7/16	0.4375	5/8	0.812	0.723	0.223	0.345	0.295	0.094	0.081	0.210	0.174
1/2	0.5000	3/4	0.875	0.775	0.223	0.354	0.299	0.106	0.091	0.216	0.176
9/16	0.5625	—	1.000	0.889	0.260	0.410	0.350	0.118	0.102	0.250	0.207
5/8	0.6250	—	1.125	1.002	0.298	0.467	0.399	0.133	0.116	0.285	0.235
3/4	0.7500	—	1.375	1.230	0.372	0.578	0.497	0.149	0.131	0.353	0.293

[1] Where specifying nominal size in decimals, zeros preceding decimal and in the fourth decimal place shall be omitted.
[2] Screws of these lengths and shorter shall have undercut heads.
[3] Tabulated values determined from formula for maximum H, Appendix V, ANSI B18.6.3-1972.
(Reprinted from The American Society of Mechanical Engineers—ANSI B18.6.3-1972)

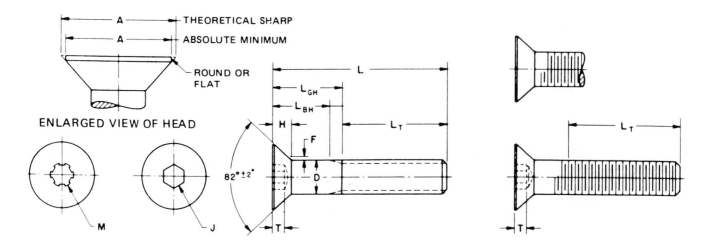

TABLE 16 DIMENSIONS OF HEXAGON AND SPLINE SOCKET FLAT COUNTERSUNK HEAD CAP SCREWS

Nominal Size or Basic Screw Diameter		D Body Dia		A Head Diameter		H Head Height		M Spline Socket Size	J Hexagon Socket Size		T Key Engagement	F Fillet Extension Above D Max
		Max	Min	Theoretical Sharp Max	Abs. Min	Reference	Flushness Tolerance		Nom		Min	Max
0	0.0600	0.0600	0.0568	0.138	0.117	0.044	0.006	0.048	0.035		0.025	0.006
1	0.0730	0.0730	0.0695	0.168	0.143	0.054	0.007	0.060	0.050		0.031	0.008
2	0.0860	0.0860	0.0822	0.197	0.168	0.064	0.008	0.060	0.050		0.038	0.010
3	0.0990	0.0990	0.0949	0.226	0.193	0.073	0.010	0.072	1/16	0.062	0.044	0.010
4	0.1120	0.1120	0.1075	0.255	0.218	0.083	0.011	0.072	1/16	0.062	0.055	0.012
5	0.1250	0.1250	0.1202	0.281	0.240	0.090	0.012	0.096	5/64	0.078	0.061	0.014
6	0.1380	0.1380	0.1329	0.307	0.263	0.097	0.013	0.096	5/64	0.078	0.066	0.015
8	0.1640	0.1640	0.1585	0.359	0.311	0.112	0.014	0.111	3/32	0.094	0.076	0.015
10	0.1900	0.1900	0.1840	0.411	0.359	0.127	0.015	0.145	1/8	0.125	0.087	0.015
1/4	0.2500	0.2500	0.2435	0.531	0.480	0.161	0.016	0.183	5/32	0.156	0.111	0.015
5/16	0.3125	0.3125	0.3053	0.656	0.600	0.198	0.017	0.216	3/16	0.188	0.135	0.015
3/8	0.3750	0.3750	0.3678	0.781	0.720	0.234	0.018	0.251	7/32	0.219	0.159	0.015
7/16	0.4375	0.4375	0.4294	0.844	0.781	0.234	0.018	0.291	1/4	0.250	0.159	0.015
1/2	0.5000	0.5000	0.4919	0.938	0.872	0.251	0.018	0.372	5/16	0.312	0.172	0.015
5/8	0.6250	0.6250	0.6163	1.188	1.112	0.324	0.022	0.454	3/8	0.375	0.220	0.015
3/4	0.7500	0.7500	0.7406	1.438	1.355	0.396	0.024	0.454	1/2	0.500	0.220	0.015
7/8	0.8750	0.8750	0.8647	1.688	1.604	0.468	0.025	...	9/16	0.562	0.248	0.015
1	1.0000	1.0000	0.9886	1.938	1.841	0.540	0.028	...	5/8	0.625	0.297	0.015
1 1/8	1.1250	1.1250	1.1086	2.188	2.079	0.611	0.031	...	3/4	0.750	0.325	0.031
1 1/4	1.2500	1.2500	1.2336	2.438	2.316	0.683	0.035	...	7/8	0.875	0.358	0.031
1 3/8	1.3750	1.3750	1.3568	2.688	2.553	0.755	0.038	...	7/8	0.875	0.402	0.031
1 1/2	1.5000	1.5000	1.4818	2.938	2.791	0.827	0.042	...	1	1.000	0.435	0.031

(Reprinted from The American Society of Mechanical Engineers—ANSI B18.3–1982)

AMERICAN NATIONAL STANDARD
MACHINE SCREWS AND MACHINE SCREW NUTS

ANSI B18.6.3—1972

TYPE I RECESS
FLAT

Type of Head

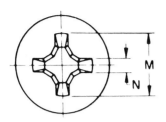

This type of recess has a large center opening, tapered wings, and blunt bottom, with all edges relieved or rounded

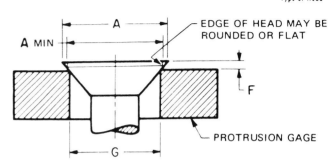

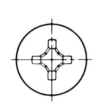

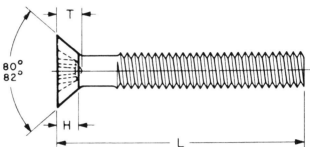

TABLE 17 DIMENSIONS OF TYPE I CROSS RECESSED FLAT COUNTERSUNK HEAD MACHINE SCREWS

Nominal Size[1] or Basic Screw Diameter		L[2] These Lengths or Shorter are Undercut	A Head Diameter		H[3] Head Height	M Recess Diameter		T Recess Depth		N Recess Width	Driver Size	Recess Penetration Gaging Depth		F[4] Protrusion Above Gaging Diameter		G[4] Gaging Diameter
			Max, Edge Sharp	Min, Edge Rounded or Flat	Ref	Max	Min	Max	Min	Min		Max	Min	Max	Min	
0	0.0600	1/8	0.119	0.099	0.035	0.069	0.056	0.043	0.027	0.014	0	0.036	0.020	0.026	0.016	0.078
1	0.0730	1/8	0.146	0.123	0.043	0.077	0.064	0.051	0.035	0.015	0	0.044	0.028	0.028	0.016	0.101
2	0.0860	1/8	0.172	0.147	0.051	0.102	0.089	0.063	0.047	0.017	1	0.056	0.040	0.029	0.017	0.124
3	0.0990	1/8	0.199	0.171	0.059	0.107	0.094	0.068	0.052	0.018	1	0.061	0.045	0.031	0.018	0.148
4	0.1120	3/16	0.225	0.195	0.067	0.128	0.115	0.089	0.073	0.018	1	0.082	0.066	0.032	0.019	0.172
5	0.1250	3/16	0.252	0.220	0.075	0.154	0.141	0.086	0.063	0.027	2	0.075	0.052	0.034	0.020	0.196
6	0.1380	3/16	0.279	0.244	0.083	0.174	0.161	0.106	0.083	0.029	2	0.095	0.072	0.036	0.021	0.220
8	0.1640	1/4	0.332	0.292	0.100	0.189	0.176	0.121	0.098	0.030	2	0.110	0.087	0.039	0.023	0.267
10	0.1900	5/16	0.385	0.340	0.116	0.204	0.191	0.136	0.113	0.032	2	0.125	0.102	0.042	0.025	0.313
12	0.2160	3/8	0.438	0.389	0.132	0.268	0.255	0.156	0.133	0.035	3	0.139	0.116	0.045	0.027	0.362
1/4	0.2500	7/16	0.507	0.452	0.153	0.283	0.270	0.171	0.148	0.036	3	0.154	0.131	0.050	0.029	0.424
5/16	0.3125	1/2	0.635	0.568	0.191	0.365	0.352	0.216	0.194	0.061	4	0.196	0.174	0.057	0.034	0.539
3/8	0.3750	9/16	0.762	0.685	0.230	0.393	0.380	0.245	0.223	0.065	4	0.225	0.203	0.065	0.039	0.653
7/16	0.4375	5/8	0.812	0.723	0.223	0.409	0.396	0.261	0.239	0.068	4	0.241	0.219	0.073	0.044	0.690
1/2	0.5000	3/4	0.875	0.775	0.223	0.424	0.411	0.276	0.254	0.069	4	0.256	0.234	0.081	0.049	0.739
9/16	0.5625	—	1.000	0.889	0.260	0.454	0.431	0.300	0.278	0.073	4	0.280	0.258	0.089	0.053	0.851
5/8	0.6250	—	1.125	1.002	0.298	0.576	0.553	0.342	0.316	0.079	5	0.309	0.283	0.097	0.058	0.962
3/4	0.7500	—	1.375	1.230	0.372	0.640	0.617	0.406	0.380	0.087	5	0.373	0.347	0.112	0.067	1.186

[1] Where specifying nominal size in decimals, zeros preceding decimal and in the fourth decimal place shall be omitted.
[2] Screws of these lengths and shorter shall have undercut heads.
[3] Tabulated values determined from formula for maximum H, Appendix V, ANSI B18.6.3—1972.
[4] No tolerance for gaging diameter is given. If the gaging diameter of the gage used differs from tabulated value, the protrusion will be affected accordingly and the proper protrusion values must be recalculated using the formulas shown in **Appendix I**, ANSI B18.6.3—1972.
(Reprinted from The American Society of Mechanical Engineers—ANSI B18.6.3—1972)

	SLOTTED
	PAN
	Type of Head

AMERICAN NATIONAL STANDARD
MACHINE SCREWS AND MACHINE SCREW NUTS

ANSI B18.6.3—1972

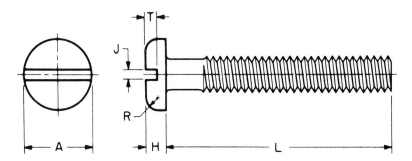

TABLE 18 DIMENSIONS OF SLOTTED PAN HEAD MACHINE SCREWS

Nominal Size[1] or Basic Screw Diameter		A Head Diameter		H Head Height		R Head Radius	J Slot Width		T Slot Depth	
		Max	Min	Max	Min	Max	Max	Min	Max	Min
0000	0.0210	0.042	0.036	0.016	0.010	0.007	0.008	0.004	0.008	0.004
000	0.0340	0.066	0.060	0.023	0.017	0.010	0.012	0.008	0.012	0.008
00	0.0470	0.090	0.082	0.032	0.025	0.015	0.017	0.010	0.016	0.010
0	0.0600	0.116	0.104	0.039	0.031	0.020	0.023	0.016	0.022	0.014
1	0.0730	0.142	0.130	0.046	0.038	0.025	0.026	0.019	0.027	0.018
2	0.0860	0.167	0.155	0.053	0.045	0.035	0.031	0.023	0.031	0.022
3	0.0990	0.193	0.180	0.060	0.051	0.037	0.035	0.027	0.036	0.026
4	0.1120	0.219	0.205	0.068	0.058	0.042	0.039	0.031	0.040	0.030
5	0.1250	0.245	0.231	0.075	0.065	0.044	0.043	0.035	0.045	0.034
6	0.1380	0.270	0.256	0.082	0.072	0.046	0.048	0.039	0.050	0.037
8	0.1640	0.322	0.306	0.096	0.085	0.052	0.054	0.045	0.058	0.045
10	0.1900	0.373	0.357	0.110	0.099	0.061	0.060	0.050	0.068	0.053
12	0.2160	0.425	0.407	0.125	0.112	0.078	0.067	0.056	0.077	0.061
1/4	0.2500	0.492	0.473	0.144	0.130	0.087	0.075	0.064	0.087	0.070
5/16	0.3125	0.615	0.594	0.178	0.162	0.099	0.084	0.072	0.106	0.085
3/8	0.3750	0.740	0.716	0.212	0.195	0.143	0.094	0.081	0.124	0.100
7/16	0.4375	0.863	0.837	0.247	0.228	0.153	0.094	0.081	0.142	0.116
1/2	0.5000	0.987	0.958	0.281	0.260	0.175	0.106	0.091	0.161	0.131
9/16	0.5625	1.041	1.000	0.315	0.293	0.197	0.118	0.102	0.179	0.146
5/8	0.6250	1.172	1.125	0.350	0.325	0.219	0.133	0.116	0.197	0.162
3/4	0.7500	1.435	1.375	0.419	0.390	0.263	0.149	0.131	0.234	0.192

[1] Where specifying nominal size in decimals, zeros preceding decimal and in the fourth decimal place shall be omitted.
(Reprinted from The American Society of Mechanical Engineers—ANSI B18.6.3—1972)

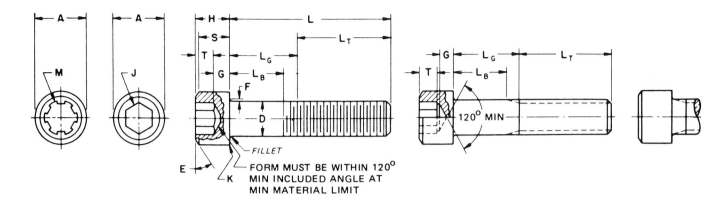

TABLE 19 DIMENSIONS OF HEXAGON AND SPLINE SOCKET HEAD CAP SCREWS (1960 SERIES)

Nominal Size or Basic Screw Diameter		D Body Diameter		A Head Diameter		H Head Height		S Head Side Height	M Spline Socket Size	J Hexagon Socket Size		T Key Engagement	G Wall Thickness	K Chamfer or Radius
		Max	Min	Max	Min	Max	Min	Min	Nom	Nom		Min	Min	Max
0	0.0600	0.0600	0.0568	0.096	0.091	0.060	0.057	0.054	0.060	0.050		0.025	0.020	0.003
1	0.0730	0.0730	0.0695	0.118	0.112	0.073	0.070	0.066	0.072	1/16	0.062	0.031	0.025	0.003
2	0.0860	0.0860	0.0822	0.140	0.134	0.086	0.083	0.077	0.096	5/64	0.078	0.038	0.029	0.003
3	0.0990	0.0990	0.0949	0.161	0.154	0.099	0.095	0.089	0.096	5/64	0.078	0.044	0.034	0.003
4	0.1120	0.1120	0.1075	0.183	0.176	0.112	0.108	0.101	0.111	3/32	0.094	0.051	0.038	0.005
5	0.1250	0.1250	0.1202	0.205	0.198	0.125	0.121	0.112	0.111	3/32	0.094	0.057	0.043	0.005
6	0.1380	0.1380	0.1329	0.226	0.218	0.138	0.134	0.124	0.133	7/64	0.109	0.064	0.047	0.005
8	0.1640	0.1640	0.1585	0.270	0.262	0.164	0.159	0.148	0.168	9/64	0.141	0.077	0.056	0.005
10	0.1900	0.1900	0.1840	0.312	0.303	0.190	0.185	0.171	0.183	5/32	0.156	0.090	0.065	0.005
1/4	0.2500	0.2500	0.2435	0.375	0.365	0.250	0.244	0.225	0.216	3/16	0.188	0.120	0.095	0.008
5/16	0.3125	0.3125	0.3053	0.469	0.457	0.312	0.306	0.281	0.291	1/4	0.250	0.151	0.119	0.008
3/8	0.3750	0.3750	0.3678	0.562	0.550	0.375	0.368	0.337	0.372	5/16	0.312	0.182	0.143	0.008
7/16	0.4375	0.4375	0.4294	0.656	0.642	0.438	0.430	0.394	0.454	3/8	0.375	0.213	0.166	0.010
1/2	0.5000	0.5000	0.4919	0.750	0.735	0.500	0.492	0.450	0.454	3/8	0.375	0.245	0.190	0.010
5/8	0.6250	0.6250	0.6163	0.938	0.921	0.625	0.616	0.562	0.595	1/2	0.500	0.307	0.238	0.010
3/4	0.7500	0.7500	0.7406	1.125	1.107	0.750	0.740	0.675	0.620	5/8	0.625	0.370	0.285	0.010
7/8	0.8750	0.8750	0.8647	1.312	1.293	0.875	0.864	0.787	0.698	3/4	0.750	0.432	0.333	0.015
1	1.0000	1.0000	0.9886	1.500	1.479	1.000	0.988	0.900	0.790	3/4	0.750	0.495	0.380	0.015
1 1/8	1.1250	1.1250	1.1086	1.688	1.665	1.125	1.111	1.012	....	7/8	0.875	0.557	0.428	0.015
1 1/4	1.2500	1.2500	1.2336	1.875	1.852	1.250	1.236	1.125	....	7/8	0.875	0.620	0.475	0.015
1 3/8	1.3750	1.3750	1.3568	2.062	2.038	1.375	1.360	1.237	....	1	1.000	0.682	0.523	0.015
1 1/2	1.5000	1.5000	1.4818	2.250	2.224	1.500	1.485	1.350	....	1	1.000	0.745	0.570	0.015
1 3/4	1.7500	1.7500	1.7295	2.625	2.597	1.750	1.734	1.575	....	1 1/4	1.250	0.870	0.665	0.015
2	2.0000	2.0000	1.9780	3.000	2.970	2.000	1.983	1.800	....	1 1/2	1.500	0.995	0.760	0.015
2 1/4	2.2500	2.2500	2.2280	3.375	3.344	2.250	2.232	2.025	....	1 3/4	1.750	1.120	0.855	0.031
2 1/2	2.5000	2.5000	2.4762	3.750	3.717	2.500	2.481	2.250	....	1 3/4	1.750	1.245	0.950	0.031
2 3/4	2.7500	2.7500	2.7262	4.125	4.090	2.750	2.730	2.475	....	2	2.000	1.370	1.045	0.031
3	3.0000	3.0000	2.9762	4.500	4.464	3.000	2.979	2.700	....	2 1/4	2.250	1.495	1.140	0.031
3 1/4	3.2500	3.2500	3.2262	4.875	4.837	3.250	3.228	2.925	....	2 1/4	2.250	1.620	1.235	0.031
3 1/2	3.5000	3.5000	3.4762	5.250	5.211	3.500	3.478	3.150	....	2 3/4	2.750	1.745	1.330	0.031
3 3/4	3.7500	3.7500	3.7262	5.625	5.584	3.750	3.727	3.375	....	2 3/4	2.750	1.870	1.425	0.031
4	4.0000	4.0000	3.9762	6.000	5.958	4.000	3.976	3.600	....	3	3.000	1.995	1.520	0.031

(Reprinted from The American Society of Mechanical Engineers—ANSI B18.3–1982)

TABLE 20 DIMENSIONS OF SQUARE HEAD AND HEADLESS SET SCREWS

Diam.	A	B	C	E	F	G	H	J	K	L	M	N	
5-.125			.023	.031	1/16	.071 .070	.053 .052	.022 .021	1/16	.083 .078	.06	.03	Max. Min.
6-.138			.025	.035	1/16	.079 .078	.056 .055	.022 .021	.069	.092 .087	.07	.03	Max. Min.
8-.164			.029	.041	5/64	.098 .097	.082 .080	.022 .021	5/64	.109 .103	.08	.04	Max. Min.
10-.190			.032	.048	3/32	.115 .113	.098 .096	.025 .023	3/32	.127 .120	.09	.04	Max. Min.
12-.216			.036	.054	3/32	.115 .113	.098 .096	.025 .023	7/64	.144 .137	.11	.06	Max. Min.
1/4	1/4	3/16	.045	.063	1/8	.149 .147	.128 .126	.032 .030	1/8	5/32	1/8	1/16	Max. Min.
5/16	5/16	15/64	.051	.078	5/32	.188 .186	.163 .161	.039 .037	11/64	13/64	5/32	5/64	Max. Min.
3/8	3/8	9/32	.064	.094	3/16	.221 .219	.190 .188	.050 .048	13/64	1/4	3/16	3/32	Max. Min.
7/16	7/16	21/64	.072	.109	7/32	.256 .254	.221 .219	.060 .058	15/64	19/64	7/32	7/64	Max. Min.
1/2	1/2	3/8	.081	.125	1/4	.298 .296	.254 .252	.068 .066	9/32	11/32	1/4	1/8	Max. Min.
9/16	9/16	27/64	.091	.141	1/4	.298 .296	.254 .252	.068 .066	5/16	25/64	9/32	9/64	Max. Min.
5/8	5/8	15/32	.102	.156	5/16	.380 .377	.319 .316	.092 .089	23/64	15/32	5/16	5/32	Max. Min.
3/4	3/4	9/16	.129	.188	3/8	.463 .460	.386 .383	.112 .109	7/16	9/16	3/8	3/16	Max. Min.
7/8	7/8	21/32			1/2	.604 .601	.509 .506	.138 .134	33/64	21/32	7/16	7/32	Max. Min.
1	1	3/4			9/16	.631 .627	.535 .531	.149 .145	19/32	3/4	1/2	1/4	Max. Min.
1 1/8	1 1/8	27/32			9/16	.709 .705	.604 .600	.168 .164	43/64	27/32	9/16	9/32	Max. Min.
1 1/4	1 1/4	15/16			5/8	.801 .797	.685 .681	.189 .185	3/4	15/16	5/8	5/16	Max. Min.
1 3/8	1 3/8	1 1/32			5/8	.869 .865	.744 .740	.207 .203	53/64	1 1/32	11/16	11/32	Max. Min.
1 1/2	1 1/2	1 1/8			3/4	.970 .966	.828 .824	.231 .227	29/32	1 1/8	3/4	3/8	Max. Min.
1 3/4					1	1.275 1.271	1.007 1.003	.298 .294	1 1/16	1 5/16	7/8	7/16	Max. Min.
2					1	1.275 1.271	1.007 1.003	.298 .294	1 7/32	1 1/2	1	1/2	Max. Min.

*Angle Y = 118° When Length Equals Diam. or Less, Y = 90° When Length Exceeds Diam.

(Reprinted from BASIC DRAFTING TECHNOLOGY by Rotmans, Horton, and Good, © 1980 Delmar Publishers Inc.)

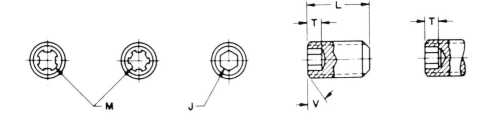

TABLE 21 DIMENSIONS OF HEXAGON AND SPLINE SOCKET SET SCREWS

Nominal Size or Basic Screw Diameter		J Hexagon Socket Size Nom		M Spline Socket Size Nom	T Min Key Engagement to Develop Functional Capability of Key		C Cup and Flat Point Diameters		R Oval Point Radius	Y Cone Point Angle 90° ±2° For These Nominal Lengths or Longer; 118° ±2° For Shorter Nominal Lengths
					Hex Socket T_H Min	Spline Socket T_S Min	Max	Min	Basic	
0	0.0600		0.028	0.033	0.050	0.026	0.033	0.027	0.045	5/64
1	0.0730		0.035	0.033	0.060	0.035	0.040	0.033	0.055	3/32
2	0.0860		0.035	0.048	0.060	0.040	0.047	0.039	0.064	7/64
3	0.0990		0.050	0.048	0.070	0.040	0.054	0.045	0.074	1/8
4	0.1120		0.050	0.060	0.070	0.045	0.061	0.051	0.084	5/32
5	0.1250	1/16	0.062	0.072	0.080	0.055	0.067	0.057	0.094	3/16
6	0.1380	1/16	0.062	0.072	0.080	0.055	0.074	0.064	0.104	3/16
8	0.1640	5/64	0.078	0.096	0.090	0.080	0.087	0.076	0.123	1/4
10	0.1900	3/32	0.094	0.111	0.100	0.080	0.102	0.088	0.142	1/4
1/4	0.2500	1/8	0.125	0.145	0.125	0.125	0.132	0.118	0.188	5/16
5/16	0.3125	5/32	0.156	0.183	0.156	0.156	0.172	0.156	0.234	3/8
3/8	0.3750	3/16	0.188	0.216	0.188	0.188	0.212	0.194	0.281	7/16
7/16	0.4375	7/32	0.219	0.251	0.219	0.219	0.252	0.232	0.328	1/2
1/2	0.5000	1/4	0.250	0.291	0.250	0.250	0.291	0.270	0.375	9/16
5/8	0.6250	5/16	0.312	0.372	0.312	0.312	0.371	0.347	0.469	3/4
3/4	0.7500	3/8	0.375	0.454	0.375	0.375	0.450	0.425	0.562	7/8
7/8	0.8750	1/2	0.500	0.595	0.500	0.500	0.530	0.502	0.656	1
1	1.0000	9/16	0.562	...	0.562	...	0.609	0.579	0.750	1 1/8
1 1/8	1.1250	9/16	0.562	...	0.562	...	0.689	0.655	0.844	1 1/4
1 1/4	1.2500	5/8	0.625	...	0.625	...	0.767	0.733	0.938	1 1/2
1 3/8	1.3750	5/8	0.625	...	0.625	...	0.848	0.808	1.031	1 5/8
1 1/2	1.5000	3/4	0.750	...	0.750	...	0.926	0.886	1.125	1 3/4
1 3/4	1.7500	1	1.000	...	1.000	...	1.086	1.039	1.312	2
2	2.0000	1	1.000	...	1.000	...	1.244	1.193	1.500	2 1/4

(Reprinted from The American Society of Mechanical Engineers—ANSI B18.3–1982)

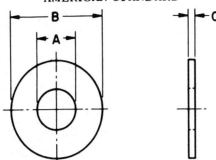

AMERICAN STANDARD

TABLE 22 DIMENSIONS OF PREFERRED SIZES OF TYPE A PLAIN WASHERS**

Nominal Washer Size***			Inside Diameter A			Outside Diameter B			Thickness C		
			Basic	Tolerance Plus	Tolerance Minus	Basic	Tolerance Plus	Tolerance Minus	Basic	Max	Min
—	—		0.078	0.000	0.005	0.188	0.000	0.005	0.020	0.025	0.016
—	—		0.094	0.000	0.005	0.250	0.000	0.005	0.020	0.025	0.016
—	—		0.125	0.008	0.005	0.312	0.008	0.005	0.032	0.040	0.025
No. 6	0.138		0.156	0.008	0.005	0.375	0.015	0.005	0.049	0.065	0.036
No. 8	0.164		0.188	0.008	0.005	0.438	0.015	0.005	0.049	0.065	0.036
No. 10	0.190		0.219	0.008	0.005	0.500	0.015	0.005	0.049	0.065	0.036
3/16	0.188		0.250	0.015	0.005	0.562	0.015	0.005	0.049	0.065	0.036
No. 12	0.216		0.250	0.015	0.005	0.562	0.015	0.005	0.065	0.080	0.051
1/4	0.250	N	0.281	0.015	0.005	0.625	0.015	0.005	0.065	0.080	0.051
1/4	0.250	W	0.312	0.015	0.005	0.734*	0.015	0.007	0.065	0.080	0.051
5/16	0.312	N	0.344	0.015	0.005	0.688	0.015	0.007	0.065	0.080	0.051
5/16	0.312	W	0.375	0.015	0.005	0.875	0.030	0.007	0.083	0.104	0.064
3/8	0.375	N	0.406	0.015	0.005	0.812	0.015	0.007	0.065	0.080	0.051
3/8	0.375	W	0.438	0.015	0.005	1.000	0.030	0.007	0.083	0.104	0.064
7/16	0.438	N	0.469	0.015	0.005	0.922	0.015	0.007	0.065	0.080	0.051
7/16	0.438	W	0.500	0.015	0.005	1.250	0.030	0.007	0.083	0.104	0.064
1/2	0.500	N	0.531	0.015	0.005	1.062	0.030	0.007	0.095	0.121	0.074
1/2	0.500	W	0.562	0.015	0.005	1.375	0.030	0.007	0.109	0.132	0.086
9/16	0.562	N	0.594	0.015	0.005	1.156*	0.030	0.007	0.095	0.121	0.074
9/16	0.562	W	0.625	0.015	0.005	1.469*	0.030	0.007	0.109	0.132	0.086
5/8	0.625	N	0.656	0.030	0.007	1.312	0.030	0.007	0.095	0.121	0.074
5/8	0.625	W	0.688	0.030	0.007	1.750	0.030	0.007	0.134	0.160	0.108
3/4	0.750	N	0.812	0.030	0.007	1.469	0.030	0.007	0.134	0.160	0.108
3/4	0.750	W	0.812	0.030	0.007	2.000	0.030	0.007	0.148	0.177	0.122
7/8	0.875	N	0.938	0.030	0.007	1.750	0.030	0.007	0.134	0.160	0.108
7/8	0.875	W	0.938	0.030	0.007	2.250	0.030	0.007	0.165	0.192	0.136
1	1.000	N	1.062	0.030	0.007	2.000	0.030	0.007	0.134	0.160	0.108
1	1.000	W	1.062	0.030	0.007	2.500	0.030	0.007	0.165	0.192	0.136
1 1/8	1.125	N	1.250	0.030	0.007	2.250	0.030	0.007	0.134	0.160	0.108
1 1/8	1.125	W	1.250	0.030	0.007	2.750	0.030	0.007	0.165	0.192	0.136
1 1/4	1.250	N	1.375	0.030	0.007	2.500	0.030	0.007	0.165	0.192	0.136
1 1/4	1.250	W	1.375	0.030	0.007	3.000	0.030	0.007	0.165	0.192	0.136
1 3/8	1.375	N	1.500	0.030	0.007	2.750	0.030	0.007	0.165	0.192	0.136
1 3/8	1.375	W	1.500	0.045	0.010	3.250	0.045	0.010	0.180	0.213	0.153
1 1/2	1.500	N	1.625	0.030	0.007	3.000	0.030	0.007	0.165	0.192	0.136
1 1/2	1.500	W	1.625	0.045	0.010	3.500	0.045	0.010	0.180	0.213	0.153
1 5/8	1.625		1.750	0.045	0.010	3.750	0.045	0.010	0.180	0.213	0.153
1 3/4	1.750		1.875	0.045	0.010	4.000	0.045	0.010	0.180	0.213	0.153
1 7/8	1.875		2.000	0.045	0.010	4.250	0.045	0.010	0.180	0.213	0.153
2	2.000		2.125	0.045	0.010	4.500	0.045	0.010	0.180	0.213	0.153
2 1/4	2.250		2.375	0.045	0.010	4.750	0.045	0.010	0.220	0.248	0.193
2 1/2	2.500		2.625	0.045	0.010	5.000	0.045	0.010	0.238	0.280	0.210
2 3/4	2.750		2.875	0.065	0.010	5.250	0.065	0.010	0.259	0.310	0.228
3	3.000		3.125	0.065	0.010	5.500	0.065	0.010	0.284	0.327	0.249

*The 0.734 in., 1.156 in., and 1.469 in. outside diameters avoid washers which could be used in coin operated devices.
**Preferred sizes are for the most part from series previously designed "Standard Plate" and "SAE." Where common sizes existed in the two series, the SAE size is designated "N" (narrow) and the Standard Plate "W" (wide). These sizes as well as all other sizes for Type A Plain Washers are to be ordered by ID, OD, and thickness dimensions.
***Nominal washer sizes are intended for use with comparable nominal screw or bolt sizes.
(Reprinted from The American Society of Mechanical Engineers—ANSI B18.22.1–1965, R|1975)

AMERICAN NATIONAL STANDARD
LOCK WASHERS

ANSI B18.21.1–1972

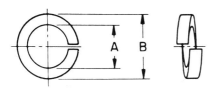
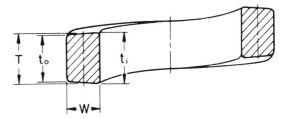

ENLARGED SECTION

TABLE 23 DIMENSIONS OF REGULAR HELICAL SPRING LOCK WASHERS[1]

Nominal Washer Size		A Inside Diameter		B Outside Diameter	T Mean Section Thickness $\left(\dfrac{t_i + t_o}{2}\right)$	W Section Width
		Max	Min	Max[2]	Min	Min
No. 4	0.112	0.120	0.114	0.173	0.022	0.022
No. 5	0.125	0.133	0.127	0.202	0.030	0.030
No. 6	0.138	0.148	0.141	0.216	0.030	0.030
No. 8	0.164	0.174	0.167	0.267	0.047	0.042
No. 10	0.190	0.200	0.193	0.294	0.047	0.042
1/4	0.250	0.262	0.254	0.365	0.078	0.047
5/16	0.312	0.326	0.317	0.460	0.093	0.062
3/8	0.375	0.390	0.380	0.553	0.125	0.076
7/16	0.438	0.455	0.443	0.647	0.140	0.090
1/2	0.500	0.518	0.506	0.737	0.172	0.103
5/8	0.625	0.650	0.635	0.923	0.203	0.125
3/4	0.750	0.775	0.760	1.111	0.218	0.154
7/8	0.875	0.905	0.887	1.296	0.234	0.182
1	1.000	1.042	1.017	1.483	0.250	0.208
1 1/8	1.125	1.172	1.144	1.669	0.313	0.236
1 1/4	1.250	1.302	1.271	1.799	0.313	0.236
1 3/8	1.375	1.432	1.398	2.041	0.375	0.292
1 1/2	1.500	1.561	1.525	2.170	0.375	0.292
1 3/4	1.750	1.811	1.775	2.602	0.469	0.383
2	2.000	2.061	2.025	2.852	0.469	0.383
2 1/4	2.250	2.311	2.275	3.352	0.508	0.508
2 1/2	2.500	2.561	2.525	3.602	0.508	0.508
2 3/4	2.750	2.811	2.775	4.102	0.633	0.633
3	3.000	3.061	3.025	4.352	0.633	0.633

[1] For use with 1960 Series Socket Head Cap Screws specified in American National Standard, ANSI B18.3.
[2] The maximum outside diameters specified allow for the commercial tolerances on cold drawn wire.
(Reprinted from The American Society of Mechanical Engineers—ANSI B18.21.1–1972)

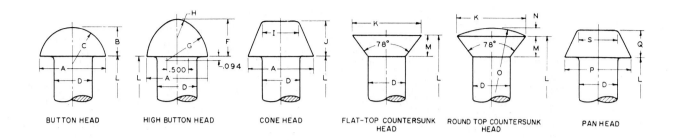

TABLE 24 AMERICAN NATIONAL STANDARD LARGE RIVETS

						Manufactured Shapes										
D nominal	A basic	B basic (min)	C	E basic	F basic	G	H	I basic	J basic (min.)	K basic	M basic (min.)	N	O	P basic	Q basic (min.)	S basic
1/2	0.875	0.375	0.443	0.781	0.500	0.656	0.094	0.469	0.438	0.905	0.250	0.095	1.125	0.800	0.381	0.500
5/8	1.094	0.469	0.553	0.969	0.594	0.750	0.188	0.586	0.547	1.131	0.312	0.119	1.406	1.000	0.469	0.625
3/4	1.312	0.562	0.664	1.156	0.688	0.844	0.282	0.703	0.656	1.358	0.375	0.142	1.688	1.200	0.556	0.750
7/8	1.531	0.656	0.775	1.344	0.781	0.937	0.375	0.820	0.766	1.584	0.438	0.166	1.969	1.400	0.643	0.875
1	1.750	0.750	0.885	1.531	0.875	1.031	0.469	0.938	0.875	1.810	0.500	0.190	2.250	1.600	0.731	1.000
1 1/8	1.969	0.844	0.996	1.719	0.969	1.125	0.563	1.055	0.984	2.036	0.562	0.214	2.531	1.800	0.835	1.125
1 1/4	2.188	0.938	1.107	1.906	1.062	1.218	0.656	1.172	1.094	2.262	0.625	0.238	2.812	2.000	0.922	1.250
1 3/8	2.406	1.031	1.217	2.094	1.156	1.312	0.750	1.290	1.203	2.489	0.688	0.261	3.094	2.200	1.009	1.375
1 1/2	2.625	1.125	1.328	2.281	1.250	1.406	0.844	1.406	1.312	2.715	0.750	0.285	3.375	2.400	1.113	1.500
1 5/8	2.844	1.219	1.439	2.469	1.344	1.500	0.938	1.524	1.422	2.941	0.812	0.309	3.656	2.600	1.201	1.625
1 3/4	3.062	1.312	1.549	2.656	1.438	1.594	1.032	1.641	1.531	3.168	0.875	0.332	3.938	2.800	1.288	1.750

TABLE 25 DRILL AND TAP SIZES

DECIMAL EQUIVALENTS AND TAP DRILL SIZES
OF WIRE GAGE LETTER AND FRACTIONAL SIZE DRILLS
(TAP DRILL SIZES BASED ON 75% MAXIMUM THREAD)

FRACTIONAL SIZE DRILLS INCHES	WIRE GAGE DRILLS	DECIMAL EQUIVALENT INCHES	TAP SIZES		FRACTIONAL SIZE DRILLS INCHES	WIRE GAGE DRILLS	DECIMAL EQUIVALENT INCHES	TAP SIZES		FRACTIONAL SIZE DRILLS INCHES	WIRE GAGE DRILLS	DECIMAL EQUIVALENT INCHES	TAP SIZES	
			SIZE OF THREAD	THREADS PER INCH				SIZE OF THREAD	THREADS PER INCH				SIZE OF THREAD	THREADS PER INCH
	80	.0135			9/64	.....	.1406			23/64	.....	.3594		
	79	.0145				27	.1440				U	.3680	7/16	14
1/64	.....	.0156				26	.1470			3/8	.....	.3750		
	78	.0160				25	.1495	10	24					
	77	.0180				24	.1520				V	.3770		
	76	.0200				23	.1540				W	.3860		
	75	.0210			5/32	.....	.1562			25/64	.....	.3906	7/16	20
	74	.0225				22	.1570				X	.3970		
	73	.0240				21	.1590	10	32		Y	.4040		
	72	.0250				20	.1610			13/32	.....	.4062		
	71	.0260				19	.1660				Z	.4130		
	70	.0280				18	.1695			27/64	.....	.4219	1/2	13
	69	.0292			11/64	.....	.1719			7/16	.....	.4375		
	68	.0310				17	.1730			29/64	.....	.4531	1/2	20
1/32	.....	.0312				16	.1770	12	24	15/32	.....	.4687		
	67	.0320				15	.1800			31/64	.....	.4844	9/16	12
	66	.0330				14	.1820	12	28	1/2	.....	.5000		
	65	.0350				13	.1850			33/64	.....	.5156	9/16	18
	64	.0360			3/16	.....	.1875			17/32	.....	.5312	5/8	11
	63	.0370				12	.1890			35/64	.....	.5469		
	62	.0380				11	.1910			9/16	.....	.5625		
	61	.0390				10	.1935			37/64	.....	.5781	5/8	18
	60	.0400				9	.1960			19/32	.....	.5937		
	59	.0410				8	.1990			39/64	.....	.6094		
	58	.0420				7	.2010	1/4	20	5/8	.....	.6250		
	57	.0430			13/64	.....	.2031			41/64	.....	.6406		
	56	.0465				6	.2040			21/32	.....	.6562	3/4	10
3/64	.....	.0469	0	80		5	.2055			43/64	.....	.6719		
	55	.0520				4	.2090			11/16	.....	.6875		
	54	.0550				3	.2130	1/4	28	45/64	.....	.7031	3/4	16
	53	.0595	1	64	7/32	.....	.2187			23/32	.....	.7187		
1/16	.....	.0625		72		2	.2210			47/64	.....	.7344		
	52	.0635				1	.2280			3/4	.....	.7500		
	51	.0670				A	.2340			49/64	.....	.7656	7/8	9
	50	.0700	2	56	15/64	.....	.2344			25/32	.....	.7812		
	49	.0730		64		B	.2380			51/64	.....	.7969		
	48	.0760				C	.2420			13/16	.....	.8125	7/8	14
5/64	.....	.0781				D	.2460			53/64	.....	.8281		
	47	.0785	3	48	1/4	E	.2500			27/32	.....	.8437		
	46	.0810				F	.2570	5/16	18	55/64	.....	.8594		
	45	.0820	3	56		G	.2610			7/8	.....	.8750	1	8
	44	.0860			17/64	.....	.2656			57/64	.....	.8906		
	43	.0890	4	40		H	.2660			29/32	.....	.9062		
	42	.0935	4	48		I	.2720	5/16	24	59/64	.....	.9219		
3/32	.....	.0937				J	.2770			15/16	.....	.9375	1	12
	41	.0960				K	.2810			61/64	.....	.9531		
	40	.0980			9/32	.....	.2812			31/32	.....	.9687		
	39	.0995				L	.2900			63/64	.....	.9844	1 1/8	7
	38	.1015	5	40		M	.2950			1	.....	1.0000		
	37	.1040	5	44	19/64	.....	.2969							
	36	.1065	6	32		N	.3020							
7/64	.....	.1094			5/16	.....	.3125	3/8	16					
	35	.1100				O	.3160							
	34	.1110				P	.3230							
	33	.1130	6	40	21/64	.....	.3281							
	32	.1160				Q	.3320	3/8	24					
	31	.1200				R	.3390							
1/8	.....	.1250			11/32	.....	.3437							
	30	.1285				S	.3480							
	29	.1360	8	32		T	.3580							
	28	.1405		36										

(Reprinted from TECHNICAL DRAWING AND DESIGN by Goetsch and Nelson, © 1986 Delmar Publishers Inc.)

TABLE 26 SPUR/PINION GEAR TOOTH PARTS

20 Degree Pressure Angles

Diametral Pitch	Circular Pitch	Circular Thickness	Addend.	Dedend.	Standard Fillet Radius
D.P.	C.P.	C.T.	A	D	
1	3.1416	1.5708	1.0000	1.2500	0.3000
2	1.5708	0.7854	0.5000	0.6250	0.1500
2.5	1.2566	0.6283	0.4000	0.5000	0.1200
3	1.0472	0.5236	0.3333	0.4167	0.1000
3.5	0.8976	0.4488	0.2857	0.3571	0.0857
4	0.7854	0.3927	0.2500	0.3125	0.0750
4.5	0.6981	0.3491	0.2222	0.2778	0.0667
5	0.6283	0.3142	0.2000	0.2500	0.0600
5.5	0.5712	0.2856	0.1818	0.2273	0.0545
6	0.5236	0.2618	0.1667	0.2083	0.0500
6.5	0.4833	0.2417	0.1538	0.1923	0.0462
7	0.4488	0.2244	0.1429	0.1786	0.0429
7.5	0.4189	0.2094	0.1333	0.1667	0.0400
8	0.3927	0.1963	0.1250	0.1563	0.0375
8.5	0.3696	0.1848	0.1176	0.1471	0.0353
9	0.3491	0.1745	0.1111	0.1389	0.0333
9.5	0.3307	0.1653	0.1053	0.1316	0.0316
10	0.3142	0.1571	0.1000	0.1250	0.0300
11	0.2856	0.1428	0.0909	0.1136	0.0273
12	0.2618	0.1309	0.0833	0.1042	0.0250
13	0.2417	0.1208	0.0769	0.0962	0.0231
14	0.2244	0.1122	0.0714	0.0893	0.0214
15	0.2094	0.1047	0.0667	0.0833	0.0200

(Reprinted from TECHNICAL DRAWING AND DESIGN by Goetsch and Nelson, © 1986 Delmar Publishers Inc.)

TABLE 27 WOODRUFF KEY SIZES FOR DIFFERENT SHAFT DIAMETERS

Shaft Diameter	5/16 to 3/8	7/16 to 1/2	9/16 to 3/4	13/16 to 15/16	1 to 1 3/16	1 1/4 to 1 7/16	1 1/2 to 1 3/4	1 13/16 to 2 1/8	2 3/16 to 2 1/2
Key Numbers	204	304 305	404 405 406	505 506 507	606 607 608 609	807 808 809	810 811 812	1011 1012	1211 1212

(Reprinted from TECHNICAL DRAWING AND DESIGN by Goetsch and Nelson, © 1986 Delmar Publishers Inc.)

TABLE 28 KEY AND KEYWAY SIZES

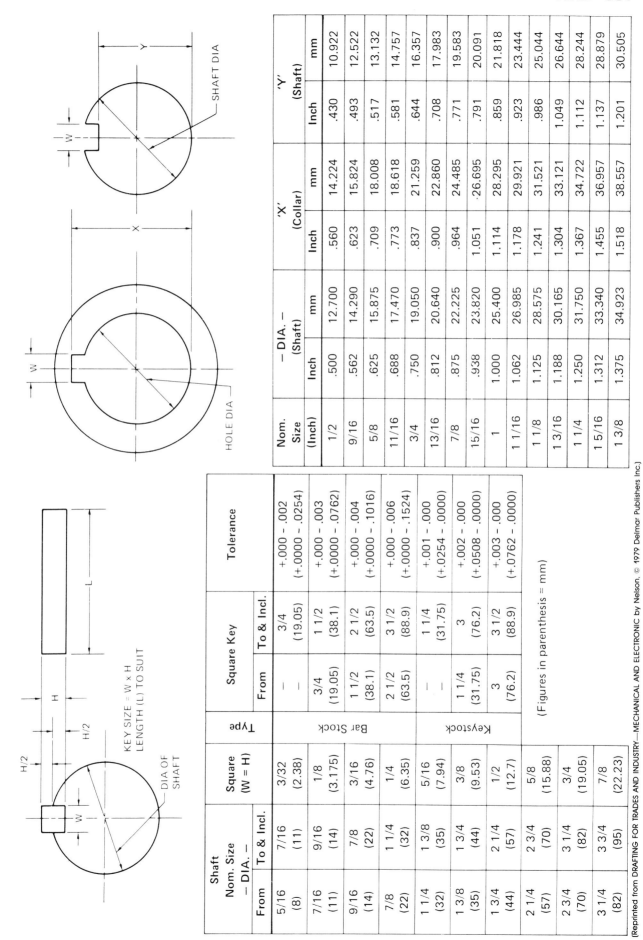

Shaft Nom. Size – DIA. –		Square (W = H)	Type	Square Key		Tolerance
From	To & Incl.			From	To & Incl.	
5/16 (8)	7/16 (11)	3/32 (2.38)	Bar Stock	—	3/4 (19.05)	+.000 -.002 (+.0000 -.0254)
7/16 (11)	9/16 (14)	1/8 (3.175)	Bar Stock	3/4 (19.05)	1 1/2 (38.1)	+.000 -.003 (+.0000 -.0762)
9/16 (14)	7/8 (22)	3/16 (4.76)	Bar Stock	1 1/2 (38.1)	2 1/2 (63.5)	+.000 -.004 (+.0000 -.1016)
7/8 (22)	1 1/4 (32)	1/4 (6.35)	Bar Stock	2 1/2 (63.5)	3 1/2 (88.9)	+.000 -.006 (+.0000 -.1524)
1 1/4 (32)	1 3/8 (35)	5/16 (7.94)	Keystock	—	1 1/4 (31.75)	+.001 -.000 (+.0254 -.0000)
1 3/8 (35)	1 3/4 (44)	3/8 (9.53)	Keystock	1 1/4 (31.75)	3 (76.2)	+.002 -.000 (+.0508 -.0000)
1 3/4 (44)	2 1/4 (57)	1/2 (12.7)	Keystock	3 (76.2)	3 1/2 (88.9)	+.003 -.000 (+.0762 -.0000)
2 1/4 (57)	2 3/4 (70)	5/8 (15.88)				
2 3/4 (70)	3 1/4 (82)	3/4 (19.05)				
3 1/4 (82)	3 3/4 (95)	7/8 (22.23)				

(Figures in parenthesis = mm)

Nom. Size (Inch)	– DIA. – (Shaft)		'X' (Collar)		'Y' (Shaft)	
	Inch	mm	Inch	mm	Inch	mm
1/2	.500	12.700	.560	14.224	.430	10.922
9/16	.562	14.290	.623	15.824	.493	12.522
5/8	.625	15.875	.709	18.008	.517	13.132
11/16	.688	17.470	.773	18.618	.581	14.757
3/4	.750	19.050	.837	21.259	.644	16.357
13/16	.812	20.640	.900	22.860	.708	17.983
7/8	.875	22.225	.964	24.485	.771	19.583
15/16	.938	23.820	1.051	26.695	.791	20.091
1	1.000	25.400	1.114	28.295	.859	21.818
1 1/16	1.062	26.985	1.178	29.921	.923	23.444
1 1/8	1.125	28.575	1.241	31.521	.986	25.044
1 3/16	1.188	30.165	1.304	33.121	1.049	26.644
1 1/4	1.250	31.750	1.367	34.722	1.112	28.244
1 5/16	1.312	33.340	1.455	36.957	1.137	28.879
1 3/8	1.375	34.923	1.518	38.557	1.201	30.505

(Reprinted from DRAFTING FOR TRADES AND INDUSTRY—MECHANICAL AND ELECTRONIC by Nelson, © 1979 Delmar Publishers Inc.)

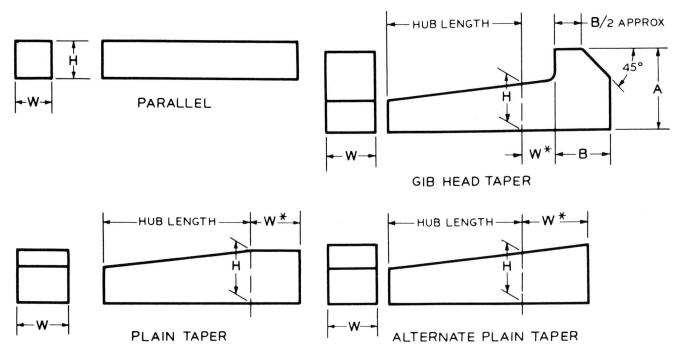

Plain and Gib Head Taper Keys Have a 1/8" Taper in 12"

TABLE 29 KEY DIMENSIONS AND TOLERANCES

KEY			NOMINAL KEY SIZE Width, W		TOLERANCE	
			Over	To (Incl)	Width, W	Height, H
Parallel	Square	Bar Stock	— 3/4 1-1/2 2-1/2	3/4 1-1/2 2-1/2 3-1/2	+0.000 −0.002 +0.000 −0.003 +0.000 −0.004 +0.000 −0.006	+0.000 −0.002 +0.000 −0.003 +0.000 −0.004 +0.000 −0.006
		Keystock	— 1-1/4 3	1-1/4 3 3-1/2	+0.001 −0.000 +0.002 −0.000 +0.003 −0.000	+0.001 −0.000 +0.002 −0.000 +0.003 −0.000
	Rectangular	Bar Stock	— 3/4 1-1/2 3 4 6	3/4 1-1/2 3 4 6 7	+0.000 −0.003 +0.000 −0.004 +0.000 −0.005 +0.000 −0.006 +0.000 −0.008 +0.000 −0.013	+0.000 −0.003 +0.000 −0.004 +0.000 −0.005 +0.000 −0.006 +0.000 −0.008 +0.000 −0.013
		Keystock	— 1-1/4 3	1-1/4 3 7	+0.001 −0.000 +0.002 −0.000 +0.003 −0.000	+0.005 −0.005 +0.005 −0.005 +0.005 −0.005
Taper	Plain or Gib Head Square or Rectangular		— 1-1/4 3	1-1/4 3 7	+0.001 −0.000 +0.002 −0.000 +0.003 −0.000	+0.005 −0.000 +0.005 −0.000 +0.005 −0.000

*For locating position of dimension H. Tolerance does not apply.
See Table 30 for dimensions on gib heads.
All dimensions given in inches.
(Reprinted from The American Society of Mechanical Engineers—ANSI B17.1–1967)

USA STANDARD

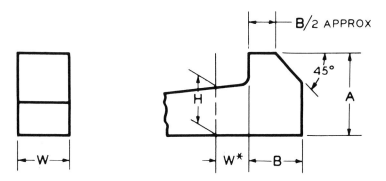

TABLE 30 GIB HEAD NOMINAL DIMENSIONS

Nominal Key Size Width, W	SQUARE			RECTANGULAR		
	H	A	B	H	A	B
1/8	1/8	1/4	1/4	3/32	3/16	1/8
3/16	3/16	5/16	5/16	1/8	1/4	1/4
1/4	1/4	7/16	3/8	3/16	5/16	5/16
5/16	5/16	1/2	7/16	1/4	7/16	3/8
3/8	3/8	5/8	1/2	1/4	7/16	3/8
1/2	1/2	7/8	5/8	3/8	5/8	1/2
5/8	5/8	1	3/4	7/16	3/4	9/16
3/4	3/4	1-1/4	7/8	1/2	7/8	5/8
7/8	7/8	1-3/8	1	5/8	1	3/4
1	1	1-5/8	1-1/8	3/4	1-1/4	7/8
1-1/4	1-1/4	2	1-7/16	7/8	1-3/8	1
1-1/2	1-1/2	2-3/8	1-3/4	1	1-5/8	1-1/8
1-3/4	1-3/4	2-3/4	2	1-1/2	2-3/8	1-3/4
2	2	3-1/2	2-1/4	1-1/2	2-3/8	1-3/4
2-1/2	2-1/2	4	3	1-3/4	2-3/4	2
3	3	5	3-1/2	2	3-1/2	2-1/4
3-1/2	3-1/2	6	4	2-1/2	4	3

*For locating position of dimension H.
For larger sizes the following relationships are suggested as guides for establishing A and B.

$$A = 1.8 H \qquad B = 1.2 H$$

All dimensions given in inches.
(Reprinted from The American Society of Mechanical Engineers—ANSI B17.1–1967)

TABLE 31 SHEET METAL AND WIRE GAGE DESIGNATION

GAGE NO.	AMERICAN OR BROWN & SHARPE'S A.W.G. OR B. & S.	UNITED STATES STANDARD	MANU-FACTURERS' STANDARD FOR SHEET STEEL	GAGE NO.
0000000	....	.500	....	0000000
000000	.5800	.469	....	000000
00000	.5165	.438	....	00000
0000	.4600	.406	....	0000
000	.4096	.375	....	000
00	.3648	.344	....	00
0	.3249	.312	....	0
1	.2893	.281	....	1
2	.2576	.266	....	2
3	.2294	.250	.2391	3
4	.2043	.234	.2242	4
5	.1819	.219	.2092	5
6	.1620	.203	.1943	6
7	.1443	.188	.1793	7
8	.1285	.172	.1644	8
9	.1144	.156	.1495	9
10	.1019	.141	.1345	10
11	.0907	.125	.1196	11
12	.0808	.109	.1046	12
13	.0720	.0938	.0897	13
14	.0642	.0781	.0747	14
15	.0571	.0703	.0673	15
16	.0508	.0625	.0598	16
17	.0453	.0562	.0538	17
18	.0403	.0500	.0478	18
19	.0359	.0438	.0418	19
20	.0320	.0375	.0359	20
21	.0285	.0344	.0329	21
22	.0253	.0312	.0299	22
23	.0226	.0281	.0269	23
24	.0201	.0250	.0239	24
25	.0179	.0219	.0209	25
26	.0159	.0188	.0179	26
27	.0142	.0172	.0164	27
28	.0126	.0156	.0149	28
29	.0113	.0141	.0135	29
30	.0100	.0125	.0120	30
31	.0089	.0109	.0105	31
32	.0080	.0102	.0097	32
33	.0071	.00938	.0090	33
34	.0063	.00859	.0082	34
35	.0056	.00781	.0075	35
36	.0050	.00703	.0067	36

(Reprinted from TECHNICAL DRAWING AND DESIGN by Goetsch and Nelson, © 1986 Delmar Publishers Inc.)

TABLE 32 AMERICAN NATIONAL STANDARDS OF INTEREST TO DESIGNERS, ARCHITECTS, AND DRAFTERS

TITLE OF STANDARD

Abbreviations	Y1.1-1972
American National Standard Drafting Practices	
Size and Format	Y14.1-1980
Line Conventions and Lettering	Y14.2M-1979
Multi and Sectional View Drawings	Y14.3-1975(R1980)
Pictorial Drawing	Y14.4-1957
Dimensioning and Tolerancing	Y14.5M-1982
Screw Threads	Y14.6-1978
Screw Threads (Metric Supplement)	Y14.6aM-1981
Gears and Splines	
Spur, Helical, and Racks	Y14.7.1-1971
Bevel and Hyphoid	Y14.7.2-1978
Forgings	Y14.9-1958
Springs	Y14.13M-1981
Electrical and Electronic Diagram	Y14.15-1966(R1973)
Interconnection Diagrams	Y14.15a-1971
Information Sheet	Y14.15b-1973
Fluid Power Diagrams	Y14.17-1966(R1980)
Digital Representation for Communication of Product Definition Data	Y14.26M-1981
Computer-Aided Preparation of Product Definition Data Dictionary of Terms	Y14.26.3-1975
Digital Representation of Physical Object Shapes	Y14 Report
Guideline — User Instructions	Y14 Report No. 2
Guideline — Design Requirements	Y14 Report No. 3
Ground Vehicle Drawing Practices	In Preparation
Chasis Frames	Y14.32.1-1974
Parts Lists, Data Lists, and Index Lists	Y14.34M-1982
Surface Texture Symbols	Y14.36-1978
Illustrations for Publication and Projection	Y15.1M-1979
Time Series Charts	Y15.2M-1979
Process Charts	Y15.3M-1979
Graphic Symbols for:	
Electrical and Electronics Diagrams	Y32.2-1975
Plumbing	Y32.4-1977
Use on Railroad Maps and Profiles	Y32.7-1972(R1979)
Fluid Power Diagrams	Y32.10-1967(R1974)
Process Flow Diagrams in Petroleum and Chemical Industries	Y32.11-1961
Mechanical and Acoustical Element as Used in Schematic Diagrams	Y32.18-1972(R1978)
Pipe Fittings, Valves, and Piping	Z32.2.3-1949(R1953)
Heating, Ventilating, and Air Conditioning	Z32.2.4-1949(R1953)
Heat Power Apparatus	Z32.2.6-1950(R1956)
Letter Symbols for:	
Glossary of Terms Concerning Letter Symbols	Y10.1-1972
Hydraulics	Y10.2-1958
Quantities Used in Mechanics for Solid Bodies	Y10.3-1968
Heat and Thermodynamics	Y10.4-1982
Quantities Used in Electrical Science and Electrical Engineering	Y10.5-1968
Aeronautical Sciences	Y10.7-1954
Structural Analysis	Y10.8-1962
Meteorology	Y10.10-1953(R1973)
Acoustics	Y10.11-1953(R1959)
Chemical Engineering	Y10.12-1955(R1973)
Rocket Propulsion	Y10.14-1959
Petroleum Reservoir Engineering and Electric Logging	Y10.15-1958(R1973)
Shell Theory	Y10.16-1964(R1973)
Guide for Selecting Greek Letters Used as Symbols for Engineering Mathematics	Y10.17-1961(R1973)
Illuminating Engineering	Y10.18-1967(R1977)
Mathematical Signs and Symbols for Use in Physical Sciences and Technology	Y10.20-1975

(Reprinted from TECHNICAL DRAWING AND DESIGN by Goetsch and Nelson, © 1986 Delmar Publishers Inc.)

ABBREVIATIONS

Lettered notations and words on working drawings are often abbreviated to save time and space. It is critical that the persons reading the working drawings will be able to interpret the abbreviations correctly.

The following are some considerations to understand when using abbreviations:

- Do not use a period after an abbreviation unless the abbreviation is a whole word.
- Use capital letters for (most) abbreviations.
- Singular and plural abbreviations may be the same.
- The same abbreviation may apply to more than one word.
- Many abbreviations are quite close in their spelling.

Using the standardized abbreviations listed here will insure that all abbreviations will be read correctly.

Accessory	ACCESS.
Accumulate	ACCUM
Adaption	ADAPT.
Addition	ADD.
Addendum	ADD.
Alteration	ALT
Alternate	ALT
Alternating current	AC
Altitude	ALT
Aluminum	AL
American Iron and Steel Institute	AISI
American Society for Testing Materials	ASTM
American Society of Mechanical Engineers	ASME
American Wire Gage	AWG
Ampere	AMP
Approved	APPD
Approximate	APPROX
Assembly	ASM
Attach	ATT
Authorize	AUTH
Auxiliary	AUX
Average	AVG
Balance	BAL
Battery	BAT.
Bearing	BRG
Bill of material	B/M
Bracket	BRKT
Brass	BRS
British thermal unit	BTU
Bronze	BRZ
Brown and Sharpe	B&S
Bushing	BUSH.
Cadmium	CAD.
Capacity	CAP.
Carbon steel	CS
Cast iron	CI
Cast steel	CS
Casting	CST
Center	CTR
Centerline	(₵) or CL
Center of gravity	CG
Centigrade	C
Centimeter	CM
Chamfer	CHAM
Change	CHG
Check	CHK
Chief	CH
Chromium	CHR
Circular pitch	CP
Circumference	CIRC
Clockwise	CW
Cold drawn steel	CDS
Cold rolled steel	CRS
Company	CO
Composition	COMP
Compression, Compressor	COMP
Concentric	CONC
Condition	COND
Conductor	COND
Conduit	CDT

Connector	CONN
Continue	CONT
Control	CONT
Copper	CU
Corporation	CORP
Correspond	CORRES
Corrosion resistant steel	CRES
Counterbalance	CBAL
Counterbore	CBORE
Counterclockwise	CCW
Counterdrill	CDRILL
Countersink	CSK
Counterweight	CTWT
Cubic	CU
Cubic centimeter	CC
Cubic feet per minute	CFM
Cubic foot	CU FT
Cubic inch	CU IN.
Cycle	CY
Dedendum	DED
Degree	(°) or DEG
Department	DEPT
Detail	DET
Develop	DEV
Deviation	DEV
Diagonal	DIAG
Diameter	DIA, θ
Diametral pitch	DP
Dimension	DIM.
Direct current	DC
Distance	DIST
Division	DIV
Drawing	DWG
Each	EA
Eccentric	ECC
Effective	EFF
Electric	ELEC
Elevation	ELEV
Engineer	ENGR
Engineering	ENGRG
Equipment	EQUIP.
Equivalent	EQUIV
Estimate	EST
Etcetera	ETC
Extension	EXT
External	EXT
Extrusion	EXTR
Fahrenheit	F
Feet per minute	FPM
Figure	FIG.
Fillister head	FIL HD
Finish	FIN.
Fitting	FTG

Flat	FL
Flat head	FHD
Flexible	FLEX.
Foot	(') or FT
Foot pounds	FT LB
Forging	FORG
Forward	FWD
Front	FRT
Gage (Gauge)	GA
Gallon	GAL
Galvanize	GALV
Gasket	GSKT
Generator	GEN
Grain	GRN
Gravity	G
Grind	GRD
Harden	HDN
Hardware	HDW
Head	HD
Heat treat	HT TR
Height	HGT
Hexagon	HEX
Horizontal	HORIZ
Hot rolled steel	HRS
Hour	HR
Hydraulic	HYD
Illustration	ILLUS
Inch	(") or IN.
Inch ounce	IN. OZ
Inch pounds	IN. LB
Inclusive	INCL
Information	INFO
Inside diameter	ID
Inside radius	IR
Inspection	INSP
Installation	INSTL
Instrument	INST
Insulation	INSL
Interchangeable	INTCHG
Intermediate	INTER.
Internal	INT
Joint	JT
Kilometer	KM
Knock out	KO
Laboratory	LAB
Left hand	LH
Length	LG
Lock washer	LWASH
Longitude, Longitudinal	LONG.

Lower	LWR
Lubricate	LUB
Machine	MACH
Magnesium	MG
Maintenance	MAINT
Malleable	MAL
Manufacture	MFR
Material	MATL
Maximum	MAX
Maximum material condition	MMC
Mechanical	MECH
Medium	MED
Memorandum	MEMO
Mercury	HG
Mile	MI
Miles per hour	MPH
Millimeter	MM
Minimum	MIN
Minute	(') or MIN
Miscellaneous	MISC
Modify	MOD
Molding	MLDG
Mounting	MTG
National	NATL
National Electrical Mfg. Association	NEMA
National Machine Tool Builders Association	NMTBA
Negative	(−) or NEG
Nickel	NI
No drawing	ND
Nominal	NOM
Nonstandard	NONSTD
Number	NO.
Obsolete	OBS
Of true position	OTP
On Center	OC
Operate	OPER
Opposite	OPP
Optional	OPT
Original	ORIG
Ounce	OZ
Outside diameter	OD
Outside radius	OR.
Oval head	OV HD
Overall	OA
Oxygen	OXY
Package, Packing	PKG
Page	P
Part	PT
Parting line	PL
Patent	PAT.
Pattern	PATT
Perpendicular	PERP

Pint	PT
Pitch	P
Pitch circle	PC
Pitch diameter	PD
Plan view	PV
Point on line	POL
Position	POSN
Positive	(+) or POS
Pound	LB
Pounds per square inch	PSI
Preliminary	PRELIM
Pressure	PRESS.
Process	PROC
Product, Production	PROD.
Quality	QUAL
Quantity	QTY
Quart	QT
Quarter	QTR
Rear view	RV
Rectangular	RECT
Reduction	RED.
Reference	REF
Regardless of feature size	RFS
Regular	REG
Reinforce	REINF
Remove	REM
Require	REQ
Required	REQD
Resistor	RES
Reverse	REV
Revision	REV
Revolution	REV
Revolutions per minute	RPM
Right hand	RH
Root diameter	RD
Round	RD
Round head	RD HD
Rubber	RUB.
Screw	SCR
Screw threads	
American National coarse	NC
American National fine	NF
American National extra fine	NEF
American National 8 pitch	8N
American National 12 pitch	12N
American National 16 pitch	16N
American Standard taper pipe	NPT
American Standard straight pipe coupling	NPSC
American Standard taper	NPTF
Unified screw thread coarse	UNC
Unified screw thread fine	UNF
Unified screw thread extra fine	UNEF
Unified screw thread 8 thread	8UN

Unified screw thread 12 thread	12UN
Unified screw thread 16 thread	16UN
Unified screw thread special	UNS
Second	(") or SEC
Section	SECT.
Serial, Series	SER
Sheet	SH
Side view	SV
Sketch	SK
Society of Automotive Engineers	SAE
Special	SPL
Specification	SPEC
Specific gravity	SP GR
Spherical	SPHER
Spot face	SF
Spring	SPR
Square	SQ
Standard	STD
Steel	STL
Support	SUPT
Switch	SW
Symbol	SYM
Symmetrical	SYM
Synthetic	SYN
Tangent	TAN.
Technical	TECH
Teeth	T
Temperature	TEMP
Tensile strength	TS
Terminal	TERM.
Theoretical	THEO
Thickness	THK
Thread	THD
Through	THRU
Tolerance	TOL

Tracer	TCR
Trade mark	TM
Transformer	TRANS
Transmission	TRANS
Transverse	TRANSV
True length	TL
True position	TP
True view	TV
Turnbuckle	TRNBKL
Typical	TYP
Ultimate	ULT
United States	US
United States of America Standards Institute	USASI
United States gage	USG
Universal	UNIV
Unless otherwise specified	UOS
Upper	UPR
Vacuum	VAC
Velocity	V
Versus	VS
Vertical	VERT
Volt	V
Volume	VOL
Water line	WL
Watt	W
Weight	WT
Wood	WD
Wood screw	WD SCR
Wrought iron	WI
Yard	YD
Year	YR

GLOSSARY

A

acute angle Any angle less than 90°.

addendum The radial distance from the pitch circle to the top of a gear tooth.

aligned section Sections which result from the offsetting or revolvement of a part to line up with the cutting plane.

aligned dimensioning The dimensions are always placed in line with the direction of the dimension lines.

allowance Acceptable clearance between mating parts.

alloy The combination of two or more metals.

alphanumeric The combination of alphabetic and numeric characters.

alphanumeric keyboard A computer keyboard that allows an operator to input letters and numerals into a computer.

AWG American Wire Gage. A standard numerical system for measuring wire sizes.

angle of thread The included angle formed between the major diameter of a thread and the depth of thread.

ANSI American National Standards Institute. The organization that coordinates and approves national standards for American industry.

arc welding Welding metals with an electric arc.

ASA American Standards Association.

assembly drawing A drawing that shows the relationship of various parts of an item when assembled.

auxiliary memory A memory device that supplements the primary memory of a computer.

auxiliary menu Lists of options available for each task listed on a master menu for computer-aided drafting.

auxiliary view A view of an object other than the normal six orthographic views. Primary auxiliary views are drawings projected directly from any of the six normal orthographic views. A secondary auxiliary view is a drawing projected from a primary auxiliary view.

axis A line around which parts rotate or are concentrically arranged.

axonometric drawing A pictorial drawing which shows the front, side, and top faces of an object in one view. The receding lines in axonometric drawings are always parallel. Isometric, dimetric, and trimetric drawings are types of axonometric drawings.

B

backing weld A weld used to add extra strength to a weld by adding additional welds to the backside of the material being welded.

baseline dimensioning Dimensioning to a common datum point or plane.

basic dimension An exact value used to describe the size, shape, and location of a feature.

basic hole size The standard size of any basic hole cutter.

basic shaft size The standard manufactured size of a shaft.

beam compass A compass used to draw circles and arcs over 10″ in diameter.

bell and spigot connections The connection between two pipes. The straight spigot end of one pipe is inserted into the flared-out end of the other pipe. The joint is sealed with caulking or a compression ring.

bend allowance The amount of extra material required to construct a bend over a radius.

bevel An inclined surface which is not constructed at a right angle to the adjoining surface.

bilateral tolerance A tolerance which allows variation in both directions from a dimension.

bill of materials A list of materials specified for a project.

block diagram Connected blocks labeled to represent sections of an electronic system.

bolt circle A circular centerline which locates centers of holes positioned about a common center point.

boot A term which means to load a software program into a computer.

bore To enlarge a hole with a hand or machine tool.

bow compass A compass used to draw circles and arcs up to 6″ in diameter.

brazing The joining of two pieces of metal with a brass alloy filler which melts about 800°F. This process is similar to soldering.

break line A line used to interrupt a drawing if an object will not fit on a drawing sheet.

broken-out section A section of an object broken away to reveal an interior feature for a sectional drawing.

buffer A device used for temporary electronic storage of information.

building line An imaginary line on a plot beyond which the building cannot extend.
bushing A hollow cylindrical sleeve bearing.
butt weld A welded joint which joins two pieces butted directly at their ends.
byte A single character of computer memory.

C

cabinet oblique drawing A pictorial drawing containing receding lines drawn at half scale, at an angle of 30° or 45° from the horizontal.
cableform diagram A drawing of groups of wires bound together into a cable or harness.
CAD Computer-aided drafting. The use of computers to aid in the preparation of technical drawings.
CADD Computer-aided drafting and design.
CAE Computer-aided engineering.
callout A drawing note connected to a drawing with a leader and arrowhead.
CAM Computer-aided manufacturing. Computer-augmented manufacturing. Computer-automated manufacturing.
cam A machine device used to transmit motion. It may change the speed, direction or type of motion.
cam follower A plunger whose reciprocating motion is produced by contact with the surface of a cam.
cartesian coordinate system A measuring system which uses rectangular grids to measure width, height, and depth (X, Y, and Z).
cavalier oblique drawing A pictorial drawing containing receding lines drawn to true scale.
center drill A drill used to drill the mounting holes in the end of a workpiece.
centerline A line that defines the common center of symmetrical objects.
chain dimensions Continuous dimensions that connect aligned features of an object without reference to a common datum point or plane.
chain line A line used to indicate a specific area.
chamfer An angle across a corner that eliminates a sharp edge.
character A symbol used to represent data in an organized system.
chordal addendum Distance from the top of a spur gear tooth to the chordal thickness line.
chordal thickness The thickness of a spur gear tooth measured on a chord on the circumference of the pitch diameter.
chord A straight line joining two points on a curve.
chuck A telescoping device that holds a rotating tool or a rotating workpiece.
CIM Computer-integrated manufacturing.
circular pitch The length of an arc along the pitch circle between the center of one gear tooth to the center of the next tooth.
circular thickness The length of an arc between the sides of a spur gear tooth measured on the pitch circle.
clearance fit A clearance space between mating parts due to the limits of the dimensions.
collar A flange or ring attached to a shaft or pipe.
column In architecture: an upright supporting member, circular in section; in engineering: a vertical structural member supporting loads.
command Instructions transmitted to a computer by an operator.
component One part of a multipart object or structure.
concave Surfaces which curve inwardly.
concentric Circles which share a common center.
construction lines Light lines used for the layout of a drawing.
convex Surfaces which curve outwardly.
counterbore To enlarge a hole to allow a screw or bolt head to be recessed below a surface.
coordinates Points of intersection of lines on a grid system.
CNC Computer-numerically controlled.
countersink A recess made to allow the tapered head of a fastener to lie below a surface.
CPU Central processing unit of a computer system.
crest The point of widest diameter of a screw thread which lies on a major diameter plane.
crosshatching Closely spaced parallel lines drawn obliquely to represent sectioned areas.
CRT Cathode ray tube.
CSI Construction Specifications Institute.
cursor A marker used to locate points on a video display screen (monitor) while working with a computer program.
cylindrical Having the form or properties of a cylinder.
cutting plane line A line drawn on a view where a cut was made in order to define the location of the imaginary sectional plane.

D

data base A stored collection of computer information.
datum A line, point, or plane from which other locations are measured.
datum dimensioning A dimensioning system in which dimensions originate from a common datum.
dedendum The radial distance from the pitch circle to the bottom of a spur gear tooth.
default Action initiated through a computer software program unless an operator specifies other actions.
degree (°) A unit of angular measurement.
design size The size of a part after clearances and tolerances are applied.
detail drawing A dimensioned working drawing of a single part.
development drawing The drawing of a surface of an object unfolded on a flat plane. It is sometimes called a stretchout pattern. Development drawings are used in

pattern drafting. Parallel line and radial line developments are used to develop patterns for objects that contain flat or single-curved surfaces. *Parallel line development* drawings are used to develop patterns for objects that contain parallel lines. *Radial line development* is used to develop patterns for objects that do not contain parallel lines. These methods cannot be used to develop patterns for warped surfaces. *Triangulation development* is a method of dividing a warped surface and other complex forms that cannot be developed with parallel or radial line development into a series of triangles, and transferring the true size of each triangle to a flat pattern.

diameter The length of a straight line which passes through the center of a circle and terminates at each end on the circumference.

diametral pitch A ratio of the number of teeth on a gear per inch of pitch diameter.

diazo A printing process that produces black, blue, or brown lines on surfaces. The process exposes a drawing with a sensitized material exposed to an ultraviolet light, then processed through an ammonia chamber.

digital circuit Electronic circuits which function through the use of binary numbers.

digital readout Devices that display numbers from the output of calculators and computers.

digitize To locate points and select computer commands with an input device.

digitizer An electronic tablet that serves as a drawing surface for the input of graphic data.

dimension Radial or linear length (width, height, or depth) labeled on a technical drawing.

dimetric drawing Pictorial drawings which contain two axes that form equal angles with the plane of projection.

direct dimensioning To be used when the distance between two points is critical to avoid tolerance buildup.

disk drive A computer input device that can read information on a rotating storage media.

displacement diagram A drawing which shows the path of a cam follower.

display The presentation of data on a computer output device.

dividers A device used to divide and transfer measurements or for scribing arcs on hard surfaces.

documentation Drawings or printed information that contains instructions for assembling, installing, operating, and servicing.

dowel A cylindrical rod or pin.

DPI Dots per inch.

drawing area The area on the display screen where the drawing will appear.

drum cam A cylinder with a groove in its surface which guides a follower through the groove to produce reciprocating motion.

drum plotter A cylindrical graphics pen plotter that accepts continuous feed paper.

dual dimensioning The use of the metric and the U.S. customary measuring systems on a drawing.

dump To transfer computer information from one system to another.

E

elbow drafter A manual drafting machine connected to the top of a drawing board.

electronics drawings Working drawings that depict electronic circuits and electronic items.

elements A distinct group of lines, shapes, and text, as defined by a computer operator; also referred to as *entities*. In development drawings, imaginary lines used to draw the pattern.

elevation drawing A drawing of the exterior or interior walls of a building. It is a perpendicular, or upright, projection from the floor plan to show vertical architectural or design details.

ellipse A closed curve in the form of a symmetrical oval.

engineering change documents Documents used to implement a specific change in a production drawing. Engineering Change Request (ECR), Engineering Change Notice (ECN), and Engineering Change Order (ECO).

engineering drawing A technical drawing in any field of engineering.

enter To insert information into a computer.

entities See elements.

equilateral A figure having sides of equal length.

ergonomics The study of human space and movement needs.

exploded view A pictorial drawing which shows all parts disassembled, but in relation to each other.

extension lines Lines used to describe the size of an object.

F

face angle The angle between the top of a gear tooth and the gear axis. Also called a bevel angle.

feature An element of a part or object.

feature control frame Contains data for each controlled specification and includes the characteristic symbol, tolerance value, and datum reference letter.

FAO Finish all over.

fastener A device used to hold two or more objects together.

fillet Curved interior intersections between two or more surfaces.

finish Applications of materials to the surface of objects.

fit The relationship between mating parts.

fittings (pipe) The parts of a piping system used for joining pipe lengths together such as bends, elbows, tees, etc.

fixtures In architecture: electrical devices that may be secured to the ceiling or walls; In manufacturing: a tool used to position and clamp an item for a machine operation.

flange A circular rim around a pipe or shaft.

flatbed plotter A pen plotter where the drawing paper is placed on a flat surface and the pen carriage moves over the surface in a manner similar to hand-held pens.

flat pattern A development or layout drawing which shows the true shape of an object before bending.

flexible curve A device used to draw irregular curves.

floppy disk A circular, thin, magnetic storage medium for computers available in 8", 5 1/4", and 3 1/2" diameters.

floor plan A horizontal, sectioned, working drawing showing all walls, windows, doors, and architectural details with the use of drawing symbols.

flow lines Lines representing the direction of a sequence of operations.

flute A groove on drills, reamers, and taps.

fold line Line of intersection between two planes in a multiview projection, also called a reference line. Representing the position of a fold to be made on flat materials such as paper, sheetmetal, and plastic.

footing An extension at the bottom of a building foundation wall which distributes the load into the ground.

force fit The joining of two mating parts which requires pressure to force the pieces permanently together.

foreshorten To show lines or objects shorter than their true lengths. Foreshortened lines are not perpendicular to the line of sight.

form tolerancing An allowable variation of a feature from the precise form as shown on a drawing.

foundation The basic structure on which a building rests.

free fit Liberal tolerances between mating parts which allows free movement without binding.

friction compass A compass used to draw circles and arcs 4" to 10" in diameter.

function keys Keys on an alphanumeric keyboard that are used in a computer program to represent specific commands.

full-divided scale A scale which contains full subdivision lines throughout the entire length of the scale.

full section A sectional drawing based on a cutting plane line which extends completely through an object.

fusion welding A welding process which results in the mixing of molten metals.

G

gage Numerical standard for the thickness of sheet metal and wire.

gas welding A welding process produced from one or more burning gases. A filler metal and pressure may or may not be used.

geometric dimensioning and tolerancing The assignment of tolerances which prescribes allowable variances in shape, angle, straightness, and position of features and parts.

gear A toothed wheel used to transmit power or motion between shafts.

general tolerance note When all tolerances are identical, the tolerance may be listed once in a note.

groove weld A weld which deposits metal in a groove between two members.

graphics tablet An electronic flat surface on which a stylus or mouse is moved to transmit information, digitize points, or pick commands.

grid size (CAD) The size of the grid or series of light dots which appears on the screen to aid the operator. The grid will not print on the final drawing.

H

half-section A sectional drawing based on a cutting plane line that cuts through one quarter of an object. A half-section reveals half of the interior and half of the exterior.

hard copy The output of a computer printed or plotted on paper or film.

hard disk A hard disk is a storage device that may be a peripheral unit or built directly into the CPU. A hard disk holds more data than a microdisk. It stores and retrieves data quickly.

hardware The physical components of a computer system.

harness drawing A drawing showing the layout or pattern for making an electrical harness or cable.

helix An evenly spaced spiral curve around a cylinder.

hems Formed along the edges of sheet metal pieces to seal in the sharp edges.

hexagon A polygon with six equal sides.

hidden lines Dashed lines which represent object lines which fall behind the object's surface.

highway diagram A point-to-point diagram used to simplify electrical interconnecting complex lines, sometimes called a trunkline diagram.

hinged planes The imaginary folded planes used in orthographic and auxiliary projections.

HVAC Heating, ventilation, and air conditioning.

I

IC Integrated circuit. An electronic circuit made on a small chip of silicon, arsenic, or germanium.

IEC International Electrotechnical Commission.

IEEE Institute of Electrical and Electronics Engineers.

IFI Industrial Fasteners Institute.

inclined Any line or plane located at an angle other than 90° from an orthographic plane.

input The insertion of information into a computer.

interchangeable Parts manufactured with consistent dimensional standards which enable use with any parent object.

interference fit When parts have negative clearances and must be forced together. Used when mating parts are designed to be permanently attached.

involute curve A spiral curve which follows a point on a string as it unwinds from a cylinder or other shapes.

ISO International Organization for Standardization

isometric drawing A pictorial drawing which contains receding planes at 30° from the horizon.

J
joystick An input device lever that controls the movement of a cursor.
jig A device which holds a workpiece and guides the cutting tool.

K
key A device embedded partially into two mating parts to prevent movement.
keyboard See alphanumeric keyboard.
keyseat A slot which holds a key.
kilo (k) Metric prefix for one thousand (10^3).

L
lay The direction of the dominant surface pattern made by machining operations.
layer Separate overlay drawings using the same base drawing.
lead The lateral distance a screw thread moves in one revolution.
leaders Used to connect a specific area on a drawing to a dimension or notation.
light pen A computer input device used by touching a video monitor with the tip of a pen to create points or lines onto the screen.
limits Acceptable variances of a dimension, or, on a CAD monitor, the size of the drawing area.
line conventions Standardization of lines used on technical drawings by line weight and style.
line of vision Imaginary lines connecting the item being drawn with the picture plane.
LMC Least material condition. The least amount of material possible in the size of a part. The LMC of an external feature is its lower limit. The LMC of an internal feature is its upper limit.
location dimensions Dimensions which show the exact location of parts of an object.
longitudinal section A full section through the length of a structure.

M
magnetic tape A magnetized strip of plastic film on which information is stored.
main menu An overall listing of tasks available in a software program.
mainframe A central processing unit networked to individual satellite terminals (workstations).
major diameter The outside diameter of the threaded portion of a fastener measured from crest to crest.
master An original drawing or intermediate print from which reproductions are produced.
melt-through welds The depth of the penetration of the weld. The depth of the melt-through is shown on the drawing's weld notation symbol. It is used with all types of groove welds.
menu Listings of specialized tasks in a software program.
metric system The international measurement system based on units in multiples of 10.
micro (μ) Metric prefix for one millionth (10^{-6}).
microcomputer A computer which uses a microprocessor as a basic CPU element. A microcomputer is a stand-alone system. It can process information without a link to a mainframe computer.
microdisks Small, hard, plastic storage medium. They are smaller than floppy disks and can store more data than a floppy disk.
microfilm Film used to store photographically reduced drawings.
microprocessor A miniaturized electronic circuit device used in personal computers.
MIL STD Military Standards.
milli Metric prefix for one thousandth (10^{-3}).
minor diameter The diameter of the thread measured from root to root, perpendicular to the axis.
MMC Maximum material condition. The most amount of material possible in the size of a part. The MMC of an external feature is its upper limit. The MMC of an internal feature is its lower limit.
modem A device used to transmit data over telephone lines.
monitor The display screen. Monitors may be monochrome (one color, usually green or amber) or may have full-color capabilities.
mouse A graphics input device which is moved across a flat surface to control the movement of a cursor on the screen.
multiview drawings Views of an object projected onto two or more orthographic planes.
mylar Polyester plastic drafting film.

N
National Coarse (NC) American Standard coarse screw thread series.
National Fine (NF) American Standard fine screw thread series.
NC Numerical control. Control of a machine process with a computer program. Sometimes called computerized numerical control (CNC).
networking When stand-alone microcomputer systems are connected to a minicomputer or mainframe system in order to use the larger system's data base.
nominal size The designation of the stock size of a standard material before machining the surfaces.
normal surface True size surfaces that are parallel or perpendicular to the plane of projection.
notations Specify materials and any other necessary manufacturing information on a drawing.

O

object line A heavy solid line used on a drawing to represent the outline of an object.

oblique drawing Pictorial drawing in which the front plane is parallel to the plane of projection.

octagon An eight-sided geometric figure with each corner forming a 135-degree angle. Each center angle is 45 degrees.

offset section A sectional drawing created by a cutting plane bent at right angles to features as though they were in the same plane.

ogee An S-shaped curve.

open-divided scale A scale where only one major unit is graduated with a full-divided unit. It is adjacent to zero.

optical disks Can store more data than any other storage device. Optical disks are used for permanent storage because laser storage data cannot be changed.

origin The zero grid point.

orthographic projection A method of representing three-dimensional objects on a plane having only length and breadth. First-angle projections include the front, left-side, and bottom views. Third angle projections include the front, right-side, and top views.

output Processed computer data transmitted to a monitor, hard copy device, storage media, a control device, or another computer.

overall dimensions Dimensions which describe the total depth, height, or width of an object.

oxyacetylene welding The welding process that uses a mixture of oxygen and acetylene for the heat source.

P

parallel Lines which will never meet if extended any distance.

parallel slide A manual drafting system where a horizontal slide is permanently attached to the drawing board. It slides up and down and functions like a T square.

pattern A drawing or a template prepared to the exact size and shape of a workpiece. Also, a model from which sand molds are made.

pattern development drawing A drawing which shows the outline of all connected surfaces of an object when laid flat.

PC Personal computer

PCB Printed circuit board. The electronic printed circuit assembly including the insulated board.

peripherals The supplemental equipment used in conjunction with, but not as part of, the computer system.

perpendicular A line or plane which is at a right angle to another line or plane.

perspective drawing A pictorial drawing which contains receding lines that converge at vanishing points on the horizon. A perspective drawing can use one, two, or three vanishing points.

pentagon A five-sided geometric figure with each corner forming a 118-degree angle. Each center angle is 72 degrees.

photodrafting The use of photographs for a working drawing.

phantom lines Lines used to show the alternate positions of an object or matching part without interfering with the main drawing.

pictorial drawing A drawing that shows the width, height, and depth of an object in one view.

picture plane Represents the vertical plane upon which the object is viewed.

pin bar The alignment bar used to align individual drawings on top of each other.

pin graphics The drafting system that separates segments of working drawings onto separate drawing formats and keeps all the segments aligned together.

pinion gear The smaller gear in a set of mated gears.

piping drawing The working drawings of a pipe system drawn in isometric or with orthographic projection. A pipe may be represented by a single line or a double line describing the thickness of both the pipe and the fitting.

pitch A uniform distance from a point on one part to a corresponding point on an adjacent part.

pitch circle An imaginary circle concentric with a gear axis which passes through the thickest point on a gear tooth.

pitch diameter (spur gear) The diameter of a pitch circle.

pixels The series of dots which make up the picture on the monitor.

plane of projection The imaginary plane between the viewer and an object onto which the object is drawn.

plat plan An architectural drawing showing the relationship of the property to the adjacent community.

plot plan An architectural drawing showing the location of all structures on the property.

point-to-point wiring diagram An electronic drawing which shows components, terminals, and connecting wires.

polar coordinate dimensioning The use of both linear and angular dimensions for locating features.

polygon A multisided closed form.

positional tolerancing The allowable variation of a feature from the precise position shown on a drawing.

primary menu The first menu which appears on a monitor when a drafter logs in on a CAD program.

printed circuit Copper electronic circuits bonded to the surface of an insulated board.

printer A device which converts computer data into printed alphanumeric or graphic images. Printers are usually dot matrix or laser operated.

prism A solid with base intersections that are parallel polygons and with sides that are parallelograms.

program Sets of instructions which instruct a computer to react to input in a specific manner.

prototype An original functional model of a product constructed prior to mass production.

prompt Instructions on the monitor.

prompt area Appears along the bottom of the display screen. It displays the last three commands used by the CAD operator.

puck A graphics input device which is moved across a graphics tablet to control the movement of the cursor on the monitor.

Q

quadrant One of the four parts that the carstesian coordinate system is divided (X and Y axes).

R

rack A flat bar containing gear teeth which engage with the teeth in a pinion gear.

radius The length of a straight line connecting the center of a circle with a point on the circumference of a circle.

RAM Random-access memory. A temporary computer memory system that includes the data input from the operator, and the software program being used.

rectangular coordinate dimensioning The location of features using linear dimensions from specified X–Y or X–Y–Z axes.

RFS Regardless of feature size. A condition which requires a tolerance of position regardless of where the feature lies within an object.

relief A groove between surfaces to provide clearance for machining.

rendering Adding embellishments to a drawing to provide a more realistic appearance.

removed section A sectional view removed from the area of the cutting plane and positioned in another location.

resistance welding Welding by using the resistance of metals to the passage of an electric current to produce fusion heat.

resolution The sharpness of the picture on the monitor.

revolution drawing A drawing of the repositioned view after the primary orthographic drawing has been rotated. The first view projected from a rotated primary orthographic view is the *primary revolution*. Rotating a primary revolution and projecting a new view is a *successive revolution*.

revolved section A sectional view that is revolved 90° and perpendicular with the plane of projection.

rise The vertical height of a roof, of a step, or a flight of stairs.

ROM Read-only memory. A permanent computer memory system that allows the computer to read the software programs.

root diameter The diameter of a circle aligned with the gear tooth roots.

root The deepest point or valley of a thread.

root opening The distance between members at the root of a weld joint.

round A curved exterior corner of two surfaces.

run The width of a step or the horizontal distance covered by a flight of stairs. This term is also used to describe the horizontal length of a rafter.

runout The area of intersection of two different shaped surfaces.

S

scale drawing A drawing prepared proportionally smaller or larger than the object it represents.

schematic diagram A drawing of an electronic circuit showing symbols for electronic components and lines representing connecting conductors.

screen menu The area on the monitor which displays the commands.

screw thread A helical thread formed on the outside of a cylindrical shaft. Screws with similar combinations of pitches and diameters into several related series: coarse thread (UNC, NC), fine thread (UNF, NF), and extra fine thread (UNEF, NEF).

seam The line of intersection between two materials.

section lines Section lines are used to represent the material through which a cut is made in order to show an interior sectional view.

sectional view A drawing which shows the interior of an object as it would appear if cut in half or quartered.

setback The allowable building distance from the property line of a lot for a building. This distance determines the area of the lot on which a structure can be built. Local zoning regulations determine these distances.

shaft A cylinder to which rotating machine parts may be attached to transmit power and motion.

SI International System of Units. The metric system of measurement as adopted by the General Conference of Weights and Measures.

single-curve surfaces All lines radiate from a common apex such as a cylinder and a cone.

site plan An architectural drawing depicting an area or plot of land with defined property lines.

slab foundation A concrete floor and foundation system poured directly on the ground.

slope diagram A drawing that indicates a comparison between the horizontal run and the rise of a roof.

snap interval The distance the cursor will jump (snap) on the display screen.

software Computer programs which contain specific instructions for the functioning of a computer system.

solder A material made of any various fusible alloy which is melted between pieces of metal in order to join them.

solid modeling The creation of a three-dimensional drawing of the exterior of an object on a CAD system with no hidden features revealed.

solid-state welding The process of heating two metal pieces to their melting point, then holding them together with pressure until the hot atoms for both pieces mix together and cool.

spacing The distances between structural members.

span The distance between structural building supports.

specification A detailed description of components, parts, or materials which includes size, color, manufacturers' number, and sometimes cost.

specify The term used when giving a measurement via the keyboard or showing a distance with a puck with a CAD system.

spline One of a number of keyways cut around a shaft. Flexible strips used for drawing curves.

spot weld A resistance weld operation that joins metal by welding individual spots rather than a continuous bead.

spread sheet The software program that keeps data aligned into columns for record keeping.

spur gear A type of gear with its teeth straight and parallel to an axis shaft.

station point Represents the location of the observer for a perspective drawing.

status line The area at the top of the screen which displays the drawing parameters.

stretchout A pattern development drawing showing the exact shape of a flat material before forming into a three-dimensional shape.

stylus pen An electronic pointer used on a graphics tablet.

subassembly An assembled part that is a part of a larger assembly.

surface texture The waviness, roughness, lay, and flaws of a surface.

surface weld The surface build-up of a metal surface using a metal filler and heat.

survey plan An architectural drawing showing the geological aspects of the property.

symbol A graphic form used to represent a standard object, part, material, or component on a drawing.

T

tabular outline dimensioning Dimensions from perpendicular datum planes listed in a table.

tangent A line that intersects a circle or an arc at a point 90° from the radius.

taper A wedge or conical shape that increases or decreases in size at a uniform rate.

taper reamer A reamer which produces tapered holes.

technical illustration A pictorial drawing used to interpret technical working drawings.

template A flat form used as a guide to draw symbols and holes.

thin wall section Used when materials are too thin to add section lining. Thin wall sections remain solid.

thread depth The distance from the crest (major diameter) to the root (minor diameter) measured perpendicular to the axis.

three-dimensional geometric forms Forms possessing height, width, and depth.

tolerance The amount of variation allowed in the dimension of an object. Plus and minus tolerancing dimensions indicate the tolerance range above and below the basic dimension.

torque Rotating or twisting force.

TPI Threads per inch.

track drafter A manual drafting machine that includes a vertical track that slides left and right on a horizontal track. A protractor head is attached to the vertical track.

train Meshed gears in a series.

trammel A drawing instrument with two adjustable points on a straight edge for drawing curves and ellipses.

transitional fit mating parts that may have either a clearance or interference fit.

translucent Material that allows light but not clear images to penetrate.

transverse section A full section through the width of a structure.

trimetric drawing A pictorial drawing in which the three principal axes are unequally foreshortened.

true length (true size) The condition which results when the line of sight is perpendicular to surfaces or lines.

true position The exact position of a feature.

truncated An object with the apex, vertex, or end removed by an angular plane.

two-dimensional geometric forms Forms possessing height and width.

two-dimensional software 2-D CAD programs which can only function on the X (horizontal) and Y (vertical) axes, and not on the Z axis.

typical A term used to describe details or conditions that are identical in many parts of a drawing.

U

undercut To allow an overhanging edge or a cut with inwardly sloping sides.

unidirectional dimensioning All isometric letters, numerals, and arrowheads are positioned vertically. A variation of this is *vertical plane dimensioning,* where all dimensions are drawn in the isometric plane.

uniform The same form or characteristics in various parts.

unilateral tolerance Tolerance permitted in only one direction from a dimension.

U.S. customary system Units based upon the inch and pound commonly used in the United States.

USMA United States Metric Association.

V

value designation The value of an electronic component expressed as volts, amperes, watts, or ohms.

vanishing point The point at which the receding lines of a perspective drawing meet.

vector A line which has value and direction and defined by its two extremes.

vellum Translucent drafting media.

vernier scale   A movable scale attached to a fixed scale which contains subdivisions of the fixed scale.
vertex   A point in the intersection of two or more sides.

W

warped surfaces   Surfaces that are curved in two directions.
whole depth   Full height of a gear tooth which is equal to the sum of the addendum and the dedendum.
wire frame   Three-dimensional computer models in which all visible and hidden edges appear as lines.
Woodruff key   Crescent-shaped flat key.
working depth   Distance a gear tooth extends into a mating space; two times the addendum.
working drawings   Drawings that provide all information needed to manufacture or construct a product.
workpiece   A piece of material which has machining operations performed on it.
workstation   A location where workers perform specific job tasks.
worm   The driver of a worm gear that contains at least one complete tooth (endless screw) around the pitch surface.
worm gears   Gears with teeth that curve to mesh with and be driven by a worm.

X

X axis   The horizontal axis in a rectangular coordinate system.
X coordinate   One coordinate of a point in a cartesian coordinate system equal to the directed distance of a point on the Y axis.

Y

Y axis   The vertical axis in a rectangular coordinate system.
Y coordinate   One coordinate of a point in a cartesian coordinate system, equal to the directed distance of a point on the X axis.

Z

Z axis   The axis in a three-dimensional cartesian coordinate system which is perpendicular to the X and Y axis.
Z coordinate   One coordinate of a point in a three-dimensional cartesian coordinate system equal to the directed distance of a point on the X and Y axis plane.
zoning   The legal restrictions on size, location, and type of structures to be built in a designated area.

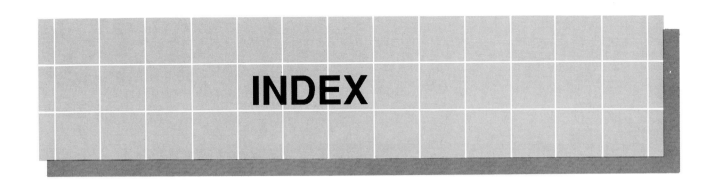

INDEX

A
American National Standards Institute (ANSI), 73
   dimensioning symbols, *138*
   drawing symbols proportions, 139
   geometric
      characteristic symbols, *164*
      tolerancing, symbol modification, 165
   metric dimension conventions, *137*
   U.S. customary dimensions, *137*
Angles
   bisecting, *90*
   dimensioning, 145
   duplicating, *90*
   lines drawn at a given, computer-aided design and, 93-94
   of thread, 285
   tolerancing, 161
Angular gears, defined, 385
Angularity, *167*
Arc
   circle instruments and, 32-34
   welding, 365
Arc tangent
   drawing a
      to two circles, *92*
      to two lines, *92*
Architectural drafting, 465-500
   computer-aided design and, 504-8
   design considerations, 473-79
      area design, 475
      doors, 478
      HVAC, 479
      lighting, 477
      room size, 475, *476*
      storage, 477
      traffic flow, 475-76
      windows, 478, *479*
   design sequence, 479
   ergonomics and, 502-3
   HVAC, *480*
   plat plan and, 466
   plot plan and, 467
   plumbing lines, *480*
   residential styles, 468-73
      American styles, later, 471-73
      contemporary styles, 473, *474*
      early American style, 468, *470*, 471
   site plan, 466
   symbols
      bathroom fixtures, *480*
      built-in components, *480*
      electrical, *479*
      kitchen and utility rooms, *480*
      wastewater facilities, *480*
   working drawings, 481-94
      7-ELEVEN convenience store, *498-501*
      detailed house sections, *489-93*
      elevations and, 482, *485-87*
      interior elevations, 482, *488-89*, 489
      pictorial, 481
      plot plans, 482
      presentation floor plans, 481
      sectional drawings, 489, 494
      site plans, 482
      support documents, 494-97
      survey plans, 482
Architectural drawings, building type and, 498
Arcs, dimensioning, 145
Assembly drawings
   working drawings and, 324, 325-30
   see also specific type of assembly drawing
Assembly sections, sectional views, 184
Auxiliary views, 192-212
   auxiliary planes and, 194-95
   completeness of, 197
   computer-aided design and, 203-7
   dimensioning and, 198
   ellipses and, 198
   foreshortening, 192-94
   hidden lines and, 198
   holes and, 198
   planes, 194-95
   primary, 196
   procedures in drawing, 199-202
      hexagonal prism, *200*
      rectangular pyramid, *200*
      truncated cylinder, *201*
      wedge stop, *199*
   projection of, 195-96
   revolutions and, 201-2
   secondary, 196
   sections, sectional views and, 185
Axis of fastener, 285
Axonometric drawings, 249, *250*

B
Backing welds, 374
Bathrooms, fixtures, architectural symbols of, *480*

556 Index

Beam compass, 32
Bell connections, pipe, 418
Belt drives, 387, *388*
Bevel gears, 386
    cutting data and formulas, *393*
Bilateral tolerancing, 159, *160*
Block diagrams, 433
Bolt, hex
    drawing isometric, *296*
    table of, *292*
Bow compass, 32
Brazing
    soldering and, 365
    symbols, 375
Broken-out, sectional views, 182

C

Cabinet drawings, 262
CAD. *See* Computer-aided design
Cams, 398-403
    computer-aided design and, 408-10
    design and drawing procedures, 402-4
    displacement diagrams, 398-99
    followers, 398
    motions of, 399-401
Cavalier drawings, 261-62
Chain
    sprocket drive and, 387, *388*
    tolerancing, 160
Chamfers, dimensioning, 145, *146*
Chassis drawings, 436, *437*
Chords, dimensioning, 145
Circles
    drawing, arc tangent to two, *92*
    drawing through three points, *93*
    isometric drawings and, 252-54
    line drawn from a point tangent, computer-aided design and, 94, *95*
    line drawn tangent to two, computer-aided design and, 94, *95*
    oblique, 262-63
Circular runout, *168*
Circularity, *166*
Class of thread fits, 285
Clearance fits, tolerancing and, 162
Combination welds, 375
Computer-aided design, 7-20, *8*
    architectural drafting and, 504-8
    auxiliary views and, 203-7
    booting the system, 63-64
    cams and, 408-10
    classifications of, 8-9
    commands, *68-70*
    data bases, 312-18
        calculating and, 312-13, *314*
        CNC machining, 316, *317-18*
        drawing status, 315
        parts list and, 316, *317*
        recording elements, 312
        time records and, 315-16
    data storage and, 15-16
    descriptive geometry and, 218-19
    development drawings, 235-39
        parallel line development and, 235, *236*
        radial line development, 237, *238-39*
    development of, 7-8
    dimensioning and, 152-54
    drafting conventions and, 81-83
    drawing
        beginning the, 64
        cartesian coordinate system, 65-67
        creating simple, 68
        editing an existing, 69
        entering entities, 67
        instruments versus, 86
        parameter setting and, 64-65
        saving and ending, 68-69
    equipment, (CAD), 43-45
    fasteners and, 298-301
    formats, 82, *83*
    gears, 404-7
    layering, 307
    multiview drawings and, 117-21
    networking of (CAD), 9-10
    pictorial drawings and, 273-77
    piping drawings, 422-27
    procedures, 63-72
    sectional views and, 186-87
    software for (CAD), 16, 19
    tolerance and, 169-73
    tool design drafting and, 458-62
    welding drawings and, 377-78
    working drawings and, 337-46
Computer-aided lettering, 57-58
Computer-aided sketching, 56-57
Computer-numerically-controlled machining (CNC), 316, *317-18*
Concentricity, *168*
Cones
    dimensioning, 147
    radial line development, 229-31
Connection diagrams, 434-36
    drawing, 440
Contour symbol, welding drawings and, 368
Curves, ogee, drawing, *93*
Cutting plane, sectional views and, 178-79
Cylinders
    dimensioning, 147
    parallel line development, 227
Cylindricity, *166*

D

Data bases
    computer-aided design, 312-18
        calculating and, 312-13, *314*
        CNC machining, 316, *317-18*
        drawing status, 315
        parts list and, 316, *317*
        recording elements, 312
        time records and, 315-16
Data storage, computer-aided design and, 15-16
Datum, 165
    reference symbols, *165*
    tolerancing, 160-61
Descriptive geometry, 213-22
    computer-aided design and, 218-19
    geometric lines and, 215-17
    geometric planes and, 214-15

points in space, 213-14
true surface size and, 217
Design assembly drawings, 325
Detail drawings, 323-24
Development drawings, 223-46
    computer-aided design and, 235-39
    parallel line development, 227-28
    pattern drawing, terminology, 225-26
    radial line development, 228-34
    surface forms, 224
    triangulation development, 231-34
Diagram assembly drawings, 330
Dimensioning, 54
    angles, 145
    arcs, 145
    auxiliary views and, 198
    chamfers, 145, *146*
    chords, 145
    computer-aided design and, 152-54
    cones, 147
    conventions, surface finishes and, 135-57
    crowded areas, 143
    cylinders, 147
    elements, 137, 139
    fillets, 145
    guidelines, 140
    machined holes, 143, *144*
    multiview drawings, 111-12
    pictorial, 147
    polar coordinate, 140-41
    prisms, 147
    pyramids, 147
    radii, 145
    rectangular coordinate, 141-42
    rounds, 145
    slots, 145
    spheres, 147
    symbols, *138*
    systems of, 137
    tabulated outline, 142
Dimetric drawings, 261
Dividers, 32, *33*
Doors, architectural drafting and, 478
Dot matrix printers, computer-aided design and, 14
Drafting, 2
    careers in, 3, 5
    electronics. *See* Electronics drafting
    fields in, 5-6
    functional, 307-10
    photo, 308
    scissor, 307, *308*
    tools of, evolution of, 7
    See also Computer-aided design
Drafting conventions, 73-85
    computer-aided design and, 81-83
    drawing formats, 77-80
    line conventions, 74-75
        types, 74-75, 81-82
        widths, 74, 81
    standardization and, 73
    standards, 73
Drafting equipment, 22-47
    arc and circle instruments, 32-34
    conventional, 22-24, 27

    paper, *23*
    economic considerations, 38
    erasers, 27
    furniture, 38
    lead holders, 24
    lettering aids, 34-35
    lighting, 38
    machines, 28-29
    pencils, 24
    protractors, 34
    scales, drafting scales, 35-37
    T squares, 30, *31*
    technical pens, 24-25
    templates, 34-35
    transfer letters and tape, 26-27
    triangles, 30, *31-32*
Drafting machines, 28-29
Drafting procedures
    functional drafting and, 307-10
    revisions, 310
Drawing
    computer-aided design and
        beginning the, 64
        booting the system, 63-64
        cartesian coordinate system, 65-67
        commands, *68-70*
        creating simple drawings, 68
        editing an existing drawing, 69
        entering entities, 67
        parameter setting and, 64-65
        saving and ending, 68-69
    history of, 1-2
    instruments, computer-aided design versus, 86
    language of, 2
    procedures, multiview drawings, 112-16
    sectional views. *See* Sectional views
    See also Drafting
Drawings
    detail, computer-aided design and, 324
    development, 223-46
        computer-aided design and, 235-39
        parallel line development, 227-28
        pattern drawing technology, 225-26
        radial line development, 228-34
        surface forms, 224
        triangulation development, 231-34
    dimensioning, computer-aided design and, 330-36
    engineering, scale selection for, *37*
    formats for, 77-80
    isometric, 250-61
        assembly, 257-58
        axis, 251
        circles, 252-54
        fillets, 254
        lines, 251
        rounds, 254
        steps in drawing, 252
    multiview, 99-134
        computer-aided design and, 117-21
        dimensioning, 111-12
        drawing procedures, 112-16
        orthographic projections, 99
        selection of, 99, 101-3
        visualization, 107-8

perspective, 264-73
    construction procedures, 270-71
    ellipses, 271
    estimating procedure, 268-70
    one-point, 265-67
    perspective grids, 271
    station point position, 268
    three-point, 271-72
    two-point, 267, 270
    vanishing point location, 268, 269
    vertical position alternatives, 267-68
pictorial, 247-83
    assembly, 324, 329
    axonometric, 249-50
    computer-aided design and, 273-77
    dimetric, 261
    isometric, 249, 250
    oblique, 261-64
    perspective, 274-73
    trimetric, 261
piping, 418-30
    computer-aided design and, 422-27
    dimensioning, 422
    symbols on, 420-22
    templates, 421
revisions, computer-aided design and, 310
sectional, 489, 494
    assembly, 325-30
    computer-aided design and, 337-46
sketches and, computer-aided design and, 48-49
storage systems and, 311
working, 323-52
    detailed house sections, 489-93
    elevations, 482, 485-87
    floor plan, 483-84
    interior elevations, 482, 488, 489
    pictorial drawings and, 481
    plot plans and, 482
    presentation floor plans, 481
    site plans, 482
    support documents for, 494-97
    survey plans and, 482
Drives
    belt, 387, *388*
    chain and sprocket, 387, *388*
Drum plotters, computer-aided design and, 13
Dual dimensioning, 137
Dusting brush, *25*

E
Elbow drafting machine, 28-29
Electrical symbols, architectural drafting of, *479*
Electronics drafting, 431-47
    computer-aided design and, 441-43
    drawing
        connection diagrams, 440
        logic diagrams, 440-41
    schematic diagrams and, 439-40
    symbols used in, 432-33
        chart, 438, *439*
    templates used in, *440*
    types of, 433-36
        block diagrams, 433
        chassis drawings, 436, *437*

connection diagrams, 434-36
logic diagrams, 436
printed circuits, 436, *438*
schematic diagrams, 433-34
Electronics industry standards, 431-32
Elevations
    architectural drawings, 482, *485-587*
    interior, 482, *488,* 489
Ellipses
    auxiliary views and, 198
    drawing, nonisometric planes and, 254-57
    perspective drawings, 271
Engineering drawings, scale selection for, *37*
Enlarged sections, sectional views, 185
Equilateral triangle, constructing an, *90*
Erasers, 27
Erection assembly drawings, 327
Ergonomics, architectural drafting and, 502-3
External thread, 285

F
Fasteners, 284-306
    axis of, 285
    computer-aided design and, 298-301
    drafting procedures, 288-90
    drive configurations of, *291*
    heads of, *291*
    major diameter, 286
    minor diameter, 286
    points, *291*
    templates for, 293-97
    threaded, 284-87
        nomenclature of, 285-87
        permanent, 297
        types of, 290, *291*
    See also Thread
Feature
    control frame, 165
    position, *168*
Field weld, welding drawings and, 369
Fillet(s)
    construction of, computer-aided design and, 95
    isometric drawings and, 254
    welds, 370
Finished method symbol, welding drawings and, 368
Fittings, pipe, 416
Flange
    connections, pipe, 417
    welds, 373
Flatbed plotters, computer-aided design and, 13
Flatness, *166*
Floor plan, *483-84*
Floppy disks, 43-44
    data storage and, 15
Foreshortening, auxiliary views, 192-94
Formats, standardized, 308, *309,* 310
French curves, 32, *33*
Friction wheels, 387, *388*
Full sections, sectional views, 182
Functional drafting, 307-10
Fusion welding, 365

G
Gears, 385-97

angular, 385
bevel, 386
    cutting data and formulas, *393*
computer-aided design and, 404-7
helical, 386, *387*
herringbone, 386, *387*
machining formulas, 391-96
nomenclature of, 389-91
parallel shaft, 385
perpendicular shaft, 385
pictorial representations of, 396-97
rack and pinion, 385-86
ring, 387, *388*
spiral bevel, 387
spur, 385, *386*
    cutting data and formulas, *391*
teeth of, 388-39
    diameter pitch, *389*
terminology, 390-91
worm, 386, *387*
    cutting data and formulas, *396*
    terminology example, *396*
General
    assembly drawings, 325
    tolerancing, 160
Geometric
    characteristic symbols, 165, 166-68
        ANSI, *165*
    tolerancing, 164-69
Geometric construction, 86-98
    computer-aided design and, 93-96
    conventional, 88, *89-93*
    forms of, 86, 88
        three-dimensional, *89*
        two-dimensional, *88*
    terminology, *87*
Geometry
    descriptive, 213-22
        geometric lines and, 215-17
        geometric planes and, 214-15
        points in space, 213-14
        true surface size and, 217
Graphics
    pin, 307, *308*
    table, computer-aided design and, 11
Grids, perspective, 271
Groove
    angles, welding drawings and, 368
    welds, 373-74

H
Half sections, sectional views, 182
Hard disk drives, data storage and, 16
Heating, ventilation and air conditioning (HVAC)
    architectural drafting and, 479
    architectural symbols of, *480*
Helical gears, 386, *387*
Helix, defined, 284
Hems, pattern drawings and, 225, *226*
Herringbone gears, 386-87
Hex bolt
    isometric, drawing, *296*
    table of, *292*
Hexagon, constructing a, *92*

Hidden lines, auxiliary views and, 198
Holes, auxiliary views and, 198

I
Inclined plane, defined, 284
Ink, 40
Inking, 39
    procedures, 40-42
    surfaces, 40
Installation assembly drawings, 330, *331*
Interference fits, tolerancing and, 162
Internal thread, 285
Isometric
    dimensioning, 258, 260
    drawings, 249, *250*
        assembly, 257-58
        axis, 251
        circles, 252-54
        fillets, 254
        lines, 251
        rounds, 254
        steps in drawing, *252*
    grids, 260-61
Isthmus, poultry, 156

J
Jamb nut, drawing, *295*
Joints, welded, 376
Joystick, computer-aided design and, 11, *12*

K
Kitchens, architectural symbols of, *480*

L
Laser printer, computer-aided design and, 14
Layout assembly drawings, 325
Lead
    defined, 284, 286
    holders, 24
    per revolution, 284-85
Least material condition, 163
Left-hand thread, 286
Lettering, 54-55
    computer-aided, 57-58
Light pens, computer-aided design and, 11, *12*
Lighting, architectural drafting and, 477
Line
    conventions, 74-75
        types, 74-75, 81-82
        widths, 74, 81
    profile, *167*
Lines
    descriptive geometry and, 214-15
        edge view, 216
        point view of a line, 216
        true line length, 216
    dividing into equal parts, computer-aided design and, 93, *94*
    drawing, arc tangent to two, *92*
    drawn at a given angle, computer-aided design and, 93-94
    isometric drawings, 251
    nonisometric, 251-52
    parallel, *89*

development drawings and, 227-28
    perpendicular, computer-aided design and, 93, *94*
    radial, development drawings and, 228-34
    tangent, *89*
    true length of, procedures to find, *226*
Logic diagrams, drawing, 440-41

M

Machine screw, drawing, *295*
Magnetic disks, data storage and, 15
Material condition, tolerance and, 163
Mating parts, 161-63
Maximum material condition, 163
Melt-through welds, 374
Metric dimensioning, 137
Microdisks, data storage and, 15
Microfilm, 311
Monitors, computer-aided design and, 13-14
Mouse, computer-aided design and, 10-11
Multiple threads fastener, 286
Multiview drawings, 99-134
    computer-aided design and, 117-21
    dimensioning, 111-12
    drawing procedures, 112-16
    orthographic projections, 99
    projection
        angles of, 106-7
        inclined and oblique, *107*, 108-11
        planes, 104-6
    selection of, 99, 101-3
    visualization, 107-8

N

Networking
    computer-aided design and, 9-10
    system components of, 10
        central processing unit, 10
        input devices, 10-12
        output devices, 12-14
Nuts, Jamb, drawing, *295*

O

Oblique
    drawings, 261-64
    sections, 264
Octagon, constructing a, *92*
Offset, sectional views, 182, *183*
Ogee curve, drawing an, *93*
One point perspective sketches, 50-51
Operation assembly drawings, 330
Optical disks, data storage and, 16
Orthographic
    multiview sketches, 49
    projections, multiview drawings and, 99
Outline assembly drawings, 329-30

P

Parallel
    line development, computer-aided design and, 235, *236*
    shaft gears, defined, 385
Parallelism, *167*
Parts, mating of, 161-63
Pattern drawing, terminology, 225-26

Pencils, 24, *25*
Pens
    cleaning, 39, *40*
    technical, 24-25
Pentagon, constructing a, *91*
Perpendicular
    bisector, *89*
    shaft gears, defined, 385
Perpendicularity, *167*
Perspective drawings, 264-73
    construction procedure, 270-71
    ellipses, 271
    estimating procedure, 268-70
    one-point, 265-67
    perspective grids, 271
    station point position, 268
    three-point, 271-72
    two-point, 267, *270*
    vanishing point location, 268, *269*
        vertical position alternatives, 267-68
Photodrafting, 308
Pictorial
    assembly drawings, 324, 329
    dimensioning, 147
    drawings, 247-83
        axonometric drawings, 249, *250*
        computer-aided design and, 273-77
        dimetric drawings, 261
        isometric drawings, 249, *250*
        oblique drawings, 261-64
        perspective drawings, 264-73
        trimetric drawings, 261
    sketches, 50-51
Pin graphics, 307, *308*
Pipe
    connections, 416-18
        cementing, 418
    standards, 418
    threads, 293, *294*
Piping drawings, 418-30
    computer-aided design, 422-27
    dimensioning, 422
    symbols on, 420-22
    templates, *421*
Pitch
    diameter of threads, 287
    of threads, 286
Planes, descriptive geometry and, 214-15
Plat plan, 466
Plot plans, 467, 482
Plotters
    drum, computer-aided design and, 13
    flatbed, computer-aided design and, 13
    pens, 44-45
        media guide, *45*
        recommendations, *45*
Plotting media, 45
Plug welds, 371
Plumbing, lines, architectural symbols of, *480*
Points, in space, 213-14
Polar coordinate dimensioning, 140-41
Polygon, construction of, computer-aided design and, 94, *95*
Printed circuits, 436, *438*

Printers, computer-aided design and, 13-14
Prisms
    dimensioning, 147
    parallel line development, 227
Projections
    angles, multiview drawings and, 106-7
    auxiliary views and, 195-96
    inclined, multiview drawings and, *107,* 108-11
    oblique, multiview drawings and, *107,* 108-11
    planes, multiview drawings and, 104-6
    welds, 372
Protractors, 34
Puck, computer-aided design and, 11, *12*
Pyramids
    dimensioning, 147
    radial line development, 229
    steps for, *228*

R
Rack and pinon gears, defined, 385-86
Radial line development, computer-aided design and, 237, *238-39*
Radii, dimensioning, 145
Rectangular coordinate dimensioning, 141-42
Reference line and arrow, welding drawings and, 366
Reformatting, 307-8
Removed sections, sectional views, 184
Resistance welding, 365
Revolved sections, sectional views, 184
Right-hand thread, 286
Right triangle, constructing a, *91*
Rivets, 297
Room size, architectural drafting and, 475, *476*
Root opening, welding drawings and, 368
Roundness, *166*
Rounds
    dimensioning, 145
    isometric drawings and, 254

S
Schedules, architectural drafting and, 494
Schematic diagrams, 433-34
Scissor drafting, 307, *308*
Screw thread series, 285
Screws, machine, drawing, *295*
Seams, pattern drawings and, 225, *226*
Section lining
    exceptions, sectional views and, 180-81
    sectional views and, 179-80
Sectional
    drawings, architectural drawings, 489, 494
    views, 178-91
        assembly sections, 184
        auxiliary sections, 185
        broken-out, 182
        computer-aided design and, 186-87
        cutting plane, 178-79
        enlarged sections, 185
        full sections, 182
        half sections, 182
        offset, 182, *183*
        removed sections, 184
        revolved sections, 184
        section lining, 179-80
        sectional lining, exceptions, 180-81
        thin wall sections, 184-85
Single-thread fastener, 286
Site plans, 466, 482
Sketches
    orthographic multiview, 49
    perspective, 50-51
    pictorial, 50-51
    working drawing, 48-49
Sketching
    computer-aided, 56-57
    dimensioning, 54
    guidelines and procedures, 51-52
    lettering, 54-55
    shading and, 52-53
Slot welds, 371
Slots, dimensioning, 145
Software, computer-aided design and, 16, 19
Soldering and brazing, 365
Solid-state welding, 364-65
Specifications, architectural drafting and, 495, *496-97*
Spheres, dimensioning, 147
Spigot connections, pipe, 418
Spiral bevel gears, 387
Spot welds, 371-72
Spur gears, 385, *386*
    cutting data and formulas, *391*
Square, constructing a, *91*
Standardization, drafting, 73
Standardized formats, 308, *309,* 310
Standards, drawing, 73
Station point position, perspective drawings and, 268
Storage systems, 307-22, 311
    computer-aided design and, 312
Straightness, *166*
Stylus pens, computer-aided design and, 11, *12*
Subassembly drawings, 327, 329
Surface
    control, 147-48
    curved, parallel line development, 227-28
    finish, symbols, *148*
    forms, development drawings, 224
    lay, 148, *150*
    profile, *167*
    projections, inclined and oblique, multiview drawings and, 108-11
    roughness
        height, *149, 151*
        tolerance, 150
        width cutoff, 148
    texture, characteristics of, *149*
    true size of, 217
    welds, 375
Survey plans, 482

T
T squares, 30, *31*
Tabs, pattern drawings and, 225
Tabulated outline dimensioning, 142
Tail symbol, welding drawings and, 366-67
Technical pens, 24-25
Templates, 34-35
    electronics, *440*
    fasteners, 293-97

piping, *421*
Thin wall sections, sectional views, 184-85
Thread
    angle of, 285
    depth, 285
    external, 285
    fits, class of, 285
    forms, *286*
        defined, 286
    internal, 285
    left-hand, 286
    metric series, 290-93
    multiple, 286
    pipe, 293, *294*
    pitch, 286
        diameter of, 287
    right-hand, 286
    single, 286
    symbolism, 287
        notations, 287-88
    See also Fasteners
Threaded connections, pipe, 417
Three-point perspective sketches, 51
Trimetric drawings, 261
Tolerance
    angle, 161
    basic sizes and, 163
    bilateral and unilateral, 159, *160*
    chain, 160
    clearance fits, 162
    computer-aided design and, 169-73
    datum, 160-61
    defined, 158
    dimensions, 158-59
    general, 160
    geometric, 164-69
    interference fits, 162
    material condition and, 163
    mating parts and, 161-63
    practices, 159
    standardized grades of, *161*
    surface roughness, 150
    transition fits, 163
Tool design drafting, 448-64
    clamps, *453,* 453
    computer-aided design and, 458-62
    drill bushings, 454, *455*
    feet and rest buttons, 451-52
    jigs and fixtures, 448
        bodies, 455
        design of, 455-58
    locators, 449-51
Total runout, *168*
Track drafting machine, 28-29
Transfer letters and tape, 26-27
Transition fits, 163
Triangles, 30, *31-32*
    constructing a, *91*
    constructing a right, *91*
    equilateral, constructing an, *90*
Triangulation development, development drawings and, 231-34
Two-point perspective sketches, 51

U
U.S. customary dimensioning, 137
Unilateral tolerancing, 159, *160*
Utility rooms, architectural symbols of, *480*

V
Valves, 420
Vanishing point location, perspective drawings, 268, *269*
Views
    auxiliary, 192-212
        auxiliary planes and, 194-95
        completeness of, 197
        computer-aided design and, 203-7
        dimensioning and, 198
        drawing procedures, 199-202
        ellipses, 198
        foreshortening, 192-94
        hidden lines and, 198
        holes and, 198
        primary, 196
        projection of, 195-96
        revolutions of, 201-2
        secondary, 196
        sectional views and, 185
    sectional, 178-91
        assembly, 185
        auxiliary, 185
        broken-out, 182
        computer-aided design and, 186-87
        cutting plane, 178-79
        enlarged, 185
        full, 182
        half, 182
        offset, 182, 183
        removed, 184
        revolved, 184
        section lining, 179-80
        section lining exceptions, 180-81
        thin wall, 184-85

W
Wastewater facilities, architectural symbols of, *480*
Welding
    drawings, 363-84
        computer-aided design and, 377-78
        data selection and, 369-70
        symbols, 365-70, 370-75
    processes, 364-65
Welds
    all around, welding drawings and, 369
    backing, 374
    combination, 375
    connections, pipe, 417
    field, welding drawings and, 369
    fillet, 370
    flange, 373
    groove, 373-74
    melt-through, 374
    pitch, welding drawings and, 368-69
    plug, 371
    projection, 372
    slot, 371
    spot, 371-72
    surface, 375

types, graphic symbols for, 366, *367*
Wheels, friction, 387, *388*
Windows, architectural drafting and, 478, *479*
Working drawings, 323-52
    architectural drafting, 481-94
        detailed house sections, *489-93*
        elevations and, 482, *485-87*
        floor plan, *483-84*
        interior elevations, 482, *488,* 489
        pictorial drawings and, 481
        plot plans and, 482
        presentation floor plans and, 481
        sectional drawings and, 489, 494
        site plans, 482
        support documents, 494-97
        survey plans and, 482
    assembly, 325, 327
        drawings and, 325-30
    computer-aided design and, 337-46
    detail drawings and, 324
    dimensions and, 330-36
    revisions, 310
    sketches, 48-49
Worm gears, 386, *387*
    cutting data and formulas, *396*
    terminology example, *396*